T0318867

MOLECULAR MACHINES IN BIOLOGY

THE concept of molecular machines in biology has transformed the medical field in a profound way. Many essential processes that occur in the cell, including transcription, translation, protein folding, and protein degradation, are all carried out by molecular machines. This volume focuses on important molecular machines whose architecture is known and whose functional principles have been established by tools of biophysical imaging (X-ray crystallography and cryo-electron microscopy) and fluorescence probing (single-molecule FRET). This edited volume includes contributions from prominent scientists and researchers who understand and have explored the structure and functions of these machines. This book is essential for students and professionals in the biological sciences and the medical field who want to learn more about molecular machines.

Dr. Joachim Frank is a Howard Hughes Medical Institute Investigator, Professor of Biochemistry and Molecular Biophysics, and Professor of Biological Sciences at Columbia University. He is a member of the National Academy of Sciences and a fellow of the American Academy of Arts and Sciences. Dr. Frank has received many awards for his research, including, with David DeRosier, the Elizabeth Robert Cole Award of the Biophysics Society. He has published more than 200 peer-reviewed articles, written numerous book chapters, and authored or edited five books, including *Electron Crystallography of Biological Macromolecules*, co-authored with Robert M. Glaeser, Kenneth Downing, David DeRosier, and Wah Chiu (2007), *Three-Dimensional Electron Microscopy of Macromolecular Assemblies* (2006), and *Electron Tomography*, Second Edition (2006).

MOLECULAR MACHINES IN BIOLOGY

Workshop of the Cell

Edited by

Joachim Frank

Columbia University

CAMBRIDGE
UNIVERSITY PRESS

CAMBRIDGE
UNIVERSITY PRESS

32 Avenue of the Americas, New York NY 10013-2473, USA

Cambridge University Press is part of the University of Cambridge.

It furthers the University's mission by disseminating knowledge in the pursuit of education, learning and research at the highest international levels of excellence.

www.cambridge.org
Information on this title: www.cambridge.org/9780521194280

© Cambridge University Press 2011

First published 2011

A catalogue record for this publication is available from the British Library

Library of Congress Cataloguing in Publication data

Molecular machines in biology : workshop of the cell / [edited by] Joachim Frank.

p.   cm.

Includes bibliographical references and index.

ISBN 978-0-521-19428-0

1. Microbiology.   2. Molecular biology.   3. Biochemistry.   4. Cell physiology.   5. Electron microscopy.

I. Frank, J. (Joachim), 1940–

QH506.M66436   2011

572 – dc23   2011015717

ISBN 978-0-521-19428-0 Hardback

Contents

Contributors

Xabier Agirrezabala
Structural Biology Unit
CIC bioGUNE
Basque Country
Spain

Karunesh Arora
Department of Chemistry and Biophysics Program
University of Michigan at Ann Arbor
Ann Arbor, MI

Sucharita Bhattacharyya
Northwestern University
Department of Molecular Biosciences
Matouschek Lab
Evanston, IL

Michael Börsch
Jena University Hospital
Friedrich Schiller University
Jena
Germany

Charles L. Brooks III
Department of Chemistry and Biophysics Program
University of Michigan at Ann Arbor
Ann Arbor, MI

Kwok-Yan Chan
Beckman Institute for Advanced Science and Technology
University of Illinois at Urbana-Champaign
Urbana, IL

Debashish Chowdhury
Department of Physics
Indian Institute of Technology
Kanpur
India

José Faraldo-Gómez
Department of Structural Biology
Max-Planck Institute of Biophysics
Frankfurt
Germany

Jingyi Fei
Department of Chemistry
Columbia University
New York, NY

Alexei V. Finkelstein
Institute of Protein Research
Russian Academy of Sciences
Pushchino, Moscow Region
Russia

Joachim Frank
Howard Hughes Medical Institute
Department of Biochemistry and Molecular
    Biophysics, and Department of Biological
    Sciences
Columbia University
New York, NY

Ruben L. Gonzalez, Jr.
Department of Chemistry
Columbia University
New York, NY

Dina Grohmann
UCL Institute for Structural and Molecular Biology
Division of Biosciences
London
United Kingdom

James Gumbart
Beckman Institute for Advanced Science and Technology
University of Illinois at Urbana-Champaign
Urbana, IL

Taekjip Ha
Department of Physics and Institute for Genomic Biology
Howard Hughes Medical Institute
University of Illinois at Urbana-Champaign
Urbana, IL

Arthur L. Horwich
Department of Genetics and HHMI
Yale School of Medicine
New Haven, CT

Daniel D. MacDougall
Department of Chemistry
Columbia University
New York, NY

Andreas Matouschek
Northwestern University
Department of Molecular Biosciences
Evanston, IL

Thomas Meier
Department of Structural Biology
Max-Planck Institute of Biophysics
Frankfurt
Germany

Helen R. Saibil
Department of Crystallography
Birkbeck College London
London
United Kingdom

Eduard Schreiner
Beckman Institute for Advanced Science and Technology
University of Illinois at Urbana-Champaign
Urbana, IL

Klaus Schulten
Beckman Institute for Advanced Science and Technology
University of Illinois at Urbana-Champaign
Urbana, IL

Xinghua Shi
Department of Physics and Institute for Genomic Biology
Howard Hughes Medical Institute
University of Illinois at Urbana-Champaign
Urbana, IL

Alexander S. Spirin
Institute of Protein Research
Russian Academy of Sciences
Pushchino, Moscow Region
Russia

Leonardo G. Trabuco
Beckman Institute for Advanced Science and Technology
University of Illinois at Urbana-Champaign
Urbana, IL

Mikel Valle
Structural Biology Unit
CIC bioGUNE
Basque Country
Spain

Finn Werner
UCL Institute for Structural and Molecular Biology
Division of Biosciences
London
United Kingdom

Shameika R. Wilmington
Northwestern University
Department of Molecular Biosciences
Matouschek Lab
Evanston, IL

Figures and Tables

TABLES

Preface

The concept of this book goes back to the Center for Molecular Machines, which I started together with similarly minded colleagues – Nilesh Banavali, April Burch, Steve Hanes, Joachim Jaeger, and Janice Pata, among others – at the Wadsworth Center in Albany back in 2005. The most visible manifestation of the Center for Molecular Machines was a monthly seminar series, which we called *Molecular Machine Shop*. The idea was to highlight some of the complicated structures at work in the cell, which were coming increasingly into view mainly through the efforts of X-ray crystallography and cryo-electron microscopy, and to bring out common features and general principles underlying these biological nanomachines. The seminars were meant to encourage interdisciplinary discourse as it was becoming increasingly clear that no single technique alone could unravel the mystery of how such machines work.

In June 2007, we organized a one-day minisymposium in Albany, which brought together experts studying molecular machines with the tools of X-ray crystallography, cryo-EM, single-molecule FRET, and molecular dynamics. In the wake of the event, Allan Ross, then senior editor at Cambridge University Press, approached me to ask if I would be interested in editing a volume on the theme. Some of the speakers of the minisymposium were receptive to the idea, and by asking other scholars working in the field I was lucky in the end to be able to assemble a team of the highest caliber.

I am grateful to all contributors for the dedicated work they have put in, and for their cooperation and patience over a considerable length of time. I would like to thank Joy Mizan, who took over the project after Alan Ross' departure from Cambridge University Press, for her dedicated work to ensure a glitch-free high-quality production, and Melissa Thomas in my lab for creating a beautiful symbolic cell as cover art, starting with a simple idea – M. C. Escher's reflecting glass ball – and with a list of deposition codes of density maps selected from the EM Data Bank.

The cover, I hope, will draw in readers with the promise of an experience that goes beyond conveying scientific facts: on the molecular level, touched by the tools of molecular graphics, life appears as a beautiful dance of colorful structures to scientists and nonscientists alike.

Joachim Frank
New York

Introduction

Joachim Frank

The term "Machine" invokes familiar images from the macroscopic world, of gears, springs, and levers engaging each other sequentially in a deterministic chain of events. Gravity and inertia are major forces to be reckoned with in the design, or are in fact utilized in the very function, of a device such as a crane, a locomotive, or a centrifuge. The world of molecules this book is concerned with is quite different in many respects. In this water-drenched nanodimension world, mobile parts are in constant jittering motion, powered by random thermal bombardment from the molecules of the aqueous solvent. The forces of gravity and inertia are dwarfed, by orders of magnitude, by those produced by non-covalent interactions, collisions with water molecules, and drag in the solvent. Complicating matters further, biological molecules and the "levers" and "actuators" within them lack the rigidity of materials like steel. Perhaps most difficult to conceive by a mind trained on the experience of the macroscopic world, however, is the disintegration of causal connections into a series of sporadic irreversible chemical events that impart directed motion, separated by stretches of "time" where the molecule has no apparent directionality.

Molecular Machines as a concept existed well before Bruce Alberts' (1998) programmatic essay in the journal *Cell*, but his article certainly helped in popularizing the term, and in firing up the imagination of students and young scientists equipped with new tools that aim to probe and depict the dynamic nature of the events that constitute life at the most fundamental level. "Machine" is useful as a concept because the molecular assemblies in this category share important properties with their macroscopic counterparts, such as processivity, localized interactions, and the fact that they perform work toward making a defined product. The concept stands in sharp contrast to the long-held view of the cell as a sack, or compendium of sacks, in which molecules engage and disengage one another more or less randomly. In invoking mechanistic imagery, it is

also, finally, a view that invites physicists to employ their tools and creative minds in developing models that take into account the reality of the nanoworld in a quantitative way. Thus Bruce Albert's is, as invitations go, the third in the past century, with Werner Schroedinger and Richard Feynman having been the first and second to usher their colleagues into tackling the grand challenges posed by biology.

To approach the subject, we must foremost know the structure of the static molecular machine at the atomic level, as a precondition for making sense of its behavior and going beyond mere phenomenological description. X-ray crystallography, a well-matured technique by now, has provided us with an ever-increasing inventory of structures – either complete structures with and without their functional ligands or, if unavailable, structures of isolated components. One of the crowning achievements of X-ray crystallography, the solution of the bacterial ribosome structure (Ban et al., 2000; Schluenzen et al., 2000; Wimberly et al., 2000; Yusupov et al., 2001), has given us unparalleled insights into the architecture of one of the most complex molecular machines in Nature.

Next, we must gather three-dimensional images showing how the structure of the molecular machine changes as a function of performing its work. Such *3-D snapshots* can be obtained by cryogenic electron microscopy (cryo-EM) of samples in which the molecular machine is imaged as an ensemble of single particles in the act of performing their work precisely as they do in solution. (In vitro systems now exist that mimic a variety of fundamental processes in the cell, such as transcription and translation). However, statistical requirements for low-dose three-dimensional imaging, which call for tens if not hundreds of thousands of projections, have the consequence that the snapshots can only be obtained for well-populated conformational states. Thus molecules that are in transition between such states defy attempts at imaging. It is possible, of course, to "trap" or stabilize the complex in additional short-lived intermediate states by some kind of intervention. Interventions that have been successfully employed in such studies include the addition of small-molecule inhibitors, use of non-hydrolyzable Guanosine triphosphate (GTP) or Adenosine triphosphate (ATP) analogs, transition state analogs, and introduction of targeted mutations. Nevertheless, of a virtual continuum of conformational states, only a relatively small number can be imaged in three dimensions. A movie generated from such a small number of samples would be filled with jarring jumps between frames of successive conformations and binding states.

Fortunately, a new powerful technique that has been developed in the past decade is rapidly filling these knowledge gaps: single-molecule fluorescence resonance energy transfer (smFRET). A donor-acceptor pair of fluorophores, placed strategically on molecular components implicated in dynamic changes, allows monitoring of distance changes in individual molecules in real time, as they perform their work.

In organizing this book, I have made the distinction between "Theory/Methods" and "Results/Biology." The distinction is not sharp because the explanation of methods entails the presentation of illustrative examples. Conversely, the description and interpretation of results is often intertwined with methodological narrative, as each system may require methods to be tweaked or refined in specific ways.

Single-molecule FRET (Chapter 1; Xinghua Shi and Takjep Ha) and Cryo-EM (Chapter 2; my own contribution) are relatively new as complementary techniques for the investigation of molecular machines, and this is why two chapters in the Methods section are devoted to them. A chapter on the theoretical description of molecular machines treated from the point of view of Statistical Mechanics (Chapter 3; Debashish Chowdhury) seemed in order, as well. Furthermore, a computational technique with predictive ambitions, Normal-Mode Analysis, often successfully employed in predicting domain motions, is featured in a contribution by prominent experts (Chapter 4; Karunesh Arora and Charles Brooks III). The degree to which a mechanistic model of a biological system lends itself to accurate predictions of the observed behavior of the system, using the principles of physics, is evidently a gold standard by which the viability of the model can be judged. (Molecular Dynamics simulations, the most prominent technique for predicting dynamics of a molecular machine, is featured in the Results section in its application to the highly complex machinery of translation.)

In the Results part of the book, I have strived to cover those molecular machines that are currently best understood through applications of a battery of biophysical probing and imaging techniques. If a preference is expressed in the selection, then it is in the limitation to globular systems and the exclusion of linear motors, which could be well covered on their own in a separate venture under a defined thematic umbrella.

The contributions in this volume toward the theme of molecular machines follow a protein from its conception as an abstract string of code; over its birth in translation as a polypeptide; its maturation, upon folding, to a fully functioning "adult"; and its function in the cell as part of a complex assembly of fellow proteins. Finally, we see the protein to its final end, as it is dissolves into its amino acid components, the universal building blocks for the next cycle.

In this logical order, we begin with the molecule instrumental for transcription of the genetic information residing on the DNA into a messenger RNA (Chapter 5: RNA polymerase; Finn Werner). The ribosome as the site of protein synthesis serves as an illustration for several lines of multidisciplinary investigation and follows with four contributions: on its dynamics as investigated either by smFRET (Chapter 6: Daniel MacDougall et al.), by cryo-EM (Chapter 7: Xabier Agirrezabala and Mikel Valle), or by molecular dynamics simulations (Chapter 8: James Gumbart et al.), and on its interpretation as a stochastic Brownian machine (Chapter 9: Alexander Spirin and Alexey Finkelstein).

Next we deal with a prominent example for a whole family of complexes involved in protein folding or refolding (Chapter 10: GroEL; Helen Saibil and Arthur Horwich). We have an instructive example for the jobs proteins are made to do once they are assembled into molecular machines performing essential functions in the cell: a protein that packages energy into the currency (ATP) universally used in the cell (Chapter 11: ATP-synthase; Thomas Meier et al.). Having said this, I should note that mature proteins are also incorporated as building blocks and catalytic agents in every single molecular machine discussed in this book. The final chapter concerns the end of a protein's life cycle and deals with the machines involved in protein degradation and recycling (Chapter 12: Andreas Matouschek).

On the whole, I believe the systems selected, while displaying the innate similarities of molecular machines I have outlined at the beginning, are different enough to let the reader appreciate the range of solutions Nature has found to achieve processivity, high fidelity, regulation, and control in a world so far removed from our own macroscopic universe.

ACKNOWLEDGMENTS

I would like to thank Ruben Gonzalez for helpful comments and suggestions. This work was supported by HHMI and NIH R01 GM29169.

REFERENCES

Alberts, B. (1998). The Cell as a Collection of Protein Machines: Preparing the next generation of molecular biologists. *Cell* 92, 291–294.

Ban, N., Nissen, P., Hansen, J., Moore, P.B., and Steitz, T.A. (2000). The complete atomic structure of the large ribosomal subunit at 2.4 Å resolution. *Science* 289, 905–920.

Schluenzen, F., Tocilj, A., Zarivach, R., Harms, J., Gluehmann, M., Janell, D., Bashan, A., Bartels, H., Agmon, I., Franceschi, F., et al. (2000). Structure of functionally activated small ribosomal subunit at 3.3 Å resolution. *Cell* 102, 615–623.

Wimberly, B.T., Brodersen, D.E., Clemons, W.M., Jr, Morgan-Warren, R.J., Carter, A.P., Vonrhein, C., Hartsch, T., and Ramakrishnan, V. (2000). Structure of the 30S ribosomal subunit. *Nature* 407, 327–339.

Yusupov, M.M., Yusupova, G.Z., Baucom, A., Lieberman, K., Earnest, T.N., Cate, J.H.D., and Noller, H.F. (2001). Crystal structure of the ribosome at 5.5 Å resolution. *Science* 292, 883–896.

CHAPTER 1

Single-Molecule FRET: Technique and Applications to the Studies of Molecular Machines

Xinghua Shi

Taekjip Ha

I. INTRODUCTION

I.1 Properties of Molecular Machines

Molecular machines are molecule-based devices, typically on the nanometer scale, that are capable of generating physical motions, for example, translocation, in response to certain inputs from the outside such as a chemical, electrical, or light stimulus. A large number of such sophisticated small devices are found in Nature, including the many biological motors discussed in this chapter, such as helicases and polymerases. These tiny nanomachines work in many ways just like an automobile on the highway, and many consume fuels on a molecular level, for instance, through the hydrolysis of adenosine-5′-triphosphate (ATP) molecules, to power their motions on their tracks. As a result, when lacking the required fuel, these nanomachines tend to slow down and even stop, same as a motor vehicle would. In addition, these biological motors often move in a directional manner with variable speeds, and their processivity characteristics can be described by how far they move on their track of a molecular highway, often formed by a biopolymer such as a nucleic acid or actin filament, before taking off at a later time. Motions of individual components within these protein machines, for example, the ribosome which is discussed in great detail throughout this book, are often nicely coordinated like in any sophisticated, larger-scaled mechanical machines. In recent years, details of the composition, stoichiometry, and three-dimensional arrangement of components within many nanomachines have become available, thanks to the ever-increasing number of high-resolution crystal structures that have been solved, which have provided valuable insights into the mechanisms of how these biological motors accomplish their tasks. In the past two decades, researchers have also brought these machines under scrutiny by a number of novel and powerful methods with ultra-high sensitivity, watching their motions one molecule at a time, and have learned a great deal of previously hidden mechanistic details about their action and

dynamics, such as the size of the fundamental steps taken by these motorized nanodevices. In a simplified view of the mechanism of action of biological motors, their strokes of physical translocation are powered by processes such as ATP hydrolysis through a modulation of their conformation, thus converting the chemical energy stored in the molecular fuel, in a stepwise fashion, into directed motions.

This chapter focuses on a particular type of single-molecule method termed Förster (fluorescence) resonance energy transfer (FRET). We start with a discussion of the technique and its developments to date and then feature a number of recent notable studies concerning the application of this technique to the investigation of a series of important molecular machines, including helicases and polymerases.

I.2 Single-Molecule Methods

By definition, single-molecule methods are based on the concept of examining the behavior of individual molecules, one at a time, therefore allowing for a direct detection of molecular heterogeneity, unsynchronizable dynamics, rare but key events, and so on. These approaches have been very successful in uncovering important mechanistic details that are otherwise hidden by ensemble averaging. As the last two decades have witnessed, single-molecule techniques and their applications to biological problems have experienced significant growth. In a survey published in 2007 of single-molecule studies (Cornish and Ha, 2007), the authors found that the number of research articles containing the words "single molecule" in the title, as obtained from a search of the PubMed database, had grown almost exponentially leading up to 2007, with a doubling time of 2.2 years. A rather dramatic prediction from this survey is that all the papers published in the biological sciences in thirty years will contain some single-molecule aspect. Although a bit bold, this prediction illustrates the ever-increasing importance and popularity of single-molecule methods in modern biological research. Note that the concept of detecting the behavior of individual molecules was not completely new even in the 1980s, as measurements on individual ion channels were already fairly common by then, but the general usefulness of the approach was not widely appreciated until the 1990s. Using single-molecule approaches, many fundamental biological problems that have fascinated scientists for a long time have started to be brought under the microscope. To use a well-known adage, seeing is believing, and many single-molecule techniques allow us to do just that (Selvin and Ha, 2008)! One day, these techniques will become an essential part of laboratory research for routine characterization of many biological phenomena, particularly those involved in the activities of molecular machines.

Among the various single-molecule methods, those based on fluorescence are particularly versatile and useful owing to their superb sensitivity, easy adaptation, and use of common instrumentation (Joo et al., 2008). From the original optical detection of individual biomolecules to the observation of their interactions in real time, and from the conventional diffraction-limited confocal or total internal reflection fluorescence (TIRF) microscopy to the recently developed super-resolution imaging microscopy, the potential of single-molecule fluorescence approaches has risen on many fronts to an unprecedented level and will, with no doubt, advance much further in the years to come (Shi, Lim, and Ha, 2010). In addition to fluorescence, force-based single-molecule tools for manipulation and measurement, such as optical and magnetic tweezers, have provided numerous insights into many important biological problems and can be largely complementary to fluorescence-based methods (Selvin and Ha, 2008). It is conceivable that a combination of different single-molecule approaches will allow researchers to best utilize the merit of each method and, in this way, to explore previously unreachable frontiers in the near future. In the rest of this chapter, we discuss mainly fluorescence-based methods, focusing on single-molecule FRET in particular.

II. SINGLE-MOLECULE FRET

II.1 Principle and Technique

Single-molecule FRET is one of the most, if not the most, general and adaptable single-molecule techniques available today. FRET is a process involving a dipole-dipole interaction between two adjacent fluorescent molecules, through which emission from the lower-energy molecule is observed upon excitation of the higher-energy molecule. As a technique that allows for direct detection of molecular interaction and its dynamics, single-molecule FRET quickly matured after it first went on stage in the mid-1990s (Ha et al., 1996) and has been employed to address

fundamental questions about helicase activities (Atkinson et al., 1997; Ha et al., 2002; Myong et al., 2005; Myong et al., 2007; Park et al., 2010), replication (Christian, Romano, and Rueda, 2009; Pandey et al., 2009; Santoso et al., 2010), transcription (Kapanidis et al., 2005; Kapanidis et al., 2006; Margeat et al., 2006), translation (Blanchard et al., 2004a; Blanchard et al., 2004b; Cornish et al., 2008; Cornish et al., 2009; Fei et al., 2008; Fei et al., 2009; Sternberg et al., 2009), reverse transcription (Rothwell et al., 2003; Liu et al., 2007; Abbondanzieri et al., 2008; Liu et al., 2008), and action of DNA-binding proteins (Joo et al., 2006; Roy et al., 2007), to name a few, and the list keeps growing rapidly. It also appears that the practitioners of this technique, originally mostly biophysicists, now encompass members of a much bigger community that includes many biochemists and molecular biologists.

As described by the Förster theory, the rate constant of energy transfer between two fluorophores, termed donor and acceptor, can be expressed as $k_{FRET} = (R_0/R)^6 k_f$, where R is the distance between the two fluorophores, R_0 is a parameter often referred to as the Förster distance, and k_f is the rate constant of the donor's fluorescence decay in the absence of the acceptor. The value of the Förster distance R_0 is determined by a few key parameters, including the orientation factor that describes the relative orientation of the transition dipoles of the donor and acceptor, the fluorescence quantum yield of the donor in the absence of the acceptor, the overlap integral depicting the degree of spectral overlap between the donor emission and acceptor absorption, and the refractive index of the medium. A more detailed discussion on this topic can be found in a study of energy transfer in yellow fluorescent proteins (Shi et al., 2007). To extract useful information about the distance R between the two fluorophores and its dynamics, one can perform fluorescence lifetime measurements (Edel, Eid, and Meller, 2007; Sorokina et al., 2009) and obtain the FRET rate and, thus, R; however, this would require a pulsed laser source with a picosecond output and sophisticated fast electronics for recording the time-resolved fluorescence signal, both being rare in a single-molecule laboratory, at least for now. As an alternative way to extract the distance information, one can obtain the time-integrated energy transfer efficiency, E, defined as $1/(1 + (R/R_0)^6)$, more conveniently from fluorescence intensity measurements. For a pair of fluorophores such as Cy3-Cy5, FRET efficiency can be approximated by the ratio of the intensity of the acceptor to the sum of intensities of the donor and acceptor, $E_{app} = I_A/(I_D + I_A)$ (Roy, Hohng, and Ha, 2008). It is worth noting that FRET efficiency becomes 50% when the distance R reaches the Förster distance R_0. Because of the sixth-power dependence on distance as shown earlier, FRET efficiency provides an extremely sensitive measure of the relative displacement between the two fluorophore probes in the range close to R_0 (Figure 1.1a) and, more importantly, the dynamics of such distance that is clearly

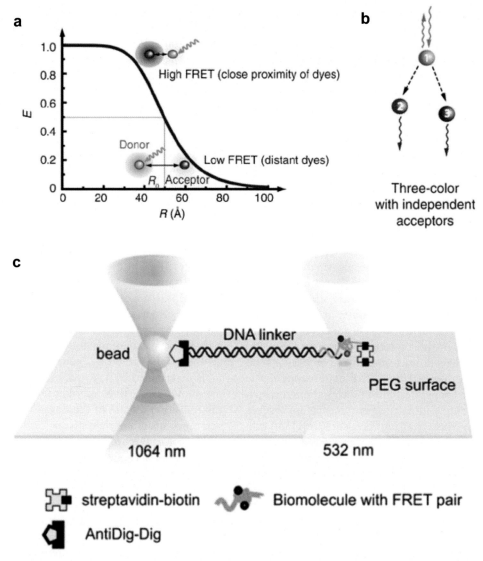

FIGURE 1.1: *Single-molecule FRET techniques. (a) Dependence of FRET efficiency on distance R, assuming $R_0 = 50$ Å (from Roy et al., 2008). (b) A two-pair, three-color FRET scheme. 1, 2, and 3 represent the donor, first, and second acceptor fluorophores, respectively (from Roy et al., 2008). (c) A FRET-tweezers hybrid scheme. The two wavelengths, 1,064 and 532 nm, correspond to the lasers for optical trapping and fluorescence detection, respectively. Reproduced with permission from (Hohng et al., 2007).*

of interest to researchers studying the motions of biological nanomachines.

As a ratiometric technique, single-molecule FRET eliminates the negative impact of instrument drift, which may not be negligible in some other single-molecule fluorescence measurements, including fluorescence imaging with one-nanometer accuracy (FIONA) (Yildiz et al., 2003) and protein-induced fluorescence enhancement (PIFE) (Myong et al., 2009). The beauty of FRET also lies in its great sensitivity to even a small change in distance in the range close to R_0, typically 3–8 nm depending on the FRET pair used, and this range clearly overlaps with the size of inter- or intra-molecular movement present in many molecular machines. It is noteworthy that, although FRET

measurements of individual molecules that are diffusing freely in solution using confocal microscopy are easier to implement, and have been very successful in revealing the presence and distribution of biomolecular subpopulations (Schuler, Lipman, and Eaton, 2002; Rothwell et al., 2003; Kapanidis et al., 2006; Michalet, Weiss, and Jager, 2006), the ability to keep track of a single molecule for an extended period of time from milliseconds to minutes greatly enriches the information content attainable through single-molecule characterization. This improvement was achieved by imaging many surface-immobilized molecules with high throughput (Zhuang et al., 2000; Ha et al., 2002) using TIRF microscopy. Such a single-molecule FRET setup can be assembled by an experienced researcher in a single day using standard, commercially available optical components that cost as much as a high-end ultracentrifuge (Roy et al., 2008), making the adoption of this technique by most laboratories straightforward. Recent extension of this technique to more than two colors (Hohng, Joo, and Ha, 2004; Clamme and Deniz, 2005) and future incorporation of novel nanotechnologies will definitely add new dimensions of capacity in obtaining useful information about the dynamics of biological nanomachines.

II.2 Comparison to FIONA

Fluorescence imaging with one-nanometer accuracy, abbreviated FIONA, is a super-precision imaging technique for determining the location of a molecule in the laboratory frame from a diffraction-limited image with a precision as high as 1.5 nanometers (Yildiz et al., 2003). This ultrahigh precision is made possible by collecting a large number of photons and curve-fitting the diffraction-limited spot to a Gaussian function, which is similar to finding the absolute peak position of a mountain of defined shape. In comparison, FRET tracks the relative displacement between two interacting objects in the molecular frame with an accuracy on the order of 0.5 nanometer (Cornish and Ha, 2007). Because of their differences in the detection scheme, FIONA would require instrument stability to be at least as good as localization precision while FRET is largely immune to microscope drift. In addition, FRET's sensitivity range depends on the fluorophores selected and is typically 3–8 nm, as previously mentioned. When a motion larger in scale is the subject of interest, for example, the long-distance transport of cargo by a kinesin protein, FIONA becomes more advantageous than FRET. From a user's point of view, the high localization precision of FIONA comes at the price of a much higher photon budget than for FRET (Cornish and Ha, 2007), making it necessary to use fairly bright and photostable fluorescent probes such as quantum dots in the former case. Although FRET is likely a more general method, FIONA can be complementary to it and can be particularly useful when it comes to studies of molecular machines.

II.3 Advances in Single-Molecule FRET

The single-molecule FRET technique discussed thus far has involved only two fluorophores, with which the information about relative displacement can be determined in real time at a high level of accuracy. This scheme has been quite successful in the studies of many molecular machines as discussed later in this chapter, focusing on one particular mode of physical motion. If the biological system of interest involves more components and thus becomes more complex, it is often necessary to acquire additional information simultaneously in order to untangle some of the intriguing riddles. To this end, Ha and coworkers extended the common two-color FRET scheme into one

with three colors (Figure 1.1b) by using three spectroscopically distinct cyanine fluorophores, including Cy3 (donor), Cy5 (acceptor 1), and Cy5.5 (acceptor 2), in two parallel energy-transfer pathways (Hohng et al., 2004). Using this multi-color scheme, correlated movements of different arms in a four-way junction structure of DNA (Hohng et al., 2004) and motions of a single-stranded binding protein along DNA (Roy et al., 2009) have been directly observed with little ambiguity. In addition to multi-color detection of fluorescence, a scheme called alternating laser excitation (ALEX) (Kapanidis et al., 2004) has been incorporated into three-color FRET by Weiss and coworkers (Lee et al., 2007), making simultaneous measurements of distances and interactions even more convenient. As a natural extension, a four-color FRET scheme has been realized recently by Hohng and coworkers, by combining multi-color fluorescence detection with pulsed, interleaved excitation. This newest version of multi-color FRET has enabled them to study a complex process of DNA strand exchange mediated by RecA proteins (Lee et al., 2010).

Another, related major technical development is the combination of single-molecule FRET with force-based manipulation and measurement tools (Shroff et al., 2005; Hohng et al., 2007; Tarsa et al., 2007). As demonstrated by Ha and coworkers, the addition of manipulation by force to single-molecule FRET detection (Hohng et al., 2007) made it possible to measure conformational changes of biomolecules through FRET as a function of the force applied by optical tweezers (Figure 1.1c). Using such a FRET-tweezers hybrid, Hohng and colleagues were able to characterize the two-dimensional reaction landscape of the four-way DNA junction structure mentioned earlier. In a related study, Liphardt and coworkers combined FRET with magnetic tweezers to construct a force sensor that provides an optical readout (Shroff et al., 2005). In addition, the importance of mechanical forces in many developmental, physiological, and pathological processes in biology has gained recent recognition and appreciation (Orr et al., 2006). Very recently, Schwartz and coworkers have used the FRET-tweezers duo-spectroscopy for the calibration, with piconewton force sensitivity, of a biosensor based on the vinculin protein that exhibits force-dependence recruitment to cell focal adhesions, which can be used to measure the force across such proteins and examine the regulation of focal adhesion dynamics (Grashoff et al., 2010). We are certain that such a combinatorial approach will be useful in the studies of important biological nanomachines in the near future.

II.4 Advances in Molecule Immobilization and Detection Schemes

As explained earlier, modern single-molecule FRET measurements are more advantageous with surface-immobilized molecules compared to freely diffusing

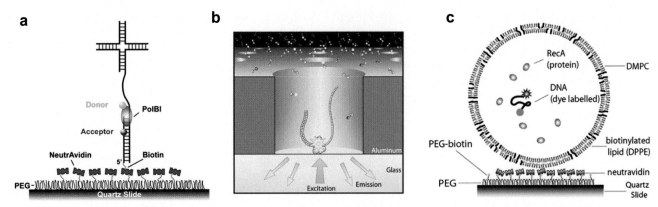

FIGURE 1.2: *Advanced molecule immobilization and detection schemes. (a) Surface passivation through PEGylation that allows for specific binding of biomolecules (Shi et al., unpublished results). (b) Illumination based on the zero-mode waveguide (ZMW) nanostructure that gives a reduction of excitation volume down to the zeptoliter (10^{-21} liter) region. Reproduced with permission from (Eid et al., 2009). (c) A porous vesicle containing DNA and proteins that is permeable to small molecules such as ATP at room temperature. Reproduced with permission from (Cisse et al., 2007).*

molecules for an extended period of observation on the same molecule. For nucleic acids, immobilization can be achieved fairly easily by using a surface that is passivated by coating with a protein called bovine serum albumin (BSA), a small fraction of which is biotinylated as an anchor for the biotin-streptavidin linkage. For proteins used in single-molecule studies, in contrast, surface coating with the BSA protein was found to be far from adequate in preventing non-specific binding of the protein being studied, and this limitation can be especially detrimental when a fluorescently labeled protein is used, which can easily overwhelm the fluorescence signal contributed by specific interactions. For this reason, a surface-passivation protocol involving the use of polyethylene glycol (PEG) was first adopted in a single-molecule study of DNA helicase unwinding activities (Figure 1.2a) (Ha et al., 2002). Ha and colleagues found that the non-specific binding of Rep helicase was 1,000-fold less on a PEG-coated surface than on a BSA-coated surface, providing a much better reproduction of the enzymatic activity at the single-molecule level. Since then, such a PEG-based passivation procedure has been employed in a large number of single-molecule FRET studies, leading to many interesting discoveries, such as repetitive shuttling of Rep helicase (Myong et al., 2005), spring-loaded DNA unwinding by HCV NS3 helicase (Myong et al., 2007), substrate-directed binding orientations of HIV reverse transcriptase (Abbondanzieri et al., 2008), spontaneous ratcheting of the ribosome (Cornish et al., 2008), and intermediates during nucleosome remodeling by ACF complex (Blosser et al., 2009). As an alternative to PEGylation, a different strategy for preparing an inert surface compatible with single-molecule experiments is passivation by a supported lipid bilayer (Graneli et al., 2006). As has been demonstrated by Greene and coworkers, many molecules can be tethered to fixed locations on such a passivated surface at once, allowing for a parallel recording of the

behavior from many individual biological motors simultaneously with a high throughput (Graneli et al., 2006).

With the surface passivated as described earlier and TIRF microscopy, one can carry out many useful single-molecule FRET studies, as long as the concentration of fluorescently labeled biomolecules in the bulk solution is limited to within the low nanomolar range, given the much higher level of fluorescence background that appears above this concentration range. However, many important biological interactions have a much weaker affinity and, thus, are governed by equilibrium dissociation constants in the micromolar range that often necessitate working at micromolar concentrations of fluorescently labeled binding partners in the bulk solution. For these studies, it is necessary to suppress the excitation volume in conventional TIRF or confocal microscopy by several orders of magnitude, as has been achieved successfully using a nanostructure-based device known as the zero-mode waveguide (ZMW) (Levene et al., 2003).

A ZMW is essentially a very tiny hole, tens of nanometers (nm) in diameter, fabricated in a 100-nm thick aluminum film that is deposited on a glass substrate (Figure 1.2b). The small, sub-wavelength size of ZMWs limits the penetration depth of visible light, typically with a wavelength between 400 and 700 nm, to about 30 nm near the bottom of the waveguide. As a result, this nanostructure-based illumination gives a reduction of excitation volume down to the zeptoliter (10^{-21} liter) region; thus, background fluorescence in the presence of an overwhelming pool of fluorescent molecules at micromolar concentrations can be suppressed almost completely. The invention of this technique has facilitated the development of a new-generation DNA sequencing instrument with high throughput and single-molecule resolution (Eid et al., 2009), which is currently being commercialized by Pacific Biosciences, Inc. In a recent study, Puglisi and coworkers demonstrated the use of ZMWs for observing translation by the ribosome in real time with single-codon resolution using fluorescently labeled aminoacyl-transfer RNA (tRNA) substrates at a physiologically relevant micromolar concentration (Uemura et al., 2010). In the same study, ZMWs were also employed in the measurements of single-molecule FRET between fluorescently labeled tRNA substrates at concentrations up to hundreds of nanomolars, which would be impossible for conventional TIRF microscopy. It is conceivable that such a FRET-ZMW combination will be a very powerful tool in studies of other important biological systems, allowing researchers to characterize weak interactions that are hard to detect otherwise.

In addition to the reduction of observation volume by using a nanostructure such as the ZMW, another promising approach is the encapsulation of single molecules within nanometer-sized compartments such as a lipid vesicle that can be tethered to the surface, which is compatible with conventional, diffraction-limited detection schemes such as TIRF microscopy. Using this strategy, heterogeneous folding pathways of the protein adenylate kinase have been studied in detail (Rhoades, Gussakovsky, and Haran, 2003). One particular feature worth noting in the use of such vesicle encapsulation techniques is that the local concentration of the encapsulated molecule can be sufficiently high for studying transient and weak interactions that are difficult to directly probe by other means.

Although it represents an effective molecule-confinement scheme, vesicle encapsulation is limited by the fact that the lipid bilayer membrane is not permeable to most ions and small molecules, including ATP, in the extra-vesicular solution, making it difficult to probe the effects of these ligands on many interesting dynamic processes. This limitation has been largely ameliorated by taking advantage of the intrinsically porous feature of lipid bilayer membranes near the gel-fluid phase transition temperature. For example, in a proof-of-principle experiment, Ha and coworkers demonstrated that vesicles containing a single-stranded DNA and the recombination protein RecA prepared by using a type of lipid molecule named dimyristoyl phospatidylcholine (DMPC) with a phase transition temperature of 23°C can be made porous at room temperature and permeable to small molecules such as ATP (Cisse et al., 2007). Enabled by the use of porous vesicles, it was possible to modulate the dynamic interactions between RecA and DNA by delivering ATP and its nonhydrolyzable analog across the lipid membrane through the leaky pores, while keeping the larger-sized protein and DNA molecules within the boundary of vesicle nanocontainers (Figure 1.2c). This approach provides a versatile way to supply desired small molecules such as ATP to the inside of the nanocontainer and should be useful for studying the weak interactions in more complex biological systems in a controllable manner.

III. APPLICATIONS

III.1 Applications to Studies of Helicases

Helicases are a class of motor proteins that couple conformational changes driven by ATP binding and hydrolysis to the unwinding of duplex nucleic acids into separate single strands. Like many other molecular machines, helicases move along the molecular track provided by the DNA or RNA substrate and consume the energy stored in ATP molecules. Prior to single-molecule FRET, several other single-molecule techniques including optical tweezers (Bianco et al., 2001) and tethered particle tracking (Dohoney and Gelles, 2001) had been used to study the processive unwinding of DNA by a helicase, *E. coli* RecBCD, with resolutions up to 100 base pairs (bp). However, many helicases exhibit a limited processivity in vitro and often fall off their nucleic acid track long before even reaching 100 bp, making it desirable to use a method with a much higher spatial resolution for studying these enzymes. For this reason, in the early 2000s, Ha and colleagues developed a single-molecule FRET assay for helicase activity that allowed them to measure the unwinding of DNA by a helicase with a resolution better than 10 bp and detect, for the first time, a number of novel helicase events such as unwinding pauses, duplex rewinding, and unwinding restarts, and facilitated a large improvement in our understanding of the underlying mechanisms (Ha et al., 2002).

In this single-molecule FRET assay, two fluorescent probes, Cy3 and Cy5, were attached to the junction between the single-stranded region and the duplex of a partial duplex DNA (Figure 1.3a), of which the unwinding by an *E. coli* helicase called Rep was observed. Because of the short distance between the two reporter fluorophores, the FRET efficiency was almost 100% to start with and decreased as the unwinding helicase proceeded along the duplex, thereby reporting the progress and dynamics of unwinding (Figure 1.3b). Ha and colleagues then observed that, for an 18-bp duplex, unwinding by Rep quickly advanced to completion, whereas for the case of a 40-bp duplex, the situation was much more complex. In the latter case, transient stalls lasting a few seconds were observed in approximately 70% of the unwinding events, followed by either a complete recovery of the initial high FRET, indicative of a duplex rewinding event, or a resumed decrease in FRET, suggesting an unwinding restart event that led to completion of the reaction (Figure 1.3b). Based on the observation that unwinding of DNA by Rep begins only upon the formation of a functional helicase involving more than one Rep monomer (Cheng et al., 2001) and the observation that an unwinding restart event would require free Rep proteins from solution (Figure 1.3c), Ha and colleagues proposed a model in which partial dissociation of the active Rep oligomer complex during unwinding leaves an inactive monomer on the DNA, giving rise to the pauses

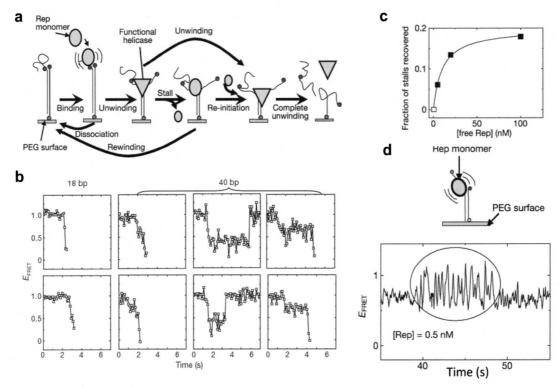

FIGURE 1.3: *A single-molecule FRET study of E. coli Rep helicase. (a) Experimental scheme and model for Rep unwinding activities. Two fluorescent probes, Cy3 and Cy5, were attached to the partial duplex DNA immobilized on the surface. (b) Typical FRET time trajectories of 18- and 40-duplex unwinding. (c) Dependence of unwinding stalls recovery on Rep concentration, suggesting the requirement of free Rep from solution for unwinding restart. (d) Rapid, ATP-dependent conformational fluctuations of a Rep monomer once in contact with the junction of the partial duplex DNA. Reproduced with permission from (Ha et al., 2002).*

observed (Figure 1.3a). Upon subsequent dissociation of the remaining Rep monomer, the duplex DNA can reform; alternatively, the unwinding can resume upon association of additional Rep monomer(s) to the one stalled on the DNA.

In addition to the mechanism of unwinding of a DNA duplex by the Rep helicase, another interesting discovery made by Ha and colleagues was a rapid, ATP-dependent conformational fluctuation of a Rep monomer, once in contact with the junction of the partial duplex DNA (Figure 1.3d), which did not lead to DNA unwinding. Such a phenomenon became the subject of a follow-up study by Ha and coworkers, as discussed below (Myong et al., 2005). The previously described early study using single-molecule FRET clearly paved the road to answering a series of outstanding mechanistic questions about helicases as molecular machines. For example, in an analogy to the quantification of fuel efficiency of an automobile, how many base pairs are unwound by a helicase per ATP molecule hydrolyzed? What is the fundamental step size involved in the helicase motion? How exactly does a helicase couple the conformational changes induced by ATP binding and hydrolysis to its translocation and DNA unwinding activities? To address these questions more directly, fluorescent

labeling of the helicase itself was then employed in single-molecule FRET studies.

Through site-directed mutagenesis, Ha and coworkers achieved site-specific labeling for a series of single-cysteine Rep mutants (Rasnik et al., 2004) and were able to confirm that all of these mutant proteins were fully functional both in vivo, using a plaque assay for bacteriophage replication (Scott et al., 1977; Cheng et al., 2002), and in vitro, as judged by single- and multiple-turnover DNA unwinding activities. As described earlier, PEG-based surface passivation was found to essentially eliminate nonspecific binding of labeled proteins to the imaging surface, allowing the detection of Rep binding to an immobilized partial duplex DNA, as visualized by an abrupt appearance of fluorescence signal in the corresponding channel. With a near-quantitative (above 90%) labeling efficiency, Ha and coworkers showed that at a sub-nanomolar protein concentration, Rep binds to DNA primarily in the monomeric form, based on counting the number of photobleaching steps observed, which could be distinguished clearly from those more complex but rare multiple-protein binding events. For each of the eight single-cysteine Rep mutants, Rasnik and colleagues studied the binding interaction between Rep and DNA in the absence of ATP by

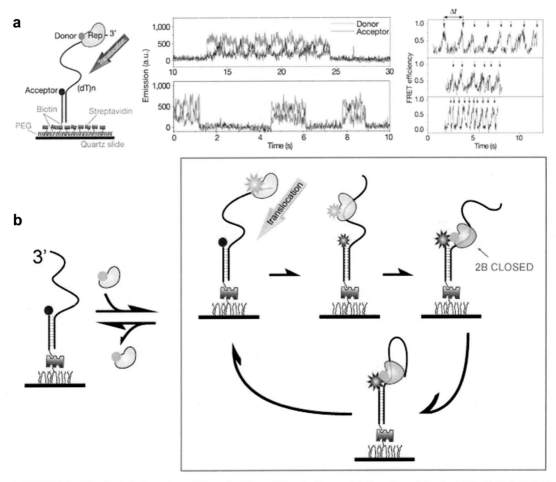

FIGURE 1.4: *Blockade-induced repetitive shuttling of Rep helicase. (a) Experimental scheme (left) and typical fluorescence (middle) time trajectories. FRET traces for 80-, 60-, and 40-nucleotide 3'-tails are shown on the right. (b) A mechanistic model for the repetitive shuttling of Rep, in which a blockade-induced Rep conformational change enhances the affinity of a secondary binding site for the 3' tail and thus promotes the formation of a transient DNA loop that leads to the rapid snapback. Reproduced with permission from (Myong et al., 2005).*

measuring the efficiency of FRET from the donor fluorophore on the protein to the acceptor fluorophore on the DNA (Rasnik et al., 2004). It was found that the relative binding orientation of Rep on single-stranded DNA, obtained using a distance-constrained triangulation procedure, was consistent with previous X-ray crystal structures. Moreover, their single-molecule FRET experiments provided evidence in support of Rep existing predominantly in the "closed" conformation when bound to a 3'-tailed DNA in solution, similar to the conformation found in the complex of PcrA and DNA (Velankar et al., 1999). Beyond these results, this work laid the foundation for using fluorescently labeled proteins in studies of many important aspects of helicase function, such as the unwinding of DNA and translocation of the helicase that is powered by ATP hydrolysis.

In a follow-up study, Ha and coworkers found that individual Rep monomers bind to the 3' tail of a partial duplex DNA and then translocate in the 3' to 5'

direction using the fuel of ATP hydrolysis, as visualized by a gradual change in FRET between the donor-labeled protein and the acceptor-labeled DNA (Figure 1.4a) (Myong et al., 2005). Interestingly, once Rep encountered a physical obstacle on the single-stranded DNA track, such as a duplex DNA, which monomeric Rep cannot unwind, or a streptavidin molecule attached to the 5' end of the single-stranded region, Rep quickly snapped back within the 15-ms time resolution to its original position near the 3' end, followed by repeated cycles of such translocation and snapback. As a result of this process, a saw-toothed pattern was observed in the FRET time trajectories; thus, this unexpected behavior of Rep was termed "repetitive shuttling" (Myong et al., 2005). Through a quantitative analysis, Myong and colleagues found that the period of repetition, defined as the time interval between adjacent snapbacks, (1) increased in proportion to the length of 3' tail; (2) was insensitive to the nucleic acid sequence content in the tail; and (3) became longer at low ATP

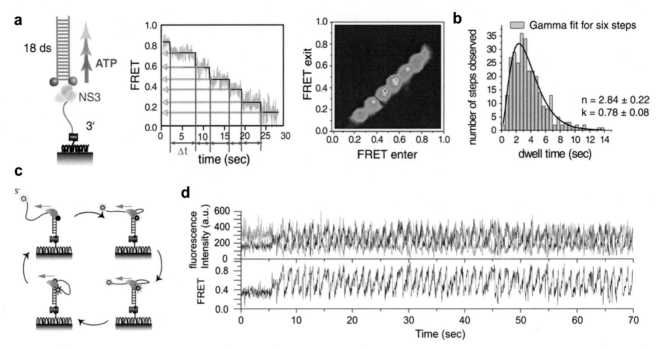

FIGURE 1.5: *Stepwise unwinding of HCV NS3 helicase and translocation of B. stearothermophilus PcrA helicase. (a) NS3 unwinds an 18-bp partial duplex DNA in six discrete steps of three base pairs. A transition density plot is shown on the right, displaying all possible connections in FRET before and after a transition. (b) Distribution in FRET plateau dwell-time exhibiting a rise followed by decay and the best fit to a Gamma function, suggesting the fundamental step size of a single base pair/nucleotide. (c) The reeling-in model for PcrA, in which PcrA translocates on the single-stranded 5' tail while maintaining contact with the duplex region, thus extruding a single-stranded DNA loop. (d) A typical single-molecule FRET time trajectory for PcrA reeling in DNA. (a) and (b) reproduced with permission from (Myong et al., 2007); (c) and (d) reproduced with permission from (Park et al., 2010).*

concentrations or when its non-hydrolyzable analog was added. Moreover, Ha and coworkers found strong evidence for a single Rep monomer shuttling repeatedly by itself, rather than multiple monomers acting in a relay. In a mechanistic model that describes the behavior of this helicase, Myong and colleagues proposed that the repetitive shuttling observed is likely caused by an obstacle-induced conformational change in Rep, which enhances the affinity of a secondary binding site for the 3' tail and thus promotes the formation of a transient single-stranded DNA loop that leads to the rapid snapback (Figure 1.4b). In addition to the evidence described earlier, a number of other observations were also in support of this model, including a periodic change in FRET efficiency between the two ends of a dual-labeled DNA, the lack of repetitive snapback behavior in the absence of an obstacle at the 5' end, and the gradual closing of the Rep 2B sub-domain upon approaching the obstacle. In an effort to relate these data to the functions of Rep in vivo (Sandler 2000; Marians 2004), Myong and colleagues also examined and confirmed the presence of repetitive shuttling on a DNA structure containing a replication fork and an analog of the Okazaki fragment, and provided evidence for Rep's interference with RecA filament formation.

Following the examinations of relatively large-scale unwinding and translocation of a helicase, recent studies have started to address the fundamental step size involved in the action of these enzymes, leading to the suggestion that the size of a single nucleotide is likely the smallest step taken by these genome-processing nanomachines. In the case of the HCV NS3 helicase, Myong and colleagues used a single-molecule FRET assay and observed that this viral helicase unwinds DNA in discrete steps of three base pairs (Myong et al., 2007). For example, for an 18-bp partial duplex DNA, six well-defined steps in FRET were observed during unwinding, which might suggest a 3-bp step for NS3 upon an initial analysis (Figure 1.5a). However, a dwell-time analysis of the detected steps in FRET efficiency did not yield an exponential distribution as would be expected if the kinetics was dominated by a single, rate-limiting step. Instead, Myong et al. found that the dwell-time histogram exhibited a rising phase, followed by a decay that is indicative of the presence of three hidden irreversible steps prior to each 3-bp unwinding step (Figure 1.5b). This result led to the suggestion that the fundamental step size of NS3 within the 3-bp kinetic step should be a single base pair/nucleotide. Based on these data and the structures of NS3 in the absence of ATP (Kim et al., 1998) as well as of closely related RNA helicases in the presence of ATP (Andersen et al., 2006; Sengoku et al., 2006), Ha and coworkers proposed a model in which NS3 translocates in three 1-bp steps on the DNA track, one per molecule of

ATP hydrolyzed, accumulating enough tension in the complex of NS3 and DNA, much like a molecular spring, which is released subsequently through the burst of 3-bp unwinding. Given the presence of very similar structural features in other helicases such as Rep, PcrA, and UvrD, it is very likely that such a physical step as observed in NS3 may be general among helicases. The single-molecule FRET technique used here thus provided a means of bridging the gap between the single-nucleotide translocation steps predicted from crystal structures and the larger unwinding steps commonly found by biochemical studies performed in solution (see below).

Very recently, Ha and coworkers have investigated the behavior of PcrA, a 3′-5′ helicase mentioned earlier (Park et al., 2010). Given this helicase's directionality, most previous studies have used a partial DNA duplex with a 3′ tail, whereas in the study by Park and colleagues, a partial DNA duplex with a 5′ tail was employed instead. Unexpectedly, PcrA was observed to translocate on the track of the single-stranded 5′ tail while maintaining contact with the duplex region; as a result, PcrA anchored itself on the duplex, reeled in the DNA single strand, and extruded a single-stranded DNA loop all at the same time (Figure 1.5c, d). As suggested by Park and colleagues, a functional consequence of this translocation-coupled looping motion was the efficient and rapid, PcrA-induced removal of a RecA filament that had been preformed on the single strand even at a low nanomolar concentration of PcrA, suggesting a potential role of PcrA in preventing the formation of potentially harmful recombination intermediates. Furthermore, an analysis of the repetitive looping behavior, similar to the one described earlier for NS3, revealed a non-exponential distribution in the period of repetition and led to the conclusion that PcrA translocates in uniform steps of one nucleotide. It is worth noting that, prior to this single-molecule study, there had been controversy over such mechanistic issues between structural and biochemical studies. Despite the agreement that, on average, one ATP molecule is consumed to move PcrA across one nucleotide, ensemble kinetic studies deduced a kinetic step size of about four nucleotides (Niedziela-Majka et al., 2007). As Ha and coworkers reasoned in their study, such an apparently large step size can be the result of persisting molecular heterogeneities in the translocation rate among different PcrA molecules in a large population, and such "static disorder" can complicate the analysis of ensemble data that relies on the measurement of variance. A molecule-by-molecule analysis, on the other hand, largely overcomes this difficulty, and we believe similar single-molecule measurements with high resolving power, including FRET, will be valuable when revisiting previous estimates of unwinding and translocation step sizes of other helicases.

Single-molecule FRET studies to date have revealed the presence of a repetitive pattern in the activity of many helicases, including the Rep shuttling and PcrA reeling motions described previously. Such a pattern has also been observed in the unwinding activity of HCV NS3 (Myong et al., 2007) and human Bloom Syndrome helicase (BLM) (Yodh et al., 2009). In the study by Myong and colleagues, upon addition of ATP, NS3 was found to unwind a dual-labeled partial DNA duplex containing a 18-bp duplex and a 3′ tail; however, once NS3 reached the end of the duplex that was immobilized on the surface, the two single strands quickly rewound again, producing the original value of FRET efficiency observed before the unwinding began. This unwinding-rewinding cycle was then repeated by the same molecule and persisted even after washing away free NS3 proteins, indicating that such repetitions were done by the same NS3 molecule instead of multiple molecules through successive binding and dissociation. In addition, Myong and colleagues also observed this repetitive pattern in the unwinding of duplexes 24 bp or longer even when the duplex end was not blocked. With all these pieces of evidence, Myong and colleagues proposed that during the unwinding, NS3 remains in contact with the strand being displaced instead of releasing this strand completely, and upon reaching a barrier or after unwinding more than 18 base pairs, NS3 returns rapidly to the junction between the single-strand region and the duplex, starting the next cycle of unwinding-rewinding motions.

A similar repetitive unwinding pattern was also observed by Ha and coworkers for the wild-type BLM helicase, as well as for a mutant containing the catalytic core but lacking the domain responsible for oligomerization (Yodh et al., 2009). Their study found that the unwinding time interval was independent of the protein concentration, suggesting that repetitive unwinding of the DNA was carried out by a single monomer of BLM. Based on the changes detected in the time trajectories of FRET efficiency, Yodh and colleagues divided the unwinding cycle into three separate stages corresponding to a gradual unwinding, a transient stall, and a fast rewinding, and performed a detailed analysis of the FRET changes, time intervals, and dependence on ATP concentration involved in these stages. From these results, Yodh and colleagues proposed a possible mechanistic model in which the BLM helicase first dissociates partially after reaching a critical depth through unwinding into the duplex that it can somehow sense, switches the motor to the strand being displaced, and then translocates on the track of this strand in the 3′ to 5′ direction. These mechanisms are likely important for BLM's ability to process stalled replication forks as well as recombination intermediates that are potentially detrimental. Future single-molecule FRET studies using a fluorescently labeled BLM will definitely provide further details about the action mechanism of this important helicase.

Single-molecule techniques based on FRET were also employed in studies of the conformational changes of DEAD-box RNA helicases on binding of the RNA substrate and ATP. For these helicases, the available crystal structures revealed a common feature in the core involving two RecA-like domains that are linked by an interdomain cleft, which led to the suggestion that an open-to-closed conformational change may be associated with the binding of RNA and ATP simultaneously. To verify the presence of such conformations and switching between them, Klostermeier and coworkers performed dual-labeling in the two RecA-like helicase domains of *B. subtillus* YxiN, a helicase involved in ribosome biogenesis, and observed the movement of the interdomain cleft by intramolecular FRET (Theissen et al., 2008). The FRET distribution of YxiN in the absence of any cofactor showed a single mid-FRET peak that is indicative of the open conformer. Interestingly, the addition of only one of the two ligands, RNA or ATP, barely affected the FRET distribution; however, upon simultaneous addition of ADPNP/ATP and RNA, the FRET efficiency distribution shifted to a high-FRET peak, suggesting the switching to the closed conformation through closure of the inter-domain cleft. In a subsequent study by the same group, Aregger and colleagues found that the YxiN helicase maintains closed conformation throughout ATP hydrolysis, indicating that the release of phosphate is probably associated with the reopening of YxiN (Aregger and Klostermeier, 2009). Such a FRET-based assay for probing conformational changes has also been used to examine the capacity of various RNA ligands to induce a conformational closure in a related RNA helicase named Hera from *T. thermophilus*, and this process can be relevant to the involvement of Hera in ribosome biogenesis (Linden, Hartmann, and Klostermeier, 2008).

The many examples discussed earlier regarding the application of single-molecule FRET to the study of helicases have clearly demonstrated the power of this technique in obtaining mechanistic details previously unattainable for this large family of biological nanomachines. Similar successes have been achieved in studies of many other important biological motors. In the rest of this chapter, we will discuss a few studies on RNA and DNA polymerases and related proteins that were made possible by this technique.

III.2 Applications to Studies of RNA and DNA Polymerases

RNA polymerases (RNAPs) are a class of enzymes that catalyze the replication of a DNA template into a complementary RNA product. At the beginning of transcription initiation, RNAP binds to the promoter and unwinds a short region of the DNA near the initiation site, followed by cycles of abortive synthesis and release of short RNA products, and upon synthesis of an RNA that is 9–11 nucleotides long, it enters into the processive mode of elongation synthesis. To understand the mechanism by which RNAP translocates relative to DNA during initial transcription, Ebright and coworkers explored three possible models, including transient excursions (Carpousis and Gralla, 1985), inchworming (Straney and Crothers, 1987; Krummel and Chamberlin, 1989), and scrunching (Carpousis and Gralla, 1985; Pal, Ponticelli, and Luse, 2005), by using a confocal scheme of single-molecule FRET with alternating laser excitation (ALEX) (Kapanidis et al., 2006). By labeling the trailing edge of RNAP with a donor fluorophore and the upstream DNA with an acceptor, Kapanidis and colleagues observed no detectable change in the FRET efficiency between the RNAP trailing edge and the DNA promoter region, thus ruling out the transient excursions model in which forward and reverse translocations of the entire RNAP would have led to a decrease in the FRET efficiency detected for the pair above (Figure 1.6a). To evaluate the possibility of RNAP expansion and contraction, Kapanidis and colleagues then monitored FRET between the leading edge of RNAP and the promoter DNA and were able to conclude that the initial transcription does not involve inchworming either, in which the part of RNAP containing the catalytic core would detach from the rest of the holoenzyme, producing a FRET decrease for this donor-acceptor pair (Figure 1.6b). Moreover, dual labeling of the DNA in the upstream and downstream regions with a donor-acceptor pair of fluorophores allowed them to observe evidence for a shortening in relative displacement between these two regions of DNA in support of the scrunching model, where the downstream DNA is pulled toward the upstream region into the RNAP holoenzyme, giving rise to an increase in FRET (Figure 1.6c). Combining these results, Ebright and coworkers concluded that a stressed intermediate complex of DNA and RNAP is involved in abortive initiation of transcription, and the elastic energy accumulated within such a complex can be used to induce promoter escape and productive initiation. Using a similar assay, Kapanidis and colleagues also studied the correlation between translocation of the RNAP complex and subunit stoichiometry of such a complex during transcription elongation and found that more than half of the mature elongation complexes retained an initiation factor (Kapanidis et al., 2005).

Apart from RNA polymerases, another major class of nucleic acid-dependent polymerases is the DNA polymerases (DNAPs), which replicate a DNA template into a complementary DNA product. To understand the mechanisms controlling the high fidelity of replicative DNAPs, Kapanidis and coworkers employed several single-molecule fluorescence techniques including FRET and fluorescence correlation spectroscopy (FCS) and investigated conformational transitions in DNAP prior to nucleotide incorporation (Santoso et al., 2010). All available crystal structures of DNAPs share a common fold that resembles the shape of a right hand, and the three sub-domains visible in this fold are often referred to as the thumb, fingers, and palm.

In the study by Santoso and colleagues, they focused particularly on the closing of the fingers sub-domain upon

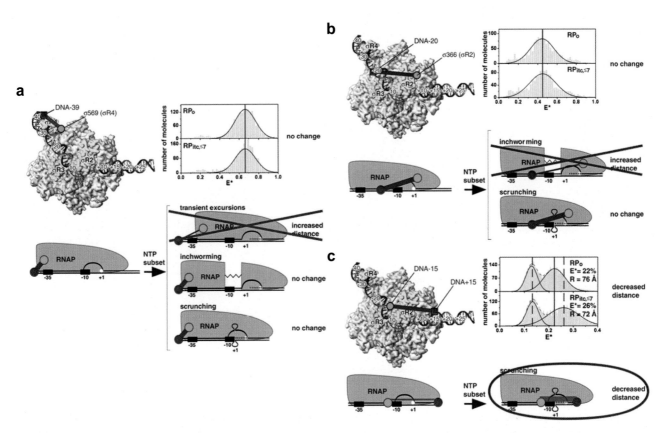

FIGURE 1.6: *Initial transcription by RNA polymerase. (a) No change detected for the FRET efficiency between the RNAP trailing edge and DNA promoter, ruling out the transient excursions model in which forward and reverse translocations of RNAP would have led to a decrease in FRET. (b) No change observed for the FRET efficiency between the RNAP leading edge and the promoter, suggesting that initial transcription does not involve inchworming either, in which the part of RNAP containing the catalytic core would slightly detach from the rest of the holoenzyme. (c) Evidence for the scrunching model, where the downstream DNA is pulled into the RNAP toward the upstream promoter, giving rise to an increase in FRET. Reproduced with permission from (Kapanadis et al., 2006).*

addition of a correct deoxyribonucleotide (dNTP), which was to be expected based on previous structural studies of DNAP. By dual-labeling the Klenow fragment of *E. coli* DNA polymerase I (Pol) with a donor-acceptor pair of fluorophores for FRET, Santoso and colleagues observed that the distributions of FRET efficiency for the binary Pol-DNA and ternary Pol-DNA-dNTP complexes were dominated by a low- and high-FRET peak, respectively, corresponding to the open and closed conformations of Pol. Surprisingly, free Pol in the absence of DNA substrate and dNTP gave a very broad distribution in FRET, which would be indicative of a high conformational flexibility. Given this broad distribution in the observed FRET efficiency and from the interpretation of a correlation analysis of the dynamics, Santoso and coworkers concluded that free Pol with no ligand bound fluctuates between the open and closed conformations on a millisecond timescale.

Furthermore, upon addition of an incorrect dNTP with a wrong base or a ribonucleotide triphosphate (NTP) with the correct base, novel FRET peaks were observed, implying the formation of new ternary complexes that are distinct from both the open and closed conformation mentioned

previously. These features are likely relevant to the mechanism by which DNAP discriminates against incorrect nucleotide subtrates for subsequent incorporation, which is critical to the high fidelity of these enzymes.

In a recent study of a DNA replisome, Patel and coworkers focused on investigating the coordination between leading- and lagging-strand DNA synthesis using single-molecule FRET and ensemble kinetics measurements (Pandey et al., 2009). Going into the study, it was already well accepted that replication of genomic DNA involves two molecules of DNAP at the replication fork, one being responsible for the continuous leading-strand synthesis and the other being in charge of replicating the lagging strand in a discontinuous fashion, through the formation of Okazaki fragment and a trombone loop (Benkovic, Valentine, and Salinas, 2001; O'Donnell, 2006). However, it was not completely clear how these two activities are synchronized with a similar apparent rate during replisome progression, taking into account the slow steps of RNA primer synthesis and subsequent transfer to the lagging-strand DNAP recycled from the previous Okazaki fragment. In contrast to a previous report (Lee et al., 2006), no pausing of the

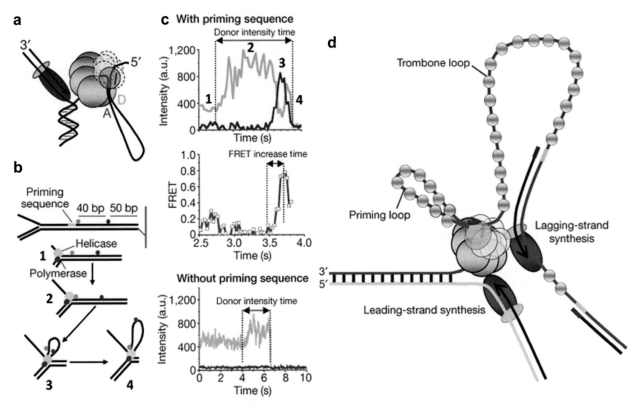

FIGURE 1.7: *Study of DNA replication coordination by the bacteriophage T7 replisome. (a) Experimental scheme for single-molecule FRET measurements. D and A represent the donor and acceptor fluorophores, respectively. (b) Different stages during the formation of priming loop. (c) A typical single-molecule fluorescence and FRET time trajectory, showing the stages depicted in (b) during the synthesis by T7 replisome. (d) A model for T7 DNA replication, in which the primase domain of T7 gp4 protein stays in contact with the priming sequence during unwinding and synthesis. This interaction leads to the formation of a priming loop and keeps the nascent RNA primer within reach of the lagging-strand polymerase. Reproduced with permission from (Pandey et al., 2009).*

bacteriophage T7 replisome was detected during the synthesis of RNA primers. By dual-labeling the lagging-strand DNA with two fluorophores at near the priming sequence and 40 base pairs downstream, Pandey and colleagues observed an increase in FRET between the two probes, suggesting the formation of a priming loop by the T7 gp4 primase/helicase protein (Figure 1.7a-c). In addition, replication of the gp2.5-coated lagging strand by T7 DNAP was found to be 30% faster than the leading-strand synthesis by the T7 replisome, consistent with the helicase being rate-limiting. These findings suggest a number of possible mechanisms by which coordination between the leading- and lagging-strand replication can be achieved. First, the replisome does not pause during primer synthesis, which would occur ahead of the ongoing lagging-strand synthesis. Second, the primase domain of the T7 gp4 protein maintains contact with the priming sequence during unwinding and synthesis, leading to formation of a priming loop and keeping the nascent RNA primer within reach of the lagging-strand polymerase (Figure 1.7d). Third, a faster synthesis by the lagging-strand polymerase allows for an overall synchronization with the leading-strand synthesis. Because the basic mechanisms of DNA

replication are highly conserved throughout evolution, the previously described feature for the T7 replication complexes are probably applicable to Bacteria and Eukaryotes as well (Benkovic et al., 2001; O'Donnell, 2006).

III.3 Application to a Study of Single-Stranded DNA-Binding Protein

As the last example in this chapter, we turn our attention to a study of the spontaneous motion of an *E. coli* single-stranded DNA-binding protein (SSB) on its DNA track recently reported by Ha and coworkers (Roy et al., 2009). This protein forms a stable homotetramer and binds to single-stranded DNA in various modes (Lohman and Ferrari, 1994). By employing two- and three-color single-molecule FRET schemes and using fluorescently labeled SSB, Roy and colleagues observed spontaneous migration of SSB along the track of single-stranded DNA, with an estimated diffusion coefficient of 270 (nucleotide)^2 per second at 37°C and an average step size of three nucleotides. According to Roy and colleagues, this observation marks the first example of any protein diffusing on single-stranded DNA and suggests a new mechanism by which proteins tightly bound to single-stranded DNA or RNA can redistribute themselves after the initial binding without fully dissociating. Moreover, Roy and colleagues found that such a diffusion-based migration can serve as a mechanism to melt short secondary structures of DNA transiently and even stimulate the formation of RecA filament on the DNA containing a hairpin structure, reflecting a possible role of this protein in DNA recombination and repair processes through coordination with other proteins for access to a common nucleic acid substrate.

IV. OUTLOOK

This chapter has shown that single-molecule FRET is becoming a standard technique in the studies of molecular machines. As the next decade arrives, we are certain that applications of this versatile method will be extended to studies of many other interesting biological nanomachines, with the scope limited only by our imagination. These applications will also benefit tremendously from the adaptation of useful, cutting-edge technologies being developed in other related fields including imaging, chemistry, material sciences, and cell biology, to name just a few. The combination of force-based manipulation and FRET measurement has already started to demonstrate its promise in the field called mechanobiology. As described earlier, incorporation of ZMWs for confinement of the illumination field in single-molecule FRET experiments will begin to be adapted in the near future, allowing for studies of more demanding interactions that are weak and transient. The development of a general, robust approach for specific, fluorescent labeling of proteins will soon appear (Shi, X., Jung, Y., Lin, L. J., Liu, C., Wu, C., et al., in preparation), further enhancing the applicability of single-molecule FRET in more sophisticated studies of molecular machines. In addition to these advances, a major emerging area of interest will be the investigation of the action mechanisms of these nanomachines in their native environments inside living cells or organisms, as their activities are carried out in real time. For example, how is leading- and lagging-strand synthesis coordinated by the replisome in vivo? This type of investigation, when achieved, will not only represent an enormous leap forward from a technical standpoint, but will greatly enhance our mechanistic understanding of these important biological nanodevices and even inspire the development of useful artificial nanomachines for biotechnological applications. We believe that the promise of single-molecule FRET will continue to grow rapidly in the many years to come.

ACKNOWLEDGMENTS

We thank the National Institutes of Health for financial support (grant GM65367) and Ruben Gonzalez and Joachim Frank for critical reading of the manuscript.

REFERENCES

Abbondanzieri, E. A., Bokinsky, G., Rausch, J. W., Zhang, J. X., Le Grice, S. F., et al. 2008. Dynamic binding orientations direct activity of HIV reverse transcriptase. *Nature*, 453, 184–9.

Andersen, C. B., Ballut, L., Johansen, J. S., Chamieh, H., Nielsen, K. H., et al. 2006. Structure of the exon junction core complex with a trapped DEAD-box ATPase bound to RNA. *Science*, 313, 1968–72.

Aregger, R. & Klostermeier, D. 2009. The DEAD box helicase YxiN maintains a closed conformation during ATP hydrolysis. *Biochemistry*, 48, 10679–81.

Atkinson, J. B., Gomperts, E. D., Kang, R., Lee, M., Arensman, R. M., et al. 1997. Prospective, randomized evaluation of the efficacy of fibrin sealant as a topical hemostatic agent at the cannulation site in neonates undergoing extracorporeal membrane oxygenation. *Am J Surg*, 173, 479–84.

Benkovic, S. J., Valentine, A. M. & Salinas, F. 2001. Replisome-mediated DNA replication. *Annu Rev Biochem*, 70, 181–208.

Bianco, P. R., Brewer, L. R., Corzett, M., Balhorn, R., Yeh, Y., et al. 2001. Processive translocation and DNA unwinding by individual RecBCD enzyme molecules. *Nature*, 409, 374–8.

Blanchard, S. C., Gonzalez, R. L., Kim, H. D., Chu, S. & Puglisi, J. D. 2004a. tRNA selection and kinetic proofreading in translation. *Nat Struct Mol Biol*, 11, 1008–14.

Blanchard, S. C., Kim, H. D., Gonzalez, R. L., Jr., Puglisi, J. D. & Chu, S. 2004b. tRNA dynamics on the ribosome during translation. *Proc Natl Acad Sci U S A*, 101, 12893–8.

Blosser, T. R., Yang, J. G., Stone, M. D., Narlikar, G. J. & Zhuang, X. 2009. Dynamics of nucleosome remodelling by individual ACF complexes. *Nature*, 462, 1022–7.

Carpousis, A. J. & Gralla, J. D. 1985. Interaction of RNA polymerase with lacUV5 promoter DNA during mRNA initiation and elongation. Footprinting, methylation, and rifampicin-sensitivity changes accompanying transcription initiation. *J Mol Biol*, 183, 165–77.

Cheng, W., Brendza, K. M., Gauss, G. H., Korolev, S., Waksman, G., et al. 2002. The 2B domain of the Escherichia coli Rep protein is not required for DNA helicase activity. *Proc Natl Acad Sci U S A*, 99, 16006–11.

Cheng, W., Hsieh, J., Brendza, K. M. & Lohman, T. M. 2001. E. coli Rep oligomers are required to initiate DNA unwinding in vitro. *J Mol Biol*, 310, 327–50.

Christian, T. D., Romano, L. J. & Rueda, D. 2009. Single-molecule measurements of synthesis by DNA polymerase with base-pair resolution. *Proc Natl Acad Sci U S A*, 106, 21109–14.

Cisse, I., Okumus, B., Joo, C. & HA, T. 2007. Fueling protein DNA interactions inside porous nanocontainers. *Proc Natl Acad Sci U S A*, 104, 12646–50.

Clamme, J. P. & Deniz, A. A. 2005. Three-color single-molecule fluorescence resonance energy transfer. *Chemphyschem*, 6, 74–7.

Cornish, P. V., Ermolenko, D. N., Noller, H. F. & HA, T. 2008. Spontaneous intersubunit rotation in single ribosomes. *Mol Cell*, 30, 578–88.

Cornish, P. V., Ermolenko, D. N., Staple, D. W., Hoang, L., Hickerson, et al. 2009. Following movement of the L1 stalk between three functional states in single ribosomes. *Proc Natl Acad Sci U S A*, 106, 2571–6.

Cornish, P. V. & Ha, T. 2007. A survey of single-molecule techniques in chemical biology. *ACS Chem Biol*, 2, 53–61.

Dohoney, K. M. & Gelles, J. 2001. Chi-sequence recognition and DNA translocation by single RecBCD helicase/nuclease molecules. *Nature*, 409, 370–4.

Edel, J. B., Eid, J. S. & Meller, A. 2007. Accurate single molecule FRET efficiency determination for surface immobilized DNA using maximum likelihood calculated lifetimes. *J Phys Chem B*, 111, 2986–90.

Eid, J., Fehr, A., Gray, J., Luong, K., Lyle, J., et al. 2009. Real-time DNA sequencing from single polymerase molecules. *Science*, 323, 133–8.

Fei, J., Bronson, J. E., Hofman, J. M., Srinivas, R. L., Wiggins, C. H., et al. 2009. Allosteric collaboration between elongation factor G and the ribosomal L1 stalk directs tRNA movements during translation. *Proc Natl Acad Sci U S A*, 106, 15702–7.

Fei, J., Kosuri, P., Macdougall, D. D. & Gonzalez, R. L., Jr. 2008. Coupling of ribosomal L1 stalk and tRNA dynamics during translation elongation. *Mol Cell*, 30, 348–59.

Graneli, A., Yeykal, C. C., Prasad, T. K. & Greene, E. C. 2006. Organized arrays of individual DNA molecules tethered to supported lipid bilayers. *Langmuir*, 22, 292–9.

Grashoff, C., Hoffman, B. D., Brenner, M. D., Zhou, R., Parsons, M., et al. 2010. Measuring mechanical tension across vinculin reveals regulation of focal adhesion dynamics. *Nature*, 466, 263–6.

Ha, T., Enderle, T., Ogletree, D. F., Chemla, D. S., Selvin, P. R., et al. 1996. Probing the interaction between two single molecules: fluorescence resonance energy transfer between a single donor and a single acceptor. *Proc Natl Acad Sci U S A*, 93, 6264–8.

Ha, T., Rasnik, I., Cheng, W., Babcock, H. P., Gauss, G. H., et al. 2002. Initiation and re-initiation of DNA unwinding by the Escherichia coli Rep helicase. *Nature*, 419, 638–41.

Hohng, S., Joo, C. & Ha, T. 2004. Single-molecule three-color FRET. *Biophys J*, 87, 1328–37.

Hohng, S., Zhou, R., Nahas, M. K., Yu, J., Schulten, K., et al. 2007. Fluorescence-force spectroscopy maps two-dimensional reaction landscape of the holliday junction. *Science*, 318, 279–83.

Joo, C., Balci, H., Ishitsuka, Y., Buranachai, C. & Ha, T. 2008. Advances in single-molecule fluorescence methods for molecular biology. *Annu Rev Biochem*, 77, 51–76.

Joo, C., Mckinney, S. A., Nakamura, M., Rasnik, I., Myong, S., et al. Real-time observation of RecA filament dynamics with single monomer resolution. *Cell*, 126, 515–27.

Kapanidis, A. N., Lee, N. K., Laurence, T. A., Doose, S., Margeat, E., et al. 2004. Fluorescence-aided molecule sorting: analysis of structure and interactions by alternating-laser excitation of single molecules. *Proc Natl Acad Sci U S A*, 101, 8936–41.

Kapanidis, A. N., Margeat, E., Ho, S. O., Kortkhonjia, E., Weiss, S., et al. 2006. Initial transcription by RNA polymerase proceeds through a DNA-scrunching mechanism. *Science*, 314, 1144–7.

Kapanidis, A. N., Margeat, E., Laurence, T. A., Doose, S., Ho, S. O., et al. 2005. Retention of transcription initiation factor sigma70 in transcription elongation: single-molecule analysis. *Mol Cell*, 20, 347–56.

Kim, J. L., Morgenstern, K. A., Griffith, J. P., Dwyer, M. D., Thomson, J. A., et al. 1998. Hepatitis C virus NS3 RNA helicase domain with a bound oligonucleotide: the crystal structure provides insights into the mode of unwinding. *Structure*, 6, 89–100.

Krummel, B. & Chamberlin, M. J. 1989. RNA chain initiation by Escherichia coli RNA polymerase. Structural transitions of the enzyme in early ternary complexes. *Biochemistry*, 28, 7829–42.

Lee, J., Lee, S., Ragunathan, K., Joo, C., Ha, T., et al. 2010. Single-Molecule Four-Color FRET. *Angew. Chem. Int. Ed.*, 49, 9922–5.

Lee, J. B., Hite, R. K., Hamdan, S. M., Xie, X. S., Richardson, C. C., et al. 2006. DNA primase acts as a molecular brake in DNA replication. *Nature*, 439, 621–4.

Lee, N. K., Kapanidis, A. N., Koh, H. R., Korlann, Y., Ho, S. O., et al. 2007. Three-color alternating-laser excitation of single molecules: monitoring multiple interactions and distances. *Biophys J*, 92, 303–12.

Levene, M. J., Korlach, J., Turner, S. W., Foquet, M., Craighead, H. G., et al. 2003. Zero-mode waveguides for single-molecule analysis at high concentrations. *Science*, 299, 682–6.

Linden, M. H., Hartmann, R. K. & Klostermeier, D. 2008. The putative RNase P motif in the DEAD box helicase Hera is dispensable for efficient interaction with RNA and helicase activity. *Nucleic Acids Res*, 36, 5800–11.

Liu, H. W., Zeng, Y., Landes, C. F., Kim, Y. J., Zhu, Y., et al. 2007. Insights on the role of nucleic acid/protein interactions in chaperoned nucleic acid rearrangements of HIV-1 reverse transcription. *Proc Natl Acad Sci U S A*, 104, 5261–7.

Liu, S., Abbondanzieri, E. A., Rausch, J. W., Le Grice, S. F. & Zhuang, X. 2008. Slide into action: dynamic shuttling of HIV reverse transcriptase on nucleic acid substrates. *Science*, 322, 1092–7.

Lohman, T. M. & Ferrari, M. E. 1994. Escherichia coli single-stranded DNA-binding protein: multiple DNA-binding modes and cooperativities. *Annu Rev Biochem*, 63, 527–70.

Margeat, E., Kapanidis, A. N., Tinnefeld, P., Wang, Y., Mukhopadhyay, J., et al. 2006. Direct observation of abortive initiation and promoter escape within single immobilized transcription complexes. *Biophys J*, 90, 1419–31.

Marians, K. J. 2004. Mechanisms of replication fork restart in Escherichia coli. *Philos Trans R Soc Lond B Biol Sci*, 359, 71–7.

Michalet, X., Weiss, S. & Jager, M. 2006. Single-molecule fluorescence studies of protein folding and conformational dynamics. *Chem Rev*, 106, 1785–813.

Myong, S., Bruno, M. M., Pyle, A. M. & Ha, T. 2007. Spring-loaded mechanism of DNA unwinding by hepatitis C virus NS3 helicase. *Science*, 317, 513–6.

Myong, S., Cui, S., Cornish, P. V., Kirchhofer, A., Gack, M. U., et al. 2009. Cytosolic viral sensor RIG-I is a 5′-triphosphate-dependent translocase on double-stranded RNA. *Science*, 323, 1070–4.

Myong, S., Rasnik, I., Joo, C., Lohman, T. M. & Ha, T. 2005. Repetitive shuttling of a motor protein on DNA. *Nature*, 437, 1321–5.

Niedziela-Majka, A., Chesnik, M. A., Tomko, E. J. & Lohman, T. M. 2007. Bacillus stearothermophilus PcrA monomer is a single-stranded DNA translocase but not a processive helicase in vitro. *J Biol Chem*, 282, 27076–85.

O'donnell, M. 2006. Replisome architecture and dynamics in Escherichia coli. *J Biol Chem*, 281, 10653–6.

Orr, A. W., Helmke, B. P., Blackman, B. R. & Schwartz, M. A. 2006. Mechanisms of mechanotransduction. *Dev Cell*, 10, 11–20.

Pal, M., Ponticelli, A. S. & Luse, D. S. 2005. The role of the transcription bubble and TFIIB in promoter clearance by RNA polymerase II. *Mol Cell*, 19, 101–10.

Pandey, M., Syed, S., Donmez, I., Patel, G., HA, T., et al. 2009. Coordinating DNA replication by means of priming loop and differential synthesis rate. *Nature*, 462, 940–3.

Park, J., Myong, S., Niedziela-Majka, A., Lee, K. S., YU, J., et al. 2010. PcrA helicase dismantles RecA filaments by reeling in DNA in uniform steps. *Cell*, 142, 544–55.

Rasnik, I., Myong, S., Cheng, W., Lohman, T. M. & Ha, T. 2004. DNA-binding orientation and domain conformation of the E. coli rep helicase monomer bound to a partial duplex junction: single-molecule studies of fluorescently labeled enzymes. *J Mol Biol*, 336, 395–408.

Rhoades, E., Gussakovsky, E. & Haran, G. 2003. Watching proteins fold one molecule at a time. *Proc Natl Acad Sci U S A*, 100, 3197–202.

Rothwell, P. J., Berger, S., Kensch, O., Felekyan, S., Antonik, M., et al. 2003. Multiparameter single-molecule fluorescence spectroscopy reveals heterogeneity of HIV-1 reverse transcriptase: primer/template complexes. *Proc Natl Acad Sci U S A*, 100, 1655–60.

Roy, R., Hohng, S. & Ha, T. 2008. A practical guide to single-molecule FRET. *Nat Methods*, 5, 507–16.

Roy, R., Kozlov, A. G., Lohman, T. M. & Ha, T. 2007. Dynamic structural rearrangements between DNA binding modes of E. coli SSB protein. *Journal of Molecular Biology*, 369, 1244–1257.

Roy, R., Kozlov, A. G., Lohman, T. M. & Ha, T. 2009. SSB protein diffusion on single-stranded DNA stimulates RecA filament formation. *Nature*, 461, 1092–7.

Sandler, S. J. 2000. Multiple genetic pathways for restarting DNA replication forks in Escherichia coli K-12. *Genetics*, 155, 487–97.

Santoso, Y., Joyce, C. M., Potapova, O., Le Reste, L., Hohlbein, J., et al. 2010. Conformational transitions in DNA polymerase I revealed by single-molecule FRET. *Proc Natl Acad Sci U S A*, 107, 715–20.

Schuler, B., Lipman, E. A. & Eaton, W. A. 2002. Probing the free-energy surface for protein folding with single-molecule fluorescence spectroscopy. *Nature*, 419, 743–7.

Scott, J. F., Eisenberg, S., Bertsch, L. L. & Kornberg, A. 1977. A mechanism of duplex DNA replication revealed by enzymatic studies of phage phi X174: catalytic strand separation in advance of replication. *Proc Natl Acad Sci U S A*, 74, 193–7.

Selvin, P. R. & Ha, T. 2008. *Single-molecule techniques: a laboratory manual.* Cold Spring Harbor, NY: Cold Spring Harbor Laboratory Press.

Sengoku, T., Nureki, O., Nakamura, A., Kobayashi, S. & Yokoyama, S. 2006. Structural basis for RNA unwinding by the DEAD-box protein Drosophila Vasa. *Cell*, 125, 287–300.

Shi, X., Jung, Y., Lin, L. J., Liu, C., Wu, C., et al. Quantitative, fluorescent labeling of aldehyde-tagged protein for single-molecule imaging. *In preparation*.

Shi, X., Basran, J., Seward, H. E., Childs, W., Bagshaw, C. R., et al. 2007. Anomalous negative fluorescence anisotropy in yellow fluorescent protein (YFP 10C): quantitative analysis of FRET in YFP dimers. *Biochemistry*, 46, 14403–17.

Shi, X., Lim, J. & Ha, T. 2010. Acidification of the oxygen scavenging system in single-molecule fluorescence studies: in situ sensing with a ratiometric dual-emission probe. *Anal Chem*, 82, 6132–8.

Shroff, H., Reinhard, B. M., Siu, M., Agarwal, H., Spakowitz, A., et al. 2005. Biocompatible force sensor with optical readout and dimensions of 6 nm3. *Nano Lett*, 5, 1509–14.

Sorokina, M., Koh, H. R., Patel, S. S. & Ha, T. 2009. Fluorescent lifetime trajectories of a single fluorophore reveal reaction intermediates during transcription initiation. *J Am Chem Soc*, 131, 9630–1.

Sternberg, S. H., Fei, J., Prywes, N., Mcgrath, K. A. & Gonzalez, R. L., JR. 2009. Translation factors direct intrinsic ribosome dynamics during translation termination and ribosome recycling. *Nat Struct Mol Biol*, 16, 861–8.

Straney, D. C. & Crothers, D. M. 1987. A stressed intermediate in the formation of stably initiated RNA chains at the Escherichia coli lac UV5 promoter. *J Mol Biol*, 193, 267–78.

Tarsa, P. B., Brau, R. R., Barch, M., Ferrer, J. M., Freyzon, Y., et al. 2007. Detecting force-induced molecular transitions with fluorescence resonant energy transfer. *Angew Chem Int Ed Engl*, 46, 1999–2001.

Theissen, B., Karow, A. R., Kohler, J., Gubaev, A. & Klostermeier, D. 2008. Cooperative binding of ATP and RNA induces a closed conformation in a DEAD box RNA helicase. *Proc Natl Acad Sci U S A*, 105, 548–53.

Uemura, S., Aitken, C. E., Korlach, J., Flusberg, B. A., Turner, S. W., et al. 2010. Real-time tRNA transit on single translating ribosomes at codon resolution. *Nature*, 464, 1012–7.

Velankar, S. S., Soultanas, P., Dillingham, M. S., Subramanya, H. S. & Wigley, D. B. 1999. Crystal structures of complexes of PcrA DNA helicase with a DNA substrate indicate an inchworm mechanism. *Cell*, 97, 75–84.

Yildiz, A., Forkey, J. N., Mckinney, S. A., Ha, T., Goldman, Y. E., et al. 2003. Myosin V walks hand-over-hand: single fluorophore imaging with 1.5-nm localization. *Science*, 300, 2061–5.

Yodh, J. G., Stevens, B. C., Kanagaraj, R., Janscak, P. & Ha, T. 2009. BLM helicase measures DNA unwound before switching strands and hRPA promotes unwinding reinitiation. *EMBO J*, 28, 405–16.

Zhuang, X., Bartley, L. E., Babcock, H. P., Russell, R., Ha, T., et al. 2000. A single-molecule study of RNA catalysis and folding. *Science*, 288, 2048–51.

CHAPTER 2

Visualization of Molecular Machines by Cryo-Electron Microscopy

Joachim Frank

I. INTRODUCTION

It is difficult nowadays to provide an introduction into cryo-EM within the space of a book chapter, given the current plethora of different methods, and the fact that there are as yet no agreed-on standards in the field. In view of this situation, the best course for the author is to provide the reader with an illustrated introduction into important concepts and strategies. However, at the same time, the focus on the molecular machine invites an expansion of scope in the most relevant section (Section IV), which concerns itself with heterogeneity, and the challenge to obtain an inventory of conformational states of a molecular machine in a single scoop.

I.1. Preliminaries: Cryo-EM as a Technique of Visualization

The transmission electron microscope (TEM) produces images that are projections of a three-dimensional object. To be more precise, the projections are line integrals of the three-dimensional Coulomb potential distribution representing the object. (For all practical purposes, especially in the resolution range down to to ~3 Å, the Coulomb potential distribution is identical to the electron density distribution "seen" by X-rays). Visualizing a molecular machine in three dimensions therefore entails the collection of multiple images showing the molecule in the same processing state (and hence, identical structure) but in different views. Thus the term "3D electron microscopy" is understood as a combination of two-dimensional imaging, following a particular data collection strategy, with three-dimensional reconstruction.

In the electron microscope, a high vacuum must be maintained that allows electrons to travel collision-free over the distances (in the order of meters) required for imaging. This requirement poses difficulties in imaging biological molecules, as they are hydrated. An initial solution to this problem was the negative staining technique, whereby heavy metal salt is added to the aqueous solution in which the molecules are suspended, and the sample is then air-dried on the grid so that each molecule is encased in a layer of heavy-metal stain. Here the contrast is mainly produced by stain exclusion, hence the finer details of the structure are lost even though the molecule is to some extent preserved in its three-dimensional aspects. Another solution to this problem is the use of glucose as an embedding medium (Unwin and Henderson, 1975), which, however, poses difficulties in its application to single molecules because the density of the medium is closely matched to the density of the protein. The third method, which addresses the problem of vacuum incompatibility without having the drawbacks of negative staining or glucose embedment, is the frozen-hydrated specimen preparation technique (Taylor and Glaeser, 1976; Dubochet et al., 1982): Here molecules are embedded in a thin layer of ice that is kept at a temperature where evaporation is negligible. In the following, we reserve the term "cryo-electron microscopy" (cryo-EM) for this particular preparation and imaging method.

I.2. Feasibility of Imaging Structure and Dynamics of Molecular Machines

In discussing the imaging of molecular machines by three-dimensional electron microscopy, we need to examine in which way the technique lends itself to the study of dynamics, or the different conformational states that either coexist or follow one another as the machine performs its work. In what has been termed *four-dimensional imaging* (e.g., Heymann et al., 2004) – the acquisition of a whole sequence of three-dimensional data as the molecule changes dynamically over time – one must keep in mind that, unlike the case of imaging with light microscopy where live processes can be analyzed relatively undisturbed, molecules are instantly incapacitated by the hostile conditions of sample preparation and the impact of the electron beam, making it impossible to collect data from the same dynamically changing molecule more than once. Thus, the "fourth dimension" must be introduced in a virtual manner, by collecting three-dimensional images from *separate* ensembles of molecules "frozen" at different states of processing and linking these states into "time lines" using additional evidence from different kinds of experiments such as single-molecule FRET. This kind of conjecture is by no means unambiguous,

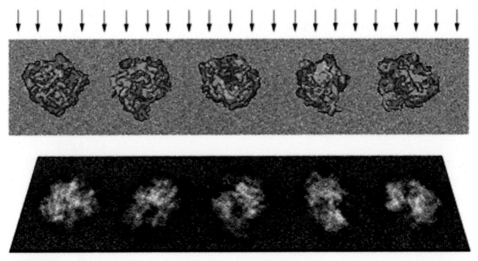

FIGURE 2.1: *Data collection for molecules in single-particle form. The molecule exists in multiple copies, all in different orientations in a matrix of vitreous (i.e., virtually amorphous) ice. The electron beam produces a set of projections, from which the molecule can be reconstructed.*

because two different states may be connected by multiple pathways, as is well known in enzyme kinetics (Boehr et al., 2006) and protein folding (Hartl and Hayer-Hartl, 2009).

Ideally we would wish to look at a molecular machine in the context of the cell, "frozen" at different time points in the act of performing its work. This type of imaging, however, is difficult to achieve, for several important reasons. One reason is that with a few exceptions, a cell's thickness exceeds the thickness that can be penetrated by electrons in the 100–300 kV range ($\sim$0.2 μm) by a large factor, necessitating the use of complicated techniques for high-pressure freezing of the cell and then sectioning the frozen cell to the required thickness (Hsieh et al., 2002; Al-Amoudi et al., 2004; see Koster and Barcena, 2006). Such slices can then be imaged by performing a tilt experiment using a series of small angular increments, collecting a set of projections from which a three-dimensional image can be reconstructed (a technique called "electron tomography" – see Frank, 2006a). The combination of three delicate techniques, especially the sectioning of the frozen specimen, is fraught with difficulties, producing a success rate of just a few percent. Another problem is that the molecule of interest is surrounded by a crowded environment of other structures, making it difficult to discern its boundaries. A third is the accumulation of dose, as many tilt images must be collected from the same area of the sample.

For all the reasons enumerated, imaging molecular machines, with the purpose of obtaining dynamic information, has to date been done almost exclusively using cryo-EM of in vitro samples combined with single-particle reconstruction. In vitro systems have been developed for a variety of fundamental processes in the cell, such as transcription, translation, protein degradation, and protein folding. Such systems, typically dependent in their action on ATP or GTP hydrolysis, may be trapped at key points of their processing path by the use of non-hydrolyzable analogs, or by small molecule inhibitors such as antibiotics – molecules that interfere with the dynamics of the process through targeted binding to specific sites, such as flexible hinge regions, which are required for mobility or for a critical binding interaction.

I.3. Cryo-EM of Molecular Machines in Single-Particle Form

Cryo-EM of molecules in single-particle form (as opposed to molecules occurring in ordered aggregates, such as two-dimensional crystals or helical bundles; Figure 2.1) is a technique of three-dimensional visualization developed over the past three decades (for recent introductory articles or reviews, see van Heel et al., 2000; Frank, 2006b; Wang and Sigworth, 2006; chapter on cryo-EM in Glaeser et al., 2007; Frank, 2009). It is suited for obtaining three-dimensional images of molecular machines in their native state, in vitro, captured in the process of performing their work.

The idea of data collection in such a case is to make use of the fact that the molecule occurs in multiple copies with (essentially) identical structure and that its orientation samples the entire angular range without leaving major gaps. Thus, instead of having to tilt the grid on which the sample is spread into multiple angles (as in electron tomography), it is then possible to take snapshots of multiple fields and, after suitable alignment, combine all projections into a density map depicting the molecule in three dimensions.

In comparison with X-ray crystallography, the absence of intermolecular contacts in the crystal means that the full range of native conformations and binding states can be observed. On the other hand, lack of a means for uniform fixation of all molecules in a definite conformation, as happens in a crystal, also has the consequence that atomic resolution is much more difficult to achieve, because normally none of the states is populated in sufficient numbers. Failure to reach atomic resolution, as we will see, is at least partially mitigated by methods of approximate atomic interpretation of the reconstructed 3D density map, mostly by flexible fitting of atomic structures obtained by X-ray crystallography.

By now, the combination of the two techniques – cryo-EM and single-particle reconstruction – has yielded spectacular results contributing insights on structure and function of many molecular machines, some of which are featured in separate chapters of this book.

II. THE BASIC TECHNIQUE

II.1. Specimen Preparation: Principle

In the basic experimental technique of cryo-EM, developed at the European Molecular Biology Laboratory (EMBL) in the early 1980s (Dubochet et al., 1982; see Dubochet et al., 1988) following the groundbreaking work by Kenneth Taylor and Robert Glaeser (1976; also see historical perspective by Taylor and Glaeser, 2008), a freeze-plunger (Figure 2.2) is employed to rapidly freeze the liquid aqueous sample, such that vitreous (amorphous) ice is formed. In this simple apparatus, an EM grid (3 mm in diameter, usually made of copper) is suspended at the tip of a pair of tweezers, which is in turn fastened to a vertically mounted, gravity-operated steel rod.

Of particular importance with regard to the imaging of molecular machines is the question to what extent the biological molecule is preserved in its native state. The transition of liquid water to a glass-like "vitreous" state induced by jet-freezing was first described by Brueggeler and Mayer (1980). Subsequently, Dubochet and coworkers demonstrated that this state could also be induced by rapid flash-freezing (or freeze-plunging) of an EM grid on which liquid sample has been deposited in a thin (~100 nm) layer, with the necessary thinning achieved by blotting off excess sample prior to plunging (Dubochet et al., 1982). The method by Dubochet et al. (1988) is now routinely used in the cryo-EM community.

According to studies of the physics of flash-freezing on small (several microns) protein crystals by Halle (2004), however, cryo-structures do not represent the equilibrium structure at the ambient temperature before the freezing, but exhibit extra heterogeneity that affects different parts of the structure differently. The author cites the example of a protein structure determined at 100° K, which might have exposed side chains sampling conformations related to 200° K substates while the same global backbone fold is maintained as at room temperature. These concerns, if

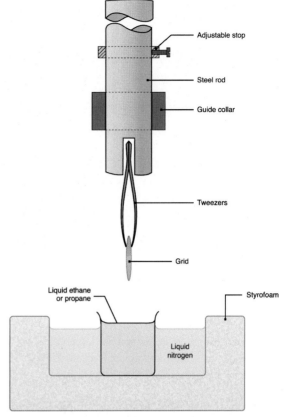

FIGURE 2.2: *Preparation of the specimen: Schematic of a freeze-plunger. A droplet of the buffer in which the molecules are suspended is applied to the electron microscope grid, which is held by tweezers. The tweezers are mounted on a gravity-operated rod held by a guide collar. Release of the rod results in the rapid immersion of the grid in the cryogen. The rapid plunge in temperature has the result that the aqueous solvent is converted into vitrified ice.*

indeed valid for cryo-EM preparation, may not carry much weight at resolutions at which side chains are not visible, but they will have to be taken into account in the interpretation as the field progresses.

The technique was originally developed for samples in which the molecules are highly ordered, such as in thin "two-dimensional" crystals or helical fibers, for which specific grid preparation methods were designed (see Glaeser et al., 2007). Indeed, ordered aggregates have two advantages: (1) They provide identical environments for the individual molecules, ensuring structural stability and uniformity; and (2) the ordered arrangement lends itself to the application of convenient Fourier processing methods. Most importantly, the structural information in images of two-dimensional crystals is concentrated in spots lying on the reciprocal lattice, making the separation of signal and noise straightforward. Similarly, in helical arrangements, structural information is concentrated in layer lines from where it can be retrieved following well-established rules. In the following, however, we deal exclusively with the sample preparation for single-molecule imaging, predicated on the assumption that the liquid sample contains large numbers of the same molecule in random orientations.

II.2. Specimen Preparation Protocol

A step-by-step protocol of the cryo-EM grid preparation has been given by Grassucci et al. (2007). At the start, a droplet of the liquid sample containing the molecules is applied to the grid. Excess liquid is blotted off using normal blotting paper, such that only a thin (ideally max. ~100 nm) liquid film remains. This step is critical, as increasing thickness leads to an increasing proportion of deleterious inelastic electron scattering and, eventually, to complete electron-opaqueness. On the other hand, a thickness close to, or smaller than, the size of the molecule will lead to the formation of damaged, partially "freeze-dried" samples. Critical parameters determining the thickness of the vitreous ice layer are temperature and humidity, and the length of time the sample is exposed to these conditions before the fast plunge. Because of the critical importance of these factors, manual operation is now increasingly replaced by use of computer-operated robots with a climate-controlled chamber (see Frederik and Storms, 2005).

Through the release of the rod (Figure 2.1), the EM grid is plunged into liquid ethane that is kept slightly above the temperature of liquid nitrogen (77.2° K) by means of a small heating coil. From that point on, the grid is kept at the liquid-nitrogen temperature, which serves to protect biological molecules from radiation damage (see Glaeser and Taylor, 1978). Essentially, the interpretation of the radiation protection is that the cleavage of bonds and fragmentation of the structure may still occur, but the low temperature "traps" or "cages" the fragments and prevents them from diffusing away. The radiation protection factor afforded by liquid nitrogen compared to room temperature is estimated as being in the range 5–10 (Glaeser et al., 2007). An additional factor may apply when going from the temperature of liquid nitrogen to that of liquid helium, depending on the nature of the specimen (crystalline or single-particle).

After the plunging, the grid is first stored in a storage box, from which it is transferred to a cryo-specimen holder or to a cartridge, depending on the type of electron microscope. In all these transfers, the grid is always kept at the temperature of liquid nitrogen. The specimen holder or cartridge is finally inserted in the EM. A step-by-step protocol is available in Grassucci et al. (2008), albeit specifically formulated for FEI (Portland, Oregon) instruments. This protocol, however, can be easily translated into a protocol for instruments of other manufacturers.

Imaging is done under low-dose conditions, again a measure designed to reduce radiation damage without compromising the image through excessive "shot noise," that is, noise due to the statistical variations of electron flux. "Low dose" is a somewhat diffuse term as used in the literature. Unwin and Henderson (1975), in their pioneering work on two-dimensional crystals of bacteriorhodopsin, used a dose of less than 1 $e^-/Å^2$, but at such a dose, molecules are virtually invisible. However, their sample was embedded in glucose and imaged at room temperature, without the radiation protection afforded by cooling to liquid nitrogen temperature, now routinely used in cryo-EM. Another factor is that higher electron voltages (200 to 300 kV versus 100 kV in Unwin and Henderson's experiment) lead to a reduced ratio of inelastic:elastic scattering (see below), hence reduced radiation damage. For these reasons, electron doses employed in the imaging of molecules are much higher now, in the range of 10 to 20 $e^-/Å^2$, enabling individual particles to be tracked down in the electron micrograph automatically by cross-correlation.

There has been a recent development in specimen preparation techniques that has the potential of revolutionizing cryo-EM imaging of molecular machines. If the objective is to image a macromolecular complex that has a particular ligand bound (for instance EF-G on the ribosome), the traditional way is to employ purification by affinity methods using a column prior to the preparation of the EM grid. The problem with this approach is that the ligand might come off in the process of the transfer, often leading to a heterogeneous mixture on the grid that must then be addressed by image classification (see Section IV.2 in this chapter). Kelly et al. (2008a; 2008b; 2010) developed a method of purification directly on the EM grid ("affinity grid"), ensuring that only molecules with the desired composition are recruited to the EM grid for imaging. The technique uses functionalized nickel-nitrilotriacetic acid (Ni-NTA) lipid monolayers deposited on the grid.

II.3. Imaging in the Electron Microscope

Electron micrographs are recorded either on film, for subsequent scanning in a microdensitometer, or by means of a charge-coupled device (CCD) camera for direct readout. Current CCD cameras that are affordable in price have a size of maximally 4,000 × 4,000 pixels, which gives a useful resolution of merely 2,000 × 2,000 independent pixels because of inter-pixel cross-talk – much less than the number of independent pixels on a film, estimated to be in the range of 6,000 × 10,000. Studies aiming at the highest spatial resolution are therefore still conducted by recording the TEM images on film (see, for instance, Seidelt et al., 2009). Electronic recording with comparable spatial resolution and high dynamic range awaits maturation and implementation of other technologies such as Complementary Metal-Oxide-Semiconductor (CMOS; see Faruqi, 2009).

Voltages being used are normally in the range of 200 to 300 kV. Voltages lower than 200 kV are avoided because of the increase in radiation damage, whereas voltages higher than 300 kV lead to diminishing contrast and inefficient electron recording. In the transmission electron

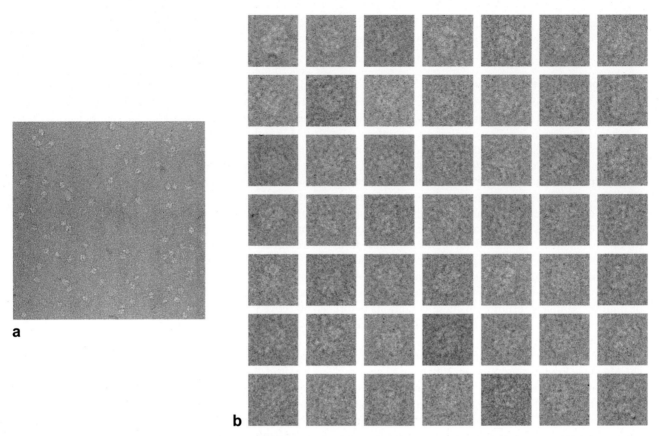

FIGURE 2.3: *Raw data collected on film with the FEI F30 electron microscope at 300 kV and a magnification of 59,000. (a) Micrograph of a ribosome complex. (b) Gallery of selected images.*

microscope, image formation is based on the scattering interaction between the electron beam and the atoms of the sample. Two types of interaction occur: elastic and inelastic scattering. The former is without dissipation of energy, such that a coherent relationship is maintained between scattered and unscattered beam, which is the origin of productive, high-resolution image formation. The latter, in contrast, is accompanied by energy loss, leading to an incoherent relationship between scattered and unscattered beams and making no productive contribution to the image. Thus the preference for higher voltages – within the limits set by the diminishing contrast – is based on the fact that the ratio between (good) elastic and (bad) inelastic scattering increases with voltage. However, going into details of image formation exceeds the scope of this chapter, and the interested reader is referred to authoritative treatments in the works by Spence (2003), Glaeser et al. (2007), and Reimer and Kohl (2008). A brief introduction is also provided in Frank (2006b).

III. IMAGE PROCESSING

There are numerous steps of image processing that bring us from the raw data to the finished product, the three-dimensional image. Extensive software packages have been written with specialization on electron microscopy, among these EMAN (Ludtke et al., 1999), SPIDER (Frank et al., 1996), FREALIGN (Grigorieff, 2007), IMAGIC (van Heel et al., 1996), MRC suite (Crowther et al., 1996), and XMIPP (Sorzano et al., 2004). (The latest special issue focused on EM-related software tools is the January 2007 issue of the *Journal of Structural Biology*). There is an increasing trend toward the use of multiple platforms, mixing and matching the modules as they have different strengths and weaknesses in the different areas of methodology. For example, Appion (Lander et al., 2009) is a pipeline designed to extend existing software applications and procedures, and provides transparent interconversions between the different file formats and angle conventions used in the various packages.

III.1. Particle Picking

In the TEM micrograph, the molecule projections are visible as very noisy "blobs" with faint contrast (Figure 2.3a). Candidate images (Figure 2.3b) are extracted from the micrographs with the aid of automated search algorithms, which normally, however, still require verification by eye. The search typically involves two-dimensional cross-correlation in some form. The cross-correlation signal is

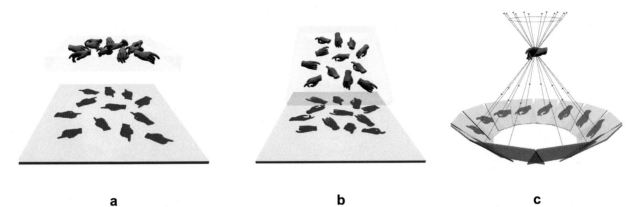

a   b   c

FIGURE 2.4: *Principle of the random-conical data collection and reconstruction. A given 3D object (here: the hand) is assumed to lie in a defined orientation on the grid, varying only in its azimuthal in-plane position. (a) When the grid is untilted, the projections of the different copies of the object remain the same, except for their rotation within the image. (b) When the grid is tilted, many different 3D projection views are realized. (c) In the coordination system of the object, the 3D projection directions form a cone. The term "random-conical" refers to the fact that the azimuths of the object are normally random, generating random placements in the conical projection geometry. Reproduced from (Frank, 1998) with permission.*

proportional to the contrast and the size of the particle. As a rule of thumb, molecules with molecular mass above 400 kD are easily recognized and processed even in the absence of symmetries.

Increasingly sophisticated methods of automated "particle picking" have been developed to replace the tedious manual selection and verification (recent example: Voss et al., 2009; see also special January 2004 issue of the

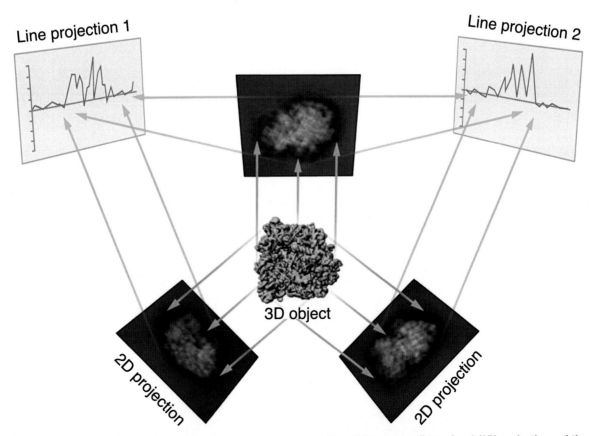

Line projection 1

Line projection 2

3D object

2D projection

2D projection

FIGURE 2.5: *Principle of the angular reconstitution technique. Two different two-dimensional (2D) projections of the same 3D object always have a one-dimensional (1D) line projection in common. By determining these pairwise common 1D line projections for a set of three or more 2D projections, one is able to determine the relative orientations of all 2D projections in a common reference system (van Heel, 1987). Adapted from (van Heel et al., 2000).*

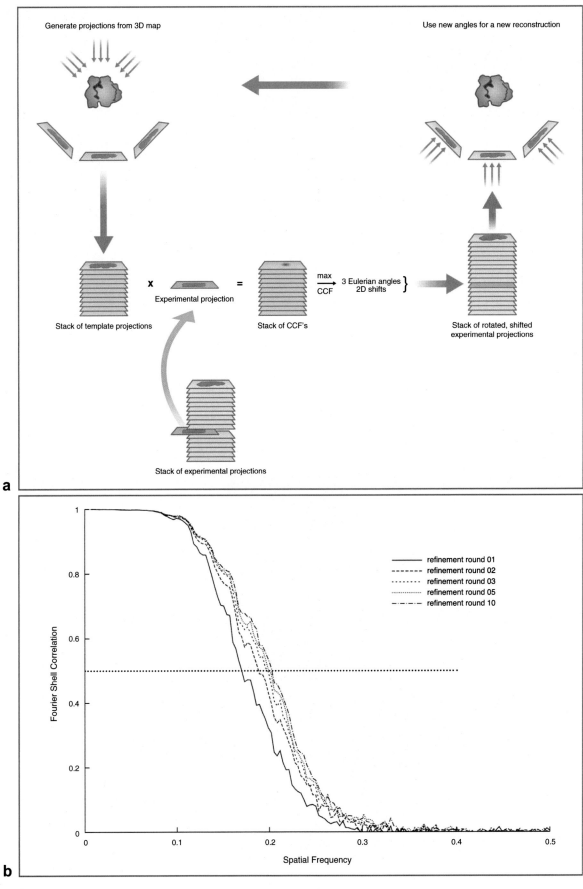

FIGURE 2.6

Journal of Structural Biology 145, Issues 1–2, devoted to this problem). Machine-learning algorithms are the latest addition to the tool set developed over time (Sorzano et al., 2009a; Langlois et al., 2011). The resulting fully verified data set may contain hundreds of thousands particles (i.e., projection images) ready to be processed (Figure 2.3b). There are two main reasons why such large numbers of images are needed: The low signal-to-noise ratio in the images – a consequence of the need to keep the electron dose low; and conformational heterogeneity, given that the minimum statistically required number of particles must be available for each conformation (see Section IV in this chapter).

III.2. Determination of Projection Angles for *Ab Initio* Reconstruction

Most of the work in single-particle reconstruction goes into the determination of projection angles, which are not *a priori* known. This problem can be compared to the indexing of X-ray diffraction patterns, a task that also places the data collected into a common three-dimensional framework. We distinguish *ab initio* methods from those that require a reference, or a preexisting model. When the structure is unknown, the project has to start with an *ab initio* reconstruction, then normally proceeds with a reference-based angular refinement using the *ab initio* reconstruction as reference. This will become clear in the following.

Two *ab initio* methods of reconstruction were developed in the 1980s: the random-conical reconstruction method (Figure 2.4; Radermacher et al., 1987; Radermacher, 1988) and the method of common lines (or "angular reconstitution") (Figure 2.5; van Heel, 1987; van Heel et al., 2000). The former uses a tilt pair to establish a coordinate system for a sub-population of the molecules that happen to lie in the same orientation on the grid, assigning a set of three Eulerian angles to each molecule. The latter *ab initio* technique uses intrinsic relationships among 2D projections of the same 3D structure – any two such projections have a central line in Fourier space in common, or, in an equivalent real-space formulation, they share a common 1D projection. For details on the mathematical principles and algorithms, the reader is referred to the original literature cited earlier. For a recent step-by-step protocol of the random-conical technique, see Shaikh et al. (2008).

A note on classification is in order here because both *ab initio* reconstruction techniques require classification of the experimental images. In the random-conical reconstruction, particles in the micrograph from the untilted grid have to be sorted into view classes, that is, groups of molecules presenting the same view. Random-conical reconstruction then proceeds separately for each view class. In the angular reconstruction technique, classification is required for a different reason: Here it is needed to boost the signal-to-noise ratio through the formation of class averages, because raw images are much too noisy to allow the comparison of common lines or 1D projections.

Classification of images is routinely done using a recipe developed by van Heel and Frank (1981): given a set of N windowed images, each containing M pixels. The images are first aligned to one another; that is, they are represented in a common two-dimensional reference system, then the whole array of $N \times M$ pixels is subjected to correspondence analysis or principal component analysis (van Heel and Frank, 1981; Lebart et al., 1984; Borland and van Heel, 1990), and some automated classification algorithms such as K-means or hierarchical ascendant classification are applied to the data represented in factor space (Frank, 1990; van Heel et al., 2000).

III.3. Three-Dimensional Reconstruction

Once the angles have been assigned by using either one of the *ab initio* methods, an initial coarse reconstruction is obtained. This is not the place to describe the methods of 3D reconstruction from projections, which has been covered in numerous articles and textbooks because of the wide range of applications; what makes single-particle reconstruction special in some way, affecting the design of algorithms, is that it is a true three-dimensional problem – unlike electron tomography with single-axis tilting – and that the angles are randomly distributed. Still, the best source specific to this subject are Michael

---

FIGURE 2.6: *(continued) Reference-based angle assignment and angular refinement. (a) Flowchart. At the outset, a library of reference or template projections is created from a given 3D reference map (left) on an initially coarse angular grid (see Figure 2.7). Each of the experimental projections is compared with all template projections by cross-correlation (denoted by the symbol X), yielding a stack of cross-correlation functions (middle). Search for the highest cross-correlation peak in the stack yields the three Eulerian angles and x,y shifts that best describe the placement of the projection in the 3D reference system. A reconstruction is performed on the whole stack of experimental data to which these transformations have been applied (right). The resulting reconstruction is the first estimate of the structure. The left-facing arrow at the top shows how this scheme is extended to form a closed loop, termed angular refinement: The previous reconstruction is used as the 3D reference in the next cycle, and so on. In each cycle, the angular spacing, initially 15 degrees, is narrowed so that finer and finer orientation-related features are captured. (b) Resolution, determined by the Fourier shell correlation, as a function of angular refinement iteration. Resolution is defined by the spatial frequency at which the Fourier shell correlation drops below 0.5. It is seen that resolution, so defined, increases steadily with each cycle, but the progress slows down. Cycle 10 is close to convergence.*

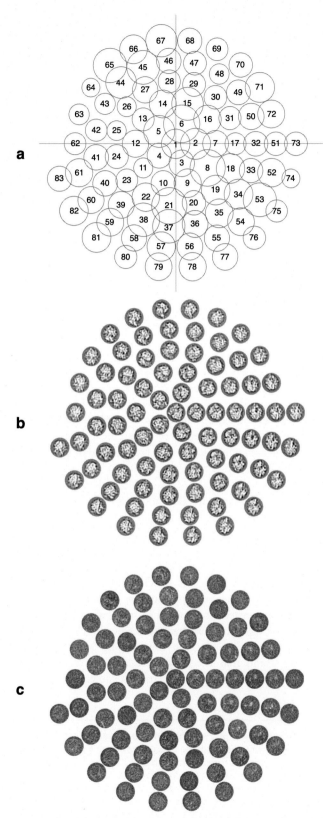

Radermacher's articles featuring weighted back-projection as the primary approach (Radermacher, 1988; Radermacher, 1991), although new approaches such as a fast gridding-based algorithm for arbitrary geometry (Penczek et al., 2004) and wavelet-based methods (Sorzano et al., 2004) have lately gained popularity.

The initial reconstruction from either *ab initio* method is typically of low resolution and may exhibit artifacts. Angular refinement is a generic term for an iterative procedure that, starting with an initial coarse reconstruction, seeks a solution that is optimally consistent with the experimental data (Penczek et al., 1994; Sorzano et al., 2009a) (Figure 2.6a). To this end, the given reconstruction is used as a 3D reference to generate an indexed library of 2D reference projections. Each experimental image is now matched with each image of the whole library to determine the rotation and shift giving maximum cross-correlation. This search yields an assignment, to the experimental projection, of three Eulerian angles and a 2D shift associated with the best match. After all experimental images have been assigned new angles, a new reconstruction can be computed, as shown on the right in Figure 2.6a. This refinement cycle is now repeated multiple times, with progressively decreasing angular increment, until some criterion of convergence has been fulfilled. A prime criterion for convergence is the progress in *resolution*.

Resolution can be estimated as follows (see recent review by Liao and Frank, 2010). In single-particle reconstruction, the absence of repeats (and symmetries, apart from certain classes of specimens such as viruses) means that the extent of the region in Fourier space where signal is present is not readily apparent – there are no diffraction spots to go by. Instead, one measures the reproducibility, in Fourier space, of reconstructions obtained from randomly drawn half-sets of the data. In other words, one measures the extent of the (usually spherical) region in Fourier space where significant correlation can be detected between the transforms of two reconstructions. The *Fourier shell correlation* curve is a plot of correlation between Fourier coefficients of the two half-set reconstructions averaged on a shell in Fourier space, as a function of shell radius. Typically, the curve falls off from the value 1 at low spatial frequencies to a value close to 0. As the angular refinement proceeds, the falloff region of the curve shifts toward higher spatial frequency (Figure 2.6b) until saturation is achieved.

Another useful criterion in monitoring progress of refinement looks at the number of particles switching angle

FIGURE 2.7: *Reference-based angle assignment in practice, as illustrated by a ribosome project. The Eulerian angles are ordered on a coarse angular grid (83 samples; 15° spacing) on a half-sphere. (a) Orientation statistics: The area of each circle is made proportional to the number of projection images assigned to it. Although there are orientational preferences, reflected by regions with larger circles, there are no real gaps (i.e., regions with circles of vanishing sizes). (b) Orientation class averages. Class averages are pasted into their angular positions in diagram (a). (c) Orientation class variances. Here two-dimensional variance patterns associated with the computations of class averages are pasted into diagram (a).*

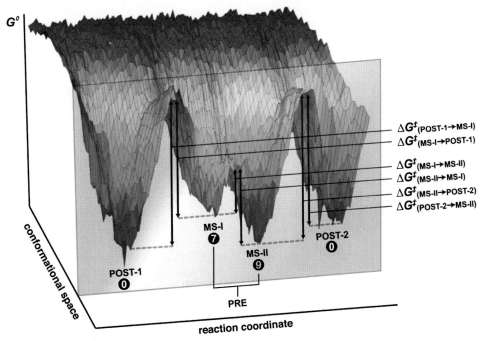

FIGURE 2.8: *Schematic diagram of the free-energy landscape for the elongation cycle of translation. The free-energy values correspond to rate constants obtained by smFRET; for details see Frank and Gonzalez (2010). Reproduced from (Frank and Gonzalez) with permission by Annual Reviews.*

assignment from one iteration to the next. This number typically stabilizes, indicating that the remaining particles have features that render them "indecisive," and that further progress cannot be expected.

IV. CRYO-EM OF HETEROGENEOUS SAMPLES AND DISCOVERY-BASED METHODS

IV.1. The Free-Energy Landscape and Inventory of States

The absence of constraints in the sample preparation of single molecules for cryo-EM implies that molecular machines can adopt all conformations and binding states compatible with the buffer conditions at the moment they are freeze-plunged. To attain the highest resolution, and to maximize the number of particles entering the reconstruction, one is interested in trapping as many molecules as possible in a single state, and indeed this was the strategy that was exclusively used in the beginning as the technique was being developed. However, as powerful classification methods have become available, opportunities for a different strategy have now opened up: to look at a sample of freely equilibrating molecules, with the idea to isolate subpopulations in all the states being present, for an exhaustive characterization of each of these subpopulations by a separate reconstruction (Connell et al., 2007; Fischer et al., 2010; Frank, 2010; Mulder et al., 2010).

This approach may be termed discovery-based because the data themselves are rich in latent "scrambled" information, which can be unscrambled by a complex process of analysis.

For further explanation, particularly in view of recent developments, it is helpful to make reference to a schematic diagram of the free-energy landscape of a molecular machine (Frank and Gonzalez, 2010), in this case the Bacterial ribosome, as inferred from smFRET (Figure 2.8). Valleys in this free-energy landscape, which correspond to our colloquial notion of "states," are well-populated, and molecules belonging to these sub-populations can be characterized by separate 3D reconstructions from the various projection classes – provided they can be separated, and provided the number of projections in each class is sufficient in statistical terms to support a reconstruction.

Depending on the ambient temperature – that is, depending on the amount of energy supplied by the thermal environment – molecules may freely interconvert among the different states, but any transient or short-lived intermediate states are too poorly populated to be captured by cryo-EM. Information on these transitions among different states is available through a complementary technique, single-molecule FRET (smFRET). Here fluorophores (a donor-acceptor pair) are placed on components whose distance is expected to change dynamically, and smFRET signals report on their distance changes in real time. smFRET has been applied to a variety of molecular machines, as described in separate chapters by Xinghua Shi and Takjep Ha (Chapter 1) and MacDougall and coworkers (Chapter 7). These smFRET results show, for instance, that the pre-translocational ribosome is constantly oscillating between two conformational states, termed macrostate I and II (or global state I and II in another terminology; see Fei et al., 2008) (Cornish et al., 2008), and possibly additional states intermediate between these conformations (Munro et al., 2007). As this example of the ribosome shows, the coexistence of multiple states giving rise to multiple conformations is a characteristic of molecular machines that are driven by Brownian motion.

Another example is the general transcription factor IID (TFIID), required for initiation of RNA polymerase II-dependent transcription at many Eukaryotic promoters. Here the binding by a protein (TATA-binding protein) induces a massive conformational change of the factor. Again, both binding states coexist in the sample, calling for a method to sort out the two states by analysis of the projection data (Elmlund et al., 2009).

IV.2. Classification: Disentangling Orientation and Conformation of the Molecules in the Projection Data

The coexistence of multiple states greatly complicates the reconstruction process in cryo-EM, because in addition to the projection *direction*, the subpopulation or *class* the projection belongs to must be determined. Given that each projection originates from a separate molecule, we are faced with an extremely difficult problem: Variations due to changes in orientation are intermingled with variations due to conformation or binding state.

Initial approaches to this problem of heterogeneity used *supervised classification*, that is, classification based on the similarity to two or more 3D references (see Valle et al., 2002). Sometimes, this scheme is hierarchically employed (Connell et al., 2007; Fischer et al., 2010). Though successful in many cases where the system is well characterized, the approach is unsatisfactory because it not only lacks generality, but also may miss conformations not anticipated, or come to incorrect conclusions if all of the references are dissimilar to the structure the experimental projections originated from. For the past few years, there has been an investment in the development of methods for *unsupervised classification*, which, by contrast, requires little or no prior information.

A position in between supervised and unsupervised methods holds those (Fu et al., 2007; Hall et al., 2007; Elad et al., 2008) that treat the conformational variability as a second-order problem, given that by far the largest variability in the data originates with the changes in orientation of the molecule. Hence, data are first sorted by orientation, then, in a second step, each orientation class is again classified into conformational classes. Finally, a third

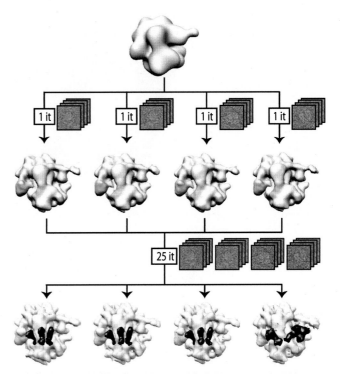

FIGURE 2.9: *Maximum likelihood (ML3D) classification of single-particle projections. A low-resolution "consensus" density map (top) is used as initial seed. The low resolution aims to minimize model bias. The data set is split into K subsets if K is the number of classes we are looking for. In the ribosome example (Scheres et al., 2007), K = 4 was chosen. A single iteration of ML3D refinement for each subset yields four density maps with random differences (second row). These maps are then used in a multi-reference ML3D refinement of the entire dataset. Bottom row presents the structures obtained after twenty-five iterations. Three of the maps prove to be virtually identical, showing the tRNA in the A, P, and E positions, whereas the fourth shows a single tRNA bound at the E site, as well as EF-G bound at the A site. Reproduced from (Scheres, 2010) with permission by Elsevier.*

step is required, which interrelates conformational classes from different orientations.

Entirely unsupervised methods fall in the following categories: maximum-likelihood classification (ML3D; Scheres et al., 2005; 2007 – see Figure 2.9), the bootstrap method (Penczek et al., 1996; Zhang et al., 2008; Hstau and Frank, 2010 – see Figure 2.10), and various methods based on multiple common lines comparisons (Herman and Kalinowski, 2007; Singer et al., 2009; Elmlund et al., 2010; Shatsky et al., 2010).

An example for the recovery of multiple conformational states from a single data set is provided in Figure 2.11. Here maximum-likelihood classification was applied to a set of 216,000 images of a pretranslocational ribosome complex, and four different conformations were extracted, three of which show the ribosome in different states of translocation.

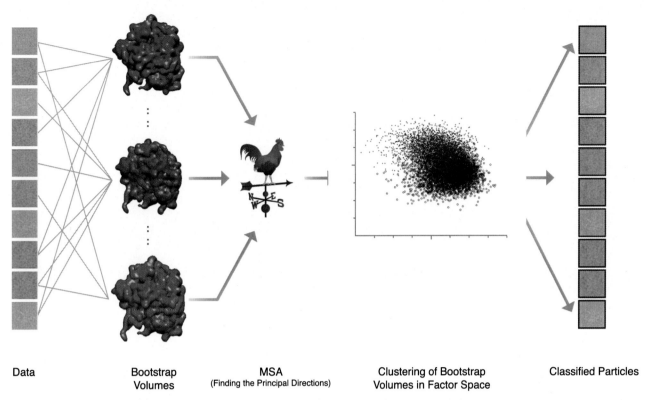

| Data | Bootstrap Volumes | MSA (Finding the Principal Directions) | Clustering of Bootstrap Volumes in Factor Space | Classified Particles |

FIGURE 2.10: Flowchart of classification based on bootstrap reconstructions. From the projection set containing N particle images, M < N samples are drawn with replacement. After each draw, a "bootstrap" reconstruction is computed. The resulting M reconstructions are analyzed by multivariate statistical analysis, resulting in a factorial representation of the bootstrap volumes in which K-means classification is performed and classes of bootstrap volumes are identified. For each particle image, its class association is finally determined.

IV.3. Time-Resolved Cryo-EM Imaging

As we have seen, the classification methods described earlier open new avenues of research by which multi-state mixtures can be studied. This technology offers the exciting opportunity to look at systems as they evolve over time; in other words, time-resolved three-dimensional imaging of molecular machines evolving from a non-equilibrium state. For sufficiently slow processes, such as virus capsid maturation (Heymann et al., 2003), ribosome biogenesis (Mulder et al., 2010), and back-translocation in translation (Fischer et al., 2010), the time for blotting and freeze-plunging the grid (in the range of several seconds) is negligible compared with the duration of the entire process, so time resolution is easily achievable by taking aliquots of the sample in regular intervals after the starting reaction for EM imaging. On the other hand, achieving time resolution on the order of a few milliseconds, such as required for studying decoding in translation (see Rodnina et al., 2002; Frank and Gonzalez, 2010), requires a radical departure from the established sample-preparation methods. Pioneering time-resolved techniques developed by Berriman and Unwin (1994) used the principle of spraying one reactant (in this instance, acetylcholine) onto a grid covered with the second reactant (acetylcholine receptor), so that the mixing

and reaction occurred on impact of the spray droplets on the target. As discussed in Lu et al. (2009), however, the unsharp definition of reaction time and possible interference of the grid with the target molecule are among the drawbacks of this method when aiming at millisecond resolution.

Specifically, the requirements may be formulated in the following way: First of all, a "zero" time point needs to be established, with little margin of error, denoting the point when the components are brought together. This calls for the design of an efficient mixer. Next, a provision needs to be made for the reaction in the fully mixed sample to take place for a defined time and under controlled conditions, in a suitable reaction chamber. Third, the reacted sample must be rapidly sprayed onto the grid, and the grid be plunged into the cryogen with a minimum of delay. The monolithic microfluidic mixing-spraying device developed by Lu et al. (2009), operated in conjunction with a traditional gravity-operated freeze-plunger, addresses these requirements successfully. Among the first applications of this novel device has been the dynamics of the association of ribosomal subunits into the 70S ribosome (Barnard et al., 2009; Shaikh et al., 2010).

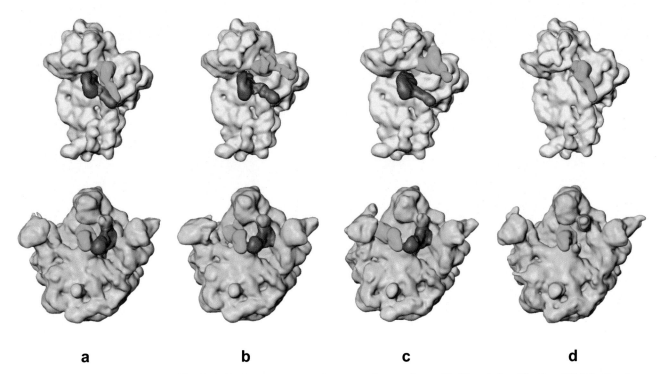

FIGURE 2.11: *Example for classification of a heterogeneous data set using maximum likelihood classification (ML3D). Specimen was a pretranslocational ribosome complex in the absence of EF-G described in Agirrezabala et al. (2008). For display of the inter-subunit space, the 70S ribosome was computationally separated into its two subunits, 30S (yellow) and 50S (blue). Supervised classification with two references identified two classes, distinguished by tRNA positions and presence versus absence of inter-subunit rotation. Number of classes specified was six. Of these, only five were found to be meaningful; the sixth class appeared to have acted as a "garbage bin" for misfits. Of the five remaining classes, one (panel d) lacks the A-site tRNA and hence is not a pretranslocational complex. Of the remaining four, two were found to be closely similar. We have therefore the result that three highly populated subpopulations (panels a, b, and c) exist, in which the positions of the tRNAs and ribosome conformation are well defined. Reproduced from (Frank, 2010, with permission by Wiley-VCH).*

IV.4. High-Throughput Data Collection and Processing

Given the fact that tens of thousands of images are required to obtain a sub-nanometer resolution reconstruction of a molecule in a single state, the diversity of coexisting states anticipated in the study of molecular machines means that the data collection and processing has to be streamlined and organized much more efficiently than currently practiced. In the quest for solutions to this problem, pioneering contributions have been made by the groups of Bridget Carragher and Clint Potter at The Scripps Institute, centered around two software tools, Leginon (Suloway et al., 2005) and Appion (Lander et al., 2009), both linked to the same database. Whereas Leginon is designed to collect data on the electron microscope automatically, Appion is a modular Python-based pipeline with interoperability with all major image-processing packages. A demonstration of efficient data collection and processing using these tools has been done by Stagg et al. (2006), who collected more than 200,000 images of GroEl and obtained a sub-nanometer reconstruction of this molecular machine in a span of a day.

Efficient organization of data storage and accelerated means of processing are obvious concerns in this context that are addressed by a variety of recent developments. Graphics Processor (GPU) implementations are now being developed to accelerate the execution of key algorithms such as projection matching significantly. All major EM-specific image-processing packages routinely make use of parallel processing, as many operations on stacks of images are parallel by nature. These new developments, however, are rapid and quite hardware-specific, precluding explicit coverage in this chapter.

V. OBTAINING AN ATOMIC MODEL

The evident goal of cryo-EM applied to a molecular machine is to obtain atomic models that show the machine in different processing states. Only in rare cases, in the presence of symmetries, cryo-EM density maps are of sufficient resolution to allow *de novo* tracing of the chains (Zhou, 2008; Lindert et al., 2009); such atomic models have been obtained for viruses (Yu et al., 2008) and for GroEl (Ludtke et al., 2008). The more typical situation is that atomic structures are available for the molecule and its functional ligands, and that these are used as the "raw material" for the composition of a comprehensive atomic model showing the conformation of the molecule and the binding interactions of its ligands. Alternatively, if the structure of the whole molecule has not been solved, at least structures of some components may be available.

A number of techniques of increasing sophistication have been developed over the years, parallel to the increase in resolution and demands for increasing fidelity of the fitting. Starting with the most straightforward approach, available structures are fitted as rigid bodies, essentially as an annotation of the (low-resolution) density map that shows putative or confirmed placements of relevant parts, giving an idea which parts of the cryo-EM density map are accounted for (see the example of the spliceosome; Stark and Lührmann, 2006). However, rigid-body docking will not do justice to the observed density map if conformational changes are present, which become noticeable with increasing resolution. Again the simplest solution is piece-wise rigid-body fitting: to break the atomic structure apart at putative flex points and place the individual pieces by eye into the EM map, to serve as an illustration for the kinds of local movements that must be invoked to bring the structure into agreement with the density. Quantitative techniques incorporating this idea, in the next advance of methodology, are SITUS (Wriggers and Chacón, 2001) and real-space refinement (RSREF; Chapman, 1995): The aim of the algorithms incorporated in these packages is to readjust the placement of pieces optimally within the local density by cross-correlation and mend their connections such that the structure is again complete, without violations of the rules of stereochemistry. An example of piece-wise flexible fitting using real-space refinement applied to the ribosome is the work by Gao et al. (2003; Gao and Frank, 2005). However, true *flexible fitting methods* go one step further and deform the structure in its entirety. One approach is normal-mode flexible fitting (NMFF; Tama et al. 2004), in which the structure is approximated by an elastic network of pseudoatoms and subjected to a normal mode analysis. Another approach, closer to the atomic nature of the structures to be fitted, is molecular dynamics flexible fitting (MDFF; Trabuco et al., 2008; see application in Villa et al., 2009; Grubisic et al., 2010).

be used, in effect, to reconstruct the free-energy landscape of the machine.

This capability invites a comparison with those of other biophysical visualization methods, nuclear magnetic resonance (NMR) and X-ray crystallography. A wide range of techniques in NMR spectroscopy allows characterization of conformational heterogeneity of macromolecules, depending critically on the timescales of inter-conversion between conformations. However, atomic resolution structure determination usually is limited to systems with a small (≤ 3) number of states (Vallurupalli et al., 2008) or to fitting of ensembles of conformations consistent with experimental data (Lange et al., 2008). Efficient transverse relaxation and spectral congestion normally limit the application of NMR spectroscopy to large molecular machines; however, TROSY techniques have enabled investigations of conformational dynamics in GroEL/GroES (Horst et al., 2005), SecA (Keramisanou et al., 2006), proteosome (Religa et al., 2010), and ribosome (Cabrita et al., 2009).

In X-ray crystallography, the state in which the molecule is visualized is defined by the energetics of crystal formation; under normal circumstances, the molecule exists in a single conformation. In some exceptional cases, two or more copies of the molecules may coexist in the unit cell in different conformations. For instance, Schuwirth et al. (2005) obtained the structure of the *E. coli* ribosome in two conformations, related by a change in small subunit head orientation, which is known to be associated with the translocation process. The same group also found ribosomes paired up in the unit cell, which were in different states of inter-subunit rotation (Zhang et al., 2009).

As a relatively new technique, cryo-EM still lacks uniform standards, data formats, and procedures. As is usually the case, such standards develop in the free marketplace of ideas. In this sense, its present state can be compared with the beginning years of protein X-ray crystallography. However, efforts on both sides of the Atlantic should be mentioned, which strive toward standardization in depositions of maps and an agreement on a common dictionary of terms (Berman et al., 2006; Lawson, 2010; Lawson et al., 2011; Velankar et al., 2010).

VI. CONCLUSIONS

Cryo-EM, making it possible to visualize molecules under native conditions, is one of the most important tools for the investigation of molecular machines. The field, as I have demonstrated here, currently experiences a widening of perspective, or a shift of paradigm: Instead of focusing on a single state, trapped by some kind of intervention, we now have the capability to make an entire inventory of all those conformational states that are well populated under the conditions of an experiment, and collect data that can

ACKNOWLEDGMENTS

This work was funded by HHMI and NIH R01 GM29169. I would like to thank Helen Saibil and Ruben Gonzalez for a critical reading and helpful comments, Bertile Halle for a discussion of possible freeze-plunge artifacts, and Art Palmer for input on the capabilities of NMR in dealing with heterogeneous samples. I further thank Lila Iino-Rubenstein and Melissa Thomas for assistance with the illustrations, and Jesper Pallesen and Hstau Liao for preparing material used in some of the figures.

REFERENCES

Agirrezabala, X. and Frank, J. (2009). Elongation in translation as a dynamic interaction among the ribosome, tRNA, and elongation factors EF-G and EF-Tu. *Q. Rev. Biophys.* 42, 139–158.

Agirrezabala, X. and Frank, J. (2010). From DNA to proteins via the ribosome: structural insights into the workings of the translational machinery. *Human Genomics* 4, 226–237.

Agirrezabala, X., Lei, J., Brunelle, J. L., Ortiz-Meoz, R. F., Green, R., and Frank, J. (2008). Visualization of the hybrid state of tRNA binding promoted by spontaneous ratcheting of the ribosome. *Mol. Cell.* 32, 190–197.

Al-Amoudi, A., Norlen, L. P., and Dubochet, J. (2004). Cryo-electron microscopy of vitreous sections of native biological cells and tissues. *J. Struct. Biol.* 148, 131–135.

Barnard, D., Lu, Z., Shaikh, T. R., Yassin, A., Mohamed, H., Agrawal, R., Lu, T.-M., and Wagenknecht, T. (2009). Time resolved cryo-electron microscopy of ribosome assembly using microfluidic mixing. *Microsc. Microan.* 15, 942–943.

Berman, H. M., Burley, S. K., Chiu, W., Sali, A., Adzhubei, A., Bourne, P. E., Bryant, S. H., Dunbrack, R. L., Jr., Fidelis, K., Frank, J., Adam Godzik, A., Henrick, K., Joachimiak, A., Heymann, B., Jones, D., Markley, J. L., Moult, J., Montelione, G. T., Orengo, C., Rossmann, M. G., Rost, B., Saibil, H., Schwede, T., Standley, D. M., and Westbrook, J. D. (2006). Outcome of a workshop on archiving structural models of biological macromolecules. *Structure* 14, 1211–1217.

Berriman, J. and Unwin, N. (1994). Analysis of transient structures by cryo-microscopy combined with rapid mixing of spray droplets. *Ultramicroscopy* 56, 241–252.

Boehr, D. D., McElheny, D., Dyson, H. J., and Wright, P. E. (2006). The dynamic energy landscape of dihydrofolate reductase catalysis. *Science* 313, 1638–1641.

Borland, L. and van Heel, M. (1990). Classification of image data in conjugate representation spaces. *J. Opt. Soc. Am.* A7, 601–610.

Brueggeler, P. and Mayer, E. (1980) Complete vitrification in pure liquid water and dilute aqueous solutions. *Nature* 288, 569–571.

Cabrita, L. D., Hsu, S. T., Launay, H., Dobson, C. M., and Christodoulou, J. (2009). Probing ribosome-nascent chain complexes produced in vivo by NMR spectroscopy. *Proc. Natl. Acad. Sci. USA* 106, 22239–22244.

Chapman, M. S. (1995). Restrained real-space macromolecular atomic refinement using a new resolution-dependent electron density function. *Acta Crystallogr.* A51, 69–80.

Connell, S. R., Takemoto, C., Wilson, D. N., Wang, H., Murayama, K., Terada, T., Shirouzu, M., Rost, M., Schüler, M., Giesebrecht, J., Dabrowski, M., Mielke, T., Fucini, P., Yokoyama, S., and Spahn, C. M. T. (2007). Structural basis for interaction of the ribosome with the switch regions of GTP-bound elongation factors. *Mol. Cell* 25, 751–764.

Cornish, P. V., Ermolenko, D. N., Noller, H. F., and Ha, T. (2008). Spontaneous intersubunit rotation in single ribosomes. *Mol Cell.* 30, 578–588.

Crowther, R. A., Henderson, R., and Smith, J. M. (1996). MRC image processing programs. *J. Struct. Biol.* 116, 9–16.

DoG Picker and Tilt Picker: software tools to facilitate particle selection in single particle electron microscopy. *J. Struct. Biol.* 166, 205–313.

Dubochet, J., Adrian, M., Chang, J., Homo, J.-C., Lepault, J., McDowell, A. W., and Schultz, P. (1988). Cryo-electron microscopy of vitrified specimens. *Q. Rev. Biophys.* 21, 129–228.

Dubochet, J., Lepault, J., Freeman, Z. R., Berriman, J. A., and Homo, J.-C. (1982). Electron microscopy of frozen water and aqueous solutions. *J. Microsc.* 128, 219–237.

Elad, N., Clare, D. K., Saibil H. R., and Orlova, E. V. (2008). Detection and separation of heterogeneity in molecular complexes by statistical analysis of their two-dimensional projections. *J. Struct. Biol.* 162, 108–120.

Elmlund, H., Baraznenok, V., Linder, T., Szilagyi, Z., Rofougaran, R., Hofer, A., Hebert, H., Lindahl, M., and Gustafsson, C. M. (2009). Cryo-EM reveals promoter DNA binding and conformational flexibility of the general transcription factor TFIID. *Structure* 17, 1442–1452.

Elmlund, D., Davis, R., and Elmlund, H. (2010). Ab initio structure determination from electron microscopic images of single molecules coexisting in different functional states. *Structure* 18, 777–786.

Faruqi, A. R. (2009). Principles and prospects of direct high-resolution electron image acquisition with CMOS detectors at low energies. *J. Phys.: Condens. Matter* 21, 314004 (9pp) doi:10.1088/0953–8984/21/31/314004.

Fei, J., Kosuri, P., MacDougall, D. D., and Gonzalez, R. L. Jr. (2008). Coupling of ribosomal L1 stalk and tRNA dynamics during translation elongation. *Mol. Cell* 30, 348–359.

Fischer, N., Konevega, A. L., Wintermeyer, W., Rodnina, M. V., and Stark, H. (2010) Ribosome dynamics and tRNA movement by time-resolved electron cryomicroscopy. *Nature* 466, 329–333.

Frank, J. (1990). Classification of macromolecular assemblies studied as single particles. *Quart. Rev. Biophys.* 23, 281–329.

Frank, J. (1998). How the ribosome works. *American Scientist* 86, 428–439.

Frank, J. (ed.) (2006a). *Electron tomography – methods for three-dimensional visualization of structures in the cell.* New York: Springer Verlag.

Frank, J. (2006b). *Three-dimensional electron microscopy of macromolecular assemblies – visualization of biological molecules in their native state.* New York: Oxford University Press.

Frank, J. (2009). Single-particle reconstruction of biological macromolecules in electron microscopy – 30 years. *Quart. Rev. Biophys.* 42, 139–158.

Frank, J. (2010). The ribosome comes alive. *Israeli J. Chem.* 50, 95–98.

Frank, J. and Gonzalez, R. (2010). Structure and dynamics of a processing Brownian motor: the translating ribosome. *Ann. Rev. Biochem.* 79, 381–412.

Frank, J., Radermacher, M., Penczek, P., Zhu, J., Li, Y., Ladjadj, M., and Leith, A. (1996). SPIDER and WEB: processing and visualization of images in 3D electron microscopy and related fields. *J. Struct. Biol.* 116, 190–199.

Frederik, P. M. and Storms, M. H. (2005). Automated robotic preparation of vitrified samples for 2D and 3D cryo electron microscopy. *Microsc. Today* 13, 32–36.

Fu, J., Gao, H., and Frank, J. (2007). Unsupervised classification of single particles by cluster tracking in multi-dimensional space. *J. Struct. Biol.* 157, 226–239.

Fu, J., Kennedy, D., Munro, J.B., Lei, J., Blanchard, S. C., and Frank, J. (2009). The P-site tRNA reaches the P/E position through intermediate positions. (Abstract) *J. Biomol. Struct. Dyn.* 26, 794–795.

Gao, H. and Frank, J. (2005). Molding atomic structures into intermediate-resolution cryo-EM density maps of ribosomal complexes using real-space refinement. *Structure* 13, 401–406.

Gao, H., Sengupta, J., Valle, M., Korostelev, A., Eswar, N., Stagg, S. M., Van Roey, P., Agrawal, R. K., Harvey, S. C., Sali, A., Chapman, M. S., and Frank, J. (2003). Study of the structural dynamics of the *E. coli* 70S ribosome using real space refinement. *Cell* 113, 789–801.

Glaeser, R. M., Downing, K. H., DeRosier, D., Chiu, W., and Frank, J. (2007). *Electron crystallography of biological macromolecules.* New York: Oxford University Press.

Glaeser, R. M. and Taylor, K. A. (1978). Radiation damage relative to transmission electron microscopy of biological specimens at low temperature: review. *J. Microsc.* 112, 127–138.

Grassucci, R. A., Taylor, D. J., and Frank, J. (2007). Preparation of macromolecular complexes for cryo-electron microscopy. *Nat. Protoc.* 2, 3239–3246.

Grassucci, R.A., Taylor, D., and Frank, J. (2008). Visualization of macromolecular complexes using cryo-electron microscopy with FEI Tecnai transmission electron microscopes. *Nat. Protoc.* 3, 330–339.

Grigorieff, N. (2007). FREALIGN: High-resolution refinement of single particle structures. *J. Struct. Biol.* 157, 117–125.

Grubisic, I., Shokhirev, M. N., Orzechowski, M., Miyashita, O., and Tama, F. (2010). Biased coarse-grained molecular dynamics simulation approach for flexible fitting of X-ray structure into cryo electron microscopy maps. *J. Struct. Biol.* 169, 95–105.

Hall, R. J., Siridechadilok, B., Nogales, E. (2007). Cross-correlation of common lines: a novel approach for single-particle reconstruction of a structure containing a flexible domain. *J. Struct. Biol.* 159, 474–482.

Halle, B. (2004). Biomolecular cryocrystallography: structural changes during flash-cooling. *Proc. Natl. Acad. Sci. USA* 101, 4793–4798.

Hartl, F. U. and Hayer-Hartl, M. (2009). Converging concepts of protein folding *in vitro* and in vivo. Nat. *Struct. Mol. Biol.* 16, 574–581.

Herman, G. T. and Kalinowski, M. (2007). Classification of heterogeneous electron microscopic projections into homogeneous subsets. *Ultramicroscopy* 108, 327–338.

Heymann, J. B., Cheng, N., Newcomb, W. W., Trus, B. L., Brown, J. C., and Steven, A. C. (2003). Dynamics of herpes simplex virus capsid maturation visualized by time-lapse cryo-electron microscopy. *Nat. Struct. Biol.* 10, 334–341.

Heymann, J. B., Conway, J. F., and Steven, A. C. (2004). Molecular dynamics of protein complexes from four-dimensional cryo-electron microscopy. *J. Struct. Biol.* 147, 291–301.

Horst, R., Bertelsen, E. B., Fiaux, J., Wider, G., Horwich, A. L., and Wüthrich, K. (2005). Direct NMR observation of a substrate protein bound to the chaperonin GroEL. *Proc. Natl. Acad. Sci. USA.* 102, 12748–12753.

Hsieh, C.-E., Marko, M., Frank, J., and Mannella, C. A. (2002). Electron tomographic analysis of frozen-hydrated tissue sections. *J. Struct. Biol.* 138, 63–73.

Kelly, D. F., Abeyrathne, P. D., Dukovski, D., and Walz, T. (2008a). The affinity grid: a prefabricated EM grid for monolayer purification. *J. Mol. Biol.* 382, 423–433.

Kelly, D. F., Dukovski, D., and Walz, T. (2008b). Monolayer purification: a rapid method for isolating protein complexes for single-particle electron microscopy. *Proc. Natl. Acad. Sci. USA* 105, 4703–4708.

Kelly, D. F., Dukovski, D. and Walz, T. (2010). Strategy for the use of affinity grids to prepare non-His-tagged macromolecular complexes for single-particle electron microscopy. *J. Mol. Biol.* 400, 675–681.

Keramisanou, D., Biris, N., Gelis, I., Sianidis, G., Karamanou, S., Economou, A., and Kalodimos, C. G. (2006). Disorder-order folding transitions underlie catalysis in the helicase motor of SecA. *Nat. Struct. Mol. Biol.* 13, 594–602.

Koster, A. J. and Barcena, M. (2006). Cryotomography: low-dose automated tomography of frozen-hydrated specimens. In: *Electron tomography – methods for three-dimensional visualization of structures in the cell* (pp. 113–161), ed. J. Frank. New York: Springer.

Lander, G. C., Stagg, S. M., Voss, N. R., Cheng, A., Fellmann, D., Pulokas, J., Yoshioka, C., Irving, C., Mulder, A., Lau, P.-W., Lyumkis, D., Potter, C. S., and Carragher, B. (2009). Appion: an integrated, database-driven pipeline to facilitate EM image processing. *J. Struct. Biol.* 166, 95–102.

Lange, O. F., Lakomek, N. A., Farès, C., Schröder, G. F., Walter, K. F., Becker, S., Meiler, J., Grubmüller, H., Griesinger, C., and de Groot, B. L. (2008). Recognition dynamics up to microseconds revealed from an RDC-derived ubiquitin ensemble in solution. *Science* 320, 1471–1475.

Langlois, R., Pallesen, J., and Frank, J. (2011). Reference-free segmentation enhanced with data-driven template matching for particle selection in cryo-electron microscopy. *J. Struct. Biol.*, in press.

Lawson, C. L., Baker, M. L., Best, C., Bi, C., Dougherty, M., Feng, P., van Ginkel, G., Devkota, B., Lagerstedt, I., Ludtke, S., Newman, R. H., Oldfield, T. J., Rees, I., Sahni, G., Sala, R., Velankar, S., Warren, J., Westbrook, J. D., Henrisck, K., Kleywegt, G. J., Berman, H. M, and Chiu, W. (2011). EMDataBank.*org: unified data resource for cryoEM.* Nucl. Acid. Res. (Database issue), 34, 287–290.

Lawson, C. L. (2010). Unified data resource for Cryo-EM. *Methods Enzymol.* 483, 73–90.

Lebart, L., Maurineau, A., and Warwick, K. M. (1984). *Multivariate descriptive statistical analysis.* New York: John Wiley.

Liao, H. and Frank, J. (2010). Definition and estimation of resolution in single-particle reconstructions. *Structure* 18, 768–775.

Lindert, S., Stewart, P. L., and Meiler, J. (2009). Hybrid approaches: applying computational methods in cryo-electron microscopy. *Curr. Opin. Struct. Biol.* 19, 218–225.

Lu, Z., Shaikh, T. R., Barnard, D., Meng, X., Mohamed, H., Yassin, A., Mannella, C. A., Agrawal, R. K., Lu, T.-M., and Wagenknecht, T. (2009). Monolithic microfluidic mixing-spraying devices for time-resolved cryo-electron microscopy. *J. Struct. Biol.* 168, 388–395.

Ludtke, S. J., Baker, M .L., Chen, D.-H., Song, J.-L., Chuang, D. T., and Wah Chiu, W. (2008). *De novo* backbone trace of GroEL from single particle electron cryomicroscopy. *Structure* 16, 441–448.

Ludtke, S. J., Baldwin, P. R., and Chiu, W. (1999). EMAN: semi-automated software for high-resolution single-particle reconstructions. *J. Struct. Biol.* 128, 82–97.

Mulder, A. M., Yoshioka, C., Beck, A. H., Bunner, A. E., Milligan, R. A., Potter, C. S., Carragher, B., and Williamson, J. R. (2010). Visualizing ribosome assembly: a structural mechanism for 30S subunit assembly. *Science* 330, 673–677.

Munro, J. B., Altman, R. B., O'Connor, N., and Blanchard, S. C. (2007). Identification of two distinct hybrid state intermediates on the ribosome. *Mol. Cell* 25, 505–517.

Munro, J. B., Sanbonmatsu, K. Y., Spahn, C. M., and Blanchard, S. C. (2009). Navigating the ribosome's metastable energy landscape. *Trends Biochem. Sci.* 34, 390–400.

Orzechowski, M., and Tama, F. (2008). Flexible fitting of high-resolution X-Ray structures into cryoelectron microscopy maps using biased molecular dynamics simulations. *Biophys. J.* 95, 5692–5705.

Penczek, P. A., Frank, J., and Spahn, C. M. T. (2006). A method of focused classification, based on the bootstrap 3D variance analysis, and its application to EF-G-dependent translocation. *J. Struct. Biol.* 154, 184–194.

Penczek, P., Grassucci, R., and Frank, J. (1994). The ribosome at improved resolution: new techniques for merging and orientation refinement in 3D cryo electron microscopy of biological particles. *Ultramicroscopy* 53, 251–270.

Penczek, P. A., Renka, R., and Schomberg, H. (2004). Gridding-based direct Fourier inversion of the three-dimensional ray transform. *J. Opt. Soc. Amer.* A21, 499–509.

Penczek, P. A., Zhu, J., and Frank, J. (1996). A common-lines based method for determining orientations for N>3 particle projections simultaneously. *Ultramicroscopy* 63, 205–218.

Radermacher, M. (1988). The three-dimensional reconstruction of single particles from random and non-random tilt series. *J. Electron Microsc. Tech.* 9, 359–394.

Radermacher, M. (1991). Three-dimensional reconstruction of single particles in electron microscopy. In: *Image Analysis in Biology* (D.-P. Haeder, ed.), pp. 219-249. CRC Press, Boca Raton, Fl.

Radermacher M., Wagenknecht T., Verschoor A., and Frank J. (1987). Three-dimensional reconstruction from a single-exposure, random conical tilt series applied to the 50S ribosomal subunit of *Escherichia coli. J. Microsc.* 146, 113–136.

Reimer, L. and Kohl, H. (2008). *Transmission electron microscopy: physics of image formation (5th Edition)*. New York: Springer Verlag.

Religa, T. L., Sprangers, R., and Kay, L. E. (2010). Dynamic regulation of archaeal proteasome gate opening as studied by TROSY NMR. *Science* 328, 98–102.

Rodnina, M. V., Daviter, T., Gromadski, K., and Wintermeyer, W. (2002). Structural dynamics of ribosomal RNA during decoding on the ribosome. *Biochimie* 84, 745–54.

Scheres, S. H. (2010). Visualizing molecular machines in action: single particle analysis with structural variability. In: *Recent advances in electron cryomicroscopy* (ed. Steve Ludtke). Advances in Protein Chemistry and Structural Biology, 81, 89–119.

Scheres, S. H., Gao, H., Valle, M., Herman, G. T., Eggermont, P. P., Frank, J., and Carazo, J. M. (2007). Disentangling conformational states of macromolecules in 3D-EM through likelihood optimization. *Nat. Methods* 4, 27–29.

Scheres, S. H., Valle, M., Nunez, R., Sorzano, C. O., Marabini, R., Herman, G. T., and Carazo, J. M. (2005). Maximum-likelihood multi-reference refinement for electron microscopy images. *J. Mol. Biol.* 348, 139–149.

Schuwirth, B .S., Borovinskaya, M. A., Hau, C. W., Zhang, W., Vila-Sanjurjo, A., Holton, J. M., and Doudna Cate, J. H. (2005). Structures of the bacterial ribosome at 3.5 Å. *Science* 310, 827–834.

Seidelt , B., Innis, C. A., Wilson, D. N., Gartmann, M., Armache, J.-P., Villa, E., Trabuco, L. G., Becker, T., Mielke, T., Schulten, K., Steitz, T. A., and Beckmann, R. (2009). Structural insight into nascent polypeptide chain–mediated translational stalling. *Science* 326, 1412–1415.

Shaikh, T. R., Gao, H., Baxter, W. T., Asturias, F. J., Boisset, N., Leith, A., and Frank, J. (2008). SPIDER image processing for single-particle reconstruction of biological macromolecules from electron micrographs. *Nat. Protoc.* 3, 1941–1974.

Shaikh, T. R., Yassin, A., Lu, Z., Barnard, D. Meng, X., Mohamed, H., Lu, T.-M., Wagenknecht, T., and Agrawal, R. K. (2010). Association of the ribosomal subunits as studied by time-resolved cryo-EM. (Abstract) *Microscopy and Microanalysis* 15 (suppl. 2), pp. 974–975.

Shatsky, M., Hall, R. J., Nogales, E., Malik, J., and Brenner, S. (2010). Automated multi-model reconstruction from single-particle electron microscopy data. *J. Struct. Biol.* 170, 98–108.

Singer, A., Coifman, R. R., Sigworth, F. J., Chester, D. W., and Shkolnisky, Y. (2009). Detecting consistent common lines in cryo-EM by voting. *J. Struct. Biol.* 169, 312–322.

Sorzano, C. O., Marabini, R., Velázquez-Muriel, J., Bilbao-Castro, J. R., Scheres, S. H., Carazo, J. M., and Pascual-Montano, A. (2004). XMIPP: a new generation of an open-source image processing package for electron microscopy. *J. Struct. Biol.* 148, 194–204.

Sorzano, C. O. S., Jonić, S., El-Bez, C., Carazo, J. M., De Carloe, S., Thévenaz P., and Unser, M. (2009a). A multiresolution approach to orientation assignment in 3D electron microscopy of single particles. *J. Struct. Biol.* 146, 381–392.

Sorzano, C. O. S., Recarte, E., Alcorlo, M., Bilbao-Castro, J. R., San-Martín, C., Marabini, R., and Carazo, J. M. (2009b). Automatic particle selection from electron micrographs using machine learning techniques. *J. Struct. Biol.* 167, 252–260.

Spence, J. (2003). *Experimental HREM.* (3rd Edition). New York: Oxford University Press.

Stagg, S. M., Lander, J. C., Pulokas, J., Fellmann, D., Cheng, A., Quispe, J. D., Mallick, S. P., Avila, R. M., Carragher, B., and Potter, C.S. (2006). Automated cryoEM data acquisition and analysis of 284 742 particles of GroEL. *J. Struct. Biol.* 155, 470–481.

Stark, H. and Lührmann, R. (2006). Cryo-Electron microscopy of spliceosomal Components. *Annu. Rev. Biophys. Biomol. Struct.* 35, 435–457.

Suloway, C., Pulokas, J., Fellmann, D., Cheng, A., Guerra, F., Quispe, J., Stagg, S., Potter, C.S., and Carragher, B. (2005). Automated molecular microscopy: the new Leginon system. *J. Struct. Biol.* 151, 41–60.

Tama, T., Miyashita, O., and Brooks III, C. L. (2004). Normal mode based flexible fitting of high-resolution structure into low-resolution experimental data from cryo-EM. *J. Struct. Biol.* 147, 315–326.

Taylor, K. A. and Glaeser, R. M. (1976). Electron microscopy of frozen-hydrated biological specimens. *J. Ultrastruct. Res.* 55, 448–456.

Taylor, K. A. and Glaeser, R. M. (2008). Retrospective on the early development of cryoelectron microscopy of macromolecules and a prospective on the opportunities for the future. *J. Struct. Biol.* 163, 214–223.

Trabucco, L. G., Villa, E., Mitra, K., Frank, J., and Schulten, K. (2008). Flexible fitting of atomic structures into electron microscopy maps using molecular dynamics. *Structure* 16, 673–683.

Unwin, P. N., Henderson, R. (1975). Molecular structure determination by electron microscopy of unstained crystalline specimens. *J. Mol. Biol.* 94, 425–440.

Valle, M., Sengupta, J., Swami, N. K., Grassucci, R. A., Burkhardt, N., Nierhaus, K. H., Agrawal, R. K., and Frank, J. (2002). Cryo-EM reveals an active role for aminoacyl-tRNA in the accommodation process. *EMBO J.* 21, 3557–3567.

Vallurupalli, P., Hansen, D. F., and Kay, L. E. (2008). Structures of invisible, excited protein states by relaxation dispersion NMR spectroscopy. *Proc. Natl. Acad. Sci. USA.* 105, 11766–11771.

van Heel, M. (1987). Angular reconstitution – a posteriori assignment of projection directions for 3-D reconstruction. *Ultramicroscopy* 21, 111–123.

van Heel, M. and Frank, J. (1981). Use of multivariate statistical analysis in analysing the images of biological macromolecules. *Ultramicroscopy* 6, 187–194.

van Heel, M., Gowen, B., Matadeen, R., Orlova, E. L., Finn, R., Pape, T., Cohen, D., Stark, H., Schmidt, R., Schatz, M., and Patwardhan, A. (2000). Single-particle electron cryomicroscopy: towards atomic resolution. *Quart. Rev. Biophys.* 33, 307–369.

van Heel, M., Harauz, G., Orlova, E. V., Schmidt, R., and Schatz, M. (1996). A new generation of the IMAGIC image processing system. *J. Struct. Biol.* 116, 17–24.

Velankar, S., Best, C., Beuth, B., Boutselakis, C. H., Cobley, N., Sousa Da Silva, A. W., Dimitropoulos, D., Golovin, A., Hirshberg, M., John, M., Krissinel, E. B., Newman, R., Oldfield, T., Pajon, A., Penkett, C. J., Pineda-Castillo, J., Sahni, G., Sen, S., Slowley, R., Suarez-Ureuena, A., Swaaminathan, J., van Ginkel, G., Vranken, W. F., Henrick, K., and Kleywegt, G. J. (2010). PDPe: protein data bank in Europe. *Nucl. Acids Res.* (Database issue) 34, 308–317.

Villa, E., Sengupta, J., Trabuco, L. G., LeBarron, J., Baxter, W. T., Shaikh, T. R., Grassucci, R. A., Nissen, P., Ehrenberg, M., Schulten, K., and Frank, J. (2009). Ribosome-induced changes in elongation factor Tu conformation control GTP hydrolysis. *Proc Natl Acad Sci USA* 106, 1063–1068.

Voss, N. R., Yoshioka, C. K., Radermacher, M., Potter, C. S., and Carragher, B. (2009). DoG Picker and TiltPicker: software tools to facilitate particle selection in single particle electron microscopy. *J. Struct. Biol.* 166, 205–313.

Wang, L. and Sigworth, F. J. (2006). Cryo-EM and single particles. *Physiology* 21, 13–18.

Wriggers, W. and Chacón, P. (2001). Modeling tricks and fitting techniques for multiresolution structures. *Structure* 9, 779–788.

Yu, X., Jin, L., and Zhou, Z. H. (2008). *3.88 Å structure of cytoplasmic polyhedrosis virus by cryo-electron microscopy.* Nature 453, 415–419.

Zhang, W., Dunkle, J. A., and Cate, J. H. (2009). Structures of the ribosome in intermediate states of ratcheting. *Science* 325, 1014–1017.

Zhang, W., Kirnmel, M., Spahn, C. M. T., and Penczek, P. A. (2008). Heterogeneity of large macromolecular complexes revealed by 3D cryo-EM variance analysis. *Structure* 16, 1770–1776.

Zhou, Z. H. (2008). Towards atomic resolution structural determination by single-particle cryo-electron microscopy. *Curr. Opin. Struct. Biol.* 18, 218–228.

CHAPTER 3

Statistical Mechanical Treatment of Molecular Machines

Debashish Chowdhury

I. INTRODUCTION

Molecular machines (Mavroidis et al. 2004) are devices that convert one form of energy into another. Just like their macroscopic counterparts, molecular machines have an "engine", an input and an output. Most of the machines I consider in this chapter are motors (Howard 2001, Kolomeisky and Fisher 2007, Schliwa 2003) which are enzymes that convert chemical energy into mechanical work.

In spite of the striking similarities, it is the differences between molecular machines and their macroscopic counterparts that makes the studies of these systems so interesting from the perspective of physicists. Biomolecular machines are usually single proteins or macromolecular complexes comprising several proteins and/or RNAs. These operate in a domain where the appropriate units of length, time, force and energy are *nano-meter, milli-second, pico-Newton* and $k_B T$, respectively (k_B being the Boltzmann constant and T is the absolute temperature). Already in the first half of the twentieth century D'Arcy Thompson, father of modern bio-mechanics, realized the importance of viscous drag and Brownian forces in this domain. He pointed out that (Thompson 1963) *"where bacillus lives, gravitation is forgotten, and the viscosity of the liquid, the resistance defined by Stokes' law, the molecular shocks of the Brownian movement, doubtless also the electric charges of the ionized medium, make up the physical environment and have their potent and immediate influence on the organism. The predominant factors are no longer those of our scale; we have come to the edge of a world of which we have no experience, and where all our preconceptions must be recast"*.

Molecular motors (i) are made of *soft matter*, (ii) have *isothermal* engines, (iii) operate under conditions far from thermodynamic equilibrium, (iv) exhibit stochastic trajectories (because of the Brownian forces), and (v) show unusual behaviours that arise from its hydrodynamics at low Reynold's number. Since inertia does not play any significant role in nano-mechnics, nature had to exploit principles quite different from those used in the macroworld for transmitting forces and for movements. It was nicely summarized by Binnig and Rohrer (1999): *"Nature's nanomechanics rests predominantly on deformation and on the transport of atoms, molecules, small entities and ionic charges, in contrast to translation and rotation in macromechanics. Simple deformations on the nanometer scale can be synthesized to create complex macromotions."*

In this chapter I present a pedagogical overview of the various approaches followed so far in developing theoretical models of molecular motors. Then, in order to illustrate the use of one particular approach, I consider two specific motors. One of these motors moves on filamentous proteins whereas the other moves on a specific nucleic acid strand.

In the context of molecular motors, the models fall into three broad categories:

(i) Modeling the operational mechanism of a single motor in terms of its structure and the coordinated dynamics of its components,

(ii) Modeling phenomena driven by a few motors in terms of their coordination, cooperation and competition,

(iii) Modeling spatio-temporal organization of systems involving a large number of interacting motors.

II. TECHNIQUES FOR THEORETICAL MODELING OF MOLECULAR MACHINES

Theory provides *understanding* and *insight*. These allow us not only to interpret the empirical observations and recognize the importance of the various ingredients but also to generalize, to create a framework for addressing the next level of question and to make predictions which can be tested in in-vivo, in-vitro or in-silico experiments.

Theorization requires a model of the system. A theoretical model is an abstract representation of the real system which helps in understanding the real system. This representation can be pictorial (for example, in terms of cartoons or graphs) or symbolical (e.g., a mathematical model). Qualitative predictions may be adequate for understanding some complex phenomena or for ruling out some plausible scenarios. But, a desirable feature of any theoretical model is that it should make quantitative predictions (Mogilner et al. 2006, Phair and Misteli 2001). The predictions of a theory, at least in principle, can be tested by in-vitro and/or in-vivo experiments in the laboratory.

The predictions of a mathematical model can be derived *analytically* in terms of abstract symbols; for specific sets of values of the model parameters, the predictions can be shown numerically or graphically. The predictions of a theoretical model can be obtained *numerically* by carrying out computer simulations (i.e., *in-silico* experiments) of the model (Andrews and Arkin 2006). Thus, simulation is not synonymous with modeling. When a model is too complicated to be formulated in abstract notations and to be treated analytically, it is called a computer model of the system. Since fully analytical treatment of a model can be accomplished exactly only in rare cases, one has to make sensible approximations so as to get results as accurate as possible. Simulation of a model also tests the validity of the approximations made in the analytical treatments of the model.

We should also make a distinction between the two different "computational methods", namely, (i) computer simulations which, as we have mentioned above, test hypotheses; and (ii) *Knowledge discovery* (or, *data mining*) which extracts hidden patterns or laws from huge quantities of experimental data, allowing hypotheses to be formulated.

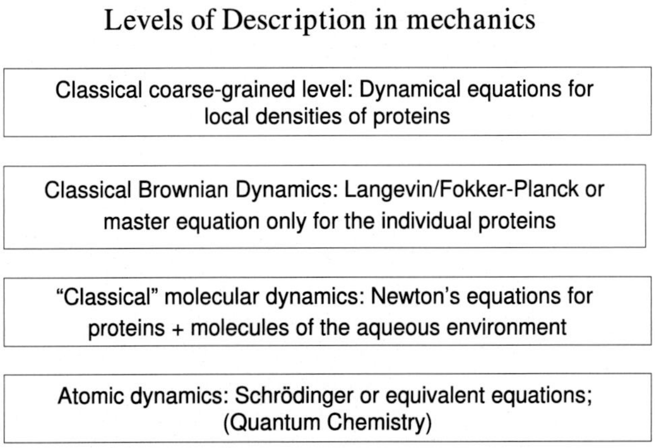

A. Computer Simulation: In-silico Experiments

Computer simulations serve several purposes:

(i) Most often, while trying to solve a mathematical model of a molecular motor analytically, one is forced to make approximations, some of which may be uncontrolled. Therefore, while comparing the predictions of an analytical calculation with experimental results, one is not sure if the discrepancies between the two arises from the inadequacy of the model, or the inaccuracies of the approximations made in the analytical calculations, or both. Computer simulation enables one to study the same model without any approximations, except for statistical and systematic errors. Therefore, this provides a method to check if the model captures the essential qualitative features of the corresponding real system and, also, to assess the validity as well as the accuracy of the approximations made in the analytical calculations.

(ii) In those circumstances where no satisfactory mathematical model of a molecular motor has yet been formulated, computer simulations of some toy models help us by providing insight as to what variables are crucially important and what are the interrelations among them. One may discover novel phenomena by carrying out new computer experiments just as new discoveries are sometimes made accidentally during laboratory experiments.

(iii) The external parameters such as temperature, external force, etc., can be varied in computer simulations just as those in laboratory experiments. Moreover, there

Levels of Description in mechanics

Classical coarse-grained level: Dynamical equations for local densities of proteins

Classical Brownian Dynamics: Langevin/Fokker-Planck or master equation only for the individual proteins

"Classical" molecular dynamics: Newton's equations for proteins + molecules of the aqueous environment

Atomic dynamics: Schrödinger or equivalent equations; (Quantum Chemistry)

FIGURE 3.1: *The hierarchy of the levels of theoretical description in mechanics.*

are other parameters, for example, the inter-molecular interaction potential, which cannot be varied continuously in a controlled systematic manner in laboratory experiments but such a variation can be achieved easily in computer experiments. Furthermore, doing such computer experiments one can calculate quantities which cannot be measured directly in any laboratory experiment, one can experiment easily with situations which are very difficult to realize in the laboratory, and one can calculate the properties of hypothetical machines, thereby "designing" machines with desirable properties. One can "switch off" and "switch on" variables and parameters in computer simulations to examine a smaller or larger set of variables and parameters when similar "switchings" are not possible in laboratory experiment on real materials. One can study models even in unphysical limits to gain insight into the physics of the problem under investigation. Even those experiments which are potentially hazardous can be conducted safely through computer simulation.

(iv) Speeding up of simulations can yield information on evolutionary time scales which cannot be done in any real experiment.

Thus, in a computer simulation we can always pose the eternal human question: "What would have happened if . . . ?" The main difficulties faced in computer simulations arise from the limited size of the available computer memory and the limitations imposed by the available CPU time.

B. Levels of Description in Mechanics

A model can be formulated at different *levels* of molecular details and dynamical complexity (see Fig. 3.1), i.e., at different physical or logical levels of resolution. The physical resolution can be spatial resolution or temporal resolution. Even at a given level, alternative, but equivalent formulations of the dynamics are possible. In principle, the dynamical equations at a higher level can be derived from one at a lower level. However, most often such a derivation is either practically impossible or lacks mathematical rigor. Moreover, even at a given level, more than one mathematical approach may be available for quantitative description of the dynamics.

Every theoretical model is intended to address a set of questions. The modeler must choose a *level of description* appropriate for this purpose, keeping in mind the phenomena that are subject of the investigation. Otherwise, the model may have either too much redundant details or it may be too coarse to provide any useful insight.

Do we need to use the formalisms of quantum mechanics for a quantitative description of the dynamics of a molecular machine? According to Planck's formula $E = hv$, where h is Planck's constant. The thermal energy $k_B T = 0.6$ kcal/mol at "room temperature" $T = 300$ K, corresponds to a frequency $v = 6.25/$ *ps*. Therefore, at room temperature, classical Newton's equations should provide a sufficiently accurate description of the dynamics of a molecular motor for all dynamical processes whose characteristic times are much longer than a picosecond.

In classical molecular dynamics, the trajectories of an individual motor, as well as those of the molecular constituents of the solvent and the fuel, can be computed by solving the corresponding Newton's equations which are, in general, coupled to each other. For demonstrating the conformational changes of the motor in each cycle, the model has to have spatial resolution that is high enough to exhibit the intra-motor movements. In other words, for the purpose of explicitly demonstrating the conformational dynamics of a motor, Newton's equations have to be formulated for the atomic constituents. For the purpose of calculating the forces exerted by the atoms on each other, one assumes some form of inter-atomic potential. These equations are *deterministic*, i.e., a unique initial condition leads to a unique trajectory. Moreover, these equations are reversible in time, i.e., the equation remains invariant under the transformation $t \rightarrow t' = -t$.

For computational accuracy and stability, the temporal resolution of a standard molecular dynamics simulation has to be sufficiently high. This imposes severe limitations on the longest simulations feasible with computational resources available at present as well as forseeable future. On the other hand, the most important conformational changes of a molecular motor take place over time intervals that are orders of magnitude longer than the longest MD simulations possible in the forseeable future. Moreover, even if such "dream machines" become available in near future, the structure of the motor at the atomic level may not be available because of the difficulties of preparing crystals of the macromolecular systems. Therefore, standard molecular dynamics cannot address the fundamental questions on the operational mechanics of molecular motors. On the other hand, shapes of many

molecular motors are now well known because of the success of cryo-electron microscopy that yields the structure at a spatial resolution that is somewhat lower than that achievable with X-ray diffraction. For motors whose shapes are known from cryo-electron microscopy, an appropriately coarse-grained model may be adequate for studying some robust features of their conformational dynamics.

A molecular machine is usually a protein or a macromolecular complex comprising several different proteins and/or RNAs. The first of the two examples which we will consider explicitly in this chapter, namely, KIF1A, is a protein. The second example, namely the ribosome, consists of several different protein and ribosomal RNA (rRNA) molecules. A molecular machine is small by macroscopic standards but is still several orders of magnitude larger than the molecules of water. On the scale of the size of a molecular machine, the surrounding fluid does not appear to be a continuum. In fact, an intracellular molecular machine "sees" that the fluid is made of molecules that constantly, but *discretely*, strike it from all sides.

For the time being, let us imagine that the engine of the machine has been switched off. The random bombardment of this machine from all sides by the molecules of water tends to *accelerate* and *decelerate* it perpetually. A single collision has a very small effect on the molecular machine; it is the *cumulative effect of a rapid and random sequence of large number of weak impulses* that would manifest as its noisy trajectory. Similar zig-zag trajectories of pollen grains observed under an optical microscope is called Brownian motion.

Now, suppose the engine of the molecular machine is switched on and supply of fuel molecules is adequate. In contrast to pollen grains, which exhibit passive Brownian motion, this molecular machine is "active" because it consumes energy from an external source. Nevertheless, its trajectory is noisy because of the random bombardment by the surrounding molecules of the aqueous medium.

Enormous reduction in the number of equations is possible by treating the aqueous medium as a reservoir (or bath) in which the molecular machine operates. We identify this level as "Brownian." At this level, the motion of the machine is governed by a differential equation that, unlike Newton's equation, is both stochastic and irreversible. The viscous damping force, which causes irreversibility, and the random force, which makes the trajectories noisy, arise from the interactions between the machine and the molecules of the reservoir.

If one is interested in capturing the intra-machine movements and conformational changes of the machine in each cycle, then the most convenient technique is the so-called normal mode analysis (NMA) (Tama and Brooks 2006; see Chapter 4). In this approach one normally uses a coarse-grained description of the machine in terms of point masses connected by harmonic springs. The dynamics of the system is assumed to be govered by a Langevin-like equation in the overdamped approximation. The eigen-states of the Hessian matrix describes the normal modes of (collective) vibration of the system which correspond to the slow conformational dynamics. NMA is a purely computational approach; to my knowledge, no analytical results have been derived so far using this technique. The advantages and the limitations of this approach has been reviewed by Ma (2005).

An alternative approach is to assume the conformational changes that the motor undergoes in a cycle, instead of demonstrating these explicitly, and to predict the transport properties of the motor. In this case, it is adequate to represent the motor by a particle with distinct "internal" states which correspond to distinct mechano-chemical states of the motor. In this approach, the number of allowed states of the system are discrete; master equation(s) are ideally suited to describe the stochastic time-evolution of the system. All the results which will be discussed in the context of the test cases in this chapter have been obtained by following this approach.

At an even coarser level, one formulates equations for the time evolution of local concentrations (densities) of the machines instead of writing equations of motion for individual machines. This approach is convenient for describing the collective behaviour of machines over extended region of space over sufficiently long period of time. But, this approach is not suitable for describing the operational mechanism of individual machines.

C. Mechanics of Molecular Motors: Noisy Power Stroke versus Brownian Ratchet

If the input energy directly causes a conformational change of the protein machine which manifests itself as a mechanical stroke of the machine, the operation of the machine is said to be driven by a "power stroke" mechanism. This is also the mechanism used by all man-made macroscopic machines. However, in case of molecular machines, the power stroke is always "noisy" because of the Brownian forces acting on the machine (Berg 1993).

Let us contrast this with the following alternative scenario: suppose, the machine exihibits "forward" and "backward" movements because of spontaneous thermal fluctuations. *On the average*, a free passive Brownian particle does not get displaced from its initial position. But, a machine is "active" and consumes input energy. If now energy input is utilized to prevent "backward" movements, but allow the "forward" movements, the system will exhibit directed, albeit noisy, movement in the "forward" direction. Note that the forward movements in this case are caused directly by the spontaneous thermal fluctuations, the input energy rectifies the "backward" movements. This alternative scenario is called the Brownian ratchet mechanism (Astumian 1997, Astumian 2001, Astumian 2007, Astumian and Hänggi 2002, Julicher et al. 1997, Reimann 2002, Vale

and Oosawa 1990). The famous Feynman-Smoluchowski ratchet-and-pawl device, working under isothermal conditions, is an example of a Brownian ratchet.

Thus, in principle, there are two idealized scenarios for a transition from a conformation A to a conformation B - one is by a power stroke and the other by a Brownian ratchet mechanism (Howard 2006). However, for a molecular motor, it is difficult to unambiguously distinguish between a power stroke and a Brownian ratchet (Wang and Oster 2002a).

III. CHEMICAL REACTIONS RELEVANT FOR MOLECULAR MACHINES

To understand molecular machines, we also have to consider *chemical reactions*, which most often supply the (free-) energy required to drive these machines. In other words, in order to understand the mechanisms of biomolecular machines, it is necessary to understand not only how these move in response to the mechanical forces but also how these are affected by generalized "chemical forces". Enzymes and ribozymes constitute two classes of biological catalysts; enzymes are proteins whereas ribozymes are RNA molecules. Molecular machines are either enzymes or ribozymes. Therefore, it is desirable to have some background knowledge in the theory of enzymatic reactions before embarking on modeling of biomolecular machines. The main aim of this section is to provide a brief summary of these essential ideas.

All chemical reactions are intrinsically *reversible* and have the general form *Reactants* $\rightleftharpoons$ *Products*. However, if the rate of the reverse reaction is very small compared to that of the forward reaction, or if the products are continuously removed from the reaction chamber as soon as these are formed, the reaction becomes, effectively, *irreversible* and takes the form *Reactants* $\rightarrow$ *Products*. Chemical kinetics is a framework for studying how fast the amounts of reactants and products change with time.

A. Levels of Description in Theories of Chemical Reactions

Just as mechanics (more appropriately, dynamics) can be formulated at different levels, the theory of chemical reactions can also be developed at several different levels depending on the purpose of the investigation (Grima and Schnell 2008).

In contrast to the section on dynamics, where we began at the molecular level, in this section on chemical reactions we begin at the opposite extreme, namely, the most coarse-grained description. At this level, the problem can be formulated as follows: suppose a macroscopically uniform mixture of S chemical species is confined in a fixed volume V and can interact through R reaction channels. If the populations (i.e., the concentrations) of all the species are given

Levels of Description for Chemical Reactions

Chemical Reaction Rate Equations
(deterministic ODEs; no spatial fluctuation: well-stirred approx.)

Reaction-Diffusion-type Equations
(deterministic PDEs; spatial variation is captured)

Chemical Master equations
(Integro-differential equations; Stochastic description)

Quantum-mechanical theories
(Quantum-chemistry)

FIGURE 3.2: *The hierarchy of the levels of theoretical description of chemical reactions.*

at some initial instant of time, what will be the corresponding molecular populations at any later arbitrary instant of time t? The traditional approach is based on ordinary differential equations, called chemical reaction rate equations, for the populations of the molecular species. The main *asumption* of the rate equation approach is that the population dynamics of the reactant and product species is a *continuous* and *deterministic* process. For a well stirred chemically reacting system, this is a reasonably good approximation.

Spatial variations in the concentrations of the reactants and products can be taken into account by replacing the ordinary differential equations (for the global concentrations) by partial differential equations (for the local concentrations). These equations describe not only chemical reactions at various spatial locations, but also the diffusion of the molecular species. Naturally, all such equations are often generically referred to as reaction-diffusion equations. However, these equations are usally deterministic.

The rate equations conceal a great deal of detailed physical processes involved in the reaction. In reality, the time evolution of the populations cannot be *continuous* because the number of molecules can change only by discrete integers. Moreover, the evolution is not *deterministic* because it is impossible to predict the exact molecular populations at an arbitrary time unless the positions and velocities of all the molecules in the system, including those in the solvent (i.e., reservoir or bath) are taken into account. Furthermore, the smaller is the population of a reacting species, the stronger are the fluctuations that make a stochastic description unavoidable. The physical reason for this stochasticity is identical to that responsible for the stochastic trajectories of Brownian particles: the dynamics is deterministic in the full phase space of the system but not in a subspace corresponding only to the reacting molecular species.

Therefore, in principle, it is possible to formulate a "chemical" Langevin equation which is analogous to the Langevin equation for a Brownian particle. Equivalently, the stochastic time evolution of the chemical species can

be described in terms of a "chemical" Fokker-Planck equation. An alternative formulation of the stochastic population dynamics of the chemical species can be developed in terms of a "chemical" Master equation. Since these stochastic theories are formulated at a more microscopic level than that of the chemical rate equations, it is possible to derive the reaction rate equations from the stochastic evolution equations.

Modelling a reaction at a more microscopic level would require a quantum-mechanical formalism to account for the electronic processes through which chemical bonds are made and broken. However, for the chemical reactions involved in the operation of molecular machines, such a detailed description would have an enormous amout of redundant information. In principle, we can ignore the electronic degrees of freedom by averaging over *length scales* smaller than the spatial extent of the molecules. Similarly, we can also average over *time scales* longer than those of electron dynamics. At this coarse-grained level, reached by such averaging, we have a theoretical framework for chemical reactions which is analogous to the stochastic (Brownian) level for describing the dynamics of molecular motors.

B. Chemical Reaction Rate Equations

Let us assume that the concentration of the s-th molecular species is denoted by a continuous, single-valued function $c_s(t)$ of time t. Then, the corresponding chemical rate equations can be expressed as

$$\frac{dc_s}{dt} = f_s(c_1, c_2, \ldots, c_s, \ldots, c_N)(s = 1, 2, \ldots, S). \quad (1)$$

The specific forms of the functions f_s are determined by the actual nature of the reactions. The coefficients of the various terms on the right hand side of these equations are called the rate constants. At the level of the chemical rate equations, the rate constants are phenomenological parameters whose numerical values are to be supplied from empirical data.

1. Stoichiometry and Reaction Rates. In general a reaction can be expressed in the general form

$$\sum_{s=1}^{S} v_s M_s = 0, \quad (2)$$

where the stoichiometric coefficients $v_i > 0(< 0)$ if M_i is a product (reactant). Any possible ambiguity in the definition of the rate of the reaction is avoided by defining the reaction rate to be $d\xi/dt = -(1/v_s)d[M_s]/dt$, where $[M_s]$ is the concentration of the s-th molecular species, so that, in general, $dM_s = v_s d\xi$.

The sign of $d\xi$ indicates whether the reaction is proceeding in the forward or the reverse direction; $d\xi > 0$ for the forward reaction whereas $d\xi < 0$ for the reverse reaction. The actual extent of a reaction depends on the amount of substance used in the reaction. A better definition of the rate of reaction is $\eta(t) = \xi(t)/V$ where V is the volume of the reaction chamber. This definition allows one to associate a single rate with the entire equation corresponding to a reaction. All practical problems of chemical kinetics can be reduced to finding how $\eta(t)$ changes with time t. From $\eta(t)$ one can calculate the time evolution of the concentrations of each chemical species involved in the reaction.

2. Energy Landscape and Free-Energy Landscape: Conformation versus Structure.
The energy landscape of a chemical reaction is a graphical way of showing how the energy of the reacting system depends on the degrees of freedom of the system which include the positions (and orientations) of all the atoms of the reactant and product molecules. Consider, for example, the reactions of the type

$$AB + C \rightarrow A + BC \tag{3}$$

For the sake of simplicity, let us assume that the reaction remains confined in essentially one-dimensional space that is collinear with the bond AB (and BC). In this case the relative orientation of the molecules AB and BC remains unaltered during the reaction. Suppose, R_{AB} and R_{BC} are the separations between A and B and that between B and C, respectively. The energy landscape of this reaction can be drawn by treating only R_{AB} and R_{BC} as the two independent variables.

For any single event of the occurrence of the reaction, the trajectory in this landscape does not necessarily proceed along the bottom of the valley, but occasionally also makes excursions up the walls of the valley. However, when averaged over large number of such trajectories, the reaction process can be described as an effective route in this landscape that corresponds to the lowest energy from the entrance to the exit over a saddle point. This average route in the multidimensional energy landscape is called the *reaction coordinate* which we will denote by the symbol ξ. Moving along this pathway alters the coordinates of all the atoms involved in the reaction; therefore, this reaction coordinate is actually a composite coordinate. The magnitude of this reaction coordinate expresses how far the reaction has progressed. Often the energy of the system is plotted against the reaction coordinate; the reactants and products correspond to two local minima separated by a maximum which corresponds to the saddle point on the multi-dimensional energy landscape. The state of maximum energy along the reaction coordinate, is called the *transition state*.

Since most of the molecular motors are proteins, we now explain how the free energy landscape for the *structural* changes of a protein can be derived from the corresponding energy landscape in the space of protein *conformations* (Howard 2001). According to our convention, a *conformational state* of a protein is given by the coordinates of all the constituent atoms. Note that we ignore the velocities of the

atoms for describing the conformational states because on the time scales longer than picosecond the velocities of the atoms average out to zero. Because of thermal fluctuations, a protein may go through a sequence of an enormously large number of conformational states (Wildman and Kern 2007). If the fluctuations in the positions of the atoms are not too large, we can regard the different conformations as small deviations about a state which is time-average of these conformations. Such a time-averaged conformational state is called a *structural state* which is obtained from X-ray crystallography. The probability of finding a protein in a conformational state with energy U_c is proportional to $\exp(-\beta U_c)$, whereas that of finding the protein in a structural state with free energy F_s is proportional to $\exp(-\beta F_s)$.

For many proteins, the conformational states segregate into two structural states, which we denote by ε_1 and ε_2. For example, ε_1 and ε_2 may correspond to the "pre-stroke" and "post-stroke" states of a motor protein. If P_1 and P_2 are the probabilities of finding the protein the structural states ε_1 and ε_2, respectively, then

$$P_2/P_1 = \exp(-\beta \Delta F) \tag{4}$$

with

$$\Delta F = F_2 - F_1 \tag{5}$$

where F_1 and F_2 are the free energies of the structural states ε_1 and ε_2, respectively.

Now consider the "reaction" (structural transition)

$$\varepsilon_1 \underset{k_r}{\overset{k_f}{\rightleftharpoons}} \varepsilon_2 \tag{6}$$

The ordinary differential equations governing the populations of proteins in these two structural states are

$$\begin{aligned} d[\varepsilon_1]/dt &= k_r[\varepsilon_2] - k_f[\varepsilon_1] \\ d[\varepsilon_2]/dt &= k_f[\varepsilon_1] - k_r[\varepsilon_2] \end{aligned} \tag{7}$$

which are usually referred to as the rate equations, and the square brackets indicate the respective concentrations. Note that the fact $d[\varepsilon_1]/dt + d[\varepsilon_2]/dt = 0$ reflects the conservation law: $[\varepsilon_1] + [\varepsilon_2] = constant$. If we deal with the number of molecules $N_1(t)$ and $N_2(t)$ of the two species ε_1 and ε_2, then

$$N_1(t) + N_2(t) = constant. \tag{8}$$

Defining $\delta N_1(t) = N_1(t) - N_1^{eq}$ and $\delta N_2(t) = N_2(t) - N_2^{eq}$, we get

$$\frac{d[\delta N_1(t)]}{dt} = -(k_f + k_r)\delta N_1(t) \tag{9}$$

and, hence,

$$\delta N_1(t) = \delta N_1(0)e^{-k_{eff}t} \tag{10}$$

with $k_{eff} = k_f + k_r$. Thus, any fluctuation in the populations of the two species decays exponentially with time with an effective time constant k_{eff}^{-1}.

In the stationary state, the time derivatives on the LHS vanish and, hence,

$$K_{eq} = \frac{[\varepsilon_2]_{eq}}{[\varepsilon_1]_{eq}} = \frac{k_f}{k_r} \qquad (11)$$

Using the equation (4), we get

$$\frac{k_f}{k_r} = \frac{[\varepsilon_2]_{eq}}{[\varepsilon_1]_{eq}} = P_2/P_1 = \exp(-\beta\Delta F) \qquad (12)$$

Equation (12) is called the *law of mass action*. If ε_1 and ε_2 are in mutual equilibrium and then some more proteins in one of the two structural forms is added to the system, then the populations of the two species will change so as to satisfy equation (12) in the new equilibrium state. Reactions for which $\Delta F < 0$, $[\varepsilon_2]_{eq} > [\varepsilon_1]_{eq}$, and $k_f > k_r$, i.e., the forward reaction is favorable. But, if $\Delta F > 0$, the forward reaction is unfavorable (or, equivalently, the reverse reaction is favorable). The condition $\Delta F < 0$ is *necessary*, but *not sufficient*, for the occurrence of a reaction at an observable average rate.

3. Effects of Temperature on Reaction Rate.

Rate constants usually depend strongly on temperature. An overwhelmingly large number of rate constants are found to vary with temperature according to the Arrhenius equation

$$k(T) = A\exp[-E_a/k_B T]. \qquad (13)$$

where E_a is called the activation energy.

In order to explain the physical origin of this temperature-dependence, let us consider the reaction (6) once again. The physical picture behind this reaction is as follows: the protein in the structural state ε_1 makes many unsuccessful attempts to overcome the free energy barrier separating it from the structural state ε_2. But, finally, when it succeeds, the actual process occurs very fast. In order to overcome the barrier through a thermally activated process, the protein must attain the transition state whose free energy exceeds that of the state ε_1 by E_a and the probability for attaining this state is proportional to $\exp[-E_a/k_B T]$.

4. Effects of Force on Reaction Rate.

Again we consider the reaction (6) for the purpose of explaining the basic concepts. Suppose an external force f is applied on the protein and the force is directed from ε_1 to ε_2. Then

$$\Delta G = \Delta G^0 - f(\Delta x) \qquad (14)$$

where ΔG^0 is the free energy difference between ε_2 and ε_1 in the absence of the external force f. Obviously, in equilibrium,

$$\frac{k_f(f)}{k_r(f)} = \frac{[\varepsilon_2]_{eq}}{[\varepsilon_1]_{eq}} = \exp(-\beta\Delta G) = K_{eq}^0 \exp(\beta f\Delta x), \quad (15)$$

i.e., the structural state ε_2 is more probable than the state ε_1.

The equation (15) implies that we can write the individual rate constants for the forward and reverse transitions as (Bustamante et al. 2004, Khan and Sheetz 1997, Tinoco and Bustamante 2002)

$$k_f(f) = k_f(0)e^{\theta\beta f(\Delta x)} \qquad (16)$$

and

$$k_r(f) = k_r(0)e^{-(1-\theta)\beta f(\Delta x)} \qquad (17)$$

where θ is a fraction of the distance Δx. The forms of f-dependence assumed in (16) and (17) is used routinely for molecular motors while deriving their force-velocity relations which are among the fundamental characteristics of each family of motors.

5. Effects of Catalysts on Reaction Rate: Enzymatic Reactions.

Enzymes are biological catalysts (Dixon and Webb 1979). These can speed up chemical reactions by a factor of 10^6 to 10^{20} and are specific in the sense that a specific catalyst speeds up a specific reaction.

- **Reaction with only one intermediate step**

Let us consider reactions of the type (Dixon and Webb 1979)

$$E + R \underset{k_{-1}}{\overset{k_1}{\rightleftharpoons}} I_1 \underset{k_{-2}}{\overset{k_2}{\rightleftharpoons}} E + P \qquad (18)$$

where E denotes the enzyme while R and P represent the reactant and products, respectively. The symbol I_1 represents the transition state of the system. The reaction rates are given by

$$\frac{dR}{dt} = -k_1 ER + k_{-1}I_1 \qquad (19)$$

$$\frac{dI_1}{dt} = k_1 ER - (k_{-1} + k_2)I_1 \qquad (20)$$

$$\frac{dP}{dt} = k_2 I_1 - k_{-2}EP \qquad (21)$$

Moreover, as the total amount of enzyme is, by definition, conserved, we must have

$$E + I_1 = E_0 = constant. \qquad (22)$$

We now make two simplifying assumptions.

Assumption 1: $k_{-2} \simeq 0$, i.e., the reverse reaction can be ignored. Then, the equation (21) simplifies to

$$\frac{dP}{dt} = k_2 I_1 \qquad (23)$$

Eliminating E from (20) and (22) we get

$$\frac{dI_1}{dt} = k_1(E_0 - I_1)R - (k_{-1} + k_2)I_1$$
$$= k_1 E_0 R - (k_{-1} + k_2 + k_1 R)I_1 \qquad (24)$$

Assumption 2: $R \gg E_0$, i.e., the reactants are in large excess, compared to the total initial amount of enzyme. If we now envisage a situation where the reaction continues to proceed in such a manner that the population of I_1 does not change with time, i.e., $dI_1/dt = 0$, then the corresponding steady population of I_1 would be given by

$$I_1 = \frac{k_1 E_0 R}{(k_{-1} + k_2 + k_1 R)}. \tag{25}$$

Assuming R to be practically constant (because there is so much excess of R), the I_1, indeed, reaches the abovementioned steady state with a relaxation time

$$\tau = \frac{1}{k_{-1} + k_2 + k_1 R} \tag{26}$$

starting from $I_1(t = 0) = 0$. Under these assumptions, the speed of the reaction is

$$V = \frac{dP}{dt} = k_2 I_1 = \frac{k_1 k_2 E_0 R}{k_{-1} + k_2 + k_1 R} \tag{27}$$

which is conventionally expressed in the form

$$V = \frac{k_2 E_0 R}{K_M + R} \tag{28}$$

where the so-called *Michaelis constant*

$$K_M = \frac{k_{-1} + k_2}{k_1} \tag{29}$$

is the ratio of the total rates of reactions *out of* I_1 and that *into* I_1.

We will now explore the physical meaning and significance of the Michaelis constant K_M. Writing $V = k_2 E_0/[1 + (K_M/R)]$, we find that $V \to V_{max} = k_2 E_0$ as $R \to \infty$, where V_{max} is the maximum possible reaction rate. Therefore, the equation (28) can be recast as

$$V = \frac{V_{max}}{1 + (K_M/R)} \tag{30}$$

From (30) we see that for $K_M = R$, $V = V_{max}/2$, i.e., K_M is the reactant concentration at which the reaction rate is half of its maximum possible value. Moreover, plotting $1/V$ against $1/R$, using the experimentally measured data, one gets a straight line with slope K_M/V_{max} and intercept $1/V_{max}$ from which both V_{max} and K_M can be extracted.

C. Stochastic Formulation of Reaction Kinetics

Stochastic chemical kinetics describes the population dynamics of the reacting species as a *discrete, stochastic* process that is assumed to evolve in a continuous time (Gillespie 2005).

1. Chemical Master Equation.

Consider S chemical species ($M_1, M_2, \ldots, M_s, \ldots M_S$), interacting through R reaction channels. Let $n_s(t) =$ Number of molecules of the *s*-th species at time t. Our goal is to obtain the state vector $\vec{n}(t) = (n_1(t), n_2(t), \ldots, n_s(t), \ldots, n_S(t))$, given the state vector $\vec{n}(0) = (n_1(0), n_2(0), \ldots, n_s(0), \ldots, n_S(0))$, at time $t = 0$. Extending the notation used in eq. (2), the reaction in the *r*-th channel can be expressed as

$$\sum_{s=1}^{S} v_{rs} M_s = 0. \tag{31}$$

Therefore, the stoichiometric coefficients form a matrix whose elements are v_{rs}. Moreover, if the *r*-th reaction takes place, the state vector $\vec{n}$ changes to $\vec{n} + \vec{v}_r$ where $\vec{v}_r = (v_{r1}, v_{r2}, \ldots, v_{rs}, \ldots v_{rS})$. We also define the *propensity function* (transition probabilities per unit time) $W_r(\vec{n})$ so that $W_r(\vec{n})\Delta t$ is the probability that the *r*-th reaction takes place in the time interval between t and $t + \Delta t$, thereby leading to a change of the molecular population $\vec{n} \to \vec{n} + \vec{v}_r$. Thus, a given reaction channel is characterized mathematically by two quantities, namely, (i) the state change vector $\vec{v}_r$, and (ii) the *propensity function* $W_r(\vec{n})$.

Let us assume that Δt is so small that no more than one reaction of any kind can take place in the interval between t and $t + \Delta t$. Then, we can write the "chemical" master equation (Gillespie 2005)

$$\frac{\partial P(\vec{n}, t)}{\partial t} = \sum_{r=1}^{R} [W_r(\vec{n} - \vec{v}_r) P(\vec{n} - \vec{v}_r, t)]$$
$$- \sum_{r'=1}^{R} [W_{r'}(\vec{n}) P(\vec{n}, t)] \tag{32}$$

for the probability $P(\vec{n}, t)$.

The rate equations for a chemically reacting system can be derived from the corresponding master equations. Let us define the average population by

$$\langle \vec{n}(t) \rangle = \sum_{\vec{n}} \vec{n}(t) P(\vec{n}, t) \tag{33}$$

It is straightforward to derive

$$\frac{d\langle \vec{n}(t) \rangle}{dt} = \sum_{r=1}^{R} v_r \langle W_r(\vec{n}(t)) \rangle \tag{34}$$

2. Reaction Speeded by a Single Enzyme Molecule: Validity of MM Theory.

If only one enzyme molecule is used to convert a macroscopic amount of reactant into product, the enzymatic reaction proceeds stochastically. The waiting time in between the completion of the enzymatic reaction becomes a random variable. The distribution of the waiting times $P_w(t)$ can be measured experimentally (English et al. 2006). The kinetics of such reactions cannot be formulated in terms of rate equations for the concentrations of the reactants and the products. Instead, one has to formulate the appropriate master equations for the probabilities of finding the different chemical states of the system. Nevertheless, the Michaelis-Menten equation still describes the average rate of the enzymatic reaction (Kou et al. 2005, Min et al. 2005). Moreover, it has been established (Min et al. 2006) that the Michaelis-Menten equation holds for the average rate of the progress of the reaction catalyzed by a single enzyme molecule under wide range of different conditions.

D. Allosterism of Enzymes and Molecular Motors

Allosterism (Changeux and Edelstein 1998, Changeux and Edelstein 2005, Koshland and Hamadani 2002, Lindsley and Rutter 2006) usually refers to the change of conformation around one location of a protein in response to binding of a ligand to another location of the same protein. Schemes for classification of the mechanisms of allostery have been proposed (Tsai et al. 2009). Allosterism is not restricted only to proteins; allosteric ribozymes are also receiving attention in recent years (Fastrez 2009). A statistical mechanical model for a cooperative mechanism of allostery has been proposed (Bray and Duke 2004).

A motor protein has separate sites for binding the fuel molecule and the track. Therefore, the mechanochemical cycle of a motor can be analyzed from the perspective of allostery (Vologodskii 2006).

E. Chemical Equilibrium and Non-Equilibrium Steady-States

Not every chemical steady state is an equilibrium state of the system. We will now show that nonequilibrium steady states (NESS) can exist only in open systems in which reactions are driven from sources to sinks between which chemical potential difference $\Delta\mu$ must exist (Qian 2006). If $\Delta\mu$ vanishes, the source and sink lose their distinct roles and the NESS reduces to an equilibrium state. As we show in this subsection, $\Delta\mu$ can be taken as a "generalized chemical force" that drives a chemical reaction, on the average, in a particular "direction" (i.e., forward or reverse).

1. Generalized Chemical Force. Let us again consider the reaction (6). Suppose, initially, there are n_1 molecules of ε_1 (each of free energy G_1) and n_2 molecules of ε_2 (each of free energy G_2). In the ideal gas appeoximation (for the molecules in solution), the initial free energy G_i is

$$G_i = n_1 G_1 + n_2 G_2 + (n_1 + n_2)k_B T$$
$$\times \left[\left(\frac{n_1}{n_1 + n_2} \right) ln \left(\frac{n_1}{n_1 + n_2} \right) \right.$$
$$\left. + \left(\frac{n_2}{n_1 + n_2} \right) ln \left(\frac{n_2}{n_1 + n_2} \right) \right] \quad (35)$$

If one molecule of ε_1 gets converted to one molecule of ε_2 by the reaction, then the new total free energy G_f is

$$G_f = (n_1 - 1)G_1 + (n_2 + 1)G_2 + (n_1 + n_2)k_B T$$
$$\times \left[\left(\frac{n_1 - 1}{n_1 + n_2} \right) ln \left(\frac{n_1 - 1}{n_1 + n_2} \right) \right.$$
$$\left. + \left(\frac{n_2 + 1}{n_1 + n_2} \right) ln \left(\frac{n_2 + 1}{n_1 + n_2} \right) \right] \quad (36)$$

Using (35) and (36), we get

$$\Delta G = \Delta G^0 + k_B T \left[n_1 ln \left(1 - \frac{1}{n_1} \right) \right.$$
$$\left. + n_2 ln \left(1 - \frac{1}{n_2} \right) + ln \left(\frac{n_2 + 1}{n_1 - 1} \right) \right] \quad (37)$$

where $\Delta G^0 = G_2 - G_1$. When n_1 and n_2 are sufficiently large,

$$\Delta G \simeq \Delta G^0 + k_B T ln \left(\frac{[\varepsilon_2]}{[\varepsilon_1]} \right) \quad (38)$$

Since Gibbs free energy per particle is also the chemical potential, we now define the "generalized chemical force" X as (Howard 2001)

$$X = \Delta\mu = \Delta G = \Delta G^0 + k_B T ln \left(\frac{[\varepsilon_2]}{[\varepsilon_1]} \right) \quad (39)$$

because, in equilibrium $X = 0$, which follows from $\frac{[\varepsilon_2]_{eq}}{[\varepsilon_1]_{eq}} = e^{-\beta\Delta G^0}$. Moreover, defining the forward and reverse fluxes $\mathcal{J}_f$ and $\mathcal{J}_r$ by the relations

$$\mathcal{J}_f = k_f[\varepsilon_1], \quad \text{and} \quad \mathcal{J}_f = k_r[\varepsilon_2] \quad (40)$$

the generalized chemical force X can be expressed as

$$X = k_B T \, ln(\mathcal{J}_r/\mathcal{J}_f) \quad (41)$$

In equilibrium state $\mathcal{J}_f = \mathcal{J}_r$ and, therefore, $X = 0$.

As an example, consider the reaction

$$ATP \rightleftharpoons ADP + P_i \quad (42)$$

where the forward reaction is called hydrolysis of ATP. For this reaction

$$X = \Delta G = \Delta G_0 - k_B T \, ln \frac{[ATP]_c}{[ADP]_c [P_i]_c} \quad (43)$$

where $\Delta G_0 = -54 \times 10^{-21} \mathcal{J}$.

F. Biochemical Cycles of Chemo-Chemical Machines: Role of Enzymes

A general cyclic reaction can be written as

$$\varepsilon_1 \rightleftharpoons \varepsilon_2 \cdots \rightleftharpoons \varepsilon_n \rightleftharpoons \varepsilon_1 \quad (44)$$

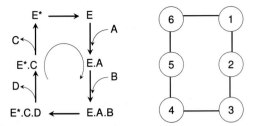 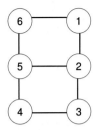

FIGURE 3.3: *The steps in a cycle of a chemo-chemical machine where absence of any direct transition between E A and E* C ensures tight mechano-chemical coupling. In each cycle, the enzyme E goes through six states that are shown explicitly in the left panel and are numbered by the sequence of integers in the right panel.*

FIGURE 3.4: *The steps in the cycle of a chemo-chemical machine which is obtained from that in Fig. 3.3 by adding a direct transition between E A and E* C.*

In this subsection we show how cyclic chemical reaction can be exploited to design a chemo-chemical machine for which both input and output are chemical energies (Hill 2005). Such machines are chemical analogues of simple mechano-mechanical machines like a simple lever. In order to motivate the design of a chemo-chemical machine, consider a reaction

$$A \to C \qquad (45)$$

which is *strongly favored* as the corresponding change of free energy $\Delta G = G_C - G_A \ll 0$. On the other hand, the reaction

$$B \to D \qquad (46)$$

is *weakly unfavored* as the corresponding $\Delta G = G_D - G_B \rangle 0$. So, given an opportunity, A molecules will spontaneously transform to C whereas D molecules will spontaneously transform into B. Is it possible to utilize the large change of free energy of the first reaction (45) to drive the second reaction (46)? If this is possible, this would be an example of "free energy transduction" and the system would operate as a chemo-chemical machine. Some of the free energy released in the reaction (45) is, then, used to pay the free energy cost required to drive the unfavourable reaction (46).

On many occasions it is hard to see how the two reactions would couple together to transduce the free energy on their own. On the other hand, free energy transduction is quite common in living cells; in these processes, usually, a large protein molecule or a macromolecular complex plays the role of a "broker" or a "middleman". In fact, most of the molecular motors we consider here fall in this category of "brokers".

1. A Chemo-Chemical Machine. To illustrate the mechanism let us consider a hypothetical (but, in principle, possible) model shown in Fig. 3.3 where E is the enzyme. Note that E exists in this model in two different conformational states, denoted by E and E^*, which are interconvertible. There is one binding site for A and another for B on the

same conformation E of the enzyme. On the other hand, in the conformational state E^* of the enzyme, these binding sites are accessible only to the molecules C and D. Therefore, once A and B bind to their respective binding sites on E, the enzyme makes a transition to the state E^* forcing A and B to make the corresponding transitions to C and D, respectively. Thus, in this model, the enzyme exists in six states numbered by the sequence of integers shown in Fig. 3.3 above. If one enzyme completes one cycle in the clockwise (CW) direction, the net effect is to convert one A molecule and one B molecule into one C molecule and one D molecule; the enzyme itself is not altered by the complete cycle. With the use of only one cycle, as shown in Fig. 3.3, there is *tight* coupling between the two reactions (45) and (46), i.e., the stoichiometry is exactly one-to-one: each complete cycle coverts exactly one A and one B into exactly one C and one D molecule.

Figure 3.4 below is a generalization of the model where possible transitions between EA and E^*C are now also included. This small extension has non-trivial consequences as we will explain below. Note that now there are three possible cycles as shown in the Figure 3.5. The possible directions are chosen arbitrarily in the CW direction in all three cycles. As explained above, cycle (a) transduces free energy. The cycle (b) runs spontaneously; but, from the point of view of free energy transduction, this cycle

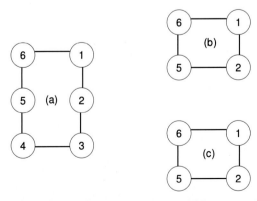

FIGURE 3.5: *The three cyclic pathways, which are possible in the kinetic scheme shown in Fig. 3.4, are shown separately in (a), (b) and (c).*

does not contribute and it simply dissipates some of the free energy of A. The cycle (c), which runs opposite to the direction of spontaneous progress of the reaction, is a wasteful cycle from the perspective of free energetics. However, if all the cycles (a), (b) and (c) occur, the cycles (b) and (c) reduce the overall efficiency of the free energy transduction. More precisely, if the transitions between EA and E^*C occur, the tight coupling of the model is lost because of the "slippage" caused by the cycles (b) and (c) converting the model into a "weak-coupling" model. Thus, for free energy transduction, the kinetic diagram must have at least one cycle that involves both free energy supply and free-energy demanding transitions.

An appropriate measure of the efficiency of the free energy transduction in any chemo-chemical machine is given by

$$\eta_{cb} = \frac{(\Delta G)_{out}}{-(\Delta G)_{in}} \tag{47}$$

For the abstract chemo-chemical machine designed above,

$$\eta_{cb} = \frac{(\Delta G)_{B \to D}}{-(\Delta G)_{A \to C}} \tag{48}$$

2. Flux and Detailed Balance in Chemo-Chemical Machines. A *transition flux* can be defined along any line ij of the kinetic diagram. The transition flux from i to j is given by $\alpha_{ji} P_i$ while the reverse flux, i.e., transition flux from j to i is given by $\alpha_{ij} P_j$. Therefore, the net transition flux in the direction from i to j is given by $\mathcal{J}_{ji} = \alpha_{ji} P_i - \alpha_{ij} P_j$. The mean rate of each cycle k in th CW and CCW directions define the cycle fluxes $\mathcal{J}_{k+}$ and $\mathcal{J}_{k-}$, respectively, and, hence, the corresponding net cycle flux in the cycle k, in the CW direction, is given by $\mathcal{J}_k = \mathcal{J}_{k+} - \mathcal{J}_-$.

At equilibrium, detailed balance holds and, we have

$$\alpha_{ji} P_i^{eq} = \alpha_{ij} P_j^{eq} \tag{49}$$

Therefore, for a cycle like $1 \rightleftharpoons 2 \rightleftharpoons 3 \rightleftharpoons 4 \rightleftharpoons 1$,

$$\frac{\alpha_{12}\alpha_{23}\alpha_{34}\alpha_{41}}{\alpha_{14}\alpha_{43}\alpha_{32}\alpha_{21}} = \frac{P_4^{eq} P_3^{eq} P_2^{eq} P_1^{eq}}{P_2^{eq} P_3^{eq} P_4^{eq} P_1^{eq}} = 1, \tag{50}$$

i.e.,

$$\alpha_{12}\alpha_{23}\alpha_{34}\alpha_{41} = \alpha_{14}\alpha_{43}\alpha_{32}\alpha_{21} \tag{51}$$

At normal conditions, the spontaneous rate of hydrolysis of ATP is extremely low. Enzymes which speed up this reaction are referred to as ATPases. A large number of motors, which we consider in this book, are also enzymes because they catalyze hydrolysis of ATP or GTP. Even for a given single motor domain, a large number of chemical states are involved in each enzymatic cycle. In principle, there are many *pathways* for the hydrolysis of ATP, i.e., there are several different sequences of states that define a complete hydrolysis cycle. Although all these pathways are

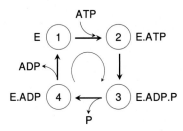

FIGURE 3.6: *A typical cyclic pathway followed by an ATPase, i.e., an enzyme E that hydrolyzes ATP into ADP and inorganic phosphate.*

allowed, some paths are more likely than others. The most likely path is identified as the *hydrolysis cycle*. The biochemical cycle shown in Fig. 3.6 is a concrete example where E denotes an ATPase, an enzyme that hydrolyzes ATP.

IV. MECHANO-CHEMISTRY OF MOLECULAR MACHINES

We combine the fundamental principles of (stochastic) nano-mechanics and (stochastic) chemical kinetics, which we elaborated in the two preceeding sections, to formulate the general theoretical framework of *mechano-chemistry* or *chemo-mechanics*. We mention a few alternative formalisms and explore the possible relations between them. We illustrate the use of these formalisms by applications to some generic models of molecular motors; these models ignore the details of the composition and structure of the track as well as those of the architectural design of the motors.

A. Free-Energy Landscape in a Mechano-Chemical State Space: FP Equation

Suppose, on the time scales relevant for the movement of molecular motors, the state of system is being described by two generalized coordinates y_1, y_2, which represent the "mechanical" and "chemical" variables, respectively. The free energy $G(y_1, y_2)$ can be represented by a landscape. In this landscape, each local minimum represents an experimentally observable kinetic state, while the passes depict the pathways for kinetic transitions. In this scenario, the current along the mechanical coordinate is the velocity of the motor in real space, whereas that along the chemical coordinate is a measure of the enzymatic turnover rate. Thus, the free-energy landscape determines the kinetic mechanism of the machine (Bustamante et al. 2001, Keller and Bustamante 2000).

Most of the tracks for the motors considered in this book are periodic in space where the period will also be the step size of the motor. The motor consumes a fixed amount of free energy in each step. Therefore, if d_1 is the step size and d_2 is the period in the chemical direction, then

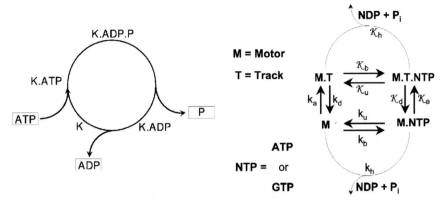

FIGURE 3.7: *A schematic representation of the generic scenario of hydrolysis of ATP by motor enzyme (a) in the absence and (b) in the presence of the corresponding cytoskeletal filament.*

in the absence of any external force (see Fig. 3.8),

$$G(y_1 + d_1, y_2) = G(y_1, y_2)$$
$$G(y_1, y_2 + d_2) = G(y_1, y_2) - |\Delta G| \tag{52}$$

where ΔG is the free energy change in "burning" of chemical fuel in each step (e.g., hydrolysis of a single molecule of ATP). Directed movement of a motor by transducing chemical energy is described by a coupling between the mechanical and chemical transitions in this landscape.

Suppose $P(y_1, y_2; t)$ denotes the probability density for finding the motor between y_1 and $y_1 + dy_1$, in the chemical state between y_2 and $y_2 + dy_2$, at time t. The equation governing the time evolution of this probability density is the Smoluchowski equation

$$\frac{\partial P(y_1, y_2; t)}{\partial t} + \sum_{i=1}^{2} \frac{\partial \mathcal{J}_{y_i}}{\partial y_i} = 0 \tag{53}$$

where the i-th component of the two-dimensional probability current $\vec{\mathcal{J}} = (\mathcal{J}_1, \mathcal{J}_2)$ is given by

$$\mathcal{J}_i = -\frac{k_B T}{\gamma_i}\left(\frac{\partial P}{\partial y_i}\right) + \frac{f_i}{\gamma_i} P \tag{54}$$

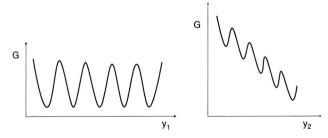

FIGURE 3.8: *Cross sections of the free-energy landscape in the mechano-chemical state space of a molecular motor where y_1 and y_2 denote the mechanical variable and the chemical variable, respectively. The panel on the left shows a typical cross section for a constant y_2 while that on the right shows a typical cross section for a constant y_1.*

with the force

$$f_i = -\frac{\partial G}{\partial y_i} + F_i \tag{55}$$

in the y_i-direction gets contribution from the external force F as well as from the "effective potential" G. Using the explicit formulae (54) and (55), the Smoluchowski equation can be written as

$$\frac{\partial P(y_1, y_2; t)}{\partial t} + \sum_{i=1}^{2}\left[-\frac{k_B T}{\gamma_i}\left(\frac{\partial^2 P}{\partial y_i^2}\right) + \frac{1}{\gamma_i}\frac{\partial}{\partial y_i}(f_i P)\right] = 0 \tag{56}$$

B. Motor Kinetics as a Jump Process in a Mechano-Chemical State Space: Master Equation

In this formulation, both the positions and "internal" (or "chemical") states of the motors are assumed to be discrete. Let $P_\mu(i, t)$ be the probability of finding the motor at the discrete position labelled by i and in the "chemical" state μ at time t. Then, the master equation for $P_\mu(i, t)$ is given by

$$\frac{\partial P_\mu(i, t)}{\partial t} = \left[\sum j \neq i P_\mu(j, t) k_\mu(j \to i)\right.$$
$$\left. - \sum j \neq i P_\mu(i, t) k_\mu(i \to j)\right]$$
$$+ \left[\sum_{\mu'} P_{\mu'}(i, t) W_{\mu' \to \mu}(i) - \sum_{\mu'} P_\mu(i, t) W_{\mu \to \mu'}(i)\right]$$
$$+ \left[\sum j \neq i \sum_{\mu'} P_{\mu'}(j, t) \omega_{\mu' \to \mu}(j \to i)\right.$$
$$\left. - \sum j \neq i \sum_{\mu'} P_\mu(i, t) \omega_{\mu \to \mu'}(i \to j)\right] \tag{57}$$

where the terms enclosed by the three different brackets [.] correspond to the purely mechanical, purely chemical and mechano-chemical transitions, respectively.

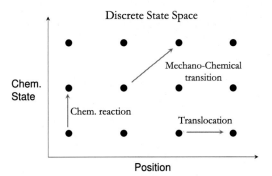

Discrete State Space

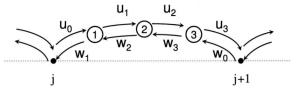

FIGURE 3.9: A schematic representation of discrete states in the mechano-chemical state space and the nature of the possible typical transitions.

FIGURE 3.10: The mechano-chemical cycle of the molecular motor in the Fisher-Kolomeisky model for m = 4. The horizontal dashed line shows the lattice which represents the track; j and j + 1 represent two successive binding sites of the motor. The circles labelled by integers denote different "chemical" states in between j and j + 1.

Utilizing an earlier result of Derrida (Derrida 1983), Fisher and Kolomeisky (see Kolomeisky and Fisher 2007 for the details) proposed a general formula for the average velocity $\langle V \rangle$ of a generic model of molecular motor where the mechano-chemical transitions form unbranched cycles. Each cycle consists of m intermediate "chemical" states in between the successive positions on the track of the motor (Fig. 3.10). The forward transitions take place at rates u_j whereas the backward transitions occur with the rates w_j. Choosing the unit of length to be the separation between the successive equispaced positions of the motor on the track, the average velocity $\langle V \rangle$ of the motor is given by (Kolomeisky and Fisher 2007)

$$V = \frac{1}{R_m}\left[1 - \prod_{j=0}^{m-1}\left(\frac{w_j}{u_j}\right)\right] \qquad (58)$$

where

$$R_m = \sum_{j=0}^{m-1} r_j = \sum_{j=0}^{m-1}\left(\frac{1}{u_j}\right)\left[1 + \sum_{k=1}^{m-1}\prod_{i=1}^{k}\left(\frac{w_{j+i}}{u_{j+i}}\right)\right] \qquad (59)$$

The diffusion constant D is a measure of fluctuations around the directed movement of the motor, on the average, in space. Fisher and Kolomeisky's general result for diffusion coefficient D is

$$D = \left[\frac{(V\,S_N + d\,U_N)}{R_N^2} - \frac{(N+2)V}{2}\right]\frac{d}{N} \qquad (60)$$

where

$$S_N = \sum_{j=0}^{N-1} s_j \sum_{k=0}^{N-1}(k+1)\,r_{k+j+1} \qquad (61)$$

and

$$U_N = \sum_{j=0}^{N-1} u_j r_j s_j \qquad (62)$$

while

$$s_j = \frac{1}{u_j}\left(1 + \sum_{k=1}^{N-1}\prod_{i=1}^{k}\frac{w_{j+1-i}}{u_{j-i}}\right) \qquad (63)$$

$$R_N = \sum_{j=0}^{N-1} r_j \qquad (64)$$

Keller and Bustamante (Keller and Bustamante 2001) wrote down the chemical rate-kinetic equations for a minimal family of models of cytoskeletal molecular motors. Only one of the steps (transitions) is assumed to generate force.

(1) Mechano-chemical *binding* model:

$$[1]_n \xrightarrow[\omega_{12}(F)]{ATP\,binding} [2]_{n+1} \xrightarrow[\omega_{23}]{ATP\,hydrolyis} [3]_{n+1} \xrightarrow[\omega_{34}]{P_i\,release} [4]_{n+1} \xrightarrow[\omega_{41}]{ADP\,release} [1]_{n+1}$$

(2) Mechano-chemical *reaction* model:

$$[1]_n \xrightarrow[\omega_{12}]{ATP\,binding} [2]_n \xrightarrow[\omega_{23}(F)]{ATP\,hydrolyis} [3]_{n+1} \xrightarrow[\omega_{34}]{P_i\,release} [4]_{n+1} \xrightarrow[\omega_{41}]{ADP\,release} [1]_{n+1}$$

(3) Mechano-chemical *release* model:

$$[1]_n \xrightarrow[\omega_{12}]{ATP\,binding} [2]_n \xrightarrow[\omega_{23}]{ATP\,hydrolyis} [3]_n \xrightarrow[\omega_{34}(F)]{P_i\,release} [4]_{n+1} \xrightarrow[\omega_{41}]{ADP\,release} [1]_{n+1}$$

(4) Mechano-chemical *trigger* model:

$$[1]_n \xrightarrow[\omega_{12}]{ATP\,binding} [2]_n \xrightarrow[\omega_{23}]{ATP\,hydrolyis} [3]_n \xrightarrow[\omega_{34}]{P_i\,release} [4]_n \xrightarrow[\omega_{41}(F)]{ADP\,release} [1]_{n+1}$$

Note that the general scheme is identical for all the four models; they differ from each other only in the step that involves movement and, hence, force-dependent rate constant. The mean position of the motor is given by

$$\langle x(t)\rangle = \sum_n n \left(\sum_{\mu=1}^{4} P_\mu(n, t)\right). \tag{65}$$

Therefore, the average speed of the motor

$$v = \frac{d\langle x(t)\rangle}{dt} = \sum_n n \left(\sum_{\mu=1}^{4} \frac{d P_\mu(n, t)}{dt}\right) \tag{66}$$

can be obtained by using the master equations for $P_\mu(n, t)$.

C. Walk in a Time-Dependent Real-Space Potential Landscape: Langevin and FP Approaches

In this subsection, we consider special situations where the changes in the chemical state of the motor are much slower than those in its position. In such situations, we can assume that the potential landscape remains unchanged while small changes in the position of the motor take place. Moreover, no mechano-chemical transition, which would alter both the position and chemical state of the motor, can take place in these situations. Suppose the allowed positions of the motor form a continuum whereas $s = 1, 2, \ldots, S$ denote the discrete chemical states of the motor in each cycle.

In the overdamped regime, the dynamics of the center of mass of the motor x obey the Langevin equation (Wang and Elston 2007, Xing et al. 2005)

$$0 = -\gamma \frac{dx}{dt} - \frac{d V_\mu(x)}{dx} + F_{ext} + \xi(t) \tag{67}$$

where $V_\mu(x)$ is the potential experienced by the motor at the position $x(t)$ when it is in the "chemical" state μ. Moreover, as usual, F_{ext} is the externally applied mechanical force (e.g., the sign of the term will be negative in case of a load force) and $\xi(t)$ is the random Brownian force. The potential $V_\mu(x)$ evolves slowly with time because of the chemical transitions. The chemical state evolves following the discrete master equation

$$\frac{\partial P_\mu(x, t)}{\partial t} = \sum_{\mu'} P_{\mu'}(x, t) W_{\mu'\to\mu}(x)$$
$$- \sum_{\mu'} P_\mu(x, t) W_{\mu\to\mu'}(x) \tag{68}$$

In order to formulate the equivalent Fokker-Planck equations (more appropriately, a hybrid of Fokker-Planck and master equations), we define the probability $P_\mu(x, t)$ that at time t the center of mass of the motor is located at x while it is in the discrete (internal) "chemical state" μ. The equation of motion governing the time evolution of $P_\mu(x, t)$ is a combination of a Fokker-Planck equation and a master equation; the Fokker-Planck part describes the dynamics in continuous space, while the Master equation accounts for the dynamics of transitions between discrete chemical states.

$$\frac{\partial P_\mu(x, t)}{\partial t}$$
$$= \frac{1}{\eta}\frac{\partial}{\partial x}\left[\{V_\mu'(x) - F\} P_\mu(x, t)\right] + \left(\frac{k_B T}{\eta}\right)\frac{\partial^2 P_\mu(x, t)}{\partial x^2}$$
$$+ \sum_{\mu'} P_{\mu'}(x, t) W_{\mu'\to\mu}(x) - \sum_{\mu'} P_\mu(x, t) W_{\mu\to\mu'}(x)$$
$$\tag{69}$$

D. Strength of Mechano-Chemical Coupling

The output of macroscopic machines is usually tightly coupled to the corresponding input; the chemical energy is converted into mechanical work via a strictly scheduled sequence of stages where in each stage there is one-to-one correspondence between the movements of the parts of the machine and the work done. On the other hand, the output of molecular machines is often loosely coupled to the input; the output work extracted from the same amount of input energy (e.g., hydrolysis of a single ATP molecule) fluctuates from one cycle to another (Oosawa 2000).

We define the strength of the coupling by

$$\alpha = \frac{(\text{average velocity of motor})}{(\text{average rate of reaction}) \times d} \tag{70}$$

where d is the step size. Note that α is the probability that the motor takes a mechanical step in space per chemical reaction. Tight coupling corresponds to $\alpha = 1$ whereas $\alpha < 1$ if the coupling is loose. Moreover, $\alpha > 1$ if the motor can take more than one mechanical step per cycle of chemical reaction.

E. Efficiency

The efficiency of molecular motors can be defined in several different ways. While one of the definitions is very similar to that of its macroscopic counterpart, the other definitions are unique to motors operating at the molecular level and characterize different aspects of its movement.

Suppose F_{ext} is the external load force opposing the movement of the motor and $\Delta\mu = \mu_{ATP} - \mu_{ADP+P}$ is the difference of the chemical potentials of ATP and the products of its hydrolysis. Let the average spatial velocity of the motor be $\langle v\rangle$ and let $\langle r\rangle$ denote the average rate of ATP hydrolysis measured in terms of the average number of ATP molecules hydrolyzed per unit time. Then, we can treat F_{ext} and $\Delta\mu$ as two *generalized forces* which are conjugate to the *generalized velocities* v and r, respectively.

The "equations of state" of the motor is given by the functional relations

$$\langle v \rangle = v(F_{ext}, \Delta\mu)$$
$$\langle r \rangle = r(F_{ext}, \Delta\mu) \tag{71}$$

The chemical energy consumed by the motor per unit time is

$$\dot{Q} = \langle r \rangle \Delta\mu \tag{72}$$

while the mechanical work performed against the external force per unit time is

$$\dot{W} = F_{ext}\langle v \rangle \tag{73}$$

An appropriate choice for the definition of the efficiency, which will be the natural counterpart of that of macroscopic motors, is the *thermodynamic efficiency* (Parmeggiani et al. 1999)

$$\eta_{td} = -\frac{F_{ext}\langle v \rangle}{\langle r \rangle \Delta\mu} \tag{74}$$

Not all molecular motors are designed to pull loads. Moreover, in contrast to the macroscopic motors, viscous drag forces not only strongly influence the function of molecular motors but are also intimately connected to the thermal fluctuations through the Einstein relation. Therefore, there is a need for a generalized definition of efficiency that does not necessarily require the application of any external load force (Wang 2005). Such a measure of efficiency, which is slightly different from the thermodynamic efficiency defined above, has also been suggested; it is called "Stokes efficiency" (Derenyi et al. 1999, Wang and Oster 2002b) because the viscous drag is calculated from Stokes law. Suppose, ΔG is the chemical free energy consumed in each reaction cycle. Then, in the absence of any load force, the Stokes efficiency η_s is defined as (Wang and Oster 2002b)

$$\eta_s = \frac{\gamma\langle v \rangle^2}{r\Delta G} \tag{75}$$

A high thermodynamic efficiency and a high Stokes efficiency have different implications. Recall that $F_{ext}^s\langle v \rangle = \langle r \rangle \Delta G$, where F_{ext}^s is the upper limit on the stall force. Therefore, as the load force approaches F_{ext}^s, the thermodynamic efficiency η_{td} approaches unity. In other words, a high thermodynamic efficiency can be achieved by slowing down the motor!

F. Modes of Operation of a Motor

Let us identify the different modes of operation of the molecular motors on the $F - \Delta\mu$ diagram (Parmeggiani et al. 1999).

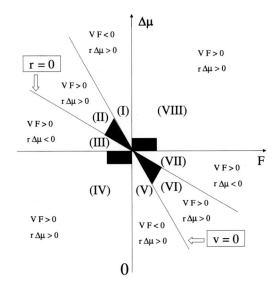

FIGURE 3.11: *Various modes of operation of a cytoskeletal molecular motor in the 2D plane spanned by the two generalized forces (adapted from Parmeggiani et al. (1999)).*

In the linear response regime one can write

$$v = \lambda_{11}F + \lambda_{12}\Delta\mu$$
$$r = \lambda_{21}F + \lambda_{22}\Delta\mu \tag{76}$$

where λ_{ij} are the phenomenological response coefficients.

In the regions II, IV, VI and VIII energy is merely dissipated. In contrast, in the regions labelled by I, III, V and VII energy transduction take place. In I the motor utilizes the ATP in excess to perform mechanical work against the load force. In V, the motor utilizes ADP in excess to perform mechanical work. In III the motor produces ATP, already in excess, from mechanical input energy. In VII the motor produces ADP, already in excess, utilizing mechanical input energy. Some of these modes of operation appear counter-intuitive and their feasibility in real system is still an open question. This conceptual framework has been extended by considering all the four generalized forces, namely, the load force and the chemical potentials of ATP, ADP and P_i (Lipowsky and Liepelt 2008).

V. KIF1A: A MONOMERIC KINESIN ON A MICROTUBULE TRACK

The cytoskeleton of a Eukaryotic cell maintains its architecture. It is a complex dynamic network that can change in response to external or internal signals. The three classes of filamentous proteins, which form the main scaffolding of the cytoskeleton, are: (a) *actin*, (b) *microtubule*, and (c) *intermediate filaments*.

Microtubules are cylindrical hollow tubes whose diameter is approximately 20 nm. The basic constituents of microtubules are globular proteins called tubulin.

Hetero-dimers, formed by α and β tubulins, assemble sequentially to form a protofilament. 13 such protofilaments form a microtubule. The length of each $\alpha - \beta$ dimer is about 8 nm. Since there is only one binding site for a motor on each dimeric subunit of MT, the minimum step size for kinesins and dyneins is 8 nm.

Although the protofilaments are parallel to each other, there is a small offset of about 0.92 nm between the dimers of the neighbouring protofilaments. Thus, total offset accumulated over a single looping of the 13 protofilaments is $13 \times 0.92 \simeq 12$ nm which is equal to the length of three tubulin monomers joined sequentially. Therefore, the cylindrical shell of a microtubule can be viewed as *three* helices of monomers. Moreover, the asymmetry of the hetero-dimeric building block and their parallel head-to-tail organization in all the protofilaments gives rise to the polar nature of the microtubules. The tip of the microtubule ending with α tubulin is called the $-$ end while that ending with β tubulin is called the $+$ end.

Filamentous actins are polymers of globular actin monomers. Each actin filament can be viewed as a double-stranded, right-handed helix where each strand is a single protofilament consisting of globular actin. The two constituent strands are half-staggered with respect to each other such that the repeat period is 72 nm.

The cytoskeleton is also responsible for intra-cellular transport of packaged molecular cargoes. The three super-families of motor proteins are: (i) *myosin* superfamily, (ii) *kinesin* superfamily, and (iii) *dynein* superfamily. Both kinesins and dyneins move on microtubules; in contrast, myosins move on actin tracks. Many cytoskeletal motors carry molecular cargo over distances which are quite long on the intracellular scale. Because of their superficial similarities with porters who carry load on their heads, these motors are often colloquially referred to as "porters".

KIF1A kinesins are members of the kinesin-3 family. These are single-headed motor proteins which move on microtubules on which the equispaced motor binding sites form a periodic linear array. A few years ago a quantitative theoretical model was developed for KIF1A motors (Greulich et al. 2007, Nishinari et al. 2005). The equispaced binding sites for KIF1A on a given protofilament of the MT are labelled by the integer index i ($i = 1, \ldots, L$). In each mechano-chemical cycle a KIF1A motor hydrolyzes one molecule of adenosine triphosphate (ATP), which supplies the mechanical energy required for its movement. The experimental results on KIF1A motors (Hirokawa et al. 2009, Kikkawa et al. 2001, Nitta et al. 2004, Okada and Hirokawa 1999, Okada and Hirokawa 2000, Okada et al. 2003) indicate that a simplified description of its mechano-chemical cycle in terms of a 2-state model (Nishinari et al. 2005) would be sufficient to understand their traffic on a MT. In the two "chemical" states labelled by the symbols S and W the motor is, respectively, strongly and weakly bound to the MT.

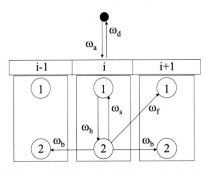

FIGURE 3.12: *Schematic description of the NOSC model of a single-headed kinesin motor that follows a Brownian ratchet mechanism.*

In the Nishinari-Okada-Schadschneider-Chowdhury (NOSC) model (see Fig. 3.12), a KIF1A molecule is allowed to attach to (and detach from) a site with rates ω_a (and ω_d). The rate constant ω_b corresponds to the unbiased Brownian motion of the motor in the state W. The rate constant ω_h is associated with the process driven by ATP hydrolysis which causes the transition of the motor from the state S to the state W. The rate constants ω_f and ω_s, together, capture the Brownian ratchet mechanism (Jülicher et al. 1997, Reimann 2002) of a KIF1A motor. In the absence of attachment and detachment of the motors, our model for a single KIF1A reduces to the Fisher-Kolomeisky multi-step chemical kinetic model of molecular motors on a single filament (see Fig. 3.10) where $m = 2$.

Often a single filamentary track is used simultaneously by many motors and, in such circumstances, the inter-motor interactions cannot be ignored. Therefore, any movement of the motor under these rules is, finally, implemented only if the target site is not already occupied by another motor.

Let us denote the probabilities of finding a KIF1A molecule in the states 1 and 2 at the lattice site i at time t by the symbols S_i and W_i, respectively. In mean-field approximation, the master equations for the dynamics of the interacting KIF1A motors in the bulk of the system are given by

$$
\begin{aligned}
\frac{dS_i}{dt} =\ & \omega_a(1 - S_i - W_i) - \omega_h S_i - \omega_d S_i \\
& + \omega_s W_i + \omega_f W_{i-1}(1 - S_i - W_i)
\end{aligned}
\tag{77}
$$

$$
\begin{aligned}
\frac{dW_i}{dt} =\ & -\omega_s W_i + \omega_h S_i - \omega_f W_i(1 - S_{i+1} + W_{i+1}) \\
& - \omega_b W_i(2 - S_{i+1} - W_{i+1} - S_{i-1} - W_{i-1}) \\
& + \omega_b(W_{i-1} + W_{i+1})(1 - S_i - W_i).
\end{aligned}
\tag{78}
$$

Similarly, the corresponding equations for the two special sites at the left boundary ($i = 1$) and right boundary ($i = L$) can be written down.

TABLE 3.1: Predicted Transport Properties in the Low-Density Limit for Four Different ATP Densities. τ is Calculated by Averaging the Intervals between Attachment and Detachment of Each KIF1A

ATP (mM)	ω_h (1/s)	v (nm/ms)	D/v (nm)	τ (s)
∞	250	0.201	184.8	7.22
0.9	200	0.176	179.1	6.94
0.3375	150	0.153	188.2	6.98
0.15	100	0.124	178.7	6.62

An important test of our model would be to check if it reproduces the single-molecule properties in the limit of extremely low density of the motors. We have already explained in our original paper (Greulich et al. 2007) how we extracted the numerical values of the various parameters involved in our model. Using those parameter sets which allow realization of the low-density of kinesins under open boundary conditions, we carried out computer simulations of the model. We chose microtubules of fixed length $L = 600$, which is the number of binding sites along a typical microtubule filament. Each run of our simulation corresponds to a duration of 1 minute of real time if each timestep is interpreted to correspond to 1 ms. The numerical results of our simulations of the model in this limit, including their trend of variation with the model parameters, are in excellent agreement with the corresponding experimental results (see Table 3.1).

Assuming *periodic* boundary conditions, the solutions $(S_i, W_i) = (S, W)$ of the mean-field equations (77), (78) in the steady-state are found to be

$$S = \frac{-\Omega_b - \Omega_s - (\Omega_s - 1)K + \sqrt{D}}{2K(1 + K)}, \quad (79)$$

$$W = \frac{\Omega_b + \Omega_s + (\Omega_s + 1)K - \sqrt{D}}{2K}, \quad (80)$$

where $K = \omega_d/\omega_a$, $\Omega_b = \omega_b/\omega_f$, $\Omega_s = \omega_s/\omega_f$, and

$$D = 4\Omega_s K(1 + K) + (\Omega_b + \Omega_s + (\Omega_s - 1)K)^2. \quad (81)$$

Thus, the density of the motors, irrespective of the internal "chemical" state, attached to the microtubule is given by

$$\rho = S + W = \frac{\Omega_b + \Omega_s + (\Omega_s + 1)K - \sqrt{D} + 2}{2(1 + K)}. \quad (82)$$

The steady-state flux of the motors along their microtubule tracks is given by

$$J = \omega_f W(1 - S - W). \quad (83)$$

Using the expressions (80) for S and W in equation (83) for the flux we get the analytical expression

$$J = \frac{\omega_f[K^2 - (\Omega_b + (1 + K)\Omega_s - \sqrt{D})^2]}{4K(1 + K)}. \quad (84)$$

VI. RIBOSOME: A MOBILE WORKSHOP ON AN mRNA TRACK

Not all motors carry cargo. And, not all of them move of filaments made of protein. Nucleic acid strands (both DNA and RNA) serve as tracks for some motors which perform some crucial roles in gene expression. As a concrete example, in this section, we consider a motor that moves on an mRNA strand and polymerizes a heteropolymer, namely, a protein.

The sequence of the amino acid subunits of a protein is dictated by that of the codons (a triplet of nucleotide subunits) on a messenger RNA (mRNA). A protein is polymerized by a machine called ribosome (Spirin 2000, Spirin 2002) that uses the corresponding mRNA as the template; this process is referred to as *translation* (of the genetic code). A ribosome hydrolyzes two molecules of Guanosine triphosphate (GTP) to elongate the protein by one amino acid and, simultaneously, to move forward by one codon on the mRNA template. Therefore, a ribosome is also regarded as a motor for which the mRNA strand serves as the track.

X-ray diffraction, cryo-electron microscopy and single-molecule experiments over the last decade have provided a deep understanding of the operational mechanism of the ribosome in terms of its structure, energetics and kinetics (Agirrezabala and Frank 2009, Blanchard 2009, Blanchard et al. 2004, Frank and Gonzalez 2010, Frank and Spahn 2006, Marshall et al. 2008, Mitra and Frank 2006, Munro et al. 2008, Munro et al. 2009, Ramakrishnan 2002, Ramakrishnan 2010, Schmeing and Ramakrishnan 2009, Steitz 2008, Steitz 2010, Tinoco and Wen 2009, Uemura et al. 2010, Wen et al. 2008, Yonath 2010). A kinetic model for translation was developed by Basu and Chowdhury (Basu and Chowdhury 2007) by capturing some of the steps in the mechano-chemical cycle of a ribosome. This model successfully accounted for the average rate of elongation of the protein as well as the distribution of times of dwell of the ribosome at the successive codons (Garai et al. 2009). Very recently this model has been extended by Sharma and Chowdhury (Sharma and Chowdhury 2010a, Sharma and Chowdhury 2010b) by identifying the states in a different manner and by allowing branched pathways which capture the possibility of translational infidelity as well as the effects of "futile" cycles caused by kinetic proofreading (Daviter et al. 2006, Ogle and Ramakrishnan 2005, Rodnina and Wintermeyer 2001, Zaher and Green 2009). This extended version also allows one of the steps, which corresponds to the so-called ratcheting (Cornish et al. 2008, Frank and Agrawal 2000, Frank et al. 2007, Horan and Noller 2007, Korostelev et al. 2008, Moazed and Noller 1989, Moran et al. 2008, Noller et al. 2002, Pan et al. 2007, Shoji et al. 2009, Valle et al. 2003, Wilson and Noller 1998), to be reversible and spontaneous (i.e., does not require GTP hydrolysis).

The transition $1 \rightarrow 2$ (see figure 3.13) accounts for the arrival of an aa-tRNA molecule. A non-cognate tRNA is

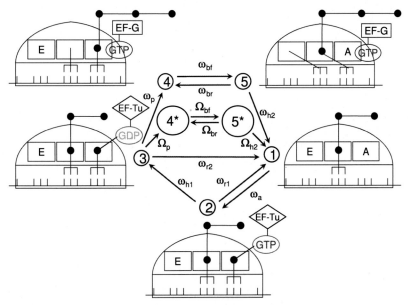

FIGURE 3.13: *Pictoral depiction of the main steps in the mechano-chemical cycle assumed in the Sharma-Chowdhury model of a single ribosome (see text for details).*

rejected on the basis of codon-anticodon mismatch; the corresponding transition being $2 \to 1$. The rejection of a near-cognate tRNA, shown by the transition $3 \to 1$, follows the hydrolysis of the GTP molecule bound to EF-Tu, which is captured by the transition $2 \to 3$. Occasionally, non-cognate and near-cognate tRNAs can escape the quality control system deployed by the ribosome. Therefore, a ribosome can follow one of the two alternative pathways $3 \to 4 \to 5 \to 1$ or $3 \to 4^* \to 5^* \to 1$ both of which elongate the protein by one amino acid; the former by a correct amino acid, whereas the latter by a wrong amino acid.

In this model the average rate of elongation of the protein is given by (Sharma and Chowdhury 2010a)

$$V = \ell_c(\omega_{h2}\mathcal{P}_5 + \Omega_{h2}\mathcal{P}_5^*) = \ell_c K_e ff \left(1 + \frac{\Omega_p}{\omega_p}\right) \quad (85)$$

where ℓ_c is the length of a codon and

$$
\begin{aligned}
\frac{1}{K_{eff}} &= \frac{1}{\omega_a}\left(1 + \frac{\omega_{r1}}{\omega_{h1}}\right)\left(1 + \frac{\omega_{r2}}{\omega_p}\right) + \frac{1}{\omega_{h1}}\left(1 + \frac{\omega_{r2}}{\omega_p}\right) \\
&+ \frac{1}{\omega_p} + \frac{1}{\omega_{bf}}\left(1 + \frac{\omega_{br}}{\omega_{b2}}\right) + \frac{1}{\omega_{b2}} \\
&+ \left(\frac{\Omega_p}{\omega_p}\right)\left[\frac{1}{\omega_a}\left(1 + \frac{\omega_{r1}}{\omega_{h1}}\right) + \frac{1}{\omega_{h1}}\right. \\
&+ \left.\frac{1}{\Omega_{bf}}\left(1 + \frac{\Omega_{br}}{\Omega_{b2}}\right) + \frac{1}{\Omega_{b2}}\right]
\end{aligned}
\quad (86)
$$

VII. SUMMARY AND CONCLUSION

In this chapter I have presented a pedagogical introduction to several interdisciplinary topics at the interface between physics, chemistry and biology. This background knowledge is essential for getting an insight into the fundamental theoretical principles and techniques involved in modeling molecular motors. I have summarized the various alternative approaches to modeling the operational mechanisms of molecular motors in terms of their structure, energetics and kinetics. Finally, to show application of one of these approaches, I have discussed two motors, namely KIF1A and the ribosome. KIF1A functions as a single-headed kinesin motor protein in vitro and moves on a microtubule, which is a filamentous protein. In contrast, the ribosome is a macromolecular complex consisting of proteins and ribosomal RNA. It moves on an mRNA strand and polymerizes a protein using the mRNA as the template. For both these motors, the models successfully predict analytical expressions for the average velocity of the motors as well as their flux when several motors move on the same track simultaneously.

ACKNOWLEDGMENT

I thank all my collaborators for their contributions in our original joint work and Joachim Frank for useful comments as well as suggestions. This work is supported, in part, by the Dr. Jag Mohan Chair Professorship of the author.

REFERENCES

Agirrezabala X and Frank J, (2009) *Elongation in translation as a dynamic interaction among the ribosome, tRNA, and elongation factors EF-G and EF-Tu*, Quart. Rev. Biophys. 42, 159–200.

Andrews S S and Arkin A P, (2006) *Simulating cell biology*, Curr. Biol. 16, R523–R527.

Astumian R D, (1997) *Thermodynamics and kinetics of a Brownian motor, Science* 276, 917–922.

Astumian R D, (2001) *Making molecules into motors, Sci. Am.* 285, 56–64.

Astumian R D and Hänggi P, (2002) *Brownian motors, Phys. Today* 55, 33–39.

Astumian R D, (2007) *Design principles of Brownian molecular machines: how to swim in molasses and walk in a hurricane, Phys. Chem. Phys.* 7, 5067–5083.

Basu A and Chowdhury D, (2007) *Traffic of interacting ribosomes: effects of single-machine mechano-chemistry on protein synthesis, Phys. Rev. E* 75, 021902.

Berg H C, (1993) Random walks in biology, (Princeton university press).

Binnig G and Rohrer H, (1999) *In touch with atoms, Rev. Mod. Phys.* 71, S324–S330.

Blanchard S C, Gonzalez Jr. R L, Kim H D, Chu S and Puglisi J D, (2004) *tRNA selection and kinetic proofreading in translation, Nat. Str. & Mol. Biol.* 11, 1008–1014.

Blanchard S C, (2009) *Single molecule observations of ribosome function, Curr. Opin. Struct. Biol.* 19, 1–7.

Bray D and Duke T, (2004) *Conformational spread: the propagation of allosteric states in large multiprotein complexes, Annu. Rev. Biophys. and Biomol. Str.* 33, 53–73.

Bustamante C, Keller D and Oster G, (2001) *The physics of molecular motors, Acc. Chem. Res.* 34, 412–420.

Bustamante C, Chemla Y R, Forde N R and Izhaky D, (2004) *Mechanical processes in biochemistry, Annu. Rev. Biochem.* 73, 705–748.

Cornish P V, Ermolenko D N, Noller H F and Ha T, (2008) *Spontaneous intersubunit rotation in single ribosomes, Molecular Cell* 30, 578–588.

Cross R A, (1997) *A protein-making motor protein, Nature* 385, 18–19.

Daviter T, Gromadski K B and Rodnina M V, (2006) *The ribosomes response to codonanticodon mismatches, Biochimie* 88, 1001–1011.

Dixon M and Webb E C, (1979) *Enzymes* (Academic Press).

English B P, Min W, van Oijen A M, Lee K T, Luo G, Sun H, Cherayil B J, Kou S C and Xie X S, (2006) *Ever-fluctuating single enzyme molecules: Michaelis-Menten equation revisited, Nat. Chem. Biol.* 2, 87–94.

Gillespie D T, (2005) *Stochastic chemical kinetics, in: Handbook of materials modeling,* ed. S. Yip (Springer).

Grima R and Schnell S, (2008) *Modelling reaction kinetics inside cells, Essays Biochem.* 45, 41–56.

Howard J, (2001) *Mechanics of motor proteins and the cytoskeleton,* (Sinauer Associates, Sunderland).

Howard J, (2006) *Protein power strokes, Curr. Biol.* 16, R517–R519.

Changeux J P and Edelstein S J, (1998) *Allosteric receptors after 30 years, Neuron* 21, 959–980.

Changeux J P and Edelstein S J, (2005) *Allosteric mechanisms of signal transduction, Science* 308, 1424–1428.

Derenyi I, Bier M and Astumian R D, (1999) *Generalized efficiency and its application to microscopic engines, Phys. Rev. Lett.* 83, 903–906.

Derrida B, (1983) *Velocity and diffusion constant of a periodic one-dimensional hopping model, J. Stat. Phys.* 31, 433–450.

Fastrez J, (2009) *Engineering allosteric regulation into biological catalysts, ChemBioChem* 10, 2824–2835.

Frank J and Agrawal R K, (2000) *A ratchet-like inter-subunit reorganization of the ribosome during translocation, Nature* 406, 318–322.

Frank J and Spahn C M T, (2006) *The ribosome and the mechanism of protein synthesis, Rep. Prog. Phys.* 69, 1383–1418.

Frank J, Gao H, Sengupta J, Gao N and Taylor D J, (2007) *The process of mRNA-tRNA translocation, PNAS* 104, 19671–19678.

Frank J and Gonzalez R L, (2010) *Structure and dynamics of a processive Brownian motor: the translating ribosome, Annu. Rev. Biochem.* 79, 381–412.

Garai A, Chowdhury D, Chowdhury D and Ramakrishnan T V, (2009) *Stochastic kinetics of ribosomes: Single motor properties and collective behavior, Phys. Rev. E* 80, 011908.

Greulich P, Garai A, Nishinari K, Schadschneider A and Chowdhury D, (2007) *Intracellular transport by single-headed kinesin KIF1A: effects of single-motor mechanochemistry and steric interactions, Phys. Rev. E* 75, 041905.

Hill T L, (2005) *Free energy transduction and biochemical cycle kinetics,* (Dover).

Hirokawa N, Nitta R and Okada Y, (2009) *The mechanisms of kinesin motor motility: lessons from the monomeric motor KIF1A, Nat. Rev. Mol. Cell Biol.* 10, 877–884, and references therein.

Horan L H and Noller H F, (2007) *Intersubunit movement is required for ribosomal translocation, PNAS* 104, 4881–4885.

Jülicher F, Ajdari A and Prost J, (1997) *Modeling molecular motors, Rev. Mod. Phys.* 69, 1269–1281.

Keller D and Bustamante C, (2000) *The mechanochemistry of molecular motors, Biophys. J.* 78, 541–556.

Khan S and Sheetz M P, (1997) *Force effects on biochemical kinetics, Annu Rev. Biochem.* 66, 785–805.

Kikkawa M, Sablin E P, Okada Y, Yajima H, Fletterick R J and Hirokawa N, (2001) *Switch-based mechanism of kinesin motors, Nature* 411, 439–445.

Kolomeisky A B and Fisher M E, (2007) *Molecular motors: a theorist's perspective, Annu. Rev. Phys. Chem.* 58, 675–695.

Korostelev A, Ermolenko D N and Noller H F, (2008) *Structural dynamics of the ribosome, Curr. Opin. Chem. Biol.* 12, 674–683.

Koshland Jr. D E and Hamadani K, (2002) *Proteomics and models for enzyme cooperativity, J. Biol. Chem.* 277, 46841–46844.

Kou S C, Cherayil B J, Min W, English B P and Xie X S, (2005) *Single-molecule Michaelis-Menten equations, J. Phys. Chem. B* 109, 19068–19081.

Lindsley J E and Rutter J, (2006) *Whence cometh the allosterome?, PNAS* 103, 10533–10535.

Lipowsky R and Liepelt S, (2008) *Chemomechanical coupling of molecular motors: thermodynamics, network representations, and balance conditions, J. Stat. Phys.* 130, 39–67.

Ma J, (2005) *Usefulness and limitations of normal mode analysis in modeling dynamics of biomolecular complexes, Structure* 13, 373–380.

Marshall R A, Aitken C E, Dorywalska M and Puglisi J D, (2008) *Translation at the single-molecule level, Annu. Rev. Biochem.* 77, 177–203.

Mavroidis C, Dubey A and Yarmush M L, (2004) *Molecular Machines, in: Annual Rev. Biomed. Engg.,* 6, 363–395.

Min W, English B P, Luo G, Cherayil B J, Kou S C and Xie X S, (2005) *Fluctuating enzymes: lessons from single-molecule studies, Acc. Chem. Res.* 38, 923–931.

Min W, Gopich I V, English B P, Kou S C, Xie X S and Szabo A, (2006) *When does the Michaslis-Menten equation hold for fluctuating enzymes?*, J. Phys. Chem. B 110, 20093–20097.

Mitra K and Frank J, (2006) *Ribosome dynamics: insights from atomic structure modeling into cryo-electron microscopy maps*, Annu. Rev. Biophys. Biomol. Struct. 35, 299–317.

Moazed D and Noller H F, (1989) *Intermediate states in the movement of transfer RNA in the ribosome*, Nature 342, 142–148.

Mogilner A, Wollman R and Marshall W F, (2006) *Quantitative modeling in cell biology: what is it good for?*, Developmental Cell 11, 279–287.

Moran S J, Flanagan J F, Namy O, Stuart D I, Brierley I and Gilbert R J C, (2008) *The mechanics of translocation: a molecular spring-and-ratchet system*, Structure 16, 664–672.

Munro J B, Vaiana A, Sanbonmatsu K Y and Blanchard S C, (2008) *A new view of protein synthesis: mapping the free-energy landscape of the ribosome using single-molecule FRET*, Biopolymers 89, 565–577.

Munro J B, Sanbonmatsu K Y, Spahn C M T and Blanchard S C, (2009) *Navigating the ribosome's metastable energy landscape*, Trends in Biochem. Sci. 34, 390–400.

Nishinari K, Okada Y, Schadschneider A and Chowdhury D, (2005) *Intracellular transport of single-headed molecular motors KIF1A*, Phys. Rev. Lett. 95, 118101.

Nitta R, Kikkawa M, Okada Y and Hirokawa N, (2004) *KIF1A alternately uses two loops to bind microtubules*, Science 305, 678–683.

Noller H F, Yusupov M M, Yusupova G Z, Baucom A and Cate J H D, (2002) *Translocation of tRNA during protein synthesis*, FEBS Lett. 514, 11–16.

Ogle J M and Ramakrishnan V, (2005) *Structural insights into translational fidelity*, Annu. Rev. Biochem. 74, 129–177.

Okada Y and Hirokawa N, (1999) *A processive single-headed motor: kinesin superfamily protein KIF1A*, Science 283, 1152–1157.

Okada Y and Hirokawa N, (2000) *Mechanism of single-headed processivity: diffusional anchoring between the K-loop of kinesin and the C terminus of tubulin*, PNAS 97, 640–645.

Okada Y, Higuchi H and Hirokawa N, (2003) *Processivity of the single-headed kinesin KIF1A through biased binding to tubulin*, Nature, 424, 574–577.

Oosawa F, (2000) *The loose coupling mechanism in molecular machines of living cells*, Genes to cells 5, 9–16.

Pan D, Kirillov S V and Cooperman B S, (2007) *Kinetically competent intermediates in the translocation step of protein synthesis*, Mol. Cell 25, 519–529.

Parmeggiani A, Jülicher F, Ajdari A and Prost J, (1999) *Energy transduction of isothermal ratchets: generic aspects and specific examples close to and far from equilibrium*, Phys. Rev. E 60, 2127–2140.

Phair R D and Misteli T, (2001) *Kinetic modelling approaches to in-vivo imaging*, Nat. Rev. Mol. Cell Biol. 2, 898–907.

Qian H, (2006) *Open-system nonequilibrium steady-state: statistical thermodynamics, fluctuations, and chemical oscillations*, J. Phys. Chem. B 110, 15063–15074.

Ramakrishnan V, (2002) *Ribosome structure and the mechanism of translation*, Cell 108, 557–572.

Ramakrishnan V, (2010) *Unraveling the structure of the ribosome*, Angew. Chem. Int. Ed. 49, 4355–4380.

Reimann P, (2002) *Brownian motors: noisy transport far from equilibrium*, Phys. Rep. 361, 57–265.

Rodnina M V and Wintermeyer W, (2001) *Fidelity of aminoacyl-tRNA selection on the ribosome: kinetic and structural mechanisms*, Annu. Rev. Biochem. 70, 415–435.

Schliwa M, (ed.) (2003) *Molecular Motors*, (Wiley-VCH).

Schmeing T M and Ramakrishnan V, (2009) *What recent ribosome structures have revealed about the mechanism of translation*, Nature 461, 1234–1242.

Sharma A K and Chowdhury D, (2010) *Quality control by a mobile molecular workshop: quality versus quantity*, Phys. Rev. E. 82, 031912.

Sharma A K and Chowdhury D, (2010) *Distribution of dwell times of a ribosome: effects of infidelity, kinetic proofreading and ribosome crowding*, Phys. Biol. 8, 026005 (2011).

Shoji S, Walker S E and Fredrick K, (2009) *Ribosomal translocation: one step closer to the molecular mechanism*, ACS Chem. Biol. 4, 93–107.

Spirin A S, (2000) *Ribosomes*, (Springer).

Spirin A S, (2002) *Ribosome as a molecular machine*, FEBS Lett. 514, 2–10.

Spirin A S, (2009) *The ribosome as a conveying thermal ratchet machine*, J. Biol. Chem. 284, 21103–21119.

Steitz T A, (2008) *A structural understanding of the dynamic ribosome machine*, Nat. Rev. Mol. Cell Biol. 9, 242–253.

Steitz T A, (2010) *From the structure and function of the ribosome to new antibiotics*, Angew. Chem. Int. Ed. 49, 4381–4398.

Tama F and Brooks III C L, (2006) *Symmetry, form, and shape: guiding principles for robustness in macromolecular machines*, Annu. Rev. Biophys. Biomol. Struct. 35, 115–133.

Thompson D'Arcy, (1963) *On Growth and Form*, vol.I reprinted 2nd edition (Cambridge University Press).

Tinoco Jr. I and Bustamante C, (2002) *The effect of force on thermodynamics and kinetics of single molecule reactions*, Biophys. Chem. 101–102, 513–533.

Tinoco Jr. I and Wen J D, (2009) *Simulation and analysis of single-ribosome translation*, Phys. Biol. 6, 025006.

Tsai C J, Sol A del and Nussinov R, (2009) *Protein allostery, signal transmission and dynamics: a classification scheme of allosteric mechanisms*, Mol. Biosyst. 5, 207–216.

Uemura S, Aitken C E, Flusberg B A, Turner S W and Puglisi J D, (2010) *Real-time tRNA transit on single translocating ribosomes at codon resolution*, Nature 464, 1012–1017.

Vale R D and Oosawa F, (1990) *Protein motors and Maxwell's demons: does mechanochemical transduction involve a thermal ratchet?*, Adv. Biophys. 26, 97–134.

Valle M, Zavialov A, Sengupta J, Rawat U, Ehrenberg M and Frank J, (2003) *Locking and unlocking of ribosomal motions*, Cell 114, 123–134.

Vologodskii A, (2006) *Energy transformation in biological molecular motors*, Phys. of Life Rev. 3, 119–132.

Wang H, (2005) *Chemical and mechanical efficiencies of molecular motors and implications for motor mechanism*, J. Phys. Condens. Matter 17, S3997–S4014.

Wang H and Oster G, (2002a) *Ratchets, power strokes and molecular motors*, Appl. Phys. A 75, 315–323.

Wang H and Oster G, (2002b) *The Stokes efficiency for molecular motors and its applications*, Europhys. Lett. 57, 134–140.

Wang H and Elston T C, (2007) *Mathematical and computational methods for studying energy transduction in protein motors*, *J. Stat. Phys.* 128, 35–76.

Wen J D, Lancaster L, Hodges C, Zeri A C, Yoshimura S H, Noller H F, Bustamante C and Tinoco Jr. I, (2008) *Following translation by single ribosomes one codon at a time*, *Nature* 452, 598–603.

Wildman K H and Kern D, (2007) *Dynamic personalities of proteins*, *Nature* 450, 964–972.

Wilson K S and Noller H F, (1998) *Molecular movements in the translational engine*, *Cell* 92, 337–349.

Xing J, Wang H and Oster G, (2005) *From continuum Fokker-Planck models to discrete kinetic models*, *Biophys. J.* 89, 1551–1563.

Yonath A, (2010) *Polar bears, antibiotics, and the evolving ribosome*, *Angew. Chem. Int. Ed.* 49, 4340–4354.

Zaher H S and Green R, (2009) *Fidelity at the molecular level: lessons from protein synthesis*, *Cell* 136, 746–762.

CHAPTER 4

Exploring the Functional Landscape of Biomolecular Machines via Elastic Network Normal Mode Analysis

Karunesh Arora

Charles L. Brooks III

I. INTRODUCTION

Proteins fold into unique three-dimensional structures predefined by their specific amino acid sequences (Brooks et al. 1998). However, proteins are not static and in their "folded state"; they can interconvert between multiple conformations as a result of thermal energy (Frauenfelder et al. 1988). In fact, protein motions display a hierarchy of timescales, with side-chain fluctuations occurring in the pico- to nanosecond timescale range whereas large-scale domain motions occur in the micro- to millisecond range or slower (Henzler-Wildman et al. 2007). Of these motions, slow-timescale (i.e., ms–μs range), large-amplitude collective motions between a relatively small number of states are of particular interest because they are linked to protein functions such as in enzyme catalysis, drug binding, signal transduction, immune response, protein folding, and protein-protein interactions (Henzler-Wildman and Kern 2007). Large-amplitude conformational changes in biomolecules are often associated with the binding or release of ligands. X-ray crystallography and NMR experiments provide crucial structural information on the conformation of the biomolecule before and after the conformation changes but reveal less about the transition dynamics between two end structures. Moreover, because proteins can sample an ensemble of conformations around the average structure, a complete understanding of protein dynamics requires the knowledge of the multi-dimensional free-energy landscape (functional landscape) that defines the relative probabilities of thermally accessible conformational states and the free-energy barriers between them. This crucial information about the system can be obtained from molecular dynamics (MD) simulations that can pinpoint the precise position and energy of each atom at any instant in time for a single structurally well-resolved protein molecule. Thus molecular dynamics simulations can connect structure and function through a wide range of thermally accessible states, and they have become an important

tool for investigating the dynamics of biological molecules (Karplus and McCammon 2002).

Rapid increases in the availability of computer power and the parallelization of molecular dynamics programs that run on massively parallel architectures have made possible simulations of small biomolecules on the microsecond timescale (Pérez et al. 2007; Freddolino et al. 2008; Maragakis et al. 2008; Shaw et al. 2010). Routine molecular dynamics simulations of large macromolecular assemblies are typically limited to few hundred nanoseconds (Freddolino et al. 2006), with a few recent exceptions (Jensen et al. 2010). However, it is clear that brute-force application of MD will not yield to the timescale of most biological processes and there is a need for algorithms that can sample the vast conformational space available to biomolecules comprehensively (Schlick 2002). The timescale limitations of standard MD have spurred interest in the development of a rich menu of advanced sampling methodologies based on MD that take advantage of clever approximations and sophisticated algorithms while still retaining the atomic-level details, to gain access to timescales relevant to functionally important large-scale conformational changes (see refs. (Arora and Brooks 2007; Arora and Brooks 2009) and references therein). Such calculations are far from routine and remain prohibitively expensive for large macromolecular machines. Coarse-grained models, which reduce the number of atoms to be simulated (typically only C_α or P atoms are considered), have also proven successful in modeling large-scale collective motions of biomolecules that occur on the timescale of a microsecond and beyond (Okazaki et al. 2006; Yoshimoto et al. 2010).

Normal-mode analysis (NMA) is an alternative computational technique for the investigation of large-scale motions of biomolecules (see (Tama and Sanejouand 2001; Tama and Brooks 2006) and references therein). NMA is based on the harmonic approximation of the potential energy in a system and provides information on its accessible equilibrium modes. Among the modes elucidated by NMA, the low-frequency modes represent collective, large-amplitude motions and describe the most facile deformations of the equilibrium structure on the multidimensional energy landscape. Consequently, relatively few low-frequency normal modes usually provide the most relevant information regarding protein function.

The low-frequency modes predicted by NMA are also robust to the choice of level of coarse-graining. As a result, different levels of coarse-graining, from atomically detailed models to models employing low-resolution structural information, can be used for studying long-timescale collective motions and the mechanical properties of extremely large macromolecular assemblies. NMA has already been proven to be extremely useful for studying collective motions of several large biological assemblies including proteins, nucleic acids, and their complexes (see for example, refs. (Duong and Zakrzewska 1997; Duong and Zakrzewska 1998; Delarue and Sanejouand 2002; Cui et al. 2004)). Interestingly, mode motions predicted by NMA match well with the conformational changes of a multitude of proteins upon ligand binding (Brooks and Karplus 1985; Marques and Sanejouand 1995; Mouawad and Perahia 1996). Thus, exploration of the normal modes of a molecular system can yield atomic-level insights into the mechanism and pathways of large-scale rearrangements of proteins that occur upon substrate binding. In recent years, NMA has also emerged as a promising tool for image reconstruction in cryo-EM experiments (Chacon et al. 2003), B-factor refinement in crystallography (Kidera and Go 1990; Kidera and Go 1992), and docking and drug discovery (Cavasotto et al. 2005).

NMA is a mature technique that has been developed over a number of years and has been applied to investigate the dynamics of a multitude of biological systems by several research groups (see (Ma 2005; Tama and Brooks 2006) and references therein). This chapter presents a brief outline of the theory of NMA and its application in our group to explore large-scale conformational changes in macromolecular machines; our focus is on presenting insights gained through the use of NMA rather than simulation details, which can be found in individual papers. First, we discuss details of the NMA technique to calculate normal modes using multi-scale structural representations, ranging from all-atom to pseudo-atomic representations. We then illustrate the applicability of NMA to explore the large-scale functional reorganization of large biomolecules, including the ribosome, an icosahedral virus capsid, a DNA helicase motor protein, and a catalytic enzyme. For some of these systems, the functional motions predicted by NMA have already been validated by independent experiments. In addition, we show that results of NMA, complemented with enhanced sampling methods, can yield a free-energy landscape description for large-scale motions

along with the mechanistic insights. Finally, we conclude by mentioning future applications of NMA and summarize general principles that have emerged from these studies in predicting functional motion, that is, the close correlation between the shape of a biomolecule and its mechanical properties.

II. METHODS

II.1 Normal-Mode Analysis

NMA is a well-established technique (Go et al. 1983; Brooks and Karplus 1985; Levitt et al. 1985) that has recently gained popularity due to algorithmic and model advances that allow applications of NMA to large biomolecular systems. The core of NMA is the diagonalization of the Hessian, which is the $3N \times 3N$ matrix of second derivatives of the potential energy (where N is the number of atoms in the system). The Hessian is constructed based on the harmonic approximation of the potential energy, that is, it is assumed that any given equilibrium system fluctuates about a single well-defined conformation, and that the nature of these thermally induced fluctuations can be calculated assuming a simple harmonic form for the potential. The solution of the Hessian matrix is a set of normal modes, each consisting of an eigenvector and its associated frequency. The eigenvector gives the direction and relative amplitude of each atomic displacement. Low-frequency eigenvectors represent collective, large-amplitude motions and describe the most facile deformations of the structure. Interestingly, the low-frequency modes often correlate well with experimentally observed conformational changes associated with the biological function of the system.

However, the diagonalization of a Hessian matrix for a large biomolecule (more than 300 amino acids) is computationally intensive. Because the application of NMA critically depends on the diagonalization of the Hessian, it can be a limiting factor in applying NMA to large macromolecular machines. To reduce the size of the Hessian, several diagonalization procedures have been developed, such as the diagonalization in mixed basis (DIMB) method (Mouawad and Perahia 1993) and the rotation-translation block (RTB) method (Tama et al. 2000). In the RTB method, first a molecular system is divided into a number of rigid blocks (n_b), each consisting of one or a few consecutive residues per block. Then, the lowest-frequency normal modes of the biological system are obtained as a linear combination of the rotations and translations of these blocks. Each block has six degrees of freedom (three translational, three rotational). The number of degrees of freedom thus reduces from $3N$ to $6n_b$, which in turn reduces the size of the matrix significantly for diagonalization while retaining the flexibility of key structural elements. The RTB method yields approximate low-frequency normal modes that very well describe the functionally important global dynamics of a biomolecule. The method has been successfully employed in NMA studies performed on several large biological machines such as the ribosome (Tama et al. 2003), RNA polymerase (Wynsberghe et al. 2004), the F1-ATPase (Tama et al. 2005), and icosahedral viruses (Tama and Brooks 2005). It has been demonstrated through these applications that the choice of rigid blocks of up to five consecutive residues does not degrade significantly the description of functionally relevant atomic displacements of a biomolecule (Tama et al. 2000). Further, in certain cases, even assigning entire proteins into rigid blocks has been shown to provide useful mechanistic information. For example, the swelling processes in multi-protein icosahedral virus particles have been studied by assigning each protein in the asymmetric subunit as a block (see detailed discussion in the Section III.4).

II.1.1 All-Atom Model of Normal-Mode Analysis.

In the standard approach for NMA, the protein model consists of classical point masses with typically one point per atom. The energy terms for interactions between atoms are defined by a fully atomistic force field. As a result, an atomic-level description of the equilibrium configuration of the structure is required. Further, since the harmonic approximation of potential energy is used in NMA, an energy minimization of the equilibrium configuration of the structure is necessary prior to performing the diagonalization of the Hessian (Goldstein 1950). The quality of the modes is dependent on the structure reaching the true minimum of the potential energy well. Therefore, thorough minimization is necessary to ensure real-valued frequencies for all modes except those related to translation or rotation of the entire molecule (Ma and Karplus 1997). However, energy minimization can become computationally prohibitive as system size increases, and this limits the size of biological systems whose dynamical properties can be investigated by NMA. One advantage of using a detailed force field is that it provides information on natural frequencies of slow normal modes. Using diagonalization techniques such as the RTB method usually shifts the frequencies to values higher than those in the full model; however, this shift occurs in a consistent manner (Tama et al. 2000; Li and Cui 2002) and can therefore be adjusted to determine the natural frequencies of these modes. From the normal-mode frequencies one can determine the vibrational contributions to thermodynamic properties (Brooks et al. 1995).

II.1.2 Coarse-Grained Elastic-Network Model.

The elastic network (EN) model is a simplified representation of the potential energy of a system and is used with NMA of biological systems (Tirion 1996). In the elastic-network model, an experimentally determined structure is represented as a three-dimensional elastic network of pseudoatoms based on the atomic coordinates of amino acid residues or base pairs. Further, amino acids of proteins or base pairs of nucleic acids can be represented in full atomic detail or at a more coarse-grained level. For example, one mass point per residue (Hinsen 1998), only C_α atoms (Atilgan et al. 2001) or more coarse-grained pseudoparticle-based models (Doruker et al. 2002) can be used to identify the junctions of the network. These junctions are representative of the mass distribution of the system and are connected together via a simple harmonic restoring force:

$$E(r_a, r_b) = \begin{cases} \frac{k}{2}(|\vec{r}_a - \vec{r}_b| - |\vec{r}_a^0 - \vec{r}_b^0|)^2 & for \; |\vec{r}_a^0 - \vec{r}_b^0| \leq R_c \\ 0 & for \; |\vec{r}_a^0 - \vec{r}_b^0| > R_c \end{cases}$$

(1)

where $\vec{r}_a - \vec{r}_b$ denotes the vector connecting pseudoatoms a and b, the zero superscript indicates the initial configuration of the pseudoatoms, and R_c is the spatial cutoff (usually ~8–10 Å) for including interactions between the particles. The harmonic spring constant k is assumed to be the same for all interactions within the cutoff distance. The choice of k does not affect the direction of the motions.

The total potential energy of the molecule is expressed as the sum of elastic strain energies:

$$E_{System} = \sum_{a,b} E(\vec{r}_a, \vec{r}_b)$$

(2)

The fact that the energy function, E, is a minimum for any chosen configuration of a system eliminates the need for minimization prior to performing NMA, unlike the all-atom approach. Consequently, NMA can be performed directly on crystallographic or NMR structures, which considerably reduces the computational expense (Wynsberghe et al. 2004).

Several computational studies have shown that the simple Hookean potential is sufficient to reproduce the low-frequency normal modes of proteins as obtained from more complete potential energy functions (Tama and Sanejouand 2001). Furthermore, recent studies of NMA based on the EN model show that the character of the low-frequency motions is unaffected even on coarse-graining the structures up to 1/40th of the number of C_α atoms (Doruker et al. 2002). Altogether, these studies show that the nature of low-frequency modes is robust and the global dynamical properties of a biomolecule can be extracted not only from its detailed structure but also from a low-resolution representation of its mass distribution. Considering that the slow modes predicted by NMA are robust to the choice of the level of coarse-graining, this property of the low-frequency modes can be exploited further to study conformational transitions of large biomolecular assemblies using low-resolution structural information from cryo-EM. As discussed in Section III.3, NMA based on a discrete representation of an electron microscopy map of the ribosome at 25 Å resolution displays motions similar to the ratchet-like reorganization observed in cryo-EM experiments (Frank and Agrawal 2000) and NMA of the

70S atomic structure (Tama et al. 2003). Thus, NMA can assist in the interpretation of inferred dynamical transitions in large biological assemblies characterized by cryo-EM.

Combining the elastic network representation of structure and connectivity with the RTB approach to reduce the size of the Hessian, one can extend the application of normal-modes methods for the exploration of conformational deformations and dynamics to macromolecular assemblies of nearly arbitrary size without substantial increase in the computational expense. In the Section III, we present several applications of methods discussed in Section II to explore dynamical properties of large macromolecular assemblies. The overall goal of these studies it to complement experimental observations by exploring functionally important rearrangements of biomolecules inferred based on the static structures of a system in different conformations and to obtain new insights into the mechanism of these transformations that are presently inaccessible to experiments.

III. BIOLOGICAL APPLICATIONS OF NORMAL-MODE ANALYSIS

III.1 Allosteric Transitions in Adenylate Kinase

Adenylate kinase (AdK) is an allosteric monomeric phosphotransferase enzyme that catalyzes the reversible transfer of a phosphoryl group from ATP to AMP and thus plays an important role in cellular energy homeostasis. Given its biological significance, AdK has been the subject of a multitude of experimental (Dahnke et al. 1992; Sinev et al. 1996; Shapiro et al. 2000; Wolf-Watz et al. 2004; Shapiro and Meirovitch 2006; Henzler-Wildman et al. 2007; Henzler-Wildman et al. 2007; Schrank et al. 2009) and theoretical/computational (Miyashita et al. 2003; Temiz et al. 2004; Maragakis and Karplus 2005; Miyashita et al. 2005; Lou and Cukier 2006; Lou and Cukier 2006; Arora and Brooks 2007; Chu and Voth 2007; Hanson et al. 2007; Whitford et al. 2007; Beckstein et al. 2009; Feng et al. 2009; Korkut and Hendrickson 2009) investigations. Structurally, AdK is composed of three main domains: the CORE (residues 1–29, 68–117 and 161–214), the ATP-binding domain, called the LID (residues 118–167), and the AMP-binding domain, called the NMP (residues 30–67) (see Figure 4.1). Several crystal structures of AdK from *E. coli* and other organisms are available, both in the free form and in complex with a variety of substrates and inhibitors (see (Vonrhein et al. 1995) and references therein). Based on structural analysis, it appears that the enzyme assumes an "open" conformation in the ligand-free state and a "closed" conformation when complexed with inhibitor AP5A (Figures 4.1a & 4.1b) (Muller and Schulz 1992; Muller and Schulz 1993). The closing of the LID and NMP domains over the substrate brings AdK to a catalytically competent state. A key question for investigation is how the binding of a substrate leads to large-scale protein motions.

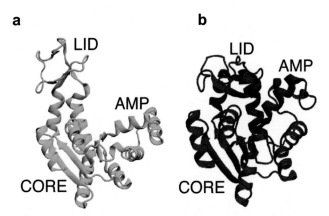

FIGURE 4.1: *Adenylate kinase in open (PDB ID: 4AKE) (left) and closed (PDB ID: 1AKE) (right) conformations (Muller and Schulz 1992; Muller and Schulz 1993).*

To explore the functional transitions of *E. coli* AdK in detail, we performed an elastic-network normal-mode analysis of the open (apo) state of AdK as well as all-atom simulations connecting open and closed endpoint configurations using enhanced sampling methods (Arora and Brooks 2007). An elastic network was constructed using 1,656 heavy atoms (214 residues) of AdK with a spatial cutoff of 8 Å for interconnections between the heavy atoms. NMA was performed using the RTB method for which two consecutive residues were assigned to a block. Thus a total of 107 blocks were considered and the matrix of size 642×642 was diagonalized to obtain the normal modes. Normal-mode analysis performed on AdK shows that the displacement of the Apo structure along the lowest frequency mode describes the functionally important closing motion (Figure 4.2). Comparison of the vector connecting the open (apo molecule) and closed (inhibitor-bound) structures with the calculated lowest-frequency mode shows that this mode contributes 68 % to the functionally important transition between the open and the closed states of AdK (Figure 4.3). Interestingly, the top-ten-ranking low-frequency modes account for 92 % of the conformational change, as shown by the cumulative overlap[1] between the NMA modes (eigenvectors) predicted for the starting conformation (open apo state) and the targeted direction (closed, inhibitor-bound state) of the structural change (Figure 4.3). Normalized mean square displacement of all C_α atoms in AdK associated with motion along the lowest-frequency mode (mode 1, after the six trivial translation and rotation modes are removed) shows that the fluctuations are dominated by LID and NMP domain dynamics that characterize the closing motion of the enzyme (Figure 4.4). Thus,

---

[1] To measure how similar a given normal mode, a_j, is to an experimentally known conformational change, $\Delta r = r_1 - r_2$, the overlap between the two vectors can be calculated as: $I_j = \frac{a_j \cdot \Delta r}{|a_j||\Delta r|}$. An overlap value of 1 indicates that the direction of the conformational change, Δr, and the direction given by the normal mode are identical.

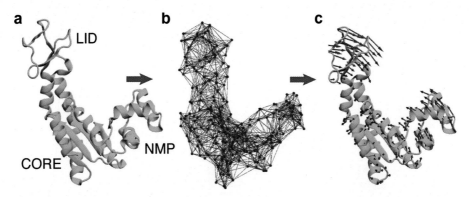

FIGURE 4.2: *Functional motions obtained from normal-mode analysis of AdK. (a) Adenylate kinase structure in open conformation (PDB ID: 4AKE). (b) Elastic network representation of the open form of AdK. NMA was carried out by establishing harmonic springs between heavy atoms less than 8 Å away from each other. Only springs between C_α atoms are depicted for clarity. (c) The arrows indicate the direction and amplitude of motions along the lowest-frequency normal mode.*

normal-mode analysis, which primarily captures the mass distribution of a structure, suggests that the architecture of AdK is such that it preferentially promotes the collective motions of the enzyme in the direction of the catalytically bound, closed conformation.

Further, a free-energy landscape depiction of these functional transitions, which occur on the ms to μs time scale (Wolf-Watz et al. 2004), was obtained from atomically detailed simulations of AdK with and without an inhibitor (Arora and Brooks 2007). We used a combination of advanced sampling methods, the nudged elastic band (NEB) method (Chu et al. 2003) and umbrella-sampling

molecular dynamics simulations (Torrie and Valleau 1977), to gain access to atomically detailed conformational transition pathways as well as free-energy profiles associated with the displacement of the LID and NMP domains.

Interestingly, our free-energy calculations reveal that in the ligand-free state, the enzyme samples near-closed conformations even in the absence of its substrate with a wide free-energy well representing the open conformation of AdK (Figure 4.5). This observation is consistent with the existence of several crystal structures of apo AdK, which differ in the degree of bending of the LID domain around hinges relative to the CORE, as well as the large fluctuations of LID domain observed upon elastic deformation along the lowest-frequency mode (Figure 4.4). However, our results show that there is a substantial free-energy barrier to reaching the completely closed state in the absence of the substrate, with the barrier lying close to the closed conformation. On the other hand, the closed conformation of AdK is energetically most favored in the ligand-bound state with a large barrier to opening (Figure 4.5). In light of these free-energy calculations, the entire conformational change process of AdK can be described as the transformation of the enzyme from the free-energy surface associated with the ligand-free form to the free-energy surface associated with the ligand-bound form following substrate binding (Arora and Brooks 2007). The emerging view of AdK conformational transitions is most consistent with a preexisting equilibrium/selected-fit ligand binding mechanism, in which the ligand simply binds to, and stabilizes, a productive binding configuration of the biomolecule (Tsai et al. 1999; Volkman et al. 2001). Our intriguing findings related to the dynamical behavior of AdK, gleaned from NMA of AdK and application of enhanced sampling methods, were subsequently corroborated by experimental studies employing a variety of sophisticated experimental techniques such as single molecule FRET and NMR (Henzler-Wildman et al. 2007; Henzler-Wildman et al. 2007).

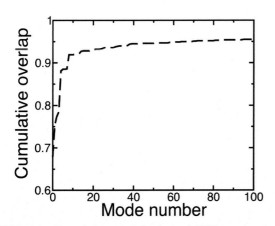

FIGURE 4.3: *Cumulative contribution of NMA modes to the structural change between the open and closed forms of AdK. The ordinate displays the cumulative overlap between the ENM modes (eigenvectors) predicted for the starting conformation (open X-ray structure) and the targeted direction (closed X-ray structure) of the structural change EN-NMA calculations were performed using the open form as the starting structure. The cumulative overlap of 0.9 is achieved by the top-ranking 10 modes (fraction of the total accessible modes). The first mode alone describes about 68% conformational change and predominantly corresponds to the fluctuations in the LID and the NMP domains of AdK.*

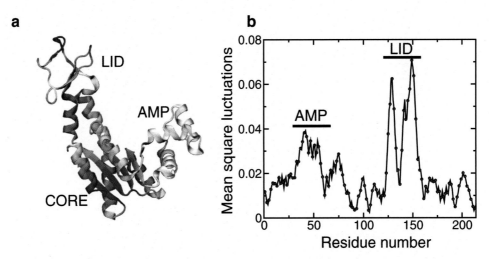

FIGURE 4.4: *Atomic motions in AdK arising from the displacement along the lowest-frequency mode. (a) Normalized mean square displacement of all C_α atoms in AdK associated with motion along the first lowest-frequency mode. (b) The AdK atoms are colored according to the amplitudes of their displacement from the equilibrium positions during the closing motion. The scale runs from red, representing the largest conformational change, to blue, indicating regions showing very little motion. Parts of the structure colored white represent intermediate-scale displacements during excursions along the mode 1.*

In summary, NMA of AdK using a simplified potential provides important data regarding the preferential direction of collective motions, and this information can be used to understand processes that are important for ligand binding and may be important for enzyme catalysis. Comparatively, mapping pathways of transitions between open and

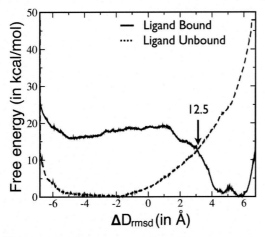

FIGURE 4.5: *Superimposed one-dimensional free-energy profile of the ligand-unbound pathway and that of the ligand-bound pathway of AdK. The difference in root mean square deviation from the open and closed states (ΔD_{rmsd}) of AdK is an order parameter for the characterization of the open-to-closed conformational transition. The intersection region of the two profiles locates the transition state of the conformational transition. The stability of the open unbound and closed bound state is assumed to be the same. For details see (Arora and Brooks 2007) (figure 5 of Arora et al.,* Proc. Natl. Acad. Sci. USA, *104:18496–18501, 2007, copyright 2007 National Academy of Sciences U.S.A)*

closed conformations using a detailed all-atom potential provides more specific information about the underlying free-energy landscape of the allosteric transitions and suggests a preexisting equilibrium ligand-binding mechanism. Thus, the combination of simplified elastic network models with more accurate physical calculations can produce a much fuller dynamical picture of the functional conformational changes occurring in enzymes.

III.2 Functional Mechanical Deformations in Hexameric Helicase

Hexameric helicases are molecular-motor proteins that utilize energy obtained from ATP hydrolysis to translocate along, and/or unwind, nucleic acids in a unidirectional manner. Despite several structural, biochemical (Wessel et al. 1992; Fouts et al. 1999; Li et al. 2003; Enemark and Joshua-Tor 2006; Erzberger and Berger 2006) and computational studies (Yu et al. 2007; Liu et al. 2009; Shi et al. 2009), there is no detailed understanding of how ATP binding and hydrolysis are coupled with conformational changes to achieve unidirectional DNA translocation. Given that the malfunctioning of helicases is linked to cancer and premature aging, it is biomedically important to understand how helicase motor proteins operate at the molecular level (van Brabant et al. 2000; Mohaghegh and Hickson 2001).

The E1 helicase of papillomavirus and Simian Virus 40 Large Tumor Antigen helicase (SV40) are two structurally well-characterized helicases that belong to the AAA + protein family (Li et al. 2003; Gai et al. 2004; Enemark and

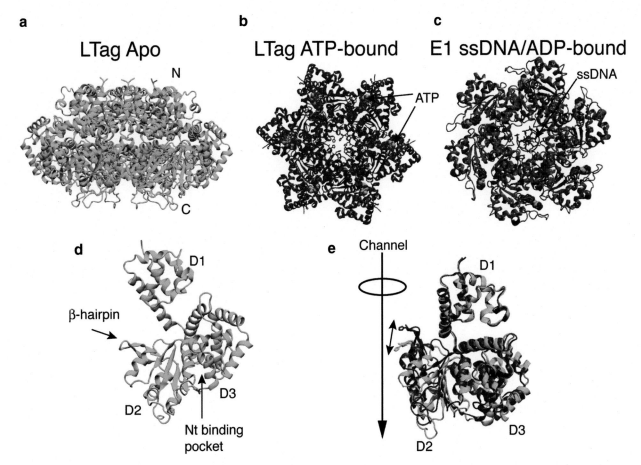

FIGURE 4.6: *The structural features of the hexameric helicase. (a) The side view of SV40 hexameric helicase in Apo conformation (PDB ID: 1SVO). (b) The top view of SV40 helicase bound to six ATP molecules (PDB ID: 1SVM), one in each subunit. (c) Top view of E1 hexameric helicase bound to ssDNA in the hexamer channel and six ADP molecules (PDB ID: 2GXA). (d) Side view of SV40 monomer structure. (e) Superposition of Apo and ATP bound monomer structures of SV40 helicase. Superposition is based on D1 domain.*

Joshua-Tor 2006). Structural studies have shown that in their functional form, these two helicases assemble to form a ring-shaped structure, with six identical protein subunits encircling the DNA (Figure 4.6). Structures of SV40 hexameric helicase in distinct nucleotide binding states (i.e., Apo, ATP-bound and ADP-bound states) and E1 hexamer structure with single-stranded DNA discretely bound within the hexamer channel and six ADP molecules at the subunit interfaces have been solved (Figures 4.6a–4.6c) (Gai et al. 2004; Enemark and Joshua-Tor 2006). These hexameric helicase structures reveal two stacked hexamer rings with a central channel (Figure 4.6a). Each subunit of the hexamer contains three structural domains, D1, D2, and D3 (Figure 4.6d). The D2 domain is a typical AAA+ domain that binds ATP. The interior of the hexamer channel is lined with β-hairpins. These hairpin loops have several positively charged residues that interact with the phosphates of the DNA backbone and account for most of the protein-DNA interactions in the E1 hexamer (Figure 4.6e).

Based on the information gleaned from these static crystal structures and bulk kinetic data, two very different mechanisms that couple ATP cycling to DNA translocation have been proposed for helicases (Gai et al. 2004; Enemark and Joshua-Tor 2006). Gai and coworkers suggested a concerted nucleotide binding and hydrolysis mechanism for the SV40 helicase (Gai et al. 2004). Alternatively, based on the crystal structure of E1 hexameric helicase bound to single-stranded DNA, a sequential ATP binding mechanism has been proposed (Enemark and Joshua-Tor 2006). A key question for investigation is whether the DNA translocation by helicases occurs via a concerted or sequential ATP binding mechanism. In addition, how is ATP binding and hydrolysis coupled to conformational changes to achieve unidirectional DNA translocation?

To decompose the functionally important mechanical process of DNA translocation in the SV40 helicase into its essential components, we employed a combination of NMA and molecular-dynamics simulations of a structure-based

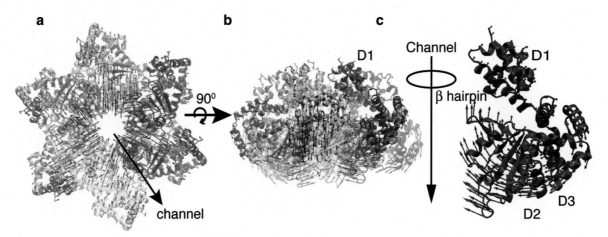

FIGURE 4.7: *Functional motions of the hexameric helicase predicted by NMA. (a) Atomic motions in the helicase arising from the displacement along a low-frequency mode 3. The arrows indicate the direction and amplitude of motions of each subunit along the low frequency mode. (b) Side view of the hexameric helicase. (c) Single subunit of the helicase depicting change in the position of β-hairpin arising from the displacement along mode 3.*

coarse-grained model (Yoshimoto et al. 2010). An elastic network was constructed from the 17,590 heavy atoms (total 2,172 residues; 332 in each subunit) of the SV40 X-ray structure in Apo conformation. Heavy atoms less than 8 Å away from each other were inter-connected via harmonic springs. NMA was performed using the RTB method, for which eleven consecutive residues were assigned to a block. Thus a total of 198 blocks were considered and the matrix of size 1188 × 1188 was diagonalized to obtain the normal modes.

The normal-mode analysis of the Apo SV40 hexameric helicase structure reveals that the motions corresponding to low-frequency mode 3 may be biologically significant. As shown in Figure 4.7, dominant elastic deformation of the helicase along low-frequency mode 3 corresponds to the longitudinal movements of the six β-hairpins along the middle of the hexameric helicase channel. This motion describes the change in the orientation between the D1 and D2/D3 domains and a large movement of β-hairpin that interacts with the ssDNA in the channel (Figure 4.7c). Likely, the six β-hairpin loops lining the interior of the hexamer channel function as a motor for pulling DNA into the SV40 double hexamer for unwinding. Comparison of the vector connecting the Apo and ATP-bound crystal hexamer structures with the calculated displacement vector along low-frequency mode 3 shows that this mode alone contributes about 70% to the functionally important transitions. Thus the direction of collective motions predicted from NMA of the Apo hexamer X-ray structure agree well with motions inferred from high-resolution crystal structures of SV40 in different nucleotide-binding states (Figure 4.6e) (Gai et al. 2004).

To investigate the coupling of nucleotide binding and release with large-scale conformational changes of DNA binding hairpin loops, we performed molecular-dynamics simulations using a structure-based coarse-grained model (Yoshimoto et al. 2010). In the coarse-grained model of the helicase, the ATP binding events (i.e., binding, hydrolysis, and product release) were introduced by imposing distance restraints between key residues in the ATP binding pockets. From the coarse-grained simulation studies we find that unidirectional DNA translocation preferentially occurs via a sequential ATP binding mechanism, and that there is a strong dependence of the hexamer motion on the sequence of ATP binding (e.g., clockwise or counter-clockwise direction around the hexamer). Furthermore, simulations of Apo hexamer with ATP bound in one of the six available binding pockets resulted in an asymmetric deformation of the empty binding pockets. This observation lends support to our hypothesis that the hexamer may have intrinsic structural properties for coordinating ATP binding in a sequential order, which can be effected by binding ATP in the active-site pocket.

Overall, coarse-grained simulation studies show that the substrate coupling, conformational change, and asymmetry in substrate binding are essential elements of translocation of the SV40 helicase along ssDNA (Yoshimoto et al. 2010). The emerging picture that is most consistent with combined normal-mode analysis and coarse-grained molecular-dynamics simulations of the SV40 helicase is that the functional motions of the helicase that are used for DNA translocation preferentially occur in the direction of low-frequency modes. The energy for the excitation of these normal modes is obtained from the strain introduced upon binding of different substrates (i.e., ATP, ADP, etc.) and is subsequently transmitted via the helicase's unique architecture into mechanical motions that are most effective for DNA translocation. Taken together,

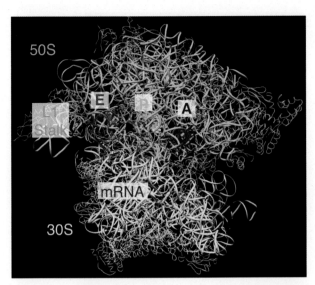

FIGURE 4.8: *A front view of the 70S ribosome architecture. The large subunit (50S) is shown in cyan and the small subunit (30S) is shown in yellow; the mRNA is shown in magenta. The position of the tRNA binding sites (A, aminoacyl; P, peptidyl; E, exit) and the L1 stalk are indicated with white overlays.*

these studies provide a near-atomic-level understanding of the mechanical properties of the helicase motor protein and possibly other AAA + proteins.

III.3 Dynamic Reorganization of the Functionally Active Ribosome

Protein synthesis from amino acids occurs in the cells of all living organisms and is one of nature's most fundamental processes. It is a multistep biochemical process that can be divided into the three stages: initiation, elongation, and termination (Berg et al. 2006). Protein synthesis is catalyzed by the ribosome, a macromolecular machine that translates the genetic information residing on mRNA into a specific sequence of amino acids. Different sets of accessory factors (i.e., initiation factors, elongation factors, and release factors) assist the ribosome at each stage of protein synthesis. Structurally, the ribosome is a huge complex of protein and RNA made up of two unequal subunits, the large (50S) and the small (30S) subunit (see Figure 4.8). The two subunits join together on an mRNA molecule to form the 70S active complex and interact via a network of intermolecular bridges. The tRNA molecules that transfer a specific active amino acid to a growing polypeptide chain during translation are located in the inter-subunit space between the 30S and 50S subunits. Both subunits contain three binding sites for tRNA: the A (aminoacyl), P (peptidyl) and E (exit) sites (Figure 4.8).

In a key mechanical step during the elongation cycle, following the formation of a new peptide bond, peptidyl-tRNA is translocated from the A site to the P site, and the deacylated tRNA from the P to the E site, respectively (Figure 4.8). Binding of elongation factor G (EF-G)

and subsequent GTP hydrolysis catalyzes the translocation process (Agrawal et al. 1999) and is believed to be accompanied by large-scale conformational rearrangements (Stark et al. 2000; Frank and Agrawal 2000). In spite of the emergence of high-resolution crystal structures and decades of biochemical and biophysical studies, the ribosome's fundamental mechanism of operation remains unclear. Key unresolved questions include: (1) How do tRNAs and mRNA move during translocation, a process that involves large molecular movements at intervals of approximately a millisecond? (2) What is the role of EF-G in that process?

Cryo-electron microscopy (Frank and Agrawal 2000) and X-ray crystallography (Ban et al. 2000; Harms et al. 2001; Yusupov et al. 2001; Ramakrishnan 2002) have provided strong evidence of the role of large dynamic motions during protein synthesis in the ribosome. Specifically, based on the cryo-EM reconstructions of certain EF-G-containing complexes, it is believed that translocation of tRNA and mRNA through the ribosome, from the A to P to E sites, involves a "ratchet-like" relative rotation of the two ribosomal subunits. In addition, other conformational changes of the ribosome that involve large displacements of the L1 stalk (Agrawal et al. 1999; Gomez-Lorenzo et al. 2000; Yusupov et al. 2001; Valle et al. 2003) region and rearrangements of the L7/L12 stalk (Agrawal et al. 1999) have also been observed and implicated as functionally relevant for protein synthesis (Ramakrishnan 2002).

However, cryo-EM reconstructions and X-ray crystallography yield only static information from which dynamics must be inferred. To connect observed and inferred functional transitions and provide atomic-level annotation of the functional transitions suggested by the lower resolution cryo-EM data, elastic network NMA was carried out (Tama et al. 2003). An elastic network model was constructed based on the 5.5 Å X-ray map of the 70S ribosome from *Thermus thermophilus*. In the elastic network model of the ribosome, phosphate and C_α atom positions were taken for the junctions of the network. Two cutoff distances to delineate the junctions within the network were considered. A cutoff of 20 Å for the P-P and P-C_α interactions and 16 Å for the C_α-C_α interactions in the X-ray structure were used. These values were based on the distance distribution functions between C_α-C_α, C_α-P and P-P positions. NMA was performed using the RTB method for which five consecutive C_α or P atoms were assigned to a block; the block boundaries were constructed such that atoms from different subunits were not included in the same blocks.

Results of NMA indicate that two low-frequency modes (mode 1 and mode 3) reproduce very well the rearrangements of ribosome subunits proposed to be functionally important based on experimental studies (Frank and Agrawal 2000). Specifically, mode 3 captures an inter-subunit rotation of the 30S relative to the 50S subunit

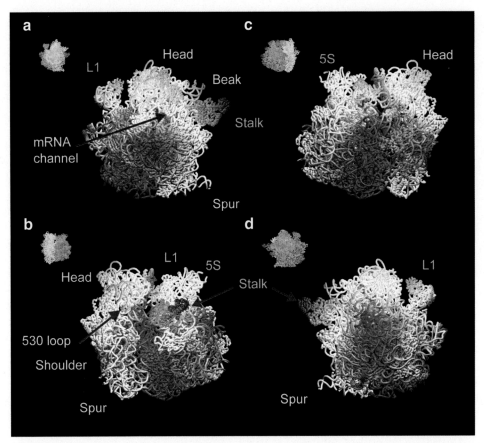

FIGURE 4.9: *Atomic motion in the ribosome arising from displacements along mode 3. (a–d) Four different views are presented, as indicated by thumbnails (yellow = 30S, blue = 50S). The atoms are colored according to the amplitudes of their displacement from the equilibrium positions during the ratchet-like motion. The scale runs from red, representing the largest conformational change, to blue, indicating regions showing very little motion. Parts of the structure colored white represent intermediate-scale displacements during excursions along the ratchet-like mode. (Figure 9 of Tama et al.,* Proc. Natl. Acad. Sci. USA, *100(16): 9319–23, 2003, copyright 2003 National Academy of Sciences U.S.A)*

(Figure 4.9). The motion along this mode agrees with a previously described "ratchet-like" reorganization of the ribosome in response to the binding of EF-G (Frank and Agrawal 2000). Notably, the ratchet-like rotation of ribosome subunits was found to be the predominant mode of motion even though NMA of the ribosome was performed in the absence of EF-G. This observation from NMA suggests that the process of ribosomal translocation involving inter-subunit rotation is most likely thermally driven, that is, the ribosome can access an alternate "ratcheted" conformation through spontaneous excursions along the low-frequency mode at small energetic cost. In support of this observation, recent single-molecule FRET (smFRET) studies show that pretranslocation ribosomes (PRE) undergo spontaneous inter-subunit rotational motion even in the absence of EF-G (Ermolenko et al. 2007; Cornish et al. 2008). Furthermore, NMA clearly indicates that the ratchet-like motion is accompanied by large conformational rearrangements in regions of the

ribosome interacting directly with the EF-G, e.g., the stalk base of the 50S subunit, and the head and shoulder of the 30S subunit (Figure 4.9). In this scenario, where the ribosome can spontaneously fluctuate between a "normal" and "ratcheted" conformation even in the absence of EF-G, the principal role of EF-G binding can be seen as stabilizing and shifting the preexisting population toward the ratcheted conformation of the ribosome rather than actively promoting the inter-subunit rotation. This picture is most consistent with a preexisting equilibrium ligand-binding mechanism, a concept derived from an energy landscape view of catalysis, as observed in the case of catalytic enzyme Adenylate kinase (see Section III.1) and likely extends to large-scale mechanical movements observed during elongation cycle of protein synthesis by the ribosome.

The ratchet-related conformational change of the ribosome is a complex motion that probably involves reconfiguration of its interactions with other structural

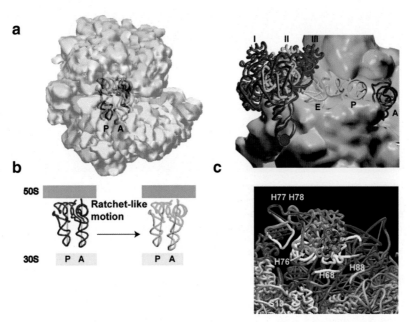

FIGURE 4.10: *Local rearrangements within the ribosome as a result of displacement along the elastic normal modes. (a) Motions of tRNAs in the A and P sites as a result of the ratchet-like motion from mode 3. (b) Motion of the L1 stalk as a result of elastic mode 1. The displacements of this structural component occur around the pivotal point denoted by the red circle and are of the extent illustrated by the outer (I) and inner (III) positions. Position II represents the state observed in the X-ray model (Yusupov et al. 2001), taken to be the equilibrium in the NMA, whereas the magnitude of displacements are taken from the structural work (Harms et al. 2001). (c) The magnitude of motions in the L1 protein arising from mode 1 is shown by coloring the atoms according to the amplitude of their displacements along this mode (red, large motion; white, intermediate-scale motion; blue, little or no motion). (Figure f10 of Tama et al., Proc. Natl. Acad. Sci. USA, 100(16): 9319–9323, 2003, copyright 2003 National Academy of Sciences U.S.A)*

components. Indeed, NMA shows that the positions of tRNAs residing in the A and P sites are also affected as a result of a ratchet-like rotation of the ribosome arising from displacement along low-frequency mode 3 (Figure 4.10). As a result of this conformational change, interactions between tRNA molecules at the A site and the ribosome are weakened, while the contacts with P-site tRNA remain intact. However, new contacts are formed between the tRNA and H69 on the ribosome subunit as a result of this motion. The change in the interaction between the tRNA molecules and the two ribosomal subunits suggests that the rotation may facilitate the movement of the tRNAs through the space between the two subunits. Thus, NMA studies provide a glimpse into the early stages of the tRNA translocation accompanied with ratchet-like movement of the ribosome subunits.

Interestingly, mode 1 describes the motion of the L1 stalk (Figure 4.11). The primary motion observed for this mode is the displacement of the L1 protein to a position away from, or close to, the inter-subunit space (Figure 4.11a). The direction of rotation of L1 stalk is in the opposite direction in relation to the rotation of 30S subunit, in

agreement with experiments (Valle et al. 2003; Agirrezabala et al. 2008; Fei et al. 2008; Fei et al. 2009). Furthermore, the motion of the L1 stalk is correlated with small rearrangements near the A, P, and E tRNA-binding sites that may play a role in effecting or regulating the removal of the E-site tRNA (Figure 4.10). The preponderance of the data available from normal-mode analysis, cryo-EM, X-ray crystallography, biochemical, and smFRET studies suggests that structural rearrangements of the ribosome at separated locations are coupled together to perform key ribosomal functions. This also lends support to the emerging view that the ribosome employs a Brownian motor mechanism for its translocation reaction (Frank and Gonzalez Jr. 2010).

In summary, NMA studies demonstrate that motions deduced from comparisons of cryo-EM maps can be well reproduced using a simplified representation of the ribosome that mainly captures the shape of the system. The key result from NMA of the ribosome's structure is that the ribosome has unique architectural features that lend themselves to ratchet-like rotational motion that is crucial for performing translocation.

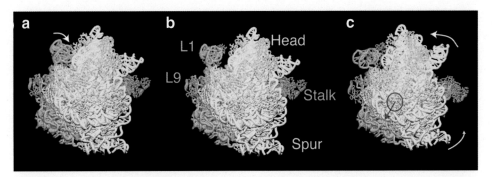

FIGURE 4.11: *Structural rearrangements of the 70S ribosome obtained from elastic network NMA (30S subunit in yellow, 50S in cyan). The ribosome is shown from the 30S solvent side. (a) The ribosome after a rearrangement along the first mode. The primary motion observed for this mode is displacement of the L1 protein to a position away from, or close to, the inter-subunit space. (b) The modified X-ray structure representing the equilibrium conformation of the ribosome (PDB ID's: 1GIX and 1GIY). (c) The ribosome after displacement along mode 3. The primary motion is a rotation of the 30S subunit relative to 50S around the axis indicated in red. The amplitude of the displacement along this mode was adjusted to match the change due to the ratchet-like motion as observed in the cryo-EM maps of the structure during translocation; however, the direction of this displacement arose as a natural consequence of the NMA. (Figure 11 of Tama et al., Proc. Natl. Acad. Sci. USA, 100(16): 9319–9323, 2003, copyright 2003 National Academy of Sciences U.S.A)*

III.4 Mechanical Properties of Icosahedral Viral Capsids

A virus is a small infectious agent that can replicate only inside a host cell it infects. A complete virus particle is made up of a protective coat of protein called a capsid that encloses nucleic acid, either DNA or RNA. Certain viruses include a tail structure that acts like a molecular syringe, attaching to the host and then injecting the viral genome into the cell. The capsid shape serves as the basis for the differentiation between virus structures, which display a wide diversity of shapes and sizes. Typically virus capsids display icosahedral symmetry (Crick and Waton 1956; Caspar and Klug 1962; Witz and Brown 2001; Tama and Brooks 2005). This requires 60 structural units (the icosahedral asymmetric unit) to complete a shell. However, very few viruses contain only 60 copies of the capsid protein. Most of the icosahedral viruses display quasi-symmetry, that is, they have 60T identical subunits in the shell, where T is the triangulation number (Caspar and Klug 1962). T reflects the selection rules for distributing identical protein subunits called capsomers (hexamers and pentamers) on a surface lattice.

An essential step in the life cycle of many viruses is the maturation of their capsid, the process by which the genome is packaged inside of the capsid, and by which the virus becomes pathogenic (Conway et al. 1995; Lata et al. 2000). This step involves a major conformational transformation of the shell of coat proteins from a non-spherical procapsid (or prohead) into a mature spherical icosahedral capsid (or head) form (Duda et al. 1995; Canady et al. 2000; Wikoff et al. 2000; Conway et al. 2001; Kuhn et al. 2002; Jiang et al. 2003). The exploration of putative pathways for the conformational changes that accompany these physical transitions can assist in understanding the mechanism associated with the maturation process by providing a description of the rearrangements of the subunits at an atomic level, and motivate hypotheses for the triggering event. Furthermore, identifying key steps in the virus maturation process can provide clues for the mechanism of disrupting this activity as well as suggesting potential targets for antiviral activity. This understanding could, of course, eventually result in the development of life-saving new drugs (Lewis et al. 1998; Zlotnick and Stray 2003). Insights into the detailed mechanism of capsid maturation could also hold promise for the development of nanotechnological devices for drug/gene therapy (Garcea and Gissmann 2004).

Primarily, X-ray crystallography and cryo-EM have provided glimpses into the structure of viral capsids and their shape-changing mechanisms (Conway et al. 2001; Helgstrand et al. 2003; Gan et al. 2004; Gertsman et al. 2009). Still, there is a lack of understanding of the atomic-level mechanisms by which viral capsids mature and change shape. To gain insights into the molecular mechanisms of capsid maturation, the fluctuation dynamics of the bacteriophage HK97 (HK97) in the procapsid, and mature capsid state was examined by performing normal-mode analysis employing the elastic-network model (Tama and Brooks 2005). HK97 was chosen as a model system to study virus capsid maturation because of the availability of a wealth of structural information for this system along the maturation pathway. Structurally, HK97 displays $T = 7$ quasi-symmetry and is made up of 420 copies of the gp5 protein which is the main building block of HK97 (Figure 4.12). The capsid structure is formed by 60 icosahedrally arranged

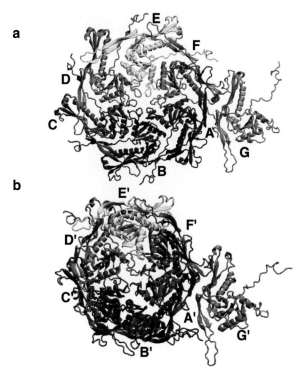

FIGURE 4.12: *Asymmetric units modeling the prohead II and head II HK97 capsid structures. (a) The prohead II asymmetric unit consists of seven identical chains (gp5 proteins), A–G, and was determined by cryo-EM (PDB ID: 1IFO) (Conway et al. 2001). The first six chains, A–F, form a pair of skewed trimers: AFE and BCD. (b) The head II asymmetric unit also contains seven gp5 proteins (A' – G') but was determined by X-ray crystallography (PDB ID: 1OHG). Here the first six chains form a fully symmetric hexamer.*

asymmetric units, each composed of seven gp5 proteins labeled A-G. The first six (A-F) form the hexamers, whereas the seventh (G) participates with five neighboring G chains in the adjacent pentamer. The structure of the HK97 capsid in both native and swollen forms is available. The 3.6 Å resolution structure of head II state of the HK97 capsid was solved using X-ray crystallography (Duda 1998; Wikoff et al. 2000), while a pseudoatomic model of its precursor, prohead II, was constructed from a cryo-EM map (Conway et al. 2001). In the head II structure, side chains from different gp5 proteins crosslink to form a highly catenated protein chain mail (Duda et al. 1995; Duda et al. 1995; Wikoff et al. 2000). Recent experimental studies suggest that the formation of these inter-subunit covalent crosslinks occurs concurrently with the conformational changes involved in the transition from prohead II to head I conformation (Gan et al. 2004).

Normal-mode analysis was performed on the prohead II and head II conformational states of HK97. For each state, an elastic network was constructed from the 107520 C_α atoms with a cutoff of 8 Å to define the C_α-C_α-based network connectivity. For the computation of the lowest-frequency normal modes using the RTB method, each

block was made of one of the proteins comprising the asymmetric subunit. Thus a total of 420 blocks was considered and the matrix to be diagonalized was 2520×2520, instead of $322{,}560 \times 322{,}560$ when diagonalization was done in the full Cartesian space. Normal-mode analysis performed on icosahedral viruses gives modes that are degenerate and non-degenerate. For the non-degenerate normal modes, the motion of each asymmetrical unit is exactly the same because this mode adheres to the icosahedral symmetry of the whole system. Experimental studies of viruses have suggested that the swollen capsids also adhere to icosahedral symmetry and to date there is no experimental evidence suggesting existence of symmetry-breaking pathways. Therefore, only the non-degenerate normal modes for which swelling pathways obey the implied symmetry were considered for further analysis.

The normal modes obtained for the two conformational states were compared with the experimentally observed conformational change. For this the cumulative sum of the squares of the overlap between the 500 modes for both states and the vector difference between the two virus states were computed. As shown in Figure 4.13, the first non-degenerate normal mode describes 65% of the conformational change, and the second non-degenerate normal mode has quite a high overlap ($\sim$0.55), which is sufficient in combination with the first to provide a nearly complete description of the conformational change (more than 90% of the overall expansion and shape transition of the capsid). This indicates that the set of low-frequency normal modes is sufficient to provide a good description of the overall conformational change between these two states of the viruses. Furthermore, each set of modes computed from the two-endpoint conformational states for the virus were also compared. The overlap between the first non-degenerate normal modes (one computed using the unswollen conformation and the other from the swollen

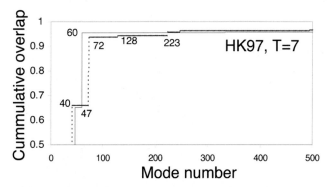

FIGURE 4.13: *The cumulative summation of the square of overlap between normal modes and the vector related to the conformational change for the two known states of the virus particle HK97. The continuous line corresponds to the cumulative summation calculated from the conformation with the smallest radius, the dotted line to the conformation with the highest radius. (Figure 13 of Tama et al., J. Mol. Biol., 345: 299–314, 2005, copyright 2005 Elsevier B.V.)*

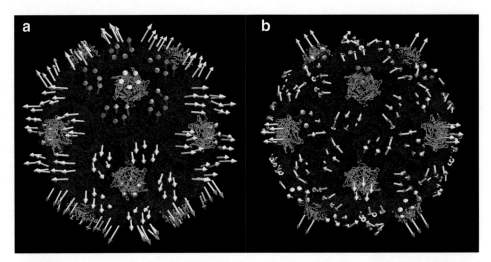

FIGURE 4.14: Amplitude and direction of motion, as indicated by the arrows, for the (a) first and (b) second non-degenerate normal modes of HK97. The G subunits forming the pentameric units are colored in green. (Figure 14 of Tama et al., J. Mol. Biol., 345: 299–314, 2005, copyright 2005 Elsevier B.V.)

conformation) is 0.96. Overall, these observations indicate that the mechanical properties of icosahedral viral capsids are well conserved for the different states.

The displacement of each asymmetric unit along the first and second non-degenerate normal modes indicates that subunit G moves very little along the first normal mode (Figure 4.14a). In addition, the neighboring subunits A and F are more constrained in the first mode compared to the other subunits. The motions of the subunits forming the hexamer consist of a radial expansion, whereas the motions for the pentameric units are limited (Figure 4.14b). On the other hand, along the second non-degenerate normal mode, the G subunit moves and the A and F subunits show larger displacements than do the other subunits. In Fig. 4.14b, we observe that along the second mode the G subunits show a motion that consists of a radial expansion, while more modest rearrangements are observed for the others subunits. Particularly, the pentamer has high flexibility to move against the other capsomers along the second normal mode. The shape transition from round to an apparent polyhedral shape seems to be achieved by moving pentamers formed by the G subunits further outside the shell as compared to the hexamers. This unique character of the dynamic properties of HK97 arises from the architecture of the capsid, and it indeed appears to be utilized in the maturation process of the virus. A substrate that could tighten the interaction between the G subunits and other units may hinder the maturation process of HK97.

Extension of this normal-mode study to a variety of viruses comprising different sizes and quasi-equivalence symmetries reveals that the pentameric units generally have higher flexibility and may move as independent units against the others capsomers (Tama and Brooks 2005). This general behavior indicates that viral capsids possess the capability to transition between conformations that might have different shape using the specific character of the pentameric unit's flexibility.

IV. CONCLUSIONS AND FUTURE OUTLOOK

Capturing large-scale, long-time conformational transitions in macromolecular assemblies is a difficult undertaking that requires the use of a number of different multi-scale methods. As demonstrated, the combination of the elastic network NMA, structure-based coarse-grained models, and advanced sampling methods provide functionally important insights into large-amplitude motions in diverse macromolecular machines, such as the ribosome, virus capsids, the DNA helicase motor, and a catalytic enzyme. These studies allow us to enter an area not currently accessible to experiment and hypothesize a complex series of structural rearrangements between different conformational states of these macromolecular machines that can help us understand their mechanistic properties.

Among the different methods that we apply to the problem, the elastic-network NMA based on a coarse-grained representation of an all-atom X-ray crystallographic image predicts structural rearrangements associated with biological function that are in close agreement with experimental observations. To obtain low-frequency vibrational modes that describe these functional rearrangements, the three-dimensional structure of a biomolecule is represented by an elastic network of pseudoatoms in which residues within a cutoff distance experience harmonic interactions. This elastic network essentially captures the shape of the molecule and the connectivity of its underlying structural framework. As we have demonstrated, these studies show that the key to understanding the function of biological systems appears to lie in the shape-dependent dynamical properties of their complex architecture. The successful application of elastic network models to predict functional motions may be another example of Nature's tendency to exploit simple structural features to achieve evolutionary important biological functions (Ma 2005; Tama and Brooks 2006).

Upon binding or releasing ligands, biomolecules often undergo large-amplitude structural transitions. Structural information on the initial and final conformation of the molecules is usually available from X-ray or NMR experiments. However, the structural transition pathway between two end structures, as well as pertinent energetic barriers, remains unresolved. Elastic-network NMA can be combined with advanced sampling methods and structure-based models to obtain this needed information.

Understanding structural transition pathways in detail can help identify transition states, design methods for controlling such transitions, and thereby modulate protein function. However, obtaining such pathways for biomolecules is not straightforward (Khavrutskii et al. 2006). The free-energy landscape of biomolecules is rugged, and during transition from one stable endpoint conformation to another, the system has to pass through multiple minima and high-energy barriers (Bolhuis et al. 2002). Moreover, multiple transition pathways may exist, which makes it difficult to find the most probable transition pathway that the system can follow. Several computational methods have been developed over a number of years to capture the most probable transition pathways between two known conformational states of molecules (see (Khavrutskii et al. 2006) and references therein). Here we used the nudged elastic band method combined with umbrella sampling to obtain conformational transition pathways as well as energetics associated with transitions in the Adenylate kinase enzyme (see Section III.1). However, the application of these methods to large macromolecular assemblies, such as the ribosome, remains challenging. Elastic-network NMA of Adenylate kinase and several other enzymes shows that biomolecular structural transitions between functional sub-states are largely captured by one or a few large-amplitude, slow modes. Given that elastic network NMA provides information on functional and robust modes and is not limited by the size of the system that can be investigated, we decided it would be reasonable to follow low-energy normal mode directions of the system between the two endpoint conformations. However, by definition, NMA is valid only in the local region surrounding a potential energy minimum; its application to describe anharmonic events such as large-scale conformational transitions between two distinct states requires a non-linear description (Ma 2005). It has been suggested that the problems arising from the harmonic approximation employed in the NMA can be ameliorated by performing the NMAs and conformational deformations in an iterative manner (Miyashita et al. 2003;

Tama et al. 2004). Instead of moving the structure from the initial to the final form directly, a sequence of small incremental deformations is used and a normal mode calculation is performed for each intermediate deformed structure. We believe that pathways obtained in this manner, by deforming structures iteratively along low-frequency modes, when used in conjunction with advanced sampling methodologies, have the potential of providing a useful energy landscape description for large macromolecular machines.

Alternative techniques for studying structural transition pathways using the elastic network model involve constructing an elastic network for each endpoint conformation. Following, the conformational transition pathway is generated by linear interpolation between the two reference elastic networks for each state (Maragakis and Karplus 2005). The transition states are defined as conformations on the cusp between the two reference states. Slight modification of this so-called double-well approach has recently been used successfully with structure-based Go models for exploring the conformational transitions of G-actin protein and Adenylate kinase enzyme (Chu and Voth 2007). These examples illustrate that there is an enormous interest in employing elastic-network models for investigating large-scale structural transition pathways, and we hope applications of these methods altogether will lead to novel insights into mechanisms of even larger macromolecular assemblies in the future.

ACKNOWLEDGMENTS

We would like to acknowledge Klaus Schulten for critically reading this manuscript. This work was supported by the National Institutes of Health through the Center for Multi-Scale Modeling Tools for Structural Biology (grant RR012255) and the National Science Foundation through the Center for Theoretical Biological Physics (PHY0216576).

REFERENCES

Agirrezabala, X., J. Lei, J. L. Brunelle, R. F. Ortiz-Meoz, R. Green and J. Frank (2008). Visualization of the hybrid state of tRNA binding promoted by spontaneous ratcheting of the ribosome. *Mol. Cell* 32: 190–197.

Agrawal, R. K., A. B. Heagle, P. Penczek, R. A. Grassucci and J. Frank (1999). EF-G-dependent GTP hydrolysis induces translocation accompanied by large conformational changes in the 70S ribosome. *Nat. Struct. Biol.* 6: 643–647.

Agrawal, R. K., P. Penczek, R. A. Grassucci, N. Burkhardt, K. H. Nierhaus and J. Frank (1999). Effect of buffer conditions on the position of tRNA on the 70 S ribosome as visualized by cryoelectron microscopy. *J. Biol. Chem.* 274: 8723–8729.

Arora, K. and C. L. Brooks (2007). Large-scale allosteric conformational transitions of adenylate kinase appear to involve a population-shift mechanism. *Proc. Natl. Acad. Sci. USA* 104: 18496–18501.

Arora, K. and C. L. Brooks (2009). Functionally Important Conformations of the Met20 Loop in Dihydrofolate Reductase are Populated by Rapid Thermal Fluctuations. *J. Am. Chem. Soc.* 131: 5642–5647.

Atilgan, A., S. Durell, R. Jernigan, M. Demirel, O. Keskin and I. Bahar (2001). Anisotropy of fluctuation dynamics of proteins with an elastic network model. *Biophys. J.* 80: 505–515.

Ban, N., P. Nissen, J. Hansen, P. B. Moore and T. A. Steitz (2000). The complete atomic structure of the large ribosomal subunit at 2.4 Å resolution. *Science* 289: 905–920.

Beckstein, O., E. J. Denning, J. R. Perilla and T. B. Woolf (2009). Zipping and unzipping of adenylate kinase: atomistic insights into the ensemble of open-closed transitions. *J. Mol. Biol.* 394: 160–176.

Berg, J. M., J. L. Tymoczko and L. Stryer (2006). *Biochemistry*. New York, W. H. Freeman and Company.

Bolhuis, P. G., D. Chandler, C. Dellago and P. L. Geissler (2002). Transition path sampling: throwing ropes over rough mountain passes, in the dark. *Ann. Rev. Phys. Chem.* 53: 291–318.

Brooks, B., D. Janezic and M. Karplus (1995). Harmonic-analysis of large systems. 1. Methodology. *J. Comput. Chem.* 16: 1522–1542.

Brooks, B. and M. Karplus (1985). Normal modes for specific motions of macromolecules: application to the hinge-bending mode of lysozyme. *Proc. Natl. Acad. Sci. USA* 82: 4995–4999.

Brooks, C. L., M. Gruebele, J. N. Onuchic and P. G. Wolynes (1998). Chemical physics of protein folding. *Proc. Natl. Acad. Sci. USA* 95: 11037–11038.

Canady, M. A., M. Tihova, T. N. Hanzlik, J. E. Johnson and M. Yeager (2000). Large conformational changes in the maturation of a simple RNA virus, nudaurelia capensis omega virus (NomegaV). *J. Mol. Biol.* 299: 573–584.

Caspar, D. and A. Klug (1962). Physical principles in the construction of regular viruses. *Cold Spring Harb Symp Quant Biol* 27: 1–24.

Cavasotto, C. N., J. A. Kovacs and R. A. Abagyan (2005). Representing Receptor Flexibility in Ligand Docking through Relevant Normal Modes. *J. Am. Chem. Soc.* 127: 9632–9640.

Chacon, P., F. Tama and W. Wriggers (2003). Mega-Dalton biomolecular motion captured from electron microscopy reconstructions. *J. Mol. Biol.* 326: 485–492.

Chu, J.-W., B. L. Trout and B. R. Brooks (2003). A super-linear minimzation scheme for the nudged elastic band method. *J. Chem. Phys.* 119: 12708–12717.

Chu, J.-W. and G. A. Voth (2007). Coarse-grained free energy functions for studying protein conformational changes: a double-well network model. *Biophys. J.* 93: 3860–3871.

Conway, J. F., R. L. Duda, N. Cheng, R. W. Hendrix and A. C. Steven (1995). Proteolytic and conformational control of virus capsid maturation: the bacteriophage HK97 system. *J. Mol. Biol.* 253: 86–99.

Conway, J. F., W. R. Wikoff, N. Cheng, R. L. Duda, R. W. Hendrix, J. E. Johnson and A. C. Steven (2001). Virus maturation involving large subunit rotations and local refolding. *Science* 292: 744–748.

Cornish, P. V., D. N. Ermolenko, H. F. Noller and T. Ha (2008). Spontaneous intersubunit rotation in single ribosomes. *Mol. Cell* 30: 578–588.

Crick, F. and J. Waton (1956). Structure of small viruses. *Nature* 177: 473–475.

Cui, Q., G. Li, J. Ma and M. Karplus (2004). A normal mode analysis of structural plasticity in the biomolecular motor F-1-ATPase. *J. Mol. Biol.* 340: 345–372.

Dahnke, T., Z. Shi, H. Yan, R. T. Jiang and M. D. Tsai (1992). Mechanism of adenylate kinase. Structural and functional roles of the conserved arginine-97 and arginine-132. *Biochemistry* 31: 6318–6328.

Delarue, M. and Y.-H. Sanejouand (2002). Simplified normal mode analysis of conformational transitions in DNA-dependent polymerases: the elastic network model. *J. Mol. Biol.* 320: 1011–1024.

Doruker, P., R. L. Jernigan and I. Bahar (2002). Dynamics of large proteins through hierarchial levels of coarse-grained structures. *J. Comput. Chem.* 23: 119–127.

Duda, R. L. (1998). Protein chainmail: catenated protein in viral capsids. *Cell* 94: 55–60.

Duda, R. L., J. Hempel, H. Michel, J. Shabanowitz, D. Hunt and R. W. Hendrix (1995). Structural transitions during bacteriophage HK97 head assembly. *J. Mol. Biol.* 247: 618–635.

Duda, R. L., K. Martincic and R. W. Hendrix (1995). Genetic basis of bacteriophage HK97 prohead assembly. *J. Mol. Biol.* 247: 636–647.

Duong, T. and K. Zakrzewska (1997). Calculation and analysis of low frequency normal modes for DNA. *J. Comput. Chem.* 18: 796–811.

Duong, T. and K. Zakrzewska (1998). Sequence specificity of bacteriophage 434 repressor-operator complexation. *J. Mol. Biol.* 280: 31–39.

Enemark, E. J. and L. Joshua-Tor (2006). Mechanism of DNA translocation in a replicative hexameric helicase. *Nature* 442: 270–275.

Ermolenko, D. N., Z. K. Majumdar, R. P. Hickerson, P. C. Spiegel, R. M. Clegg and H. F. Noller (2007). Observation of intersubunit movement of the ribosome in solution using FRET. *J. Mol. Biol.* 370: 530–540.

Erzberger, J. P. and J. M. Berger (2006). Evolutionary relationships and structural mechanisms of AAA$^+$ proteins. *Ann. Rev. Biophys. Biomol. Struct.* 35: 93–114.

Fei, J., J. E. Bronson, J. M. Hofman, R. L. Srinivas, C. H. Wiggins and R. L. Gonzalez (2009). Allosteric collaboration between elongation factor G and the ribosomal L1 stalk directs tRNA movements during translation. *Proc. Natl. Acad. Sci. USA* 106: 15702–15707.

Fei, J., P. Kosuri, D. D. MacDougall and R. L. Gonzalez (2008). Coupling of ribosomal L1 stalk and tRNA dynamics during translation elongation. *Mol. Cell* 30: 348–359.

Feng, Y., L. Yang, A. Kloczkowski and R. L. Jernigan (2009). The energy profiles of atomic conformational transition intermediates of adenylate kinase. *Proteins* 77: 551–558.

Fouts, E. T., X. Yu, E. H. Egelman and M. R. Botchan (1999). Biochemical and electron microscopic image analysis of the hexameric E1 helicase. *J. Biol. Chem.* 274: 4447–4458.

Frank, J. and R. K. Agrawal (2000). A ratchet-like inter-subunit reorganization of the ribosome during translocation. *Nature* 406: 318–322.

Frank, J. and R. L. Gonzalez Jr. (2010). Structure and Dynamics of a Processive Brownian Motor: The Translating Ribosome. *Annu. Rev. Biochem.* 79: 381–412.

Frauenfelder, H., F. Parak and R. D. Young (1988). Conformational substates in proteins. *Annu. Rev. Biophys. Biophys. Chem.* 17: 451–479.

Freddolino, P. L., A. S. Arkhipov, S. B. Larson, A. McPherson and K. Schulten (2006). Molecular dynamics simulations of the complete satellite tobacco mosaic virus. *Structure* 14: 437–449.

Freddolino, P. L., F. Liu, M. Gruebele and K. Schulten (2008). Ten-microsecond molecular dynamics simulation of a fast-folding WW domain. *Biophys. J.* 94: L75–L77.

Gai, D., R. Zhao, D. Li, C. V. Finkielstein and X. S. Chen (2004). Mechanisms of conformational change for a replicative hexameric helicase of SV40 large tumor antigen. *Cell* 119: 47–60.

Gan, L., J. F. Conway, B. A. Firek, N. Cheng, R. W. Hendrix, A. C. Steven, J. E. Johnson and R. L. Duda (2004). Control of crosslinking by quaternary structure changes during bacteriophage HK97 maturation. *Mol. Cell* 14: 559–569.

Garcca, R. L. and L. Gissmann (2004). Virus like particles as vaccines and vessels for the delivery of small molecules. *Curr. Opin. Biotech.* 15: 513–517.

Gertsman, I., L. Gan, M. Guttman, K. Lee, J. A. Speir, R. L. Duda, R. W. Hendrix, E. A. Komives and J. E. Johnson (2009). An unexpected twist in viral capsid maturation. *Nature* 458: 646–650.

Go, N., T. Noguti and T. Nishikawa (1983). Dynamics of a small globular proteins in terms of low frequency vibrational modes. *Proc. Natl. Acad. Sci. USA* 80: 3696–3700.

Goldstein, H. (1950). *Classical Mechanics.* Reading, MA, Addison-Wesley.

Gomez-Lorenzo, M. G., C. M. Spahn, R. K. Agrawal, R. A. Grassucci, P. Penczek, K. Chakraburtty, J. P. Ballesta, J. L. Lavandera, J. F. Garcia-Bustos and J. Frank (2000). Three-dimensional cryo-electron microscopy localization of EF2 in the Saccharomyces cerevisiae 80S ribosome at 17.5 Å resolution. *EMBO J.* 19: 2710–2718.

Hanson, J., K. Duderstadt, L. Watkins, A. Bhattacharyya, J. Brokaw, J. Chu and H. Yang (2007). Illuminating the mechanistic roles of enzyme conformational dynamics. *Proc. Natl. Acad. Sci. U S A* 104: 18055–18060.

Harms, J., F. Schluenzen, R. Zarivach, A. Bashan, S. Gat, I. Agmon, H. Bartels, F. Franceschi and A. Yonath (2001). High resolution structure of the large ribosomal subunit from a mesophilic eubacterium. *Cell* 107: 679–688.

Helgstrand, C., W. R. Wikoff, R. L. Duda, R. W. Hendrix, J. E. Johnson and L. Liljas (2003). The refined structure of a protein catenane: the HK97 bacteriophage capsid at 3.44 Å resolution. *J. Mol. Biol.* 334: 885–899.

Henzler-Wildman, K. and D. Kern (2007). Dynamic personalities of proteins. *Nature* 450: 964–972.

Henzler-Wildman, K., V. Thai, M. Lei, M. Ott, P. G. Vendruscolo, T. Fenn, E. Pozharski, R. Venkatramani, G. A. Pedersen, M. Karplus, C. Hübner and D. Kern (2007). Intrinsic motions along an enzymatic reaction trajectory. *Nature* 450: 838–844.

Henzler-Wildman, K. A., M. Lei, V. Thai, S. J. Kerns, M. Karplus and D. Kern (2007). A hierarchy of timescales in protein dynamics is linked to enzyme catalysis. *Nature* 450: 913–916.

Hinsen, K. (1998). Analysis of domain motions by approximate normal mode calculations. *Proteins* 33: 417–429.

Jensen, M. O., D. W. Borhani, K. Lindorff-Larsen, P. Maragakis, V. Jogini, M. P. Eastwood, R. O. Dror and D. E. Shaw (2010). Principles of conduction and hydrophobic gating in K$^+$ channels. *Proc Natl Acad Sci U S A* 107: 5833–5838.

Jiang, W., Z. Li, Z. Zhang, M. L. Baker, P. E. Prevelige and W. Chiu (2003). Coat protein fold and maturation transition of bacteriophage P22 seen at subnanometer resolutions. *Nat. Struc. Biol.* 10: 131–135.

Karplus, M. and J. A. McCammon (2002). Molecular dynamics simulations of biomolecules. *Nat. Struct. Biol.* 9: 646–652.

Khavrutskii, I. V., K. Arora and C. L. Brooks (2006). Harmonic Fourier beads method for studying rare events on rugged energy surfaces. *J. Chem. Phys.* 125: 174108–174115.

Kidera, A. and N. Go (1990). Refinement of protein dynamic structure: normal mode refinement. *Proc. Natl. Acad. Sci. USA* 87: 3718–3722.

Kidera, A. and N. Go (1992). Normal mode refinement: crystallographic refinement of protein dynamic structure. 1. Theory and test by simulated diffraction data. *J. Mol. Biol.* 225: 457–475.

Korkut, A. and W. A. Hendrickson (2009). Computation of conformational transitions in proteins by virtual atom molecular mechanics as validated in application to adenylate kinase. *Proc. Natl. Acad. Sci. USA* 106: 15673–15678.

Kuhn, R. J., W. Zhang, M. G. Rossmann, S. V. Pletnev, J. Corver, E. Lenches, C. T. Jones, S. Mukhopadhyay, P. R. Chipman, E. G. Strauss, T. S. Baker and J. H. Strauss (2002). Structure of dengue virus: implications for flavivirus organization, maturation, and fusion. *Cell* 108: 717–725.

Lata, R., J. Conway, N. Cheng, R. Duda and R. Hendrix (2000). Maturation dynamics of a viral capsid: visualization of transitional intermediate states. *Cell* 100: 253–263.

Levitt, M., C. Sander and P. S. Stern (1985). Protein normal-mode dynamics: trypsin inhibitor, crambin, ribonuclease and lysozyme. *J. Mol. Biol.* 181: 423–447.

Lewis, J. K., B. Bothner, T. J. Smith and G. Siuzdak (1998). Antiviral agent blocks breathing of the common cold virus. *Proc. Natl. Acad. Sci. U S A* 95: 6774–6778.

Li, D., R. Zhao, W. Lilyestrom, D. Gai, R. Zhang, J. A. DeCaprio, E. Fanning, A. Jochimiak, G. Szakonyi and X. S. Chen (2003). Structure of the replicative helicase of the oncoprotein SV40 large tumour antigen. *Nature* 423: 512–518.

Li, G. and Q. Cui (2002). A coarse-grained normal mode approach for macromolecules: an efficient implementation and application to Ca(2 +)-ATPase. *Biophys. J.* 83: 2457–2474.

Liu, H., Y. Shi, X. S. Chen and A. Warshel (2009). Simulating the electrostatic guidance of the vectorial translocations in hexameric helicases and translocases. *Proc. Natl. Acad. Sci. USA* 106: 7449–7454.

Lou, H. and R. I. Cukier (2006). Molecular dynamics of apo-adenylate kinase: a distance replica exchange method for the free energy of conformational fluctuations. *J. Phys. Chem. B* 110: 24121–24137.

Lou, H. and R. I. Cukier (2006). Molecular dynamics of apo-adenylate kinase: a principal component analysis. *J. Phys. Chem. B* 110: 12796–12808.

Ma, J. (2005). Usefulness and limitations of normal mode analysis in modeling dynamics of biomolecular complexes. *Structure* 13: 373–380.

Ma, J. and M. Karplus (1997). Ligand-induced conformational changes in ras p21: a normal mode and energy minimization analysis. *J. Mol. Biol.* 274: 114–131.

Maragakis, P. and M. Karplus (2005). Large amplitude conformational change in proteins explored with a plastic network model: adenylate kinase. *J. Mol. Biol.* 352: 807–822.

Maragakis, P., K. Lindorff-Larsen, M. P. Eastwood, R. O. Dror, J. L. Klepeis, I. T. Arkin, M. Jensen, H. Xu, N. Trbovic, R. A. Friesner, A. G. P. III and D. E. Shaw (2008). Microsecond molecular dynamics simulation shows effect of slow loop dynamics on backbone amide order parameters of proteins. *J. Phys. Chem. B* 112: 6155–6158.

Marques, O. and Y. Sanejouand (1995). Hinge-bending motion in citrate synthase arising from normal mode calculations. *Proteins* 23: 557–560.

Miyashita, O., J. N. Onuchic and P. G. Wolynes (2003). Nonlinear elasticity, proteinquakes, and the energy landscapes of functional transitions in proteins. *Proc. Natl. Acad. Sci. USA* 100: 12570–12575.

Miyashita, O., P. G. Wolynes and J. N. Onuchic (2005). Simple energy landscape model for the kinetics of functional transitions in proteins. *J. Phys. Chem. B* 109: 1959–1969.

Mohaghegh, P. and I. Hickson (2001). DNA helicase deficiencies associated with cancer predisposition and premature ageing disorders. *Hum. Mol. Genet.* 10: 741–746.

Mouawad, L. and D. Perahia (1993). Diagonalization in a mixed basis: a method to compute low-frequency normal-modes for large macromolecules. *Biopolymers* 33: 569–611.

Mouawad, L. and D. Perahia (1996). Motions in hemoglobin studied by normal mode analysis and energy minimization: evidence for the existence of tertiary T-like, quaternary R-like intermediate structures. *J. Mol. Biol.* 258: 393–410.

Muller, C. W. and G. E. Schulz (1992). Structure of the complex between adenylate kinase from Escherichia coli and the inhibitor Ap5A refined at 1.9 Å resolution. A model for a catalytic transition state. *J. Mol. Biol.* 224: 159–177.

Muller, C. W. and G. E. Schulz (1993). Crystal structures of two mutants of adenylate kinase from Escherichia coli that modify the Gly-loop. *Proteins* 15: 42–49.

Okazaki, K., N. Koga, S. Takada, J. N. Onuchic and P. G. Wolynes (2006). Multiple-basin energy landscapes for large-amplitude conformational motions of proteins: Structure-based molecular dynamics simulations. *Proc. Natl. Acad. Sci. USA* 103: 11844–11849.

Pérez, A., F. J. Luque and M. Orozco (2007). Dynamics of B-DNA on the microsecond time scale. *J. Amer. Chem. Soc.* 129: 14739–14745.

Ramakrishnan, V. (2002). Ribosome structure and the mechanism of translation. *Cell* 108: 557–572.

Schlick, T. (2002). *Molecular Modeling and Simulation: An Interdisciplinary Guide.* New York, Springer-Verlag.

Schrank, T. P., D. W. Bolen and V. J. Hilser (2009). Rational modulation of conformational fluctuations in adenylate kinase reveals a local unfolding mechanism for allostery and functional adaptation in proteins. *Proc. Natl. Acad. Sci. USA* 106: 16984–16989.

Shapiro, Y. E. and E. Meirovitch (2006). Activation energy of catalysis-related domain motion in E. coli adenylate kinase. *J. Phys. Chem. B* 110: 11519–11524.

Shapiro, Y. E., M. A. Sinev, E. V. Sineva, V. Tugarinov and E. Meirovitch (2000). Backbone dynamics of escherichia coli adenylate kinase at the extreme stages of the catalytic cycle studied by (15)N NMR relaxation. *Biochemistry* 39: 6634–6644.

Shaw, D. E., P. Maragakis, K. Lindorff-Larsen, S. Piana, R. O. Dror, M. P. Eastwood, J. A. Bank, J. M. Jumper, J. K. Salmon, Y. Shan and W. Wriggers (2010). Atomic-level characterization of the structural dynamics of proteins. *Science* 330: 341–346.

Shi, Y., H. Liu, D. Gai, J. Ma and X. S. Chen (2009). A computational analysis of ATP binding of SV40 large tumor antigen helicase motor. *PLoS Comput Biol* 5: e1000514.

Sinev, M. A., E. V. Sineva, V. Ittah and E. Haas (1996). Domain closure in adenylate kinase. *Biochemistry* 35: 6425–6437.

Stark, H., M. V. Rodnina, H. J. Wieden, M. van Heel and W. Wintermeyer (2000). Large-scale movement of elongation factor G and extensive conformational change of the ribosome during translocation. *Cell* 100: 301–309.

Tama, F. and C. L. Brooks (2005). Diversity and identity of mechanical properties of icosahedral viral capsids studied with elastic network normal mode analysis. *J. Mol. Biol.* 345: 299–314.

Tama, F. and C. L. Brooks (2006). Symmetry, Form, and Shape: Guiding Principles for Robustness in Macromolecular Machines. *Ann. Rev. Biophys. Biomol. Struct.* 35: 115–133.

Tama, F., M. Feig, J. Liu, C. L. Brooks and K. A. Taylor (2005). The requirement for mechanical coupling between head and S2 domains in smooth muscle myosin ATPase regulation and its implications for dimeric motor function. *J. Mol. Biol.* 345: 837–854.

Tama, F., F. X. Gadea, O. Marques and Y. H. Sanejouand (2000). Building-block approach for determining low-frequency normal modes of macromolecules. *Proteins* 41: 1–7.

Tama, F., O. Miyashita and C. L. Brooks (2004). Normal mode based flexible fitting of high-resolution structure into low-resolution experimental data from cryo-EM. *J. Struct. Biol.* 147: 315–326.

Tama, F. and Y. Sanejouand (2001). Conformational change of proteins arising from normal mode calculations. *Protein. Eng.* 14: 1–6.

Tama, F., M. Valle, J. Frank and C. L. Brooks (2003). Dynamic reorganization of the functionally active ribosome explored by normal mode analysis and cryo-electron microscopy. *Proc. Natl. Acad. Sci. USA* 100: 9319–9323.

Temiz, N. A., E. Meirovitch and I. Bahar (2004). Escherichia coli adenylate kinase dynamics: comparison of elastic network model modes with mode-coupling (15)N-NMR relaxation data. *Proteins* 57: 468–480.

Tirion, M. (1996). Large amplitude elastic motions in proteins from a single-parameter atomic analysis. *Phys. Rev. Lett.* 77: 1905–1908.

Torrie, G. M. and J. P. Valleau (1977). Nonphysical sampling distribution in Monte Carlo free-energy estimation: Umbrella sampling. *J. Comp. Phys.* 23: 187–199.

Tsai, C. J., S. Kumar, B. Ma and R. Nussinov (1999). Folding funnels, binding funnels, and protein function. *Protein Sci* 8: 1181–1190.

Valle, M., A. Zavialov, J. Sengupta, U. Rawat, M. Ehrenberg and J. Frank (2003). Locking and unlocking of ribosomal motions. *Cell* 114: 123–134.

van Brabant, A. J., R. Stan and N. A. Ellis (2000). DNA helicases, genomic instability, and human genetic disease. *Annu. Rev. Genomics Hum. Genet.* 1: 409–459.

Volkman, B. F., D. Lipson, D. E. Wemmer and D. Kern (2001). Two-state allosteric behavior in a single-domain signaling protein. *Science* 291: 2429–2433.

Vonrhein, C., G. J. Schlauderer and G. E. Schulz (1995). Movie of the structural changes during a catalytic cycle of nucleoside monophosphate kinases. *Structure* 3: 483–490.

Wessel, R., J. Schweizer and H. Stahl (1992). Simian virus 40 T-antigen DNA helicase is a hexamer which forms a binary complex during bidirectional unwinding from the viral origin of DNA replication. *J. Virol.* 66: 804–815.

Whitford, P. C., O. Miyashita, Y. Levy and J. N. Onuchic (2007). Conformational transitions of adenylate kinase: switching by cracking. *J. Mol. Biol.* 366: 1661–1671.

Wikoff, W. R., L. Liljas, R. L. Duda, H. Tsuruta, R. W. Hendrix and J. E. Johnson (2000). Topologically linked protein rings in the bacteriophage HK97 capsid. *Science* 289: 2129–2133.

Witz, J. and F. Brown (2001). Structural dynamics, an intrinsic property of viral capsids. *Arch. Virol.* 146: 2263–2274.

Wolf-Watz, M., V. Thai, K. Henzler-Wildman, G. Hadjipavlou, E. Z. Eisenmesser and D. Kern (2004). Linkage between dynamics and catalysis in a thermophilic-mesophilic enzyme pair. *Nat. Struc. Mol. Biol.* 11: 945–949.

Wynsberghe, A. V., G. Li and Q. Cui (2004). Normal-mode analysis suggests protein flexibility modulation throughout RNA polymerase's functional cycle. *Biochemistry* 43: 13083–13096.

Yoshimoto, K., K. Arora and C. L. Brooks (2010). Hexameric helicase deconstructed: interplay of conformational change and substrate coupling. *Biophys. J.* 98: 1449–1457.

Yu, J., T. Ha and K. Schulten (2007). How directional translocation is regulated in a DNA helicase motor. *Biophys J* 93: 3783–3797.

Yusupov, M., G. Yusupova, A. Baucom, K. Lieberman, T. Earnest, J. Cate and H. Noller (2001). Crystal structure of the ribosome at 5.5 Å resolution. *Science* 292: 883–896.

Zlotnick, A. and S. J. Stray (2003). How does your virus grow? Understanding and interfering with virus assembly. *Trends Biotechnol.* 21: 536–542.

CHAPTER 5

Structure, Function, and Evolution of Archaeo-Eukaryotic RNA Polymerases – Gatekeepers of the Genome

Finn Werner

Dina Grohmann

I. PREFACE

RNA polymerases (RNAPs) are essential to all life forms and responsible for the regulated and template DNA-dependent transcription of all genetic information. A plethora of basal and gene-specific transcription factors interact physically and functionally with RNAP, which results in the execution of a highly fine-tuned genetic program that is at the very heart of biology. RNAPs come in a range of flavors, but notably all RNAPs responsible for the transcription of cellular genomes are evolutionary related and are thus derived from one common ancestor. Recent technological advances have given us unprecedented insights into the function and mechanisms of RNAPs. This book chapter serves to describe our modern understanding of the structure, function, and evolution of RNAPs in the three principal domains of life: the Bacteria, Archaea, and Eukarya.

II. TRANSCRIPTION IN THE INFORMATION-PROCESSING CIRCUITRY OF LIFE

Since Francis Crick phrased the "central dogma" of molecular biology in the mid-1950s – according to which *DNA-makes-RNA-makes-protein* – scientists from a broad range of backgrounds have investigated the flow of genetic information in biological systems (Watson and Crick, 1953). According to this traditional view, the DNA template-dependent synthesis of DNA is referred to as *replication*, the DNA template-dependent synthesis of RNA is *transcription*, and RNA in turn is *translated* into proteins (Figure 5.1). Soon after the discovery that nucleic acids not only encode the genetic information, but are also instrumental in translating it into proteins in the form of ribosomes (e.g., rRNA) and their ligands (e.g., tRNA), it became apparent that this assumed unidirectional flow of information is anything but simple, nor is it unidirectional (Figure 5.1).

For example, a plethora of viral genomes are made of RNA, which is *reverse transcribed* into DNA. The perpetuation of RNA genomes is facilitated by RNA-template-dependent RNA synthesis – transcription – implying that replication and transcription can be the same process. What initially was perceived as a quaint catalytic property of self-splicing introns (Cech and Bass, 1986) and processes involved in tRNA maturation (Altman and Robertson, 1973) rapidly led to the discovery of enzymes that were entirely made of RNA with a whole range of activities, *ribozymes*. In the last decade, a large number of small noncoding RNA molecules have emerged as potent regulators of replication, transcription, translation, mRNA folding, and stability. Thus, RNA molecules are involved in all stages of the information processing in extant biology; they encode information and provide structural, regulatory and catalytic properties. What is the reason for this pervasive omnipresence of RNA in life – is it due to the versatile properties of RNA alone, or is it a relic of the ancient past of our biosphere? Currently no theoretical models provide satisfying and unequivocal answers to the origins of life (Schuster, 2010). However, both the perpetuation of genetic information and the ability to alter the information by mutation or recombination was necessarily required for a hypothetical ancestor to be subject to Darwinian evolution. The "RNA world" hypothesis provides such a model, and RNA polymerases play a fundamental role in this scenario (Joyce, 1999).

III. RIBOZYME RNA POLYMERASES AND THE RNA WORLD

The RNA world hypothesis proposes that prior to the emergence of DNA and protein synthesis, RNA served both as genome and catalyst, in the form of ribozymes (Joyce, 1999). DNA and proteins emerged later in evolution and, due to their competitive properties, "displaced" RNA from many of its biological roles. Thus, DNA could have been a better storage medium for genetic information due to its higher stability, whereas proteins were better

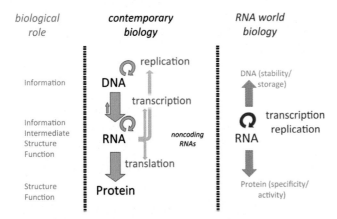

FIGURE 5.1: *Information-processing circuitry in biology. The vast majority of organisms on earth store their genetic information in the format of DNA, which is transcribed into RNA, and the RNA is translated into protein (green arrows). In the hypothetical RNA-world era more than 4 billion years ago, RNA molecules both encoded information and carried out catalysis (highlighted in red). DNA as storage medium and protein as catalyst emerged much later in evolution (gray arrows) and adapted to resemble the flow of information we know from extant life. Note that small non-coding RNA species regulate all three fundamental processes (thin gray arrows).*

catalysts due to the chemical resourcefulness of their side chains.

The "RNA world" hypothesis depends on a hypothetical ribozyme replicase that was responsible for the perpetuation of the genetic information and itself able to evolve – in effect, a ribozyme RNAP capable of synthesizing copies of itself and other RNA molecules. Even though no additional cofactors were required, the binding of small amino-acid-like molecules or even short proto-peptides to the ribozyme could have increased its thermal and chemical stability, specificity, and interaction properties that, for example, favored binding of substrates or release of products. Notably no naturally occurring ribozyme RNA polymerases have been discovered yet. However, in vitro evolution approaches starting from a pool of random sequences and driven by reiterated cycles of sequence alteration (mutation) and selection have generated both ribozyme RNA ligases and (relatively non-processive) RNAPs (Ekland et al., 1995; Johnston et al., 2001; Lincoln and Joyce, 2009). This demonstrates that it is possible for ribozymes to catalyze phosphodiester bond formation in a template-dependent fashion, and at least in theory supports the possibility of a natural ribozyme RNAP. One of the main objections raised against the putative ribozyme RNAPs is the apparent difficulty of synthesizing their nucleotide triphosphate substrates in the absence of proteinaceous enzymes. In an alternative scenario, the protein synthesis machinery (proto-ribosomes) evolved prior to nucleic acid polymerization; proto-enzymes were catalyzing the formation of nucleotides and their phosphates, and nucleic acid polymerases evolved without any ribozyme ancestors. This hypothesis, however, fails to address the

coding of the proto-enzymes, which would have been required for the necessary inheritability and principles of Darwinian evolution. Recently the X-ray structure of the catalytic core of a synthetic, in vitro-evolved, RNA ligase has been solved at 3.0 Å (Shechner et al., 2009). The structure suggests that the RNA catalytic mechanism closely resembles that of proteinaceous RNAP, in which two magnesium ions facilitate an SN_2 nucleophilic substitution reaction. In the ribozyme the magnesium ions are chelated by phosphate moieties of the RNA chain. In proteinaceous RNAPs, the two magnesium ions are coordinated by side chain carboxylates and the phosphate moieties of the NTP substrate (Figure 5.2).

The two-metal mechanism is conserved not only in all RNAPs, but also in most, if not all, RNA-mediated phosphoryl transfer reactions including self-splicing introns and RNase P (Steitz, 1998; Steitz and Steitz, 1993). According to this model, an extremely simplified role of both RNA and amino acid side chains of the RNAP active site is to position two magnesium ions 3.9 Å apart in the accurate spatial orientation to the RNA 3′ OH-terminus and a NTP substrate, thereby facilitating phosphodiester bond formation. However, *processive* RNA polymerization requires a structural framework that enables the active site to translocate along the template (or vice versa) to generate long RNA polymers. In all naturally occurring RNAPs, this structural framework is provided by proteins.

IV. EMERGENCE OF PROTEINACEOUS RNAP

All multi-subunit RNAPs encompass homologs of Bacterial RNAP subunits: alpha, beta, beta′ and omega (Werner, 2008). The two large beta and beta′ subunits contribute the bulk of the enzyme and harbor the catalytic center at their interface (Ruprich-Robert and Thuriaux, 2010). Interestingly beta and beta′ are paralogs – that is, they are derived from a common ancestor, which has given rise to the hypothesis that they descended from a homodimeric RNA-binding protein (Iyer et al., 2003). Figure 5.3 depicts a model for the evolution of multi-subunit RNAPs from a ribozyme replicase precursor. A ribozyme polymerase acquired a homodimeric protein cofactor (Figure 5.3, stage 1), which protected it against thermal or chemical denaturation and thereby gave a selective advantage to the resulting ribonucleoprotein polymerase. Subsequently the gene encoding the homodimeric cofactor underwent duplication and speciation, thereby generating a heterodimeric cofactor, which represents the progenitor of the two largest subunits of contemporary multi-subunit RNAPs (stage 2). The role of the cofactor became more eminent and changed from assisting catalysis to taking over the catalytic mechanism entirely. In effect, the active site was transferred (stage 3) from the RNA ribozyme to the heterodimeric protein cofactors. The now functionally obsolete RNA component was lost (step 4), the subunit complexity of the RNAP

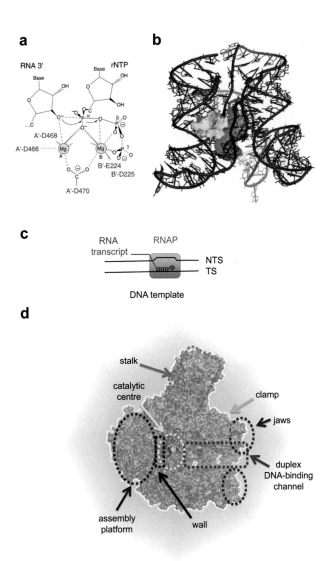

increased (stage 5), and the template specificity switched from RNA to DNA (stage 6).

It should be pointed out that the above model is highly speculative, because neither ribozyme RNAP nor any ribonucleoprotein particle RNAP intermediates have been identified in nature yet. This in itself is worth contemplating. If ribosomes are undergoing a similar evolution from ribozymes through ribonucleoprotein particle (RNPs) intermediates to proteinaceous ribosomes, why have they *not* lost their RNA components, when RNAPs are made of protein only? RNAPs need, like ribosomes, to interact dynamically with nucleic acid templates and substrates, and to catalyze a nucleophilic substitution reaction between activated substrates and a growing chain of a biological polymer. In principle, ribosomal proteins could usurp all ribosome functions that are currently carried out by rRNA, and ribosomes, like RNAPs, could be made entirely of protein – maybe in the distant future they will. In this context, it is interesting that the average protein-to-RNA ratio of ribosomes from a range of organisms increases with increasing evolutionary complexity, and that the mitochondrial ribosome – which is considered to have evolved at the fastest rate – has the highest protein content of all. It is quite possible that we are witnessing this process as it happens. However, it should be noted that there is no evidence that the RNA components of *all* RNP enzymes are fading out but instead in some cases have added novel functionality (Cech, 2009).

V. EXPANSION OF RNAP SUBUNIT REPERTOIRE

Bacteria employ one type of RNAP with four distinct subunits; Archaea also use one type of RNAP but are made of 11–13 subunits; and most Eukaryotes have at least three distinct classes of RNAP containing 12–17 subunits (Figure 5.4). How can we explain this variation in complexity?

One of the more plausible theories explaining the increase of RNAP complexity from Bacteria, through Archaea and to Eukaryotes is "transcription factor capture" (Carter and Drouin, 2009). During evolution, regulatory proteins that initially were reversibly associated with RNAP could over time become stably incorporated into the enzyme (Werner, 2008). However, even though the increase in RNAP complexity from Bacteria to Archaea and Eukarya by transcription factor capture is appealing (see block arrow in Figure 5.4), it cannot be ruled out that all subunits were present in

FIGURE 5.2: The universally conserved two-metal catalytic mechanism of phosphodiester bond catalysis. (a) Catalytic mechanism illustrated by the active center of the M. jannaschii RNAP. Two magnesium ions facilitate an SN₂ nucleophilic attack of the RNA 3'-OH group on the alpha phosphate group of a nucleotide triphosphate subtrate by stabilizing the pentavalent transition state of the reaction. The reaction products are a pyrophosphate and an RNA-chain that has been extended by one nucleotide. Metal A, which is stably bound by RNAP, is coordinated by three invariant aspartate residues of the largest RNAP subunit. Alanine substitution of any of these residues (D466/468/470 in M. jannaschii) renders the RNAP catalytically inactive (Werner and Weinzierl, 2002). Metal B, which enters the active site with the NTP substrate and leaves it bound to pyrophosphate, is chelated via invariant aspartate and glutamate residues in the second largest RNAP subunit (E224/D225 in M. jannaschii). (b) The phosphodiester bond formation reaction in synthetic ribozyme ligases (pdb 3IVK) and ribozyme RNA polymerases is highly reminiscent of proteinaceous RNAPs. The ligation junction is highlighted in green and the active site in red. (c) The basic function of RNAP is to synthesize an RNA transcript in a DNA template-dependent manner. RNAPs engage with duplex DNA templates and separate the two DNA strands – only the template strand is loaded into the active site. NTP substrates enter the active site and anneal with the template DNA strand via Watson-Crick basepairing in a sequence-specific manner.

(d) Structural layout of contemporary Archaeal and Eukaryotic RNA polymerases. Important features discussed in this chapter are highlighted with color-coded dashed circles, including the assembly platform (dark blue), the stalk (orange), the clamp (green), the jaws (red), the DNA-binding channel (blue), the wall (black), and the active site (yellow). The two magnesium ions in the active site are highlighted as magenta spheres.

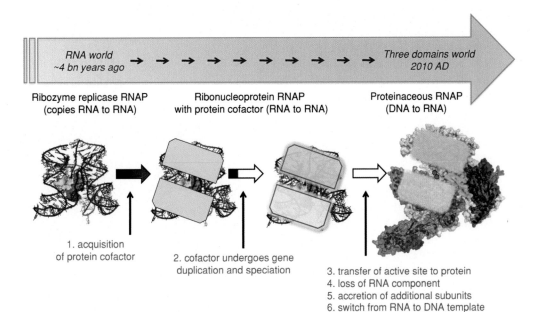

FIGURE 5.3: *The evolution of proteinaceous RNAP from a ribozyme ancestor. This model describes the transition of a hypothetical ribozyme RNA polymerase through ribonucleoprotein RNAP intermediates to the proteinaceous RNAPs found in contemporary organisms. Because of the lack of any naturally occurring ribozyme RNAPs, this model is highly speculative, yet intriguing. The ribozyme ancestor illustrated in the figure in dark blue is the in vitro evolved ribozyme ligase [pdb 3IVK] (Shechner et al., 2009), which is not to be confused with the "real" ancestor of RNAP. The active site of the ribozyme is highlighted in red, whereas the two active site magnesium ions in the proteinaceous RNAP are highlighted with magenta spheres.*

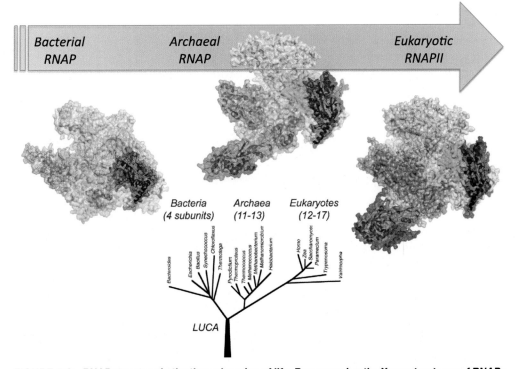

FIGURE 5.4: *RNAP structure in the three domains of life. By comparing the X-ray structures of RNAPs from the three domains of life, Bacteria (T. aquaticus, pdb 1I6V), Archaea (S. solfataricus, 2PMZ) and Eukarya (S. cerevisae, pdb 1NT9), their evolutionary relationship becomes strikingly apparent. All multisubunit RNAPs have a universally conserved core; in addition, the Archaeal and Eukaryotic enzymes display a high degree of structural homology. The arrow refers to an increase in complexity, not necessarily a temporal direction.*

LUCA (Last Universal Common Ancestor) and the additional subunits, and transcription factors (Figure 5.4) were "ablated" or streamlined in the Bacterial lineage. Carter and Drouin have reasoned that overall, including multiple classes of Eukaryotic RNAPs, the subunit accretion theory of RNAP expansion requires the smallest number of evolutionary changes (i.e., parsimony), because more independent events are required to explain loss than gain of subunits (Carter and Drouin, 2009). However, whereas the early origin of multi-subunit RNAPs remains in the realm of speculation, given that it is not amenable to experimental validation, the availability of experimental systems of RNAPs from extant Bacteria, Archaea, and Eukarya has allowed us to investigate them in great detail and has given us detailed mechanistic insights into transcription. By comparing the properties of RNAPs from all three domains of life we can study the generic differences between the transcription machineries, and develop hypotheses on their evolution. Do their structural and functional characteristics reflect distinct demands for the successful execution of the genetic programs of the host organisms?

VI. RNAP SUBUNIT ARCHITECTURE AND FUNCTION

Despite a relatively low sequence identity, the two main types of multi-subunit RNAPs, exemplified by the Bacterial and Archaeal/Eukaryotic enzymes, respectively, display an impressive degree of structural homology (Figure 5.5; homologous RNAP subunits are color-coded). The strictly conserved residues cluster around the RNAP active site (Figure 5.5, red circle) including the bridge (Figure 5.5, orange) and trigger helices, form the NTP entry pore, and are involved in the handling of the template and non-template DNA- and RNA strands, or encode flexible motifs including the RNAP clamp (Figure 5.5, green circle) and the switch regions (Ruprich-Robert and Thuriaux, 2010). How does the active site of multi-subunit RNAPs work? X-ray structures of the Bacterial and yeast enzymes have captured conformational intermediates of the NTP addition cycle that correspond to distinct functional states of RNAPs, which in combination are the molecular basis for the physical translocation of RNAPs along the template gene (Brueckner et al., 2009; Vassylyev et al., 2007a; Vassylyev et al., 2007b). Two structural motifs, the bridge and trigger helices, are crucial to the mechanism. The RNAP-DNA-RNA elongation complex is in equilibrium between pre- and post-translocated states, where the latter corresponds to the RNAP having moved one basepair in the downstream direction. The NTP substrate is inserted into the RNAP active site in the post-translocated "preinsertion" state, the active site "closes" by a structural rearrangement of the trigger motif from a loop to a helical structure, which results in the formation of a trihelix bundle with the bridge helix – the "insertion" state. In this

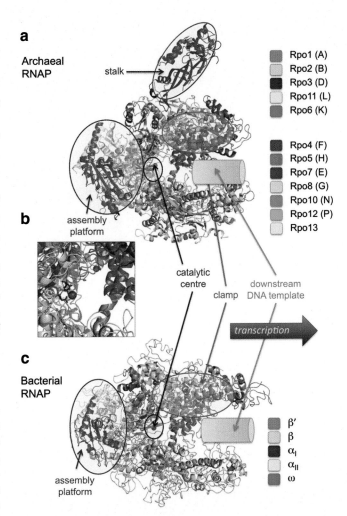

FIGURE 5.5: *Multi-subunit RNAPs share a common structural framework. The structural conserved core of the Bacterial* T. aquaticus *RNAP (c, pdb 1I6V) and Archaeal S. shibatae RNAP (a, pdb 2WAQ) can easily be recognized by close inspection of their X-ray structures. Important functional features such as the active site (b, metal A shown as pink sphere), the bridge helix (highlighted in orange), and the main DNA-binding channel are well conserved. The homologous RNAP subunits are color-coded according to the key in the figure.*

post-translocated insertion-state conformation, the two magnesium ions are ideally positioned to render the active site competent for catalysis, and a new phosphodiester bond is formed. Pyrophosphate leaves the active site, and the elongation complex is rendered in the pre-translocated state. Transition of the pre- into post-translocated state involves an "opening" of the active site into the preinsertion state, that is, a structural rearrangement of the trihelix bundle into trigger loop and bridge helix – ready for the subsequent NTP binding event and the next nucleotide addition cycle. Many of the RNAP core functions including the one described previously have recently been reviewed in a special issue of *Current Opinion in Structural Biology* (Cramer and Arnold, 2009), and rather than elaborating on these, this chapter henceforth focuses on recent and novel

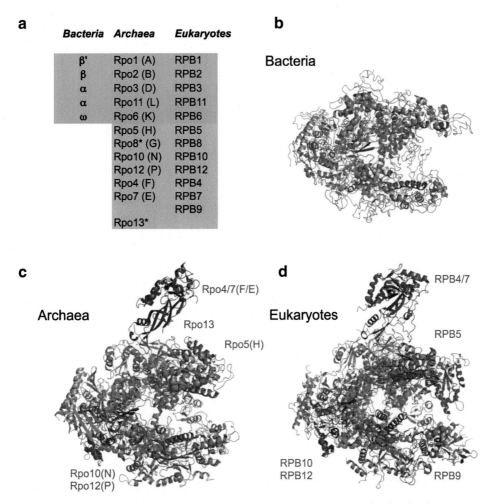

FIGURE 5.6: *Archaeal and Eukaryotic RNAPs contain a complement of subunits that are not present in the Bacterial enzyme. Panel (a) tabulates the RNAP subunit composition in the three domains of life (Eukaryotes refers to RNAPII). The X-ray structures of the Bacterial T. aquaticus RNAP (b, 1I6V)), the Archaeal S. shibatae (c, 2WAQ) RNAP, and Eukaryotic S. cerevisiae RNAPII (d, 1NT9) are shown with universally conserved RNAP subunits colored marine blue, whereas the Archaeal-Eukaryote specific subunits are colored magenta. RPB8 and 13 are only found in some archaeal species (Koonin et al., 2007; Korkhin et al., 2009).*

insights into the molecular mechanisms of RNAP and in particular concentrate on the subunits that are specific for the Archaeal and Eukaryotic enzyme (Werner, 2008).

The Archaeal/Eukaryote-specific RNAP subunits – which have no homologs in Bacterial RNAPs (Figures 5.6c and 5.6d, highlighted in magenta) – interact with many of the universally conserved subunits (Figure 5.5b, highlighted in blue) and are not clustered at one particular site of the enzyme. The nomenclature for the eukaryotic RNAPII subunits is RPB1–12 (from largest to smallest *S. cerevisiae* polypeptide), whereas the Archaeal RNAP subunits are named Rpo1–13 (Figure 5.6a) or are designated with a letter code in the older literature (A, B, D, E, F, G, H, K, L, N and P). Four subunits, Rpo3/10/11/12, are required for the efficient assembly of RNAPs; they form the aptly named assembly platform (Werner and

Weinzierl, 2002). Rpo3/11 is homologous to the Bacterial alpha homodimer, which is sufficient for Bacterial RNAP assembly (Werner et al., 2000) (Figures 5.5a and 5.5c). Subunits Rpo10 (N)- and 12 (P) fill concave depressions in the second largest RNAP subunit (Rpo2 [B]) and thereby act as molecular adaptors between Rpo2 and 3 (RPB2 and 3) (Figures 5.6c and 5.6d), which explains at least in part their role during RNAP assembly. However, Rpo10 and 12 have additional functions beyond RNAP assembly. Thus, the Archaeal homolog of RPB12, Rpo12 (P) has been shown to play a role during transcription initiation by promoting DNA melting and stabilizing the open complex (Reich et al., 2009). Similarly, Rpo5 (H) is instrumental in DNA melting and early transcription (Grunberg et al., 2010). RPB5 consists of two discrete domains; a Eukaryote-specific N-terminal domain that interacts with the basal initiation factor TFIIB (Cheong et al., 1995; Lin et al., 1997) (Figures 5.6c and 5.6d). The C-terminal domain of RPB5, which corresponds to the full-length Archaeal homolog Rpo5 (H), makes intricate contacts with the C-terminus of the largest RNAP subunit (Rpo1 [A″], Figures 5.5 and 5.6). Rpo5 (H) and a fragment of Rpo1 form the lower jaw domain of RNAP, which is more extended than its Bacterial counterpart (Hirata et al., 2008; Korkhin et al., 2009) (Figure 5.5).

The jaw interacts with the downstream duplex DNA and has been reported to undergo substantial conformational changes between the initiation and elongation phase of transcription (Bartlett et al., 2004; Grunberg et al., 2010). RPB8/Rpo8 is located at the underside of the RNAP between the assembly platform and the NTP entry pore. It encodes a nucleic-acid binding OB-fold (Kang et al., 2006) and possibly interacts with the 3′ end of the nascent transcript in backtracked elongation complexes, which is extruded through the NTP entry pore (Komissarova and Kashlev, 1998). Yeast RPB8 is essential for cell viability, but its precise function during transcription is not clear (Briand et al., 2001). RPB9 is probably the only subunit found exclusively in Eukaryotic RNAPs. RPB9 is not essential for cell viability in yeast, but has been shown to influence interactions with the basal factor TFIIF, transcription start site selection and fidelity (Walmacq et al., 2009; Ziegler et al., 2003). Rpo13 is the only Archaea-specific subunit; its function is unknown, and it is only present in a subset of Archaeal genomes (Korkhin et al., 2009).

The most pronounced difference between Archaeal-Eukaryotic and Bacterial RNAPs is the stalk, which is located at the periphery of the enzyme and is comprised of Rpo4 and 7 (F and E) (Figure 5.5). The Rpo4/7 (F/E) complex is well characterized and its multiple functions are discussed in greater detail later in this chapter.

VII. MOLECULAR MECHANISMS OF TRANSCRIPTION INITIATION

Promoter-directed transcription requires sequence-specific recruitment of RNAP to the promoter, initiation of RNA polymerization in a primer-independent fashion, and efficient escape from the promoter. The Archaeal RNAP and Eukaryotic RNAPII have identical minimal requirements for basal transcription initiation factors, the TATA-binding protein, TBP, and transcription factor IIB, TF(II)B. The molecular mechanisms of Archaeal RNAP during transcription initiation are illustrated in Figures 5.7a to 5.7d. TBP triggers a recruitment cascade by binding to the promoter TATA element with intermediate affinity (Kd~10–100nM), which results in a bending of the promoter DNA by 50–80° (Kim et al., 1993, Whittington et al., 2008). Next TFB binds to the TBP-TATA complex by sequence-specific recognition of the TFB-responsive element (BRE), and this stabilizes

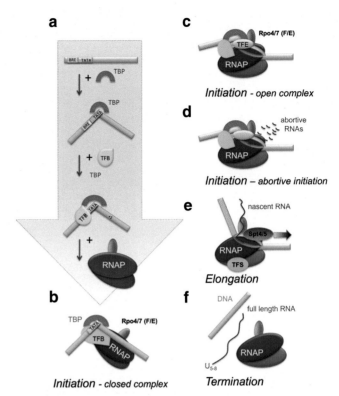

FIGURE 5.7: *Molecular mechanisms of Archaeal RNAP during the transcription cycle. Transcription is initiated by the step-wise assembly (a) of the preinitiation complex, resulting in the "closed" complex (b). After the DNA strands are separated to form the "open" complex (c), RNAPs enter the abortive initiation phase during which short RNA transcripts are synthesized and ejected while the RNAP is still engaged with the promoter (d). Following promoter escape, RNAP enters the elongation phase of transcription (e), during which elongation factors such as Spt4/5 (red-orange) and TFS (green) associate with RNAP and modulate its activities. Eventually RNAP terminates upon transcription of short U_{5-8} stretches that serve as termination signals (f). The Rpo4/7 (F/E) complex is involved in multiple steps including open complex formation during initiation (b/c), high processivity during elongation (e), and efficient transcription termination (f). The basal initiation factors TBP, TFB, and TFE are highlighted in red, turquoise, and orange, respectively.*

the preinitiation complex (Bell et al., 1999). RNAP is subsequently recruited to the DNA-TBP-TFB complex (Werner and Weinzierl, 2005) (Figure 5.7a). However, TF(II)B serves as more than a simple molecular adaptor between promoter DNA and RNAP.

TF(II)B consists of two discrete domains, which are connected by a flexible linker (Figure 5.8c). The C-terminal core domain of TF(II)B binds to the TATA-TBP complex and makes contacts with the DNA both upstream and downstream of the TATA element. Partial crystal structures combined with biochemical proximity probing (cross-linking and cleavage) have been useful to build structural models of the minimal Eukaryotic DNA-TBP-TF(II)B-RNAP initiation complex (Chen and Hahn, 2003; Chen

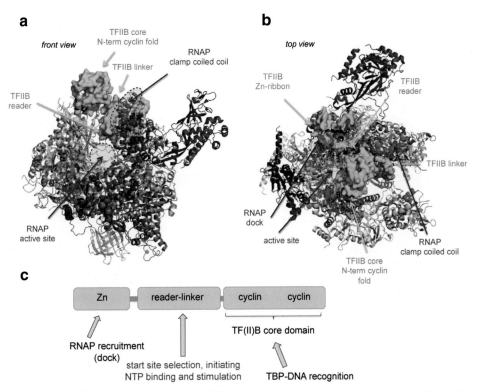

FIGURE 5.8: *Molecular mechanisms of TFIIB during transcription initiation. Transcription initiation in Archaea and Eukaryotes is facilitated by the intimate interplay between RNAP and TF(II)B in the preinitiation complex. This figure shows the X-ray structure of the yeast RNAPII-TFIIB complex (pdb code 3K1F) in front (a) and top (b) views, and a schematic of the TFIIB domain organization (c). Part of TF(II)B penetrates deep into the active site and modulates the properties of RNAP such as NTP binding, transcription start site selection, and polymerization. Only one of the two cyclin repeats could be resolved in the X-ray structure.*

and Hahn, 2004; Kostrewa et al., 2009). According to our current understanding of initiation complex architecture, neither the Archaeal nor Eukaryotic RNAPs is in extensive direct contact with the DNA in the closed complex (Figure 5.7b). Instead, RNAP is anchored to the promoter via multiple interactions with TF(II)B. The N-terminal Zn-ribbon domain of TF(II)B interacts with the dock domain of RNAP (Figure 5.8b), and the C-terminal core domain is positioned across the DNA binding channel (Figures 5.8a and 5.8b). In addition to the RNAP dock domain, the clamp coiled-coil motif is an important binding site for TF(II)B (Figure 5.8; highlighted with a red dashed circle) (Kostrewa et al., 2009). The highly flexible linker region that connects the TF(II)B domains (B-reader helix and B-linker) penetrates deep into the active center of RNAP (Figures 5.8a and 5.8b). The linker can be cross-linked to the template DNA strand (Renfrow et al., 2004) and is displaced by the growing RNA transcript (longer than 5 nt), which aids the promoter escape (Bushnell et al., 2004). Thus, RNAPs form a "composite" active site that is complemented by TF(II)B, which possibly alters the affinity for the initiating nucleotide substrate, affects the transcription start site selection of RNAPII (Kostrewa et al.,

2009), and stimulates catalysis of Archaeal RNAP (Werner and Weinzierl, 2005).

Following recruitment to the promoter, all RNAPs enter a non-productive phase of transcription called abortive initiation (Figure 5.7d), during which the downstream DNA template repeatedly is "reeled" in (Kapanidis et al., 2006), small (3–9 nt) transcripts are synthesized and released without the RNAP disengaging from the promoter (Goldman et al., 2009). The molecular reason for abortive initiation is still unclear but is likely to be caused by the energy barrier of breaking interactions within the initiation complex – between RNAP and TF(II)B. Within the transcription initiation complex, multiple interactions with individual low affinities combine to form a stable complex (Kostrewa et al., 2009; Liu et al., 2010; Murakami and Darst, 2003). This dynamic interaction network enables both efficient recruitment of RNAP to the promoter and the dissociation of the complex by small conformational changes, during which the individual contacts are broken in a stepwise manner. In addition to TBP and TF(II)B, other basal factors contribute to transcription initiation in Archaea (TFE) and Eukaryotes (TFIIE, TFIIF, and TFIIH). TFE and TFIIE interact with the RNAP clamp;

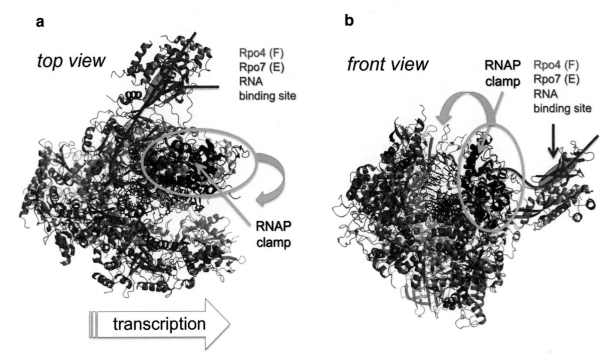

FIGURE 5.9: *Rpo4/7 (F/E) modulates RNAP activities during the elongation and termination phase of transcription by two mechanisms. This model of the ternary elongation complex (TEC) in top view (a) and front view (b) is based on S. cerevisiae RNAPII (pdb 1Y1W). Rpo4/7 (F/E) increases processivity and efficient transcription termination of RNAP. Both activities depend on two mechanisms involving (1) RNA binding (RNA binding site is highlighted as red semitransparent oval) and (2) allostery of the RNAP clamp (highlighted in green), which is speculated to close over the major DNA-binding channel (indicated with a green block arrow).*

they are not strictly required for initiation but stimulate DNA melting and stabilize the open complex (Holstege et al., 1996; Naji et al., 2007; Werner and Weinzierl, 2005).

VIII. STRUCTURE AND FUNCTION OF RNAP SUBUNITS RPO4/7 (F/E)

The most prominent structural feature that distinguishes Archaeo-Eukaryotic RNAPs from their Bacterial counterparts is a stalk-like protrusion located at the periphery of the roughly ellipsoidal shape of the enzyme in proximity of the RNA exit channel (Figures 5.5 and 5.6 [Werner, 2007]). This "signature" module of Archaeo-Eukaryotic RNAPs is a heterodimeric complex consisting of RNAP subunits Rpo4/7 (F/E), which are homologous to RPB4/7 in the *Saccharomyces cerevisiae* RNAPII (Todone et al., 2001). In addition to the Archaeal RNAP, all five types of Eukaryotic RNAPs (RNAPI, II, III, IV, and V) harbor homologs of Rpo4/7 (F/E), which suggests that they are important for RNAP function (Ream et al., 2009). Rpo4/7 (F/E) is a highly versatile RNAP module that plays multiple roles during the transcription cycle (Figure 5.7 [Hirtreiter et al., 2010; Grohmann et al., 2009; Grohmann, 2010; Werner and Weinzierl, 2005]).

VIII.1 Transcription Initiation: Rpo4/7 (F/E) is Involved in DNA Melting During the Initiation Phase of Transcription

During open complex formation, the two DNA strands of the promoter are separated ("melted") and the template DNA strand is loaded into the active site. Rpo4/7 (F/E) facilitates DNA melting (Figures 5.7b and 5.7c) and is in addition required for the function of a third basal factor, TFE (Naji et al., 2007; Werner and Weinzierl, 2005). TFE increases the stability of the complex and, like Rpo4/7 (F/E), assists DNA strand separation. Similarly RPB4/7, the yeast homolog of Rpo4/7 (F/E), plays an important role during transcription initiation of *S. cerevisiae* RNAPII (Edwards et al., 1991). The molecular mechanisms of Rpo4/7 (F/E) during "open complex" formation are unclear, but crystallographic data have suggested that Rpo4/7 (F/E) modulates the position of the RNAP clamp, which in turn leads to DNA strand separation (Armache et al., 2005). The tip of RPB7 forms a wedge that is inserted at the base of the RNAP clamp and thereby closes it over the main DNA binding channel (Armache et al., 2005) (Figure 5.9). RPB4/7 dynamically associates with yeast RNAPII during the transcription cycle (Sampath and Sadhale, 2005). This hypothesis is mostly based on indirect evidence: RNAPII purified from exponentially growing yeast is sub-saturated with respect to RPB4/7, whereas RNAPII isolated from stationary phase cells contain RPB4/7 in stoichiometric amounts (Kolodziej et al., 1990; Woychik and Young, 1989). Archaeal Rpo4/7 (F/E) is stably incorporated into RNAP and is not in a dynamic equilibrium with Rpo4/7 (F/E) complexes in solution (Grohmann et al., 2009). This result implies that Rpo4/7 (F/E) is present in the elongating form of RNAP, and suggest that it is instrumental in transcript elongation (Hirtreiter et al., 2010a).

VIII.2 Transcription Elongation: Rpo4/7 (F/E) Increases Processivity

Transcription elongation complexes have a remarkable processivity; RNAPII molecules routinely transcribe several million basepairs (e.g. the 2.4 Mb dystrophin gene) during which they remain associated with the template DNA for as long as sixteen hours (Tennyson et al., 1995). This stability is crucial for the elongation process, because RNAP cannot re-elongate a partial RNA transcript after the enzyme has disengaged from the template, and consequently RNAP has to initiate transcription from the promoter *de novo*. Despite the requirement of forming stable elongation complexes, RNAP cannot bind too tightly to its template or its product because this would impede efficient transcription elongation. Therefore transcribing RNAPs have to interact dynamically with the nucleic acid scaffold of the elongation complex (Figure 5.7e). Crystal structures of elongating RNAPs (e.g., pdb 1I6H, 1Y1W) have resolved the downstream duplex DNA template and a 9-basepair (bp) DNA-RNA hybrid but have so far failed to resolve the upstream duplex DNA or the nascent RNA (Gnatt et al., 2001; Kettenberger et al., 2004). An elegant study by the Michaelis group probed the location of the upstream DNA employing a fluorescence resonance energy transfer system called the nanopositioning system (NPS) (Andrecka et al., 2009). However, the same approach failed to determine the path of the nascent RNA transcript beyond 23 nt (Andrecka et al., 2008), and thus there is no hard structural information about the path and secondary structure of the nascent RNA.

The nascent RNA emerges from the main body of the RNAP through the RNA exit channel and is directed toward Rpo4/7 (F/E) (Figure 5.9). Indeed, both Rpo4/7 (F/E) and RPB4/7 bind RNA in vitro and in vivo (Meka et al., 2005; Orlicky et al., 2001; Ujvari and Luse, 2006). This interaction is apparently sequence-independent, of intermediate affinity (Rpo4/7 (F/E)-RNA $K_d = 0.34 \pm 0.06$ mM) (Grohmann et al., 2010), and covers approximately 15 nt of RNA between position $+ 26$ to $+ 41$ relative to the active site (Ujvari and Luse, 2006). RNAPs lacking Rpo4/7 (F/E) suffer a severe elongation failure in the absence of the non-template strand (NTS) and display a marked reduction in processivity compared to wild-type RNAPs when using the more biological relevant scaffold containing the NTS (Hirtreiter et al., 2010a). Furthermore,

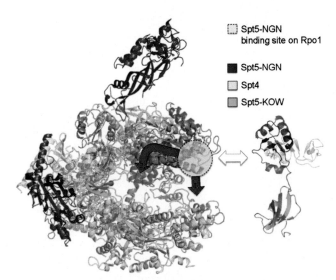

Spt5-NGN binding site on Rpo1

Spt5-NGN

Spt4

Spt5-KOW

FIGURE 5.10: *The transcription elongation factor Spt4/5 stimulates RNAP processivity. The NGN domain (colored firebrick red) of the universally conserved transcription elongation factors Spt5 (NusG) interacts with the tip of the RNAP clamp coiled coil (yellow circle) and increases transcription processivity. This activation could be facilitated by altering interactions with the DNA template via a closure of the RNAP clamp domain (red arrow labeled 1.) and/or by an allosteric signal through the RNAP clamp to the bridge helix (highlighted in orange) in the catalytic center (red arrow labeled 2.).*

the elongation activity of Rpo4/7 (F/E) correlates with its RNA-binding activity, which implies that interactions with the transcript directly increase processivity (Hirtreiter et al., 2010a). Interestingly, even the most severe RNA binding-deficient mutant variants of Rpo4/7 (F/E) are still able to stimulate elongation at low levels. This effect could be attributed to a residual RNA binding by Rpo4/7 (F/E) in the context of the elongation complex, or alternatively Rpo4/7 (F/E) could stimulate elongation allosterically by inducing a conformational change in RNAP, such as the closure of the RNAP clamp (Armache et al., 2005). An inward movement of the RNAP clamp would narrow the main DNA binding channel, stabilize the elongation complex, and increase processivity (see above and Figure 5.9b).

VIII.3 Transcription Termination: Rpo4/7 (F/E) Augments Termination from Weak U₅-Signals

Transcription termination is one of the least understood mechanisms of RNAP. In the Archaeal system, transcription of short poly-U stretches (5–8 U-residues) leads to termination in vitro and in vivo (Santangelo et al., 2009; Santangelo and Reeve, 2006; Spitalny and Thomm, 2008). The transcription of a poly-U stretch serves as a universal pause signal for all RNAPs. In the Bacterial RNAP, additional events are required for termination; either a stable RNA hairpin (*intrinsic* terminator) or the action of the rho termination factor triggers conformational changes in the RNAP active site (Figure 5.10, bridge helix highlighted in orange) and the RNAP clamp (Figure 5.5), which results in the dissociation of the elongation complex and termination (Epshtein et al., 2007; Epshtein et al., 2010). This model and the working hypotheses concerning allostery during initiation and elongation (see earlier discussion) have suggested a role of Rpo4/7 (F/E) in transcription termination (Hirtreiter et al., 2010a). Whereas Rpo4/7 (F/E) is not strictly required for termination, the termination efficiency at weak termination signals (e.g. comprised of five rather than eight U-residues) is significantly increased by Rpo4/7 (F/E) in a manner that is largely dependent on F/E-RNA interactions (Hirtreiter et al., 2010a). How does Rpo4/7 (F/E) enhance transcription termination? In the Archaeal system, the pausing induced by transcription of a poly-U tract is sufficient to trigger termination. The efficiency of termination correlates with the duration of the pause (Landick, 2006), which in turn is likely to be affected by the stability of the elongation complex – which could be increased by Rpo4/7 (F/E)-RNA interactions. In addition, Rpo4/7 (F/E) could directly lead to the conformational changes within the RNAP that result in the dissociation of the elongation complex, possibly via the RNAP clamp or the trigger/bridge helices (Figures 5.9 and 5.10). The role and distribution of termination signals at the 3′ end of Archaeal transcription units, and their relative strength in vivo (mainly determined by the number of U residues) is far from clear (Santangelo and Reeve, 2006; Santangelo et al., 2009). The effect of Rpo4/7 (F/E) on termination is likely to have an even greater impact in vivo than in vitro, because efficient and accurate transcription termination is of crucial importance in Archaea, which typically have small genomes, very short intergenic regions and transcription-translation are coupled (Santangelo et al., 2008; Santangelo et al., 2009; Santangelo and Reeve, 2006).

IX. BOTH TRANSCRIPTION-ELONGATION FACTORS AND RNAP SUBUNITS ACTIVATE TRANSCRIPTION THROUGH THE RNAP CLAMP

The transcription elongation properties of RNAPs are influenced by their subunit composition (i.e. ± Rpo4/7 [F/E]) and regulated by exogenous transcription factors including Spt5, the only known RNAP-associated transcription factor that is universally conserved in evolution. The X-ray structure and function of the archaeal *M. jannaschii* Spt4/5 (Hirtreiter, 2010b) reveal an astonishing degree of structural conservation, conservation of RNAP interaction sites and elongation-enhancing properties (Hirtreiter, 2010b). Archaeal Spt5 is, like its Bacterial homolog NusG, comprised of two domains: the NGN domain (*NusG* N-terminal domain) and a KOW domain (*Kyrpidis, Ouzounis* and *Woese* domain). A deletion analysis has revealed that Spt5-NGN is the effector domain of Spt5 which mediates the dimerization with Spt4, the binding to RNAP, and is required for the elongation activity

of Spt4/5. The last two features are reliant on an interaction between a deep hydrophobic cavity in the Spt5-NGN domain and the tip of the RNAP clamp coiled coil, a surface-exposed structural feature that is conserved in all multi-subunit RNAPs (Figure 5.10). What are the molecular mechanisms by which Spt4/5 stimulates elongation? The Spt5-NGN binding site on RNAP is approximately 70 Å distant from the active site, which suggests an allosteric mechanism of stimulation (Figure 5.10).

It is significant that both Rpo4/7 (F/E) and Spt4/5 are located in close proximity of the RNAP clamp and that both affect transcription elongation in a similar manner: by increasing the processivity in a fashion that is not dependent on the NTS. The latter result makes it unlikely that Rpo4/7 (F/E) and Spt4/5 function solely by interacting with the NTS, or act by affecting downstream DNA-strand separation, or upstream DNA-strand joining. Rather it has been proposed that Spt4/5 induces a conformational change in the RNAP clamp, which is translated into the RNAP interior and results in increased translocation efficiency (Figure 5.10, red block arrow 2). This mechanism is reminiscent of the Bacterial Spt5 variant, NusG and its paralogue RfaH, which have been proposed to stimulate elongation by stabilizing the forward-translocated state of the RNAP-active site (Belogurov et al., 2007). Alternatively, Spt4/5 could bridge the gap over the main DNA-binding channel of RNAP and "lock" the RNAP clamp in a closed position, which would result in an increased elongation complex stability and enhanced processivity (Figure 5.10, red block arrow 1).

X. ELONGATION GONE WRONG: RESCUE BY TRANSCRIPT CLEAVAGE

Whereas reversibly paused ternary elongation complexes (TECs) can either spontaneously recover or be aided by Spt5/NusG-like factors, they can also become irreversibly arrested. Paused RNAPs have a tendency to move "backward" (3' to 5') along the DNA template (backtracking). During backtracking, the RNA 3' end is extruded from the RNAP through the NTP entry pore and RNA polymerization cannot occur. RNAPs are only able to overcome this impediment by cleaving the transcript internally, releasing short (3–18 nt) RNA 3'-cleavage products and creating a new RNA 3'-OH that is aligned in the active site and conducive to catalysis (Deighan and Hochschild, 2006; Sigurdsson et al., 2010). This endonucleolytic cleavage activity of RNAP is prominent at elevated pH and stimulated by transcript cleavage factors under physiological conditions. In Eukaryotes and Archaea, TF(II)S stimulates transcript cleavage (Hausner et al., 2000), and the structure of the yeast RNAPII-TFIIS complex provides insights into its molecular mechanism (Kettenberger et al., 2004).

TFIIS is recruited to the TEC via interactions between TFIIS domain II and the RNAP jaw, whereas domain III

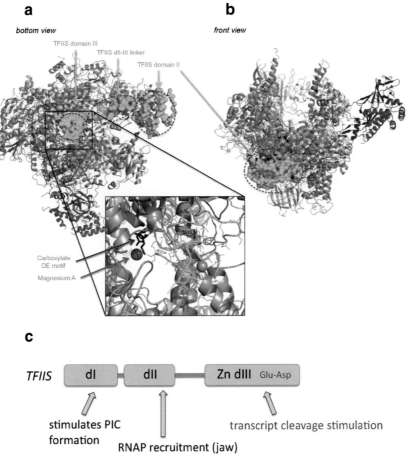

FIGURE 5.11: *Rescue of stalled transcription elongation complexes. The X-ray structure of the RNAPII-TFIIS complex reveals the molecular mechanism by which TFIIS stimulates transcript cleavage. All multi-subunit RNAPs possess a "tunable" active site that can catalyze either the formation (RNA polymerization) or cleavage (endonucleolysis) of RNA phosphodiester bonds, depending on the geometry of two magnesium ions in the active site. TFIIS inserts two carboxylate groups into the active site and thereby retunes it from polymerization to cleavage. TFIIS is colored turquoise, metal A is highlighted as pink sphere, and the "DE" motif of TFIIS domain III is shown as red stick model in the insert.*

is inserted into the active site through the NTP entry pore (Figure 5.11). Domain II forms a Zn-ribbon domain with a thin protruding beta hairpin, which complements the active site without denying access of NTP substrates to the active site, or blocking the extrusion of the transcript through the pore. Two invariant acidic residues on the tip of the hairpin are essential for the stimulatory effect of TFIIS on transcript cleavage (Figure 5.11, highlighted in red in the insert) (Jeon et al., 1994). They are brought into close proximity of the magnesium ions in the active site and likely alter the binding characteristics of the metal ions and/or modulate the structure and location of the RNA or DNA-RNA hybrid in the active site in a manner that stimulates endonucleolysis (Figure 5.4) (Kettenberger et al., 2004). Given that most, if not all, genes contain frequent pause sites, TFIIS regulates RNAP activity by increasing its processivity and thereby the overall transcription elonga-

tion rate. The mechanisms of TF(II)B and TF(II)S bear some similarity because these factors invade the catalytic center of RNAP either via the major DNA-binding channel or the NTP entry pore, complement the active site, alter the interactions with nucleic acids, magnesium ions, and NTP substrates, and thereby modulate the catalytic properties of RNAP.

XI. CONCLUDING REMARKS

RNAPs are complex molecular machines that carry out RNA polymerization and cleavage, in a highly regulated manner by interacting with basal transcription factors. A combination of cutting-edge structural and biophysical techniques with more traditional biochemical and molecular biological approaches has given us intriguing insights into the *modus operandi* of transcription complexes.

We hope this chapter will encourage the reader to delve deeper into this fascinating subject by reading the primary research literature listed in the References section.

ACKNOWLEDGMENTS

Finn Werner would like to thank the Wellcome Trust and BBSRC for funding research at the RNAP laboratory.

REFERENCES

Altman, S. & Robertson, H. D. 1973. RNA precursor molecules and ribonucleases in E. coli. *Mol Cell Biochem*, 1, 83–93.

Andrecka, J., Lewis, R., Bruckner, F., Lehmann, E., Cramer, P. & Michaelis, J. 2008. Single-molecule tracking of mRNA exiting from RNA polymerase II. *Proc Natl Acad Sci USA*, 105, 135–40.

Andrecka, J., Treutlein, B., Arcusa, M. A., Muschielok, A., Lewis, R., Cheung, A. C., Cramer, P. & Michaelis, J. 2009. Nano positioning system reveals the course of upstream and nontemplate DNA within the RNA polymerase II elongation complex. *Nucleic Acids Res*, 37, 5803–9.

Armache, K. J., Mitterweger, S., Meinhart, A. & Cramer, P. 2005. Structures of complete RNA polymerase II and its subcomplex, Rpb4/7. *J Biol Chem*, 280, 7131–4.

Bartlett, M. S., Thomm, M. & Geiduschek, E. P. 2004. Topography of the euryArchaeal transcription initiation complex. *J Biol Chem*, 279, 5894–903.

Bell, S. D., Kosa, P. L., Sigler, P. B. & Jackson, S. P. 1999. Orientation of the transcription preinitiation complex in Archaea. *Proc Natl Acad Sci USA*, 96, 13662–7.

Belogurov, G. A., Vassylyeva, M. N., Svetlov, V., Klyuyev, S., Grishin, N. V., Vassylyev, D. G. & Artsimovitch, I. 2007. Structural basis for converting a general transcription factor into an operon-specific virulence regulator. *Mol Cell*, 26, 117–29.

Briand, J. F., Navarro, F., Rematier, P., Boschiero, C., Labarre, S., Werner, M., Shpakovski, G. V. & Thuriaux, P. 2001. Partners of Rpb8p, a small subunit shared by yeast RNA polymerases I, II and III. *Mol Cell Biol*, 21, 6056–65.

Brueckner, F., Ortiz, J. & Cramer, P. 2009. A movie of the RNA polymerase nucleotide addition cycle. *Curr Opin Struct Biol*, 19, 294–9.

Bushnell, D. A., Westover, K. D., Davis, R. E. & Kornberg, R. D. 2004. Structural basis of transcription: an RNA polymerase II-TFIIB cocrystal at 4.5 Angstroms. *Science*, 303, 983–8.

Carter, R. & Drouin, G. 2009. The increase in the number of subunits in eukaryotic RNA polymerase III relative to RNA polymerase II is due to the permanent recruitment of general transcription factors. *Mol Biol Evol.*, 27, 1035–1043.

Cech, T. R. 2009. Crawling out of the RNA world. *Cell*, 136, 599–602.

Cech, T. R. & Bass, B. L. 1986. Biological catalysis by RNA. *Annu Rev Biochem*, 55, 599–629.

Chen, H. T. & Hahn, S. 2003. Binding of TFIIB to RNA polymerase II: mapping the binding site for the TFIIB zinc ribbon domain within the preinitiation complex. *Mol Cell*, 12, 437–47.

Chen, H. T. & Hahn, S. 2004. Mapping the location of TFIIB within the RNA polymerase II transcription preinitiation complex: a model for the structure of the PIC. *Cell*, 119, 169–80.

Cheong, J. H., YI, M., Lin, Y. & Murakami, S. 1995. Human RPB5, a subunit shared by eukaryotic nuclear RNA polymerases, binds human hepatitis B virus X protein and may play a role in X transactivation. *Embo J*, 14, 143–50.

Cramer, P. & Arnold, E. 2009. Proteins: how RNA polymerases work. *Curr Opin Struct Biol*, 19, 680–2.

Deighan, P. & Hochschild, A. 2006. Conformational toggle triggers a modulator of RNA polymerase activity. *Trends Biochem Sci*, 31, 424–6.

Edwards, A. M., Kane, C. M., Young, R. A. & Kornberg, R. D. 1991. Two dissociable subunits of yeast RNA polymerase II stimulate the initiation of transcription at a promoter in vitro. *J Biol Chem*, 266, 71–5.

Ekland, E. H., Szostak, J. W. & Bartel, D. P. 1995. Structurally complex and highly active RNA ligases derived from random RNA sequences. *Science*, 269, 364–70.

Epshtein, V., Cardinale, C. J., Ruckenstein, A. E., Borukhov, S. & Nudler, E. 2007. An allosteric path to transcription termination. *Mol Cell*, 28, 991–1001.

Epshtein, V., Dutta, D., Wade, J. & Nudler, E. 2010. An allosteric mechanism of Rho-dependent transcription termination. *Nature*, 463, 245–9.

Gnatt, A. L., Cramer, P., Fu, J., Bushnell, D. A. & Kornberg, R. D. 2001. Structural basis of transcription: an RNA polymerase II elongation complex at 3.3 A resolution. *Science*, 292, 1876–82.

Goldman, S. R., Ebright, R. H. & Nickels, B. E. 2009. Direct detection of abortive RNA transcripts in vivo. *Science*, 324, 927–8.

Grohmann, D., Hirtreiter, A. & Werner, F. 2009. The RNAP subunits F/E (RPB4/7) are stably associated with Archaeal RNA polymerase – using fluorescence anisotropy to monitor RNAP assembly in vitro. *Biochem J.*, 421, 339–343.

Grohmann, D., Klose, D., Klare, J. P., Kay, C. M., Steinhoff, H.-J., Werner, F. 2010. RNA-binding to Archaeal RNAP subunits F/E- a DEER and FRET study. *JACS*, 132, 5954–5955.

Grunberg, S., Reich, C., Zeller, M. E., Bartlett, M. S. & Thomm, M. 2010. Rearrangement of the RNA polymerase subunit H and the lower jaw in Archaeal elongation complexes. *Nucleic Acids Res*, 38, 1950–63.

Hausner, W., Lange, U. & Musfeldt, M. 2000. Transcription factor S, a cleavage induction factor of the Archaeal RNA polymerase. *J Biol Chem*, 275, 12393–9.

Hirata, A., Klein, B. J. & Murakami, K. S. 2008. The X-ray crystal structure of RNA polymerase from Archaea. *Nature*, 451, 851–4.

Hirtreiter, A., Damsma, F., Cheung, A., Klose, D., Grohmann, D., Vojnic, E., Martin, C. R., Cramer, P., Werner, F. 2010b. Spt4/5 Stimulates Transcription Elongation through the RNA Polymerase Clamp Coiled Coil Motif. *NAR*, 38, 4040–4051.

Hirtreiter, A., Grohmann, D. & Werner, F. 2010a. Molecular mechanisms of RNA polymerase – the F/E (RPB4/7) complex is required for high processivity in vitro. *Nucleic Acids Res*, 38, 585–96.

Holstege, F. C., Van Der Vliet, P. C. & Timmers, H. T. 1996. Opening of an RNA polymerase II promoter occurs in two distinct steps and requires the basal transcription factors IIE and IIH. *EMBO J*, 15, 1666–77.

Iyer, L. M., Koonin, E. V. & Aravind, L. 2003. Evolutionary connection between the catalytic subunits of DNA-dependent RNA polymerases and eukaryotic RNA-dependent RNA polymerases and the origin of RNA polymerases. *BMC Struct Biol*, 3, 1.

Jeon, C., Yoon, H. & Agarwal, K. 1994. The transcription factor TFIIS zinc ribbon dipeptide Asp-Glu is critical for stimulation of elongation and RNA cleavage by RNA polymerase II. *Proc Natl Acad Sci USA*, 91, 9106–10.

Johnston, W. K., Unrau, P. J., Lawrence, M. S., Glasner, M. E. & Bartel, D. P. 2001. RNA-catalyzed RNA polymerization: accurate and general RNA-templated primer extension. *Science*, 292, 1319–25.

Joyce, G. F., Orgel, L. E. 1999. *The RNA World*, New York, Cold Spring Harbor Laboratory Press.

Kang, X., Hu, Y., Li, Y., Guo, X., Jiang, X., Lai, L., Xia, B. & Jin, C. 2006. Structural, biochemical, and dynamic characterizations of the hRPB8 subunit of human RNA polymerases. *J Biol Chem*, 281, 18216–26.

Kapanidis, A. N., Margeat, E., Ho, S. O., Kortkhonjia, E., Weiss, S. & Ebright, R. H. 2006. Initial transcription by RNA polymerase proceeds through a DNA-scrunching mechanism. *Science*, 314, 1144–7.

Kettenberger, H., Armache, K. J. & Cramer, P. 2004. Complete RNA polymerase II elongation complex structure and its interactions with NTP and TFIIS. *Mol Cell*, 16, 955–65.

Kim, Y., Geiger, J. H., Hahn, S. & Sigler, P. B. 1993. Crystal structure of a yeast TBP/TATA-box complex. *Nature*, 365, 512–20.

Kolodziej, P. A., Woychik, N., Liao, S. M. & Young, R. A. 1990. RNA polymerase II subunit composition, stoichiometry, and phosphorylation. *Mol Cell Biol*, 10, 1915–20.

Komissarova, N. & Kashlev, M. 1998. Functional topography of nascent RNA in elongation intermediates of RNA polymerase. *Proc Natl Acad Sci USA*, 95, 14699–704.

Koonin, E. V., Makarova, K. S. & Elkins, J. G. 2007. Orthologs of the small RPB8 subunit of the eukaryotic RNA polymerases are conserved in hyperthermophilic Crenarchaeota and "Korarchaeota." *Biol Direct*, 2, 38.

Korkhin, Y., Unligil, U. M., Littlefield, O., Nelson, P. J., Stuart, D. I., Sigler, P. B., Bell, S. D. & Abrescia, N. G. 2009. Evolution of Complex RNA Polymerases: The Complete Archaeal RNA Polymerase Structure. *PLoS Biol*, 7, e102.

Kostrewa, D., Zeller, M. E., Armache, K. J., Seizl, M., Leike, K., Thomm, M. & Cramer, P. 2009. RNA polymerase II-TFIIB structure and mechanism of transcription initiation. *Nature*, 462, 323–30.

Landick, R. 2006. The regulatory roles and mechanism of transcriptional pausing. *Biochem Soc Trans*, 34, 1062–6.

Lin, Y., Nomura, T., Cheong, J., Dorjsuren, D., Iida, K. & Murakami, S. 1997. Hepatitis B virus X protein is a transcriptional modulator that communicates with transcription factor IIB and the RNA polymerase II subunit 5. *J Biol Chem*, 272, 7132–9.

Lincoln, T. A. & Joyce, G. F. 2009. Self-sustained replication of an RNA enzyme. *Science*, 323, 1229–32.

Liu, X., Bushnell, D. A., Wang, D., Calero, G. & Kornberg, R. D. 2010. Structure of an RNA polymerase II-TFIIB complex and the transcription initiation mechanism. *Science*, 327, 206–9.

Meka, H., Werner, F., Cordell, S. C., Onesti, S. & Brick, P. 2005. Crystal structure and RNA binding of the Rpb4/Rpb7 subunits of human RNA polymerase II. *Nucleic Acids Res*, 33, 6435–44.

Murakami, K. S. & Darst, S. A. 2003. Bacterial RNA polymerases: the whole story. *Curr Opin Struct Biol*, 13, 31–9.

Naji, S., Grunberg, S. & Thomm, M. 2007. The RPB7 orthologue E' is required for transcriptional activity of a reconstituted Archaeal core enzyme at low temperatures and stimulates open complex formation. *J Biol Chem*, 282, 11047–57.

Orlicky, S. M., Tran, P. T., Sayre, M. H. & Edwards, A. M. 2001. Dissociable Rpb4-Rpb7 subassembly of rna polymerase II binds to single-strand nucleic acid and mediates a post-recruitment step in transcription initiation. *J Biol Chem*, 276, 10097–102.

Ream, T. S., Haag, J. R., Wierzbicki, A. T., Nicora, C. D., Norbeck, A. D., Zhu, J. K., Hagen, G., Guilfoyle, T. J., Pasa-Tolic, L. & Pikaard, C. S. 2009. Subunit compositions of the RNA-silencing enzymes Pol IV and Pol V reveal their origins as specialized forms of RNA polymerase II. *Mol Cell*, 33, 192–203.

Reich, C., Zeller, M., Milkereit, P., Hausner, W., Cramer, P., Tschochner, H. & Thomm, M. 2009. The Archaeal RNA polymerase subunit P and the eukaryotic polymerase subunit Rpb12 are interchangeable in vivo and in vitro. *Mol Microbiol*, 71, 989–1002.

Renfrow, M. B., Naryshkin, N., Lewis, L. M., Chen, H. T., Ebright, R. H. & Scott, R. A. 2004. Transcription factor B contacts promoter DNA near the transcription start site of the Archaeal transcription initiation complex. *J Biol Chem*, 279, 2825–31.

Ruprich-Robert, G. & Thuriaux, P. 2010. Non-canonical DNA transcription enzymes and the conservation of two-barrel RNA polymerases. *Nucleic Acids Res*, 38, 4559–4569.

Sampath, V. & Sadhale, P. 2005. Rpb4 and Rpb7: a sub-complex integral to multi-subunit RNA polymerases performs a multitude of functions. *IUBMB Life*, 57, 93–102.

Santangelo, T. J., Cubonova, L., Matsumi, R., Atomi, H., Imanaka, T. & Reeve, J. N. 2008. Polarity in Archaeal operon transcription in Thermococcus kodakaraensis. *J Bacteriol*, 190, 2244–8.

Santangelo, T. J., Cubonova, L., Skinner, K. M. & Reeve, J. N. 2009. Archaeal intrinsic transcription termination in vivo. *J Bacteriol*, 191, 7102–8.

Santangelo, T. J. & Reeve, J. N. 2006. Archaeal RNA polymerase is sensitive to intrinsic termination directed by transcribed and remote sequences. *J Mol Biol*, 355, 196–210.

Schuster, P. 2010. Origins of Life: Concepts, data and debates. *Complexity*, 15, 7–10.

Shechner, D. M., Grant, R. A., Bagby, S. C., Koldobskaya, Y., Piccirilli, J. A. & Bartel, D. P. 2009. Crystal structure of the catalytic core of an RNA-polymerase ribozyme. *Science*, 326, 1271–5.

Sigurdsson, S., Dirac-Svejstrup, A. B. & Svejstrup, J. Q. 2010. Evidence that transcript cleavage is essential for RNA polymerase II transcription and cell viability. *Mol Cell*, 38, 202–10.

Spitalny, P. & Thomm, M. 2008. A polymerase III-like reinitiation mechanism is operating in regulation of histone expression in Archaea. *Mol Microbiol*, 67, 958–70.

Steitz, T. A. 1998. A mechanism for all polymerases. *Nature*, 391, 231–2.

Steitz, T. A. & Steitz, J. A. 1993. A general two-metal-ion mechanism for catalytic RNA. *Proc Natl Acad Sci USA*, 90, 6498–502.

Tennyson, C. N., Klamut, H. J. & Worton, R. G. 1995. The human dystrophin gene requires 16 hours to be transcribed and is cotranscriptionally spliced. *Nat Genet*, 9, 184–90.

Todone, F., Brick, P., Werner, F., Weinzierl, R. O. & Onesti, S. 2001. Structure of an Archaeal homolog of the eukaryotic RNA polymerase II RPB4/RPB7 complex. *Mol Cell*, 8, 1137–43.

Ujvari, A. & Luse, D. S. 2006. RNA emerging from the active site of RNA polymerase II interacts with the Rpb7 subunit. *Nat Struct Mol Biol*, 13, 49–54.

Vassylyev, D. G., Vassylyeva, M. N., Perederina, A., Tahirov, T. H. & Artsimovitch, I. 2007a. Structural basis for transcription elongation by Bacterial RNA polymerase. *Nature*, 448, 157–62.

Vassylyev, D. G., Vassylyeva, M. N., Zhang, J., Palangat, M., Artsimovitch, I. & Landick, R. 2007b. Structural basis for substrate loading in Bacterial RNA polymerase. *Nature*, 448, 163–8.

Walmacq, C., Kireeva, M. L., Irvin, J., Nedialkov, Y., Lubkowska, L., Malagon, F., Strathern, J. N. & Kashlev, M. 2009. Rpb9 subunit controls transcription fidelity by delaying NTP sequestration in RNA polymerase II. *J Biol Chem*, 284, 19601–12.

Watson, J. D. & Crick, F. H. 1953. A structure for deoxyribose nucleic acid. *Nature*, 171, 737–8.

Werner, F. 2007. Structure and function of Archaeal RNA polymerases. *Mol Microbiol*, 65, 1395–404.

Werner, F. 2008. Structural evolution of multi-subunit RNA polymerases. *Trends Microbiol*, 16, 247–50.

Werner, F., Eloranta, J. J. & Weinzierl, R. O. 2000. Archaeal RNA polymerase subunits F and P are bona fide homologs of eukaryotic RPB4 and RPB12. *Nucleic Acids Res*, 28, 4299–305.

Werner, F. & Weinzierl, R. O. 2002. A recombinant RNA polymerase II-like enzyme capable of promoter-specific transcription. *Mol Cell*, 10, 635–46.

Werner, F. & Weinzierl, R. O. 2005. Direct modulation of RNA polymerase core functions by basal transcription factors. *Mol Cell Biol*, 25, 8344–55.

Whittington, J. E., Delgadillo, R. F., Attebury, T. J., Parkhurst, L. K., Daugherty, M. A. & Parkhurst, L. J. 2008. TATA-binding protein recognition and bending of a consensus promoter are protein species dependent. *Biochemistry*, 47, 7264–73.

Woychik, N. A. & Young, R. A. 1989. RNA polymerase II subunit RPB4 is essential for high- and low-temperature yeast cell growth. *Mol Cell Biol*, 9, 2854–9.

Ziegler, L. M., Khaperskyy, D. A., Ammerman, M. L. & Ponticelli, A. S. 2003. Yeast RNA polymerase II lacking the Rpb9 subunit is impaired for interaction with transcription factor IIF. *J Biol Chem*, 278, 48950–6.

CHAPTER 6

Single-Molecule Fluorescence Resonance Energy Transfer Investigations of Ribosome-Catalyzed Protein Synthesis

Daniel D. MacDougall

Jingyi Fei

Ruben L. Gonzalez, Jr.

I. INTRODUCTION

Protein synthesis, or translation, is an inherently dynamic process in which the ribosome traverses the open reading frame of a messenger RNA (mRNA) template in steps of precisely one triplet-nucleotide codon, catalyzing the selection of aminoacyl-transfer RNA (aa-tRNA) substrates and polymerization of the nascent polypeptide chain, while simultaneously coordinating the sequential binding of exogenous translation factors. The complexity of this process is mirrored by the intricate molecular architecture of the ribosome itself, highlighted in atomic detail by recent X-ray crystallographic structures that reveal an elaborate network of RNA-RNA, RNA-protein, and protein-protein interactions (Korostelev and Noller, 2007; Steitz, 2008). This high degree of intra- and inter-molecular connectivity suggests that allosteric mechanisms may regulate the activity and coordinate the timing of biochemical events catalyzed by spatially distal ribosomal functional centers. Large-scale conformational dynamics of the ribosome have similarly been implicated as a means by which to regulate the biochemical steps of protein synthesis and to power forward progression through the kinetic steps of the translation process.

Comparison of X-ray crystallographic structures of ribosomal subunits as well as the intact ribosome in the absence and presence of translation factors (reviewed in Schmeing and Ramakrishnan [2009]), together with the analysis of cryogenic electron microscopy (cryo-EM) reconstructions of the ribosome trapped at various functional states during protein synthesis (see Chapter 7), has allowed visualization of large-scale conformational rearrangements of the translational machinery. Through such comparative structural analysis, mobile ribosomal domains have been identified and specific conformational changes have been inferred. However, these static structural images lack information regarding the timescales of the inferred

conformational changes, and the kinetic and thermodynamic parameters underlying the corresponding ribosomal motions. Such dynamic information has recently been uncovered through the application of single-molecule fluorescence resonance energy transfer (smFRET) to studies of protein synthesis. This technique has proven to be particularly well-suited for monitoring and characterizing large-scale conformational dynamics of the ribosome and its tRNA and translation factor ligands, which often occur on length scales (~tens of Å) and time scales (~ms to s) that are well matched with the spatio-temporal resolution of current smFRET methodologies (see Chapter 1). Guided by the structural data, numerous donor-acceptor fluorophore labeling schemes have already been developed, each capable of monitoring specific conformational changes of the translational machinery in real time.

In this chapter, we will first briefly discuss experimental considerations pertaining to the design and implementation of smFRET investigations of highly purified *in vitro* translation systems (Section II). We then describe some of the major findings from smFRET studies of Bacterial protein synthesis, highlighting emergent themes and single-molecule-specific insights that have been gleaned (Sections III–V). A majority of the literature to date has focused on events occurring during the elongation phase of translation, and primarily during the aa-tRNA selection and translocation steps of the translation elongation cycle; accordingly, we confine the bulk of our discussion to the conformational dynamics of the translational machinery that are pertinent to aa-tRNA selection and translocation. Specifically, in Section III, we discuss pre–steady state and steady-state smFRET measurements of aa-tRNA selection, which have allowed observation and characterization of the conformational trajectory of aa-tRNA as it is selected and accommodated into the ribosomal A site, revealing a crucial intermediate that had previously evaded biochemical detection and has thus far eluded structural characterization due to its transient nature. Section IV focuses on steady-state investigations of translocation-relevant conformational equilibria, which have led to the discovery that many of the conformational rearrangements associated with translocation occur spontaneously and reversibly upon peptide bond formation, with the ribosome possessing the intrinsic capability of accessing functionally relevant conformational states through thermal fluctuations alone. The ability of translation factors and antibiotics to modulate these equilibria – through the manipulation of transition rates and stabilization/destabilization of particular conformational states – can be directly observed and correlated with their ability to promote or inhibit translocation, respectively. Finally, in Section V, we discuss recent smFRET investigations that have extended these ideas to the initiation, termination, and ribosome recycling stages of protein synthesis, providing evidence that modulation of intrinsic ribosomal dynamics and conformational equilibria represents a common regulatory mechanism used by translation factors during all stages of protein synthesis. A dynamic picture of the translating ribosome emerges in which thermal fluctuations drive spontaneous ribosome and tRNA motions that form the basis for ribosome function. Addition of translation factors to this mechanistic foundation provides a means by which to modulate, accelerate, guide the directionality, and increase the efficiency of these intrinsic processes, thereby accomplishing highly regulated and tightly controlled protein synthesis.

II. DESIGN OF smFRET EXPERIMENTS

II.1 Site-Specific Fluorescent Labeling of Translation Components

Preparation of fluorescently labeled translation components is the starting point for any smFRET investigation of ribosome conformational dynamics (Fei et al., 2010). Donor and acceptor fluorophore pairs can be conjugated to tRNAs, translation factors, the ribosome, or any combination thereof, with the choice of labeled components depending on the particular molecular interaction(s) or conformational rearrangement(s) to be monitored. The design of a mechanistically informative labeling scheme relies heavily on X-ray crystallographic and cryo-EM structural models, which are used to choose labeling positions that will allow sensitive detection of the conformational change of interest without perturbing biochemical activity. Site-specific labeling of translation components is critical for being able to interpret changes in FRET efficiency in molecular detail as corresponding to movement of a particular ribosomal domain or the formation of a particular inter-molecular interaction. Consequently, numerous labeling strategies have been developed, which have helped to increase the scope of smFRET studies of ribosome

conformational dynamics, allowing researchers to probe various structural transitions.

Donor and acceptor fluorophores have been covalently attached to tRNA species at naturally occurring, post-transcriptionally modified nucleotides within the molecule's central elbow region, or to the amino acid linked to the 3'-terminal aminoacyl acceptor stem of the tRNA (Sytnik et al., 1999; Blanchard et al., 2004b). Translation factors can be fluorescently labeled at unique cysteine residues or unnatural amino acids that have been incorporated at appropriate positions on the molecule's surface (Wang et al., 2007; Munro et al., 2009b; Sternberg et al., 2009). Fluorescent labeling of the ribosome itself has been achieved by two major approaches. In the first, purified ribosomal proteins (r-proteins) are fluorescently labeled and subsequently reconstituted in vitro with ribosomal subunits (Hickerson et al., 2005; Fei et al., 2009). In the second approach, helical extensions engineered into ribosomal RNA (rRNA) stem-loops are hybridized to a complementary fluorescently labeled oligonucleotide (Dorywalska et al., 2005).

II.2 Surface Immobilization of Ribosomal Complexes and smFRET Imaging

Once fluorescently labeled translation components have been prepared and their biochemical activities have been confirmed to be unimpaired by labeling, smFRET imaging typically requires immobilization of ribosomal complexes on the surface of a polymer-passivated microscope slide. Quartz microscope slides can be passivated with a mixture of polyethylene glycol (PEG) and biotin-PEG (Ha et al., 2002), thereby allowing specific attachment of biotinylated ribosomal complexes through a biotin-streptavidin-biotin linkage (Blanchard et al., 2004b). Most frequently, ribosomal complexes are assembled on a 5'-biotinylated mRNA molecule (or a 3'-biotinylated oligonucleotide hybridized to the mRNA 5' end), which serves as the attachment point between ribosome and surface (Figure 6.1). An alternative approach has been reported whereby the 3' end of the large 50S subunit 23S rRNA can be oxidized, biotinylated, and used as the anchor point (Stapulionis et al., 2008).

Stable attachment of fluorescently labeled ribosomal complexes to the slide surface permits acquisition of smFRET versus time trajectories from single ribosomes, with an observation time (seconds to minutes) that is often limited by photobleaching of the organic fluorophores. Total internal reflection (TIR) illumination is generally used for excitation of donor fluorophores within single ribosomal complexes; when combined with wide-field imaging, this approach allows acquisition of smFRET versus time data from hundreds of ribosomal complexes simultaneously (see Chapter 1). Importantly, the biochemical activities of ribosomes immobilized using the methods described

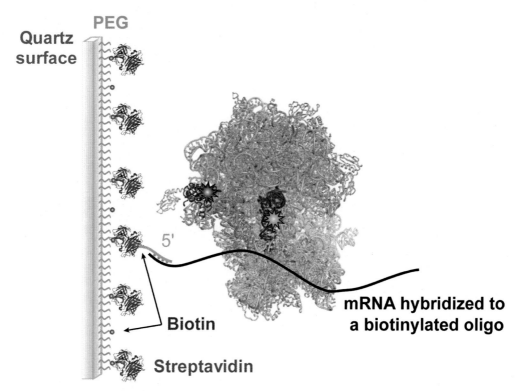

FIGURE 6.1: *Surface immobilization strategy. Quartz flow cells are first passivated with a mixture of PEG and biotin-PEG. This passivated flow cell is incubated with streptavidin prior to use. Fluorescently labeled ribosomal complexes are immobilized on the surface via a biotin-streptavidin-biotin interaction.*

in the previous paragraph remain intact; surface-tethered ribosomes have been demonstrated to be active in the individual steps of translation initiation (Marshall et al., 2008; Marshall et al., 2009), elongation (Blanchard et al., 2004b; Stapulionis et al., 2008), termination, and ribosome recycling (Sternberg et al., 2009).

III. AMINOACYL-TRNA SELECTION

III.1 Selection of aa-tRNA by the Ribosome

During each elongation cycle in protein synthesis, an aa-tRNA is delivered to the ribosome in a "ternary complex" with elongation factor Tu (EF-Tu) and GTP. Based on biochemical experiments, a kinetic model has been formulated that details the stepwise progression of aa-tRNA into the ribosomal A site during aa-tRNA selection, culminating in accommodation of aa-tRNA into the peptidyl transferase center and peptide bond formation (reviewed in Rodnina et al. [2005]). One of the first applications of smFRET to the study of protein synthesis allowed direct visualization of aa-tRNA selection by the ribosome, using fluorophore-labeled tRNAs as FRET probes to follow the conformational trajectory of the incoming aa-tRNA in real time (Blanchard et al., 2004a). This study added important mechanistic details to our understanding of how the ribosome rapidly and efficiently selects the correct

aa-tRNA, thereby ensuring faithful incorporation of the mRNA-encoded amino acid into the growing polypeptide chain, and highlighted the role of aa-tRNA dynamics in the selection process.

Selection of the cognate (correct) aa-tRNA from a pool of competitor near-cognate (one mismatch at a non-wobble position) and non-cognate (at least two mismatches) aa-tRNAs is performed rapidly and efficiently by the ribosome, with an error rate of approximately 1 out of every 1,000 to 10,000 amino acids incorporated into the polypeptide (Parker, 1989). Such a high level of discrimination between correct and incorrect aa-tRNAs, which can differ by as little as a single Watson-Crick base pair within the codon-anticodon duplex, cannot be explained based solely on differences in the free energy of codon-anticodon formation (Grosjean et al., 1978). Biochemical experiments have shed light on the mechanisms through which the ribosome can achieve such a high degree of selectivity. A kinetic proofreading strategy is exploited whereby the ribosome discriminates in favor of cognate aa-tRNAs at two independent selection steps termed "initial selection" and "proofreading," which are separated by the chemical step of GTP hydrolysis (Hopfield, 1974; Thompson and Stone, 1977). Furthermore, during both initial selection and proofreading, induced-fit mechanisms act to preferentially select for the cognate aa-tRNA (reviewed in Daviter et al. [2006]). Correct base pairing of the codon-anticodon duplex within

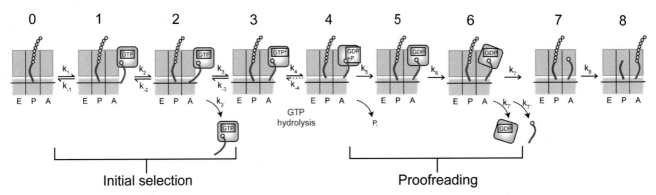

FIGURE 6.2: *Kinetic model for aa-tRNA selection. Step 0→1: Initial binding of the EF-Tu(GTP)aa-tRNA ternary complex to the ribosome via interactions between EF-Tu and the L7/L12 stalk. Step 1→2: Formation of codon-anticodon interaction at the decoding center on the 30S subunit. Step 2→3: GTPase activation. At this stage, the ternary complex can either dock into the GTPase center on the 50S subunit (k_3) or be rejected from the ribosome (k_3'). Step 3→4: GTP hydrolysis by EF-Tu. Step 4→5: Release of P_i from EF-Tu. Step 5→6: EF-Tu conformational change from its GTP-bound form to its GDP-bound form. Step 6→7: aa-tRNA is released from EF-Tu, which dissociates from the ribosome (k_7'). aa-tRNA can either accommodate into the peptidyl transferase center (k_7) or be rejected and dissociate from the ribosome (k_7''). Step 7→8: Peptidyl transfer between P- and A-site tRNAs (k_8). Figure adapted from Frank and Gonzalez (2010), copyright © 2010 Annual Reviews.*

the small 30S ribosomal subunit's decoding center induces specific conformational rearrangements of the tRNA and ribosome that accelerate its forward progression through the reaction pathway compared to non- or near-cognate aa-tRNAs.

The detailed kinetic model for aa-tRNA selection by the ribosome is depicted schematically in Figure 6.2 (for a review, see Rodnina and Wintermeyer [2001]). The ternary complex initially binds to the ribosome through protein-protein interactions between EF-Tu and the ribosomal L7/L12 stalk (k_1/k_{-1}), followed by formation of the codon-anticodon interaction within the 30S ribosomal subunit's decoding center (k_2/k_{-2}). Subsequent GTPase activation of EF-Tu (k_3), which is rate-limiting for GTP hydrolysis (k_4), is selectively accelerated in response to recognition of a cognate codon-anticodon interaction through an induced-fit mechanism. Non- and near-cognate aa-tRNAs, in contrast, have a lower probability of advancing past this initial selection step as a result of a lower rate of GTPase activation (slower k_3) as well as an increased rate of ternary complex dissociation (faster k_3'), owing to weaker interactions with the ribosome. These effects lead to near-complete discrimination against non-cognate aa-tRNAs during initial selection. Following GTPase activation and GTP hydrolysis by EF-Tu, inorganic phosphate (P_i) is released (k_5), EF-Tu undergoes a conformational change to its GDP-bound form (k_6), and ultimately dissociates from the ribosome (k_7'). The GDP-bound form of EF-Tu has a low affinity for aa-tRNA; consequently, the 3'-terminus of the aa-tRNA is released, and the aa-tRNA may either be accommodated into the peptidyl-transferase center of the 50S ribosomal subunit (k_7) to form a peptide bond (k_8) or be rejected from the ribosome (k_7''). An induced-fit mechanism operates during this proofreading step of aa-tRNA selection by accelerating the rate of accommodation in response to a

cognate, but not a near-cognate, codon-anticodon interaction. In addition, the rate of dissociation (k_7'') is faster for more weakly bound near-cognate aa-tRNAs, further decreasing the probability that they will be accommodated into the peptidyl transferase center and allowed to participate in peptide bond formation.

III.2 Real-Time smFRET Observation of aa-tRNA Selection

smFRET studies of aa-tRNA selection were designed and interpreted within the biochemical framework described in the previous section. The selection and incorporation of aa-tRNA into single ribosomes was followed by monitoring the time evolution of smFRET upon delivery of acceptor-labeled EF-Tu(GTP)Phe-tRNA^{Phe} ternary complex (labeled with a Cy5 acceptor fluorophore at the acp^{3}U47 residue within tRNA^{Phe}) to surface-immobilized ribosomal initiation complexes bearing donor-labeled fMet-tRNA^{fMet} (labeled with a Cy3 donor fluorophore at the s^{4}U8 position within tRNA^{fMet}) in the P site (Figure 6.3a) (Blanchard et al., 2004a). FRET generated between donor and acceptor fluorophores on the P site-bound fMet-(Cy3)tRNA^{fMet} and incoming EF-Tu(GTP)Phe-(Cy5)tRNA^{Phe} showed rapid progression from low to high FRET upon ternary complex binding to the ribosome, with the final FRET value of ~0.75 corresponding to full accommodation of Phe-(Cy5)tRNA^{Phe} into the peptidyl-transferase center (Figures 6.3b and 6.3c). Dynamic fluctuations in the smFRET signal following accommodation were observed and attributed to tRNA dynamics after peptide bond formation (see Section IV).

The real-time evolution of smFRET observed during aa-tRNA selection contains a wealth of information concerning the conformational states through which the

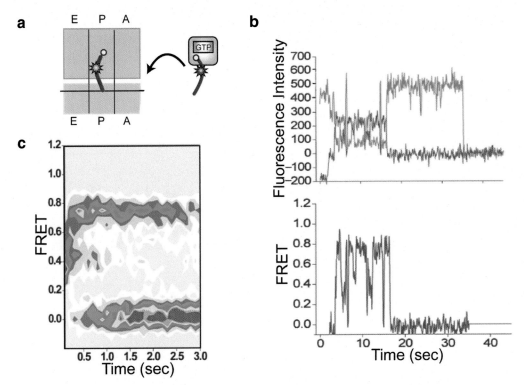

FIGURE 6.3: *Selection of cognate aa-tRNA studied using tRNA-tRNA smFRET signal. (a) Cartoon representation of tRNA-tRNA smFRET signal. Cognate EF-Tu(GTP)Phe-(Cy5)tRNA^Phe is delivered to a surface-immobilized initiation complex carrying fMet-(Cy3)tRNA^fMet. (b) Sample Cy3 and Cy5 emission intensity versus time trajectories are shown in green and red, respectively (top). The corresponding smFRET versus time trajectory, FRET = $I_{Cy5}/(I_{Cy3} + I_{Cy5})$, is shown in blue (bottom). (c) Contour plot of the time evolution of population FRET, generated by superimposing the individual smFRET versus time traces and post-synchronizing to the first observation of FRET ≥ 0.25. Contours are plotted from tan (lowest population) to red (highest population). Molecules in the ~0-FRET state arise from photobleaching, blinking of Cy5, and dissociation of tRNA from the ribosome. Figure adapted from Blanchard et al. (2004a), copyright © 2004 Nature Publishing Group, with permission from Macmillan Publishers Ltd.*

aa-tRNA transits as it is selected by the ribosome; structural parameters (i.e., the relative distance between P- and A-site tRNAs) as well as kinetic parameters (i.e., the transition rates between different states) can be extracted from the smFRET versus time trajectories. Assignment of intermediate FRET states to particular conformational states was facilitated through the use of small-molecule inhibitors of protein synthesis that stall the aa-tRNA selection process at particular steps, and by programming the 30S subunit's A site with a near-cognate codon. In the presence of the antibiotic tetracycline or a near-cognate A-site codon, transient sampling of a low (~0.35) FRET state was observed, which was identified as the codon recognition state in which the codon-anticodon interaction is formed in the 30S subunit's decoding center (Figures 6.4a and 6.4b). The non-hydrolyzable GTP analog GDPNP and the antibiotic kirromycin were used to stall the ternary complex immediately before and after GTP hydrolysis, respectively, generating a mid-(~0.5) FRET state where EF-Tu is docked at the 50S subunit's GTPase-associated center (referred to hereafter simply as the GTPase

center) (Figures 6.4c and 6.4d). The transition from low to mid FRET, therefore, represents GTPase activation of EF-Tu. At this stage, aa-tRNA adopts the A/T configuration, first characterized structurally by chemical probing (Moazed and Noller, 1989a) and later by cryo-EM (see Chapter 7). Finally, the high- (~0.75) FRET state, achieved during uninhibited delivery of cognate EF-Tu(GTP)Phe-(Cy5)tRNA^Phe, corresponds to successful accommodation of Phe-(Cy5)tRNA^Phe into the peptidyl transferase center and peptide bond formation (Figure 6.3).

The low-FRET codon recognition state was found to represent a critical branchpoint in the mechanism used to preferentially select cognate over near-cognate aa-tRNAs. smFRET allowed direct observation of this intermediate – which had not been resolved in bulk biochemical experiments or in structural studies due to its transient nature – and permitted real-time observation of the frequencies and rates with which it is traversed by cognate versus near-cognate aa-tRNAs. When the 30S subunit's A site was programmed with the cognate UUU codon, the majority of EF-Tu(GTP)Phe-(Cy5)tRNA^Phe progressed rapidly

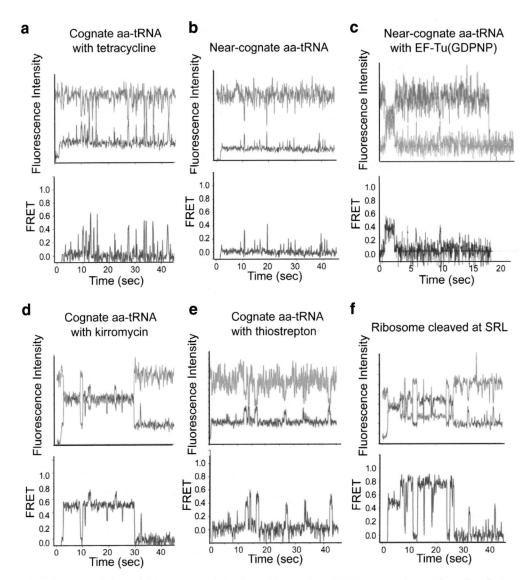

FIGURE 6.4: *Single-molecule fluorescence intensities and smFRET versus time trajectories for aa-tRNA selection under various conditions. Phe-(Cy5)tRNA^{Phe} (in ternary complex with EF-Tu and GTP or GDPNP) was stopped-flow delivered to ribosomal initiation complexes carrying fMet-(Cy3)tRNA^{fMet} at the P site. Cy3 and Cy5 emission intensity versus time trajectories are shown in green and red, respectively (top). The corresponding smFRET versus time trajectories, FRET = I_{Cy5}/(I_{Cy3} + I_{Cy5}), are shown in blue (bottom). (a) Stopped-flow delivery of cognate ternary complex in the presence of 100 μM tetracycline. (b) Stopped-flow delivery of near-cognate ternary complex. (c) Stopped-flow delivery of near-cognate aa-tRNA as a ternary complex with EF-Tu(GDPNP). (d) Stopped-flow delivery of cognate ternary complex in the presence of 200 μM kirromycin. (e) Stopped-flow delivery of cognate ternary complex in the presence of 50 μM thiostrepton. (f) Stopped-flow delivery of cognate ternary complex to ribosomal complexes cleaved at the sarcin-ricin loop (SRL). Figures (a), (b), (d) and (f) are reproduced from Blanchard et al. (2004a), copyright © 2004 Nature Publishing Group, with permission from Macmillan Publishers Ltd., Figure (c) is reproduced from Lee et al. (2007), copyright © 2007 National Academy of Sciences, U.S.A., and Figure (e) is reproduced from Gonzalez et al. (2007), copyright © 2007 RNA Society.*

through the codon recognition state (low FRET) en route to GTPase activation (mid FRET) and accommodation (high FRET) (Figure 6.3). In the presence of a near-cognate CUU codon, however, the majority of incoming ternary complexes are unable to progress past the codon recognition state, instead only transiently sampling this state before dissociating from the ribosome (Figure 6.4b). For near-cognate ternary complexes, sampling of the low-FRET codon recognition state was followed by a transition to higher FRET values only 11% of the time, versus 65% of the time for cognate ternary complexes. Furthermore, analysis of rates exiting the codon recognition state demonstrated that near-cognate ternary complexes have both a higher rate of dissociation compared with cognate ternary complexes ($k_{low \to 0}$; 16.2 sec^{-1} versus 6.4 sec^{-1}, respectively) and also a slower rate of transit to the GTPase-activated state ($k_{low \to mid}$; 2 sec^{-1} versus 11.8 sec^{-1}, respectively). These observations shed light on the induced-fit mechanism that acts to stabilize the binding of a cognate ternary complex and to accelerate its GTPase activation. Formation of the cognate codon-anticodon interaction specifically accelerates transit from the low- to the mid-FRET state. Therefore, the allosteric mechanism linking cognate codon-anticodon recognition in the decoding center to enhanced rates of GTPase activation by EF-Tu involves a movement of aa-tRNA toward the P site, allowing productive interactions to be made between the ternary complex and the 50S subunit's GTPase center that stimulate EF-Tu's GTP hydrolysis activity.

III.3 Thermally Driven Fluctuations of the Ribosome-tRNA Complex Permit Sampling of Conformational States Along the aa-tRNA Selection Pathway

The kinetic barrier separating the low-FRET codon recognition state from the mid-FRET GTPase-activated state is overcome through large thermal fluctuations of the ternary complex-bound ribosomal complex. This feature of initial selection was highlighted by higher-time-resolution smFRET measurements of the delivery of EF-Tu(GDPNP)Phe-(Cy5)tRNA^{Phe} to ribosomal initiation complexes carrying fMet-(Cy3)tRNA^{fMet} in the P site (Lee et al., 2007). In the presence of GDPNP, both cognate and near-cognate aa-tRNAs were found to fluctuate reversibly between the low- and mid-FRET states, reporting on attempts by the ternary complex to form stabilizing contacts with the GTPase center of the 50S subunit (Figure 6.4c). Stabilization of a long-lived mid-FRET state was interpreted to correspond to successful docking of the ternary complex at the GTPase center, where all stabilizing contacts between the ribosome and ternary complex required for GTPase activation have been formed. Specifically, interactions of the ternary complex with ribosomal proteins L10, L7/L12, L11 and its associated 23S rRNA, and the sarcin-ricin loop of 23S rRNA presumably play important roles in this stabilization. Short-lived excursions of the ternary complex to mid FRET, with lifetimes less than 100 ms were also observed (Figure 6.4c). These were interpreted as unsuccessful attempts to dock at the GTPase center, in which only a subset of the requisite stabilizing interactions are formed; this short-lived mid-FRET state was termed the pseudo-GTPase-activated state. Both cognate and near-cognate aa-tRNAs fluctuate rapidly into and out of the pseudo-GTPase-activated state before successful stable binding to the GTPase center. However, detailed kinetic analysis revealed that cognate aa-tRNAs fluctuate to mid FRET more often than near-cognate

aa-tRNAs (27 attempts s^{-1} versus 8 attempts s^{-1}, respectively). Additionally, fluctuations to mid FRET were more likely to result in successful docking for cognate as compared with near-cognate aa-tRNA; on average, cognate aa-tRNAs underwent two attempts per every successful docking event, compared to four for near-cognate aa-tRNA. These findings imply that the induced-fit rearrangements of the ribosomal complex triggered by cognate codon-anticodon interactions position the cognate ternary complex in a favorable orientation, such that fluctuations to the GTPase-activated state can occur more readily and with a higher probability of success. These results, in addition to highlighting the role of thermal fluctuations in aa-tRNA selection, emphasized the dynamic and inherently reversible nature of kinetic steps in the early stages of the aa-tRNA selection pathway.

Dynamic fluctuations of the tRNA-tRNA smFRET signal, corresponding to transient sampling of conformational states in the aa-tRNA selection pathway, have also been observed for ternary complexes stalled at mid FRET before and after GTP hydrolysis using GDPNP and the antibiotic kirromycin, respectively (Blanchard et al., 2004a). Kirromycin binds directly to EF-Tu and permits GTPase activation and GTP hydrolysis but blocks the subsequent conformational change of EF-Tu to its GDP-bound form. In the presence of either GDPNP or kirromycin, residency at the mid-FRET state is interrupted by brief excursions to both the low- and high-FRET states (Figure 6.4d). This behavior hints at the ability of the aa-tRNA to sample relevant conformational states of the reaction pathway even in the absence of GTP hydrolysis or EF-Tu's conformational change. It is tempting to speculate, then, that EF-Tu and GTP hydrolysis may not be strictly required for incorporation of aa-tRNA into the ribosomal A site. Perhaps the requisite aa-tRNA-ribosome interactions can be made, and the relevant conformational states sampled, even without EF-Tu. In support of this notion, factor-free translation from a poly(U) template can occur in vitro, albeit at a much slower rate than in the presence of translation factors and GTP (Pestka, 1969; Gavrilova and Spirin, 1971). In this view, aa-tRNA selection by the primordial ribosome may have predated the evolution of translation factors, and EF-Tu may have evolved later to increase the speed, directionality, and fidelity of this process.

III.4 Regulation of aa-tRNA Selection by Antibiotics, Ribosome Structural Elements, and Amino Acid–tRNA Pairing

As described in the previous sections, the tRNA-tRNA smFRET signal allows direct observation of aa-tRNA's stepwise movement through the various conformational states that comprise the aa-tRNA selection process. As such, it provides a powerful experimental framework for investigating the effects of ribosome-targeting antibiotics that interfere with aa-tRNA selection. This approach has proved useful in identifying the particular stage at which antibiotics act, as well as the kinetic mechanism by which they interfere with protein synthesis. As already discussed, the antibiotic tetracycline, whose primary binding site is located near the 30S subunit's A site (Brodersen et al., 2000; Pioletti et al., 2001), was shown to block progression of the ternary complex from the low-FRET codon recognition state to the mid-FRET GTPase-activated state. Upon sampling of a cognate or near-cognate codon-anticodon interaction in the presence of tetracycline, ternary complexes dissociate rapidly from the ribosome (Figure 6.4a). In contrast, thiostrepton, a thiazole antibiotic that binds to ribosomal protein L11 and the associated rRNA helices H43 and H44 of the 50S subunit's GTPase center (Harms et al., 2008), exerts its inhibitory action at the mid-FRET state (Figure 6.4e) (Gonzalez et al., 2007). This drug does not affect progression of the EF-Tu(GTP)Phe-(Cy5)tRNA^{Phe} ternary complex through the codon recognition state, but instead prevents stable binding of the ternary complex at the GTPase center, with an observed mid-FRET lifetime of ~26 ms. Therefore, thiostrepton likely acts by blocking stabilizing contacts between the ternary complex and the L11 protein and/or L11-associated rRNA. Consequently, the ternary complex is unable to progress past the mid-FRET state, instead being rejected from the GTPase center and retracing its steps back through the codon recognition state before dissociating from the ribosome.

Comparison of these results with smFRET data collected using ribosomes with a cleaved sarcin-ricin loop has aided in the assignment of specific functional roles to distinct structural components of the ribosome's GTPase center. The sarcin-ricin loop represents an important component of the GTPase center, which has been shown to interact with EF-Tu's guanine nucleotide-binding domain (Schmeing et al., 2009; Villa et al., 2009). Like binding of thiostrepton, cleavage of the sarcin-ricin loop blocks progression of the EF-Tu(GTP)Phe-(Cy5)tRNA^{Phe} ternary complex past the mid-FRET state, but through an entirely different mechanism (Blanchard et al., 2004a). In the case of a cleaved sarcin-ricin loop, ternary complexes transit to the GTPase center but become trapped there (Figure 6.4f), with a lifetime of ~8–12 s, in contrast to the transient (~26 ms) excursions to the GTPase center observed in the presence of thiostrepton. A model thus emerges in which the L11 region mediates stable binding of the ternary complex to the GTPase center, whereas the sarcin-ricin loop stimulates EF-Tu's GTP hydrolysis activity (Gonzalez et al., 2007).

Another, recent application of the tRNA-tRNA smFRET signal has been to explore the role of amino acid identity and amino acid–tRNA pairing in the selection process (Effraim et al., 2009). Using a ribozyme capable of misacylating tRNAs with non-native amino acids, misacylated Ala-tRNA^{Phe} and Lys-tRNA^{Phe} were prepared and shown to be efficiently selected by the ribosome, capable of participating in peptide bond formation to nearly the same extent as correctly charged Phe-tRNA^{Phe} in a dipeptide formation assay. However, competition experiments in which ribosomal initiation complexes were presented with an equimolar mixture of correctly charged and misacylated tRNA^{Phe} demonstrated that the ribosome is capable of discriminating between the two species, leading to a slight preferential selection of tRNA^{Phe} charged with its native amino acid (Phe-tRNA^{Phe} was selected 3.7- and 2.2-fold more efficiently than Ala-tRNA^{Phe} and Lys-tRNA^{Phe}, respectively). This indicated that during aa-tRNA selection, the ribosome is sensitive to not only the codon-anticodon pairing, but also to the amino acid's identity and/or the specific amino acid–tRNA pairing.

Experiments utilizing the tRNA-tRNA smFRET signal revealed the molecular basis of this subtle discrimination. Sub-population analysis led to the classification of smFRET trajectories into two categories: trajectories exhibiting productive binding events that result in accommodation and peptide bond formation, and trajectories in which multiple A site sampling events were observed, none of which lead to full accommodation of aa-tRNA during the observation period. An increase in the latter sub-population was observed for both Ala-(Cy5)tRNA^{Phe} and Lys-(Cy5)tRNA^{Phe} compared with Phe-(Cy5)tRNA^{Phe} (this sub-population accounts for 52% and 44% of the trajectories, respectively, compared with 16% for Phe-(Cy5)tRNA^{Phe}), which closely mirrored the 3.7- and 2.2-fold enhanced selection efficiency of Phe-tRNA^{Phe} compared to the misacylated species in the biochemical competition experiments. Therefore, the increased frequency of unproductive A-site sampling events points to the ribosome's capacity to sense amino acid identity and/or amino acid–tRNA pairing at an early stage in the aa-tRNA selection process, thereby discriminating against certain incorrectly charged tRNAs. This application of the tRNA-tRNA smFRET signal provides an important starting point for mechanistic studies of the translational machinery's response to tRNAs charged with unnatural amino acids that are poorly incorporated into proteins. A more detailed mechanistic understanding of how the ribosome discriminates against unnatural amino acids could facilitate biomedically relevant protein-engineering applications by aiding in the design of unnatural amino acid-tRNA pairings and, ultimately, mutant ribosomes that yield increased incorporation efficiencies.

IV. MRNA-TRNA TRANSLOCATION

IV.1 Transit of mRNA and tRNAs Through the Ribosome

After accommodation of aa-tRNA into the A site, peptide bond formation occurs rapidly, thereby transferring

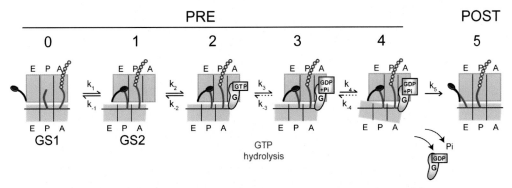

FIGURE 6.5: Kinetic model for translocation. Step 0→1: Following peptide bond formation, the PRE complex exists in a dynamic equilibrium between Global State 1 (GS1) and Global State 2 (GS2). Spontaneous conformational changes characterizing the GS1→GS2 transition include movement of the acceptor stems of A- and P-site tRNAs to the P and E sites on the 50S subunit, closing of the L1 stalk (dark blue), and a rotational movement of the 30S subunit relative to the 50S subunit. Step 1→2: Binding of EF-G(GTP) stabilizes GS2. Step 2→3: GTP hydrolysis by EF-G. Step 3→4: Further conformational rearrangements of the ribosome and EF-G that occur subsequent to GTP hydrolysis and facilitate translocation (Savelsbergh et al., 2003). Step 4→5: Translocation of mRNA and tRNAs on the 30S subunit and release of EF-G(GDP) and P_i. Adapted from Frank and Gonzalez (2010), copyright © 2010 Annual Reviews.

the nascent polypeptide to the A-site tRNA and deacylating the P-site tRNA. The resulting complex, referred to as the pretranslocation (PRE) complex, is the substrate for elongation factor G (EF-G)-catalyzed translocation of the mRNA-tRNA complex through the ribosome by precisely one codon. This translocation event moves A-site and P-site tRNAs into the P and E sites, respectively, and places the next mRNA codon into the decoding center so that it may be recognized by a new ternary complex in the next elongation cycle. The resulting ribosomal complex, bearing a peptidyl-tRNA in the P site and an empty A site, is referred to as a post-translocation (POST) complex (see Figure 6.5, which summarizes the kinetic steps of the translocation process). Conformational dynamics within the PRE and POST complexes has been the subject of intensive investigation by smFRET, and the results from these studies have enhanced our understanding of the mechanism and regulation of translocation. In particular, detailed smFRET investigations of conformational rearrangements of the PRE complex have provided direct evidence that translation factors and antibiotics are able to accelerate or impede translocation through specific modulation of the ribosome's dynamic conformational equilibria.

IV.2 Conformational Rearrangements of the Pretranslocation Complex Required for Translocation

Large-scale conformational rearrangements of PRE complexes were initially identified through biochemical, ensemble FRET, and cryo-EM structural studies. Chemical probing experiments led to the discovery – subsequently corroborated by ensemble FRET measurements – that upon peptide bond formation, tRNAs spontaneously transition into intermediate "hybrid" configurations on the

ribosome, in which the 3′-terminal acceptor ends of the A- and P-site tRNAs occupy the large subunit P and E sites, respectively, while their anticodon stem loops remain bound at the small subunit A and P sites (termed A/P and P/E hybrid states, respectively) (Moazed and Noller, 1989b; Odom et al., 1990). Subsequent movement of the tRNA anticodon stems with respect to the 30S subunit, coupled with movement of the associated mRNA, is catalyzed by EF-G(GTP).

Cryo-EM reconstructions of PRE complex analogs containing vacant A sites (PRE^{-A} complexes) and stabilized through the binding of EF-G(GDPNP) allowed visualization of the P-site tRNA bound in the P/E hybrid configuration and led to the discovery of additional large-scale conformational rearrangements of the PRE complex possibly associated with hybrid state formation (Frank and Agrawal, 2000; Valle et al., 2003). Comparison of cryo-EM reconstructions of PRE^{-A} complexes in the presence and absence of EF-G(GDPNP) revealed three major conformational changes, highlighted in Figure 6.6. These were: (1) the aforementioned movement of deacylated P-site tRNA from the classical P/P to the hybrid P/E binding configuration; (2) movement of the universally conserved L1 stalk domain of the 50S E site ~20 Å toward the inter-subunit space, thereby establishing an intermolecular interaction with the elbow of the P/E tRNA; and (3) a counter-clockwise ratchet-like rotation of the 30S subunit with respect to the 50S subunit (when viewed from the solvent side of the 30S subunit). The global conformational states of the PRE^{-A} complex observed by cryo-EM in the absence and presence of EF-G(GDPNP) will be referred to here as Global State 1 (GS1) and Global State 2 (GS2), respectively (Fei et al., 2008); we note that the analogous terms Macrostate I (MSI) and Macrostate II

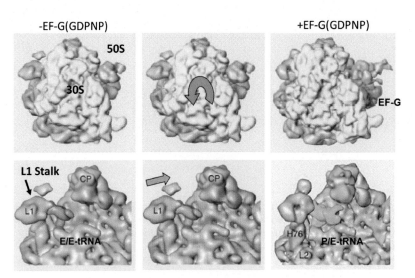

FIGURE 6.6: *Conformational rearrangements within the PRE complex inferred from cryo-EM reconstructions. Images of the PRE complex analog (PRE^{-A}) in the absence (left panel) and in the presence (right panel) of EF-G(GDPNP) reveal conformational rearrangements (middle panel), which include transition of the P-site tRNA from the classical to the hybrid binding configuration, counter-clockwise rotation of the 30S subunit relative to the 50S subunit (middle panel, top), and closing of the L1 stalk (middle panel, bottom). Adapted from Valle et al. (2003), copyright © 2003 Cell Press, with permission from Elsevier.*

(MSII) are also in frequent use throughout the literature (Frank et al., 2007). The conformational changes characterizing the GS1→GS2 transition likely play a major role in facilitating the translocation reaction. Indeed, biochemical evidence lends support to the notion that GS2 represents an authentic on-pathway translocation intermediate (Dorner et al., 2006; Horan and Noller, 2007). It had been suggested early on that large-scale conformational rearrangements of PRE complexes – in particular a relative movement of the subunits, inferred from the ribosome's universally conserved two-subunit architecture (Bretscher, 1968; Spirin, 1968) – underlie the translocation of mRNA and tRNAs through the ribosome, and the cryo-EM data provided important validation of this idea.

IV.3 Spontaneous and Reversible Conformational Fluctuations of the Pretranslocation Complex are Thermally Driven

Numerous fluorophore-labeling strategies have been designed to investigate the conformational changes of PRE and PRE^{-A} complexes by smFRET, a subset of which will be discussed here (Figure 6.7). smFRET between elbow-labeled A- and P-site tRNAs was shown early on to report on the occupancy of the classical (∼0.74 FRET) or hybrid (∼0.45 FRET) states (Blanchard et al., 2004b). L1 stalk movement from an open to a closed conformation has been tracked through smFRET between fluorophores attached to ribosomal proteins L1 and L9 (Fei et al., 2009) (an L1-L33 smFRET signal has also been used for this

purpose in an independent study [Cornish et al., 2009]). Based on cryo-EM reconstructions of the L1 stalk in the open and closed states, an L1-L9 smFRET state centered at ∼0.56 FRET was assigned to the open L1 stalk conformation, while a second state centered at ∼0.34 FRET was assigned to the closed conformation. In the closed conformation, the L1 stalk can form inter-molecular contacts with the elbow region of P/E hybrid tRNA; smFRET signals between fluorophore-labeled L1 and P-site tRNA were developed to report on the formation (high FRET, ∼0.84) and disruption (low FRET, ∼0.21) of these contacts (Fei et al., 2008; Munro et al., 2009a). Finally, inter-subunit rotation has been monitored through smFRET between dye-labeled ribosomal proteins reconstituted with the small and large subunits. Results obtained with an S6(Cy5)-L9(Cy3) construct will be described below, in which smFRET states centered at ∼0.56 and ∼0.4 FRET were assigned to the non-rotated and rotated conformations, respectively (Ermolenko et al., 2007a; Cornish et al., 2008).

PRE and PRE^{-A} complexes are prepared via peptidyl transfer from the P-site tRNA to either aa-tRNA or the antibiotic puromycin at the A site, respectively (Fei et al., 2008; Munro et al., 2009a)). Puromycin mimics the 3′-terminal acceptor stem of aa-tRNA, participating in peptide bond formation to deacylate the P-site tRNA before dissociating from the ribosome (Traut and Monro, 1964). Steady-state smFRET measurements of PRE and PRE^{-A} complexes using the previously described donor-acceptor labeling schemes yield the striking observation of

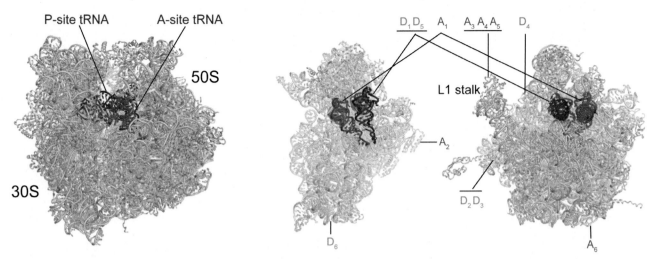

FIGURE 6.7: *Positions of the translational machinery labeled with donor (D)/acceptor (A) fluorophore pairs in smFRET studies of translocation. The 70S ribosome (PDB ID: 2J00 and 2J01) carrying A- and P-site tRNAs (in purple and red, respectively) (left panel) is split into 30S and 50S subunits (in tan and lavender, respectively), which are viewed from the inter-subunit space (right panel). D_1: s^4U8 of P-site tRNA^{fMet}; A_1: acp^{3}U47 of A-site tRNA^{Phe}. D_2: position 11 within N11C single-cysteine mutant of r-protein L9; A_2: position 41 within D41C single-cysteine mutant of r-protein S6. D_3: position 18 within Q18C single-cysteine mutant of r-protein L9; A_3: position 202 within T202C single-cysteine mutant of r-protein L1. D_4: position 29 within T29C single-cysteine mutant of r-protein L33; A_4: position 88 within A88C single-cysteine mutant of r-protein L1. D_5: acp^{3}U47 of P-site tRNA^{Phe}; A_5: position 202 within T202C single-cysteine mutant of r-protein L1, or position 55 within S55C single-cysteine mutant of L1. D_6: helix 44 of 16S rRNA (nucleotides 1450-1453); A_6: Helix 101 of 23S rRNA (nucleotides 2853-2864). Figure adapted from Frank and Gonzalez (2010).*

spontaneous and stochastic conformational fluctuations, corresponding to dynamic and reversible exchange between classical and hybrid configurations of the tRNAs, open and closed conformations of the L1 stalk, formation and disruption of L1-tRNA interactions, and rotated and non-rotated inter-subunit orientations (Figure 6.8). These large-scale conformational rearrangements, which require extensive remodeling of RNA-RNA, protein-protein, and RNA-protein interactions, are observed to occur spontaneously, driven solely by thermal energy.

Each smFRET signal is consistent with a specific conformational change associated with the GS1→GS2 transition characterized by cryo-EM; taken together, the smFRET signals, therefore, imply spontaneous fluctuations of the entire PRE complex between the GS1 and GS2 conformational states. These results suggest that transition to GS2 – and thus forward progression along the translocation reaction coordinate – can occur in the absence of EF-G and GTP hydrolysis. Indeed, full rounds of spontaneous translocation have been observed *in vitro* in a factor-free environment, in which the ribosome moves slowly but directionally along the mRNA template to generate polypeptides of defined length (Gavrilova et al., 1976). It seems, therefore, that many, if not all, of the conformational rearrangements required for translocation can be accessed with the input of thermal energy alone. Fluctuations of the PRE complex observed by smFRET represent dynamic events likely important for promoting mRNA and tRNA movement during translocation; these

fluctuations may thus increase the probability that spontaneous translocation will occur.

Conformational fluctuations within the PRE complex appear to be triggered by deacylation of the P-site peptidyl-tRNA, an "unlocking" event that prepares the ribosome for movement of the mRNA-tRNA complex by one codon (Valle et al., 2003). Following translocation, the ribosome is converted back to a "locked" state in which large-scale dynamics appear to be largely suppressed (with the exception of L1 stalk dynamics, which are important for E-site tRNA release and will be discussed in depth in Section IV.6). This suppression of conformational dynamics is evident from smFRET interrogation of POST complexes containing a peptidyl-tRNA at the P site. Structural features characteristic of GS1 appear to predominate in POST complexes, with a large majority of ribosomes observed to be fixed in a non-rotated state with a strong preference for the classical tRNA configuration (Ermolenko et al., 2007a; Cornish et al., 2008).

Multiple cycles of ribosome locking and unlocking during translation elongation have been observed using an inter-subunit smFRET signal consisting of fluorescently labeled oligonucleotides hybridized to helical extensions engineered into h44 of 16S rRNA within the 30S subunit and H101 of 23S rRNA within the 50S subunit (Figure 6.7) (Marshall et al., 2008; Aitken and Puglisi, 2010). Using the H101(Cy5)-h44(Cy3) labeling scheme, peptide bond formation and ribosome unlocking are signaled by a high→low FRET transition, whereas

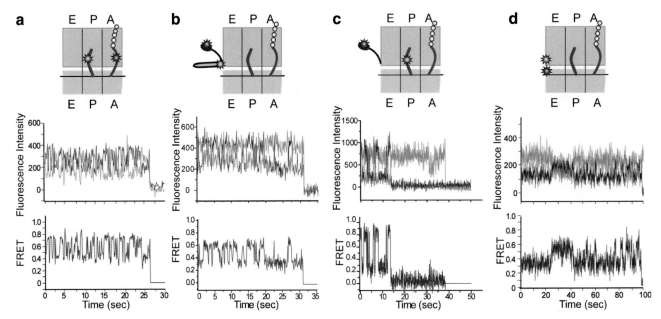

FIGURE 6.8: *Conformational dynamics within PRE complexes studied using different smFRET probes. Cartoon representation (top row) of PRE complexes labeled at different positions (see Figure 6.7) for studying (a) tRNA dynamics using D_1/A_1; (b) L1 stalk dynamics using D_3/A_3; (c) formation and disruption of intermolecular contacts between the L1 stalk and P-site tRNA using D_5/A_5; and (d) inter-subunit rotation using D_2/A_2. Cy3 and Cy5 emission intensities are shown in green and red, respectively (middle). The corresponding smFRET traces, $FRET = I_{Cy5}/(I_{Cy3} + I_{Cy5})$, are shown in blue (bottom). (a) adapted from Blanchard et al. (2004b), copyright 2004 National Academy of Sciences, U.S.A.; (b) adapted from Fei et al. (2009), copyright 2009 National Academy of Sciences, U.S.A.; (c) adapted from Fei et al. (2008), copyright 2008 Cell Press, with permission from Elsevier; (d) adapted from Cornish et al. (2008), copyright 2008 Cell Press, with permission from Elsevier.*

subsequent translocation and re-locking of the ribosome result in a low→high FRET transition (Figure 6.9). Accordingly, the lifetime of the high-FRET locked state was shown to decrease as a function of increasing ternary complex concentration, whereas the lifetime of the low-FRET unlocked state decreased as a function of increasing EF-G(GTP) concentration. The locked conformation of POST complexes may help prevent ribosome slipping and

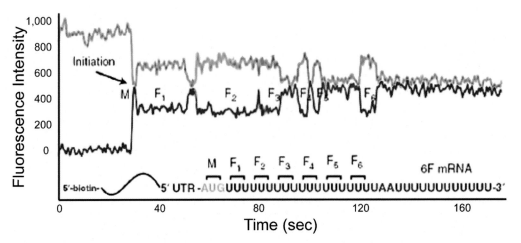

FIGURE 6.9: *Direct observation of locking and unlocking during multiple rounds of the elongation cycle using the H101(Cy5)-h44(Cy3) smFRET signal. Locking and unlocking events during each round of elongation were observed using the D_6/A_6 smFRET probes shown in Figure 6.7. A ribosomal 30S initiation complex was immobilized via an mRNA coding for six phenylalanines (6F). The arrival of FRET corresponds to 50S subunit joining during initiation and is followed by multiple cycles of high-low-high FRET, each reporting on ribosome unlocking and locking during one round of elongation. Reproduced from Aitken and Puglisi (2010), copyright © 2010 Nature Publishing Group, with permission from Macmillan Publishers Ltd.*

thus act to preserve the correct reading frame following translocation. Additionally, the locked conformation may facilitate selection of the next aa-tRNA and its precise positioning in the peptidyl transferase center. Following peptide bond formation, the POST complex is unlocked and converted to a dynamic PRE complex in which thermal fluctuations power conformational rearrangements that are required for translocation.

IV.4 Conformational Dynamics within Pretranslocation Complexes are Modulated by Magnesium Ion Concentration, tRNA Identity and Acylation State, and Antibiotics

Single-molecule FRET versus time trajectories that report on conformational fluctuations of the PRE complex contain a wealth of mechanistic information, allowing determination of the number of states sampled, their equilibrium population distributions (K_{eq}), and the transition rates between states (i.e., $k_{state(i) \rightarrow state(j)}$). This information, which would be masked in bulk measurements due to the stochastic and asynchronous nature of the conformational fluctuations, is uniquely accessible to single-molecule techniques. Both the equilibrium distribution of states and the transition rates between states for the various conformational equilibria characterizing GS1$\rightleftarrows$GS2 transitions were found to be highly sensitive to experimental conditions, including the concentration of Mg^{2+} ions, the absence, presence, identity, and acylation state of the tRNA ligands, the absence or presence of translation factors, and the absence or presence of ribosome-targeting antibiotics. Assuming that conformational changes associated with the GS1$\rightarrow$GS2 transition are a fundamental part of the translocation process, these observations suggest that specific control over ribosome dynamics within the PRE complex, through the acceleration/deceleration of conformational change and the stabilization/destabilization of specific conformational states, could provide an effective means for regulating the rate of translocation. In this view, ribosomal ligands may function by promoting or inhibiting conformational dynamics that are intrinsic to the ribosomal complex. Indeed, as described in the following paragraphs, the effect of changes in experimental conditions on the rate of translocation is often correlated with the effect of those changes on PRE complex dynamics.

The dynamic exchange of tRNAs between classical and hybrid configurations necessarily requires the disruption and formation of multiple tRNA-rRNA and tRNA-ribosomal protein interactions; this suggests that the classical$\rightleftarrows$hybrid tRNA equilibrium may be modulated by the concentration of Mg^{2+} ions in solution because Mg^{2+} is known to play a crucial role in the folding and stabilization of RNA structures (Draper, 2004). Examination of the classical$\rightleftarrows$hybrid tRNA equilibrium over a range of Mg^{2+} concentrations (3.5 to 15 mM) within a PRE complex

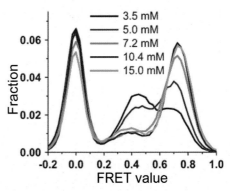

FIGURE 6.10: *Mg^{2+}-dependence of tRNA classical$\rightleftarrows$hybrid dynamic equilibrium tRNA dynamics were observed using the D_1/A_1 smFRET probes shown in Figure 6.7. The distribution of FRET values is plotted as a function of Mg^{2+} concentration. FRET states centered at ~0.4 and ~0.75 FRET correspond to hybrid and classical tRNA configurations, respectively. The FRET state at ~0 FRET arises from Cy5 blinking and photobleaching. Reproduced from Kim et al. (2007), copyright © 2007 Cell Press, with permission from Elsevier.*

carrying N-acetyl-Phe(Cy5)tRNA[Phe] at the A site and deacylated (Cy3)tRNA[fMet] at the P site revealed a Mg^{2+}-dependent shift in the equilibrium distribution of classical and hybrid configurations (Kim et al., 2007). Specifically, at low concentrations (3.5 mM) the hybrid configuration is favored. However, the equilibrium fraction of the classical configuration increases with increasing concentration of Mg^{2+}, with the classical and hybrid configurations becoming equally populated at ~4 mM Mg^{2+} (Figure 6.10). Lifetime analysis revealed that this shift occurs primarily through a Mg^{2+}-dependent stabilization of the classical configuration, whose lifetime increases as a function of Mg^{2+}, whereas the lifetime of the hybrid configuration is unaffected. In structural terms, this is interpreted to mean that classically bound tRNAs form a more extensive and compact network of Mg^{2+}-stabilized tRNA-rRNA and/or tRNA-ribosomal protein interactions. At high Mg^{2+} concentrations (~7 mM and above), the classical configuration is almost exclusively favored on account of a decreased rate of classical$\rightarrow$hybrid transitions. These results offer a mechanistic explanation for the known inhibitory and stimulatory effects, respectively, of high and low Mg^{2+} concentration on the rate of translocation. At very high Mg^{2+} concentrations (~30 mM), translocation is blocked almost entirely, even in the presence of EF-G(GTP) (Spirin, 1985), which can be rationalized by a Mg^{2+}-induced stalling of the classical$\rightarrow$hybrid tRNA transition evidenced by smFRET. At the other extreme of low Mg^{2+} (~3 mM), spontaneous translocation can proceed rapidly (Spirin, 1985), an effect presumably linked to the accelerated rate of the classical$\rightarrow$hybrid transition under low-Mg^{2+} conditions. Thus, smFRET evidence suggests that the rate of the classical$\rightarrow$hybrid tRNA transition is closely linked with the rate of translocation, implying that under certain conditions, movement of tRNAs into their hybrid configuration may represent a rate-limiting step for translocation of mRNA and tRNAs through the ribosome.

Changes in the acylation state and identity of the P- and A-site tRNAs within PRE complexes have similarly been found to influence the energetics of its conformational fluctuations. As discussed earlier, the presence of a peptide on the P-site tRNA (i.e., in a POST complex) correlates with a locked ribosome in which ribosome and tRNA dynamics are suppressed, whereas ribosomes bearing a deacylated P-site tRNA (i.e., in a PRE complex) are unlocked and exhibit pronounced dynamic behavior. In addition, ribosome dynamics have been shown to be sensitive to the identity of the P-site tRNA. For example, a comparison of inter-subunit rotation dynamics within four different PRE^{-A} complexes differing only in the identity of the deacylated P-site tRNA (tRNA^{fMet}, tRNA^{Phe}, tRNA^{Tyr}, and tRNA^{Met} were used) revealed distinct thermodynamic and kinetic parameters underlying reversible inter-subunit rotation (Cornish et al., 2008). Different P-site tRNA species, therefore, make sufficiently unique contacts with the ribosome to influence large-scale structural rearrangements at the subunit interface in a characteristic way.

Similarly, the presence and acylation state of the A-site tRNA appears to dictate thermodynamic and kinetic behavior of conformational equilibria monitored by the individual smFRET signals. For example, the presence of A-site dipeptidyl-tRNA versus aa-tRNA increases the population of the hybrid configuration by increasing the rate of classical→hybrid tRNA transitions, as monitored by the tRNA-tRNA smFRET signal (Blanchard et al., 2004b). Likewise, using the L1-tRNA smFRET signal, addition of aa-tRNA to PRE complexes caused a slight increase in the rate with which the L1 stalk-P/E tRNA interaction is formed, with minimal effect on the rate with which this interaction is disrupted. Occupancy of the A site by a peptidyl-tRNA increased the forward rate by an additional six-fold, again with minimal effect on the reverse rate (Fei et al., 2008). Finally, the presence of a peptidyl-tRNA at the A site of PRE complexes shifts the equilibrium from the open to the closed L1 stalk conformation, as monitored by the L1-L9 smFRET signal, primarily by accelerating the rate of open→closed L1 stalk transitions (Fei et al., 2009).

From the data discussed in the previous paragraphs, a picture begins to emerge in which large-scale conformational rearrangements of the entire PRE complex can be allosterically controlled through even subtle and highly localized changes in interactions between the ribosome and its ligands (i.e., the presence of peptidyl- versus aa-tRNA at the A site). This feature of the PRE complex has been exploited by ribosome-targeting antibiotics, which, as described in Section III for drugs that inhibit aa-tRNA selection, often function by inhibiting the dynamics of

the translational machinery. Indeed, smFRET studies have provided evidence that translocation inhibitors specifically interfere with the conformational dynamics of PRE complexes. One example is the potent translocation inhibitor viomycin, which binds at the interface of the 30S and 50S subunits between helix 44 within the 16S rRNA and helix 69 within the 23S rRNA (Yamada et al., 1978; Stanley et al., 2010). Viomycin halts inter-subunit rotation dynamics and causes a net stabilization of the rotated state (Ermolenko et al., 2007b; Cornish et al., 2008). In addition, viomycin has been shown to slow classical⇌hybrid tRNA fluctuations (Kim et al., 2007; Feldman et al., 2010), although there are conflicting reports regarding the question of whether the drug stabilizes the classical or the hybrid configuration (Ermolenko et al., 2007b; Kim et al., 2007; Feldman et al., 2010).

smFRET investigations of PRE complexes were also conducted in the presence of a collection of aminoglycoside antibiotics (Feldman et al., 2010), drugs that bind to helix 44 within the 16S rRNA, stabilizing a conformation of the universally conserved 16S rRNA nucleotides A1492 and A1493 in which they are displaced from helix 44, adopting extrahelical positions that allow them to interact directly with the codon-anticodon helix at the decoding center (Carter et al., 2000). The aminoglycosides were shown to suppress tRNA dynamics, in general decreasing the rate of transition out of the classical tRNA binding configuration and causing a net stabilization of the classical state. The magnitude of these effects elicited by each of the aminoglycosides tested, although modest, correlated with the reduction in translocation rate observed in the presence of each drug (results from kanamycin, gentamycin, paromomycin, and neomycin are shown in Figure 6.11) (Feldman et al., 2010). Therefore, stabilization of the classical state and inhibition of transitions into the hybrid state represents a general mechanism for translocation inhibition by aminoglycosides, with subtle differences in antibiotic chemical structure dictating the degree of inhibition. Taken together, the results presented in this section illustrate that inhibition of ribosome and/or tRNA dynamics within the PRE complex represents a general inhibition strategy leveraged by a variety of ribosome-targeting antibiotics.

IV.5 Regulation of Pretranslocation Complex Dynamics by EF-G

Perhaps the most dramatic effect on ribosome dynamics within the PRE complex is elicited by EF-G, the elongation factor responsible for catalysis of full mRNA-tRNA translocation. smFRET analysis revealed that binding of EF-G(GDPNP) to PRE^{-A} complexes leads to stabilization of all conformational features characterizing the GS2 ribosome: ribosomal subunits are stabilized in their rotated conformation, the L1 stalk strongly favors the closed conformation, and the P-site tRNA is stabilized in the P/E

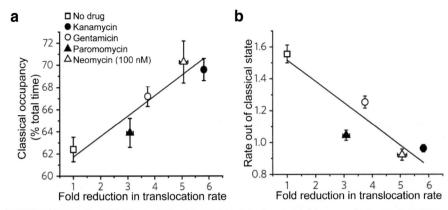

FIGURE 6.11: *Stabilization of the classical state is strongly correlated with inhibition of translocation by decoding site-binding aminoglycosides. (a) A strong correlation is observed between time-averaged classical state occupancy and fold-reduction of the single-step translocation rate of wild-type ribosomes in the presence of drug (20 μM, unless otherwise noted). (b) A strong correlation is also observed between translocation rates and the rate constant of transitioning from the classical to hybrid states. Reproduced from Feldman et al. (2010), copyright © 2010 Nature Publishing Group, with permission from Macmillan Publishers Ltd.*

hybrid state where it forms a long-lived inter-molecular interaction with the L1 stalk (Ermolenko et al., 2007a; Cornish et al., 2008; Cornish et al., 2009; Fei et al., 2009). Particularly remarkable is the stabilization of the closed state of the L1 stalk, which demonstrates that binding of EF-G(GDPNP) to the ribosome's GTPase center can allosterically regulate L1 stalk dynamics ~175 Å away at the ribosomal E site. A major role of EF-G, therefore, appears to be to bias intrinsic conformational fluctuations of the ribosome toward the on-pathway translocation intermediate GS2. In accord with the ability of the ribosome to translocate in the absence of translation factors (Pestka, 1969; Gavrilova et al., 1976; Bergemann and Nierhaus, 1983), one of EF-G's main mechanistic functions may be to stabilize GS2, preventing backward fluctuations along the translocation reaction coordinate and thus guiding the directionality of a process that the ribosome is inherently capable of coordinating on its own. This model finds strong support from biochemical experiments demonstrating that EF-G(GDPNP) stimulates the rate of translocation ~1,000-fold relative to uncatalyzed, spontaneous translocation, and that GTP hydrolysis in the EF-G(GTP)-catalyzed reaction provides an additional rate enhancement of only ~50-fold (Rodnina et al., 1997; Katunin et al., 2002). GTP hydrolysis, which, based on fast kinetics measurements, precedes movement of the mRNA-tRNA duplex on the small subunit, likely leads to conformational changes in EF-G and the ribosome that promote the second step of translocation (Rodnina et al., 1997; Taylor et al., 2007).

As previously discussed, a full round of mRNA-tRNA translocation converts the PRE complex into a POST complex in which non-rotated subunits and classical tRNA configurations characteristic of GS1 prevail and ribosome and tRNA dynamics are suppressed (Ermolenko et al., 2007a;

Cornish et al., 2008; Fei et al., 2008). This effect could be observed in real time through stopped-flow delivery of EF-G(GTP) to PRE complexes labeled with the inter-subunit S6(Cy5)-L9(Cy3) smFRET pair (Cornish et al., 2008). The PRE complex exhibits rotated⇌non-rotated inter-subunit fluctuations until the delivery of EF-G(GTP), which binds to the PRE complex and catalyzes full translocation, thereby rectifying inter-subunit dynamics and locking the ribosome in the post-translocation, non-rotated state (Figure 6.12).

IV.6 Conformational Dynamics of the L1 Stalk Before, During, and After Translocation

In contrast to the predominately static behavior of the inter-subunit smFRET signal in POST complexes, the L1-L9 smFRET signal demonstrates a persistence of L1 stalk dynamics. The L1 stalk, as suggested by its conservation throughout all kingdoms of life (Nikulin et al., 2003), represents an important structural component of the ribosome that likely plays a crucial role in both the translocation event and subsequent release of deacylated tRNA from the E site (Valle et al., 2003; Andersen et al., 2006). Results from a study in which tRNA dynamics were probed using a tRNA-tRNA smFRET signal hint at the role of the L1 stalk in promoting translocation. In this study, stabilization of the classical tRNA configuration was observed in PRE complexes formed with L1-depleted mutant ribosomes, an effect that was correlated with the slowed rate of translocation observed in the absence of L1 (Subramanian and Dabbs, 1980; Munro et al., 2007). Further single-molecule evidence underscoring the functional importance of the L1 stalk was achieved by monitoring L1-tRNA interactions during real-time EF-G-catalyzed

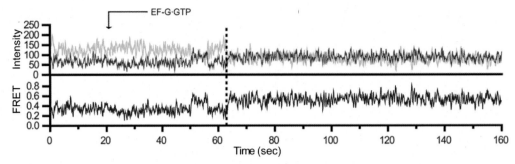

FIGURE 6.12: *EF-G(GTP)-catalyzed translocation rectifies inter-subunit rotation dynamics and converts the ribosome into the non-rotated state. The D_2/A_2 smFRET probes shown in Figure 6.7 were used to report on inter-subunit rotation. EF-G(GTP) (300 nM EF-G, 250 μM GTP) was added at ~20 s (arrow) to PRE complexes containing deacylated tRNA^{fMet} bound to the P site and N-Ac-Phe-tRNA^{Phe} bound to the A site. Translocation is observed as the transition to the stable high-FRET state (vertical dashed line). Reproduced from Cornish et al. (2008), copyright © 2008 Cell Press, with permission from Elsevier.*

translocation reactions (Fei et al., 2008). Stopped-flow delivery of Lys-tRNA^{Lys} and EF-G(GTP) to a POST complex bearing L1(Cy5) and fMet-Phe-(Cy3)tRNA^{Phe} in the P site leads to peptidyl transfer followed by EF-G(GTP)-catalyzed translocation. The smFRET versus time trajectories exhibit a sharp transition from low to high FRET upon peptidyl transfer (corresponding to the formation of inter-molecular contacts between L1 and the tRNA's elbow region), followed by stable occupancy of the high-FRET state until fluorophore photobleaching (Figure 6.13b). This is in contrast to the analogous experiment performed in the absence of EF-G(GTP), where the initial transition from low to high FRET is followed by fluctuations between the two FRET states (corresponding to repetitive formation and disruption of L1-tRNA contacts) (Figure 6.13a). These results suggest that during EF-G(GTP)-catalyzed translocation, inter-molecular interactions formed between the L1 stalk and P/E-tRNA are maintained during the movement of the deacylated tRNA from the hybrid P/E configuration into the classical E/E configuration. Formation and maintenance of these interactions provides a molecular rationale to help explain how the L1 stalk

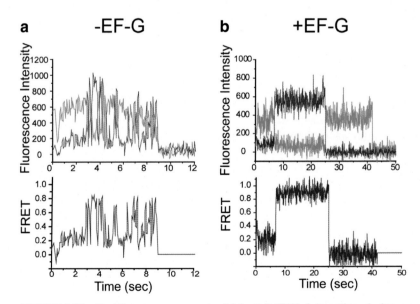

FIGURE 6.13: *Real-time measurement of L1 stalk-tRNA interaction during a full elongation cycle. The D_5/A_5 smFRET probes shown in Figure 6.7 were used to study the L1 stalk-tRNA interaction. Stopped-flow delivery of 100 nM EF-Tu(GTP)Lys-tRNA^{Lys} in the absence (a) and presence (b) of 1 μM EF-G(GTP) to surface-immobilized POST complexes bearing L1(Cy5) and fMet-Phe-(Cy3)tRNA^{Phe} at the P site. Reproduced from Fei et al. (2008), copyright © 2008 Cell Press, with permission from Elsevier.*

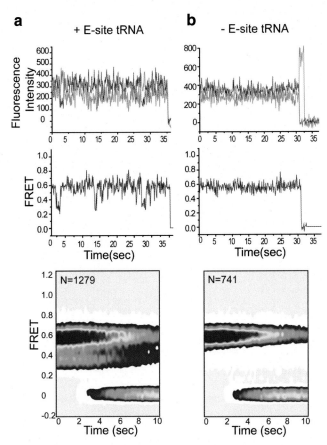

a + E-site tRNA

b - E-site tRNA

FIGURE 6.14: L1 stalk conformational dynamics within POST complexes. The intrinsic conformational dynamics of the L1 stalk within POST complexes were studied using the D_3/A_3 smFRET probes shown in Figure 6.7. (a) In an authentic POST complex containing an E-site tRNA, the L1 stalk undergoes fluctuations between open (~0.56 FRET) and half-closed (~0.34 FRET) conformations. (b) In a POST complex with a vacant E site, the L1 stalk predominately occupies the open conformation. Reproduced from Fei et al. (2009), copyright © 2009 National Academy of Sciences, U.S.A.

facilitates the translocation reaction (Subramanian and Dabbs, 1980).

Following translocation, L1 stalk dynamics within the POST complex (in this case containing deacylated tRNA^{Phe} at the E site and fMet-Phe-Lys-tRNA^{Lys} at the P site) may actively promote the deacylated tRNA's dissociation from the ribosome. The majority of smFRET versus time trajectories collected using the L1-L9 smFRET signal with this POST complex exhibited fluctuations between low- and high-FRET states (Fei et al., 2009) (Figure 6.14a). The low-FRET state within POST complexes likely corresponds to a "half-closed" conformation of the L1 stalk identified by Cornish et al. (2009), whereas the high-FRET state reports on the open stalk conformation. The observed fluctuations likely originate from a sub-population of complexes whose E-site tRNA has not yet been released, because this dynamic sub-population largely disappears for POST complexes from which E-site tRNA was quantitatively

dissociated prior to smFRET measurements. For these POST complexes with a vacant E site, the majority of smFRET trajectories instead correspond to a stable open conformation of the L1 stalk (Figure 6.14b). In contrast to the dynamic fluctuations of the L1-L9 signal, the L1-tRNA signal yields a stable high-FRET value in the analogous POST complex (Figure 6.13b). This observation strongly suggests that inter-molecular interactions between L1 and the E-site tRNA are maintained while the stalk fluctuates between open and half-closed conformations, thus implying a reconfiguration of the tRNA between at least two different configurations within the E site. Analogous to the L1 stalk's role in directing translocation of tRNA from the P to the E site, it is likely that maintenance of L1-tRNA contacts during the tRNA's residency at the E site allows opening of the stalk to guide the release trajectory of deacylated tRNA from the POST complex. Interestingly, the rate of stalk opening was found to be ~ten-fold faster than the rate of E-site tRNA release, implying that multiple fluctuations of the L1 stalk/E-site tRNA complex may occur prior to ejection. Therefore, opening of the L1 stalk, though presumably required for release of E-site tRNA from the POST complex, may not constitute the rate-limiting step for this process.

V. BEYOND ELONGATION: smFRET INVESTIGATIONS OF TRANSLATION INITIATION, TRANSLATION TERMINATION, AND RIBOSOME RECYCLING

V.1 Translation Factor–Mediated Modulation of Ribosome Dynamics as a Unifying Theme During All Stages of Protein Synthesis

In Sections III and IV, we described how conformational dynamics of the ribosome and its tRNA substrates are modulated during the elongation phase of protein synthesis, providing a regulatory mechanism that is exploited by EF-Tu and EF-G to promote aa-tRNA selection and translocation, respectively, as well as by ribosome-targeting antibiotics that impede these processes. Although the majority of smFRET studies to date have focused on elongation, recent smFRET investigations into initiation, termination, and ribosome recycling have provided evidence that translation factors serve an analogous function at the beginning and end of each round of protein synthesis. Modulation of the ribosome's global architecture through factor-dependent shifts in the translational machinery's conformational equilibria may serve as a general paradigm for translation regulation throughout all stages of protein synthesis.

V.2 Translation Initiation

During initiation of protein synthesis, a ribosomal initiation complex is assembled from its component parts in

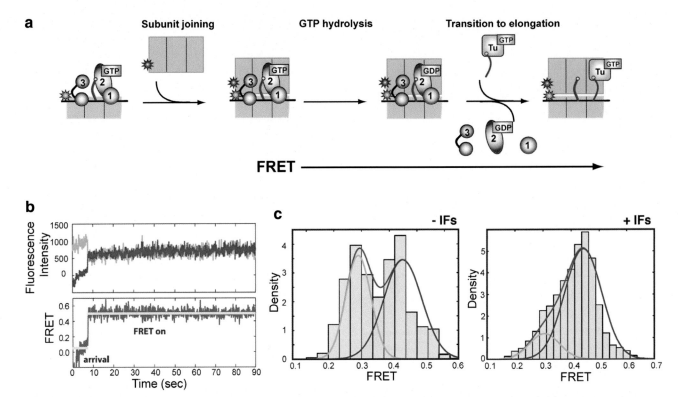

FIGURE 6.15: *Regulation of ribosomal conformational dynamics by initiation factors during translation initiation. The D_6/A_6 smFRET probes shown in Figure 6.7 were used to study the inter-subunit conformation of the ribosome during translation initiation. (a) Surface immobilization of Cy3-labeled 30S initiation complexes containing 30S subunits, initiation factors, fMet-tRNA^fMet, and mRNA followed by delivery of Cy5-labeled 50S subunits results in formation of a 70S initiation complex and establishment of a FRET signal sensitive to inter-subunit conformation. (b) Representative Cy3/Cy5 emission intensities and smFRET versus time trace. Upon stopped-flow delivery of Cy5-50S, an initial dwell time is observed followed by a burst of FRET. The FRET signal is stable, with the observation time often limited by fluorophore photobleaching. (c) FRET distribution histograms for 70S complexes formed in the absence (left) and presence (right) of initiation factors. Adapted from Marshall et al. (2009), copyright © 2009 Cell Press, with permission from Elsevier.*

a coordinated process that is directed by three initiation factors: IF1, IF2, and IF3. Following the initiation factor–mediated assembly of a 30S initiation complex (30S IC) bearing a P-site initiator fMet-tRNA^fMet at an AUG start codon, joining of the 50S subunit is catalyzed by IF2 in its GTP-bound form (Laursen et al., 2005). This process, which results in the formation of a 70S initiation complex (70S IC) with fMet-tRNA^fMet in the P site, has been studied using smFRET with the H101(Cy5)-h44(Cy3) inter-subunit labeling scheme described in Section IV (Figure 6.7) (Marshall et al., 2009). Delivery of H101(Cy5)50S subunits to h44(Cy3)30S ICs allowed real-time observation of subunit joining, an event signaled by a sharp transition from 0 to high FRET (Figures 6.15a and 6.15b). The mean FRET arrival time, which reports on the kinetics of 50S subunit joining, was 31.3 ± 7.3 s in the absence of initiation factors, and 5.6 ± 0.9 s in the presence of IF1, IF2, and IF3 (at 15 mM Mg^{2+}). In addition to enhancing the rate of 70S IC formation, initiation factors were found to influence the conformational state in which the 70S IC is assembled. A bimodal FRET distribution with peaks centered at

~0.30 and ~0.44 FRET was observed (Figure 6.15c), with the ~0.44-FRET state representing the ribosomal conformation that is competent to bind the first EF-Tu(GTP)aa-tRNA ternary complex and enter into the elongation phase of translation. In the absence of initiation factors, little preference is shown toward assembly of the 70S IC in the ~0.44- versus the ~0.30-FRET state. However, addition of IF1, IF2, and IF3 leads to preferential formation of the 70S IC in the ~0.44-FRET state (Figure 6.15c), implying a role for initiation factors in guiding proper assembly of the 70S IC into its elongation-competent conformation during initiation.

V.3 Translation Termination

In a similar manner, release factors (RFs) and the ribosome recycling factor (RRF) have been shown by smFRET to promote preferential population of particular conformations of the translational machinery, thereby guiding forward progression through the termination and recycling stages of protein synthesis (Sternberg et al., 2009). The

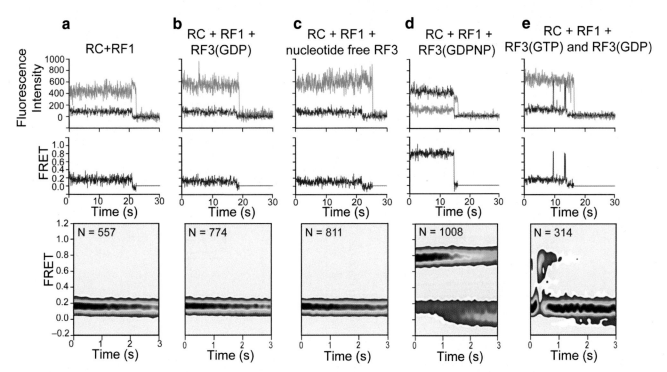

FIGURE 6.16: *Ribosomal conformational dynamics regulated by release factors during translation termination. The D₅/A₅ smFRET probes shown in Figure 6.7 were used to report on the dynamic equilibrium between GS1 and GS2 in ribosomal release complexes (RCs). (a) RC in the presence of 1 µM RF1. Binding of RF1 blocks the GS1→GS2 transition and stabilizes GS1. (b) RF1-bound post-hydrolysis RC in the presence of 1 µM RF3(GDP). (c) RF1-bound post-hydrolysis RC in the presence of 1 µM nucleotide-free RF3. (d) Puromycin-treated RC in the presence of 1 µM RF3(GDPNP). (e) RF1-bound post-hydrolysis RC with 1 µM RF3 in the presence of a mixture of 1 mM GTP and 1 µM GDP. Only those trajectories exhibiting fluctuations between GS1 and GS2 (42%) make up the time-synchronized contour plot (bottom row), which was generated by post-synchronizing the onset of the first GS1→GS2 transition event in each trajectory to time = 0.5 sec. Adapted from Sternberg et al. (2009), copyright © 2009 Nature Publishing Group, with permission from Macmillan Publishers Ltd.*

termination of protein synthesis is signaled by the translocation of a stop codon into the A site, which is recognized by a class I RF (RF1 or RF2 in *E. coli*) that subsequently catalyzes hydrolysis of the nascent polypeptide from the P-site peptidyl-tRNA (Petry et al., 2008). This biochemical step generates a post-hydrolysis release complex (RC) with a deacylated tRNA at the P site. As described in Section IV, deacylation of P-site peptidyl-tRNA during elongation triggers large-scale fluctuations of the PRE complex. The presence of deacylated tRNA in the P site following RF1-catalyzed peptide release implies that the post-hydrolysis RC is intrinsically capable of analogous conformational fluctuations. The role of release and recycling factors in regulating these dynamics was assessed by monitoring their effect on the L1-tRNA smFRET signal, interpreted to report on transitions between the GS1 (low FRET, no L1-tRNA contact) and GS2 (high FRET, formation of L1-P/E-tRNA contact) global conformations of the ribosome.

In contrast to deacylation of the P-site tRNA via peptidyl transfer to puromycin or aa-tRNA during elongation, which results in stochastic fluctuations between GS1 and GS2, deacylation via RF1-catalyzed peptide release during termination generates post-hydrolysis RCs that are

locked in GS1. This finding suggests that RF1 binding blocks the GS1→GS2 transition. Indeed, addition of RF1 to puromycin-treated RCs suppresses fluctuations between GS1 and GS2, and shifts the GS1⇌GS2 equilibrium predominately toward GS1 (Figure 6.16a). RF1 thus prevents fluctuations of the post-hydrolysis RC, which would otherwise occur spontaneously, and locks the ribosome in GS1 in anticipation of binding of the GTPase class II RF, RF3.

Biochemical experiments have suggested that RF3(GDP) binds to the RF1-bound RC, and that subsequent GDP-to-GTP exchange by ribosome-bound RF3 catalyzes the dissociation of RF1; GTP hydrolysis by RF3(GTP) then leads to its own dissociation from the RC (Zavialov et al., 2001). The role of the GS1→GS2 transition in this process has been demonstrated by a sequence of smFRET experiments using RF1, RF3, and various guanine nucleotides and their analogs (Figure 6.16). Neither addition of RF3(GDP) (Figure 6.16b) nor nucleotide-free RF3 (Figure 6.16c) to the RF1-bound, post-hydrolysis RC was capable of eliciting the GS1→GS2 transition. However, adding RF3(GDPNP) to a puromycin-treated RC traps it in GS2 (Figure 6.16d), suggesting that the GS1→GS2 transition occurs

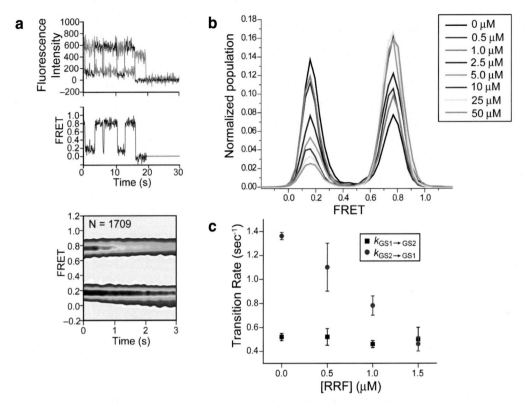

FIGURE 6.17: *Ribosome recycling factor fine-tunes the GS1 ⇌ GS2 equilibrium within the post-termination complex. The D_5/A_5 probes shown in Figure 6.7 were used to report on the dynamic equilibrium between GS1 and GS2 in post-termination complexes (PoTCs). (a) Cy3/Cy5 emission intensities, smFRET versus time trace, and contour plot of the time evolution of population FRET for PoTCs in the presence of 1 μM RRF. (b) FRET distribution histograms of PoTCs as a function of RRF concentration. Low- and high-FRET states correspond to GS1 and GS2, respectively. (c) $k_{GS1→GS2}$ and $k_{GS2→GS1}$ as a function of RRF concentration. Adapted from Sternberg et al. (2009), copyright © 2009 Nature Publishing Group, with permission from Macmillan Publishers Ltd.*

concomitantly with, or subsequent to, GTP binding, but prior to GTP hydrolysis by RF3. In experiments where RF3(GDP) was added to RF1-bound post-hydrolysis RCs in a background of 10 μM GDP, 1 mM GTP, and 1 μM RF1, multiple, short transitions from GS1 to GS2 could be observed in the smFRET versus time trajectories (Figure 6.16e), consistent with multiple rounds of RF3-catalyzed RF1 release followed by GTP hydrolysis/dissociation of RF3 and re-binding of RF1. Taken together, these results demonstrate that the GS1→GS2 transition occurs only upon GTP binding to RF3. This change in ribosome global structure might occur spontaneously following RF1 dissociation, or alternatively, could serve as the driving force for RF1 release.

V.4 Ribosome Recycling

RF3-catalyzed release of RF1, followed by RF3 dissociation subsequent to GTP hydrolysis, generates a post-termination complex (PoTC) that is initially recognized by ribosome recycling factor (RRF). Subsequent splitting

of the PoTC into its constituent 30S and 50S subunits is catalyzed by the joint action of RRF and EF-G in a GTP-dependent reaction (Petry et al., 2008). The PoTC contains a deacylated tRNA in the P site and therefore is expected to exhibit dynamic GS1 ⇌ GS2 fluctuations. Indeed, monitoring of PoTC dynamics using the L1-tRNA smFRET signal clearly demonstrates spontaneous transitions between GS1 and GS2. Adding RRF to a PoTC was found to exert a subtle effect on the GS1 ⇌ GS2 equilibrium, tilting occupancy toward GS2 in an RRF concentration-dependent manner (Figure 6.17). At concentrations near the equilibrium dissociation constant of RRF for GS2 ($K_{d,GS2}$), this effect could be explained by a decrease in the rate of GS2→GS1 transitions with increasing RRF concentration (Figure 6.17c). Considered in light of EF-G's known preference for GS2, this RRF-promoted shift toward higher fractional occupancy of GS2 should favor EF-G(GTP) binding and thus splitting of the ribosomal subunits. This suggests that spatial or temporal variations in intracellular RRF concentrations could provide a means for regulating the efficiency of ribosome recycling in vivo. Regulation of the GS1 ⇌ GS2 equilibrium thus appears to provide a common mechanism used by release and recycling factors to coordinate sequential biochemical events during the termination and recycling stages of protein synthesis. More generally, regulation of the ribosome's global state by translation factors serves to organize factor-binding events and biochemical steps over the course of the entire protein synthesis cycle.

VI. CONCLUSIONS AND FUTURE DIRECTIONS

Investigation of the translational machinery by smFRET has allowed direct observation of large-scale conformational rearrangements of this universally conserved macromolecular machine. Through site-specific attachment of donor and acceptor fluorophores to the ribosome and its tRNA and translation factor ligands, specific conformational processes such as tRNA movements through the ribosome, inter-subunit rotation, and movements of the L1 stalk have been monitored in real time during the initiation, elongation, termination, and ribosome recycling phases of translation. Analysis of smFRET versus time trajectories collected under both pre–steady state and equilibrium conditions has allowed a detailed characterization of the kinetic and thermodynamic parameters underlying the dynamics of the translational machinery.

Collectively, these studies highlight the stochastic nature of individual steps within the mechanism of translation, in which thermal fluctuations of the ribosome and its tRNA substrates permit sampling of meta-stable conformational states on a complex multi-dimensional free-energy landscape (Munro et al., 2009c; Frank and Gonzalez, 2010). The preferred modes of thermally driven ribosomal motion, programmed into the ribosome's modular two-subunit architecture, may have been harnessed by the primordial ribosome to catalyze the essential reactions of protein synthesis – aa-tRNA selection, peptide bond formation, and translocation – long before the evolution of translation factors. Indeed, the contemporary ribosome can perform all of these functions, albeit slowly, to direct protein synthesis from an mRNA template in factor-free in vitro systems (Pestka, 1969; Gavrilova and Spirin, 1971; Gavrilova et al., 1976). The smFRET studies described in this chapter illustrate the ability of translation factors to regulate and direct conformational equilibria of the ribosome and its tRNA substrates during all stages of protein synthesis. Through the stabilization/destabilization of particular conformational states and the acceleration/deceleration of particular conformational transitions, a major mechanistic role of translation factors appears to be to guide the directionality of conformational processes intrinsic to the ribosome-tRNA complex. A particularly well-studied example is the ability of EF-G(GDPNP) to rectify stochastic conformational fluctuations of PRE complexes and to stabilize the

on-pathway translocation intermediate GS2, thereby facilitating and accelerating translocation. In an analogous way, smFRET characterization of the effect of ribosome-targeting antibiotics on the translational machinery's conformational dynamics has revealed that these drugs exert their inhibitory activities through the inhibition of the large-scale structural rearrangements that are required to drive protein synthesis.

Moving forward, many opportunities exist to apply the techniques described in this chapter to mobile ribosomal domains and conformational changes suggested by structural work, but not yet probed by smFRET. For example, an enhanced understanding of the function of the L7/L12 protein stalk of the 50S GTPase center, thought to recruit translation factors to the ribosome and facilitate biochemical steps such as GTP hydrolysis and P_i release (Mohr et al., 2002; Savelsbergh et al., 2005), would be greatly facilitated through characterization of the nature and timescale of its movements with respect to the ribosome, as well as the timing of its interactions with translation factors during the various stages of protein synthesis. Similarly, smFRET provides a means by which to characterize the kinetic and thermodynamic underpinnings of putative movements of the small subunit's head domain, which have been suggested to play an important role in regulating events during translation initiation (Carter et al., 2001), aa-tRNA selection (Ogle et al., 2002), and translocation (Spahn et al., 2004; Ratje et al., 2010). Efforts to obtain a complete mechanistic understanding of the conformational dynamics of the translating ribosome will benefit from the emergence of new technologies and experimental platforms. Recent advances, such as probing multiple conformational changes simultaneously using three-fluorophore labeling to investigate the degree of conformational coupling (Hohng et al., 2004; Munro et al., 2009a; Munro et al., 2009b), new illumination strategies permitting single-molecule detection in the presence of freely diffusing dye-labeled ligands at physiologically relevant micromolar concentrations (Levene et al., 2003; Uemura et al., 2010), and new data analysis algorithms permitting increasingly unbiased analysis of smFRET versus time trajectories (Bronson et al., 2009), will allow ever more complex mechanistic questions to be addressed. These techniques should prove particularly useful in the extension of smFRET techniques from the studies of Bacterial protein synthesis described here to the more complex and highly regulated translational machinery of higher organisms.

ACKNOWLEDGMENTS

Work in the Gonzalez laboratory is supported by a Burroughs Wellcome Fund CABS Award (CABS 1004856), an NSF CAREER Award (MCB 0644262), an NIH-NIGMS grant (GM 084288–01), and an American Cancer Society Research Scholar Grant (RSG GMC-117152) to R.L.G.

We thank Joachim Frank, Dmitri Ermolenko, and Ilya Finkelstein for critically reading the chapter and providing valuable comments.

REFERENCES

Aitken, C. E. & Puglisi, J. D. (2010) Following the intersubunit conformation of the ribosome during translation in real time. *Nat Struct Mol Biol*, 17, 793–800.

Andersen, C. B., Becker, T., Blau, M., Anand, M., Halic, M., Balar, B., Mielke, T., Boesen, T., Pedersen, J. S., Spahn, C. M., Kinzy, T. G., Andersen, G. R. & Beckmann, R. (2006) Structure of eEF3 and the mechanism of transfer RNA release from the E-site. *Nature*, 443, 663–8.

Bergemann, K. & Nierhaus, K. H. (1983) Spontaneous, elongation factor G independent translocation of Escherichia coli ribosomes. *J Biol Chem*, 258, 15105–13.

Blanchard, S. C., Gonzalez, R. L., Kim, H. D., Chu, S. & Puglisi, J. D. (2004a) tRNA selection and kinetic proofreading in translation. *Nat Struct Mol Biol*, 11, 1008–14.

Blanchard, S. C., Kim, H. D., Gonzalez, R. L., JR., Puglisi, J. D. & Chu, S. (2004b) tRNA dynamics on the ribosome during translation. *Proc Natl Acad Sci USA*, 101, 12893–8.

Bretscher, M. S. (1968) Translocation in protein synthesis: a hybrid structure model. *Nature*, 218, 675–7.

Brodersen, D. E., Clemons, W. M., Jr., Carter, A. P., Morgan-Warren, R. J., Wimberly, B. T. & Ramakrishnan, V. (2000) The structural basis for the action of the antibiotics tetracycline, pactamycin, and hygromycin B on the 30S ribosomal subunit. *Cell*, 103, 1143–54.

Bronson, J. E., Fei, J., Hofman, J. M., Gonzalez, R. L., Jr. & Wiggins, C. H. (2009) Learning rates and states from biophysical time series: a Bayesian approach to model selection and single-molecule FRET data. *Biophys J*, 97, 3196–205.

Carter, A. P., Clemons, W. M., Brodersen, D. E., Morgan-Warren, R. J., Wimberly, B. T. & Ramakrishnan, V. (2000) Functional insights from the structure of the 30S ribosomal subunit and its interactions with antibiotics. *Nature*, 407, 340–8.

Carter, A. P., Clemons, W. M., Jr., Brodersen, D. E., Morgan-Warren, R. J., Hartsch, T., Wimberly, B. T. & Ramakrishnan, V. (2001) Crystal structure of an initiation factor bound to the 30S ribosomal subunit. *Science*, 291, 498–501.

Cornish, P. V., Ermolenko, D. N., Noller, H. F. & Ha, T. (2008) Spontaneous intersubunit rotation in single ribosomes. *Mol Cell*, 30, 578–88.

Cornish, P. V., Ermolenko, D. N., Staple, D. W., Hoang, L., Hickerson, R. P., Noller, H. F. & Ha, T. (2009) Following movement of the L1 stalk between three functional states in single ribosomes. *Proc Natl Acad Sci USA*, 106, 2571–6.

Daviter, T., Gromadski, K. B. & Rodnina, M. V. (2006) The ribosome's response to codon-anticodon mismatches. *Biochimie*, 88, 1001–11.

Dorner, S., Brunelle, J. L., Sharma, D. & Green, R. (2006) The hybrid state of tRNA binding is an authentic translation elongation intermediate. *Nat Struct Mol Biol*, 13, 234–41.

Dorywalska, M., Blanchard, S. C., Gonzalez, R. L., Kim, H. D., Chu, S. & Puglisi, J. D. (2005) Site-specific labeling of the ribosome for single-molecule spectroscopy. *Nucleic Acids Res*, 33, 182–9.

Draper, D. E. (2004) A guide to ions and RNA structure. *Rna*, 10, 335–43.

Effraim, P. R., Wang, J., Englander, M. T., Avins, J., Leyh, T. S., Gonzalez, R. L., Jr. & Cornish, V. W. (2009) Natural amino acids do not require their native tRNAs for efficient selection by the ribosome. *Nat Chem Biol*, 5, 947–53.

Ermolenko, D. N., Majumdar, Z. K., Hickerson, R. P., Spiegel, P. C., Clegg, R. M. & Noller, H. F. (2007a) Observation of intersubunit movement of the ribosome in solution using FRET. *J Mol Biol*, 370, 530–40.

Ermolenko, D. N., Spiegel, P. C., Majumdar, Z. K., Hickerson, R. P., Clegg, R. M. & Noller, H. F. (2007b) The antibiotic viomycin traps the ribosome in an intermediate state of translocation. *Nat Struct Mol Biol*, 14, 493–7.

Fei, J., Bronson, J. E., Hofman, J. M., Srinivas, R. L., Wiggins, C. H. & Gonzalez, R. L., Jr. (2009) Allosteric collaboration between elongation factor G and the ribosomal L1 stalk directs tRNA movements during translation. *Proc Natl Acad Sci USA*, 106, 15702–7.

Fei, J., Kosuri, P., Macdougall, D. D. & Gonzalez, R. L., Jr. (2008) Coupling of ribosomal L1 stalk and tRNA dynamics during translation elongation. *Mol Cell*, 30, 348–59.

Fei, J., Wang, J., Sternberg, S. H., Macdougall, D. D., Elvekrog, M. M., Pulukkunat, D. K., Englander, M. T. & Gonzalez, R. L., Jr. (2010) A highly purified, fluorescently labeled in vitro translation system for single-molecule studies of protein synthesis. *Methods Enzymol*, 472, 221–59.

Feldman, M. B., Terry, D. S., Altman, R. B. & Blanchard, S. C. (2010) Aminoglycoside activity observed on single pre-translocation ribosome complexes. *Nat Chem Biol*, 6, 54–62.

Frank, J. & Agrawal, R. K. (2000) A ratchet-like inter-subunit reorganization of the ribosome during translocation. *Nature*, 406, 318–22.

Frank, J., Gao, H., Sengupta, J., Gao, N. & Taylor, D. J. (2007) The process of mRNA-tRNA translocation. *Proc Natl Acad Sci USA*, 104, 19671–8.

Frank, J. & Gonzalez, R. L. (2010) Structure and dynamics of a processive Brownian motor: the translating ribosome. *Annual Review of Biochemistry*, 79, 9.1–9.32.

Gavrilova, L. P., Kostiashkina, O. E., Koteliansky, V. E., Rutkevitch, N. M. & Spirin, A. S. (1976) Factor-free ("non-enzymic") and factor-dependent systems of translation of polyuridylic acid by Escherichia coli ribosomes. *J Mol Biol*, 101, 537–52.

Gavrilova, L. P. & Spirin, A. S. (1971) Stimulation of "non-enzymic" translocation in ribosomes by p-chloromercuribenzoate. *FEBS Lett*, 17, 324–6.

Gonzalez, R. L., Jr., Chu, S. & Puglisi, J. D. (2007) Thiostrepton inhibition of tRNA delivery to the ribosome. *Rna*, 13, 2091–7.

Grosjean, H. J., De Henau, S. & Crothers, D. M. (1978) On the physical basis for ambiguity in genetic coding interactions. *Proc Natl Acad Sci USA*, 75, 610–4.

Ha, T., Rasnik, I., Cheng, W., Babcock, H. P., Gauss, G. H., Lohman, T. M. & Chu, S. (2002) Initiation and re-initiation of DNA unwinding by the Escherichia coli Rep helicase. *Nature*, 419, 638–41.

Harms, J. M., Wilson, D. N., Schluenzen, F., Connell, S. R., Stachelhaus, T., Zaborowska, Z., Spahn, C. M. & Fucini, P.

(2008) Translational regulation via L11: molecular switches on the ribosome turned on and off by thiostrepton and micrococcin. *Mol Cell*, 30, 26–38.

Hickerson, R., Majumdar, Z. K., Baucom, A., Clegg, R. M. & Noller, H. F. (2005) Measurement of internal movements within the 30 S ribosomal subunit using Forster resonance energy transfer. *J Mol Biol*, 354, 459–72.

Hohng, S., Joo, C. & Ha, T. (2004) Single-molecule three-color FRET. *Biophys J*, 87, 1328–37.

Hopfield, J. J. (1974) Kinetic proofreading: a new mechanism for reducing errors in biosynthetic processes requiring high specificity. *Proc Natl Acad Sci USA*, 71, 4135–9.

Horan, L. H. & Noller, H. F. (2007) Intersubunit movement is required for ribosomal translocation. *Proc Natl Acad Sci USA*, 104, 4881–5.

Katunin, V. I., Savelsbergh, A., Rodnina, M. V. & Wintermeyer, W. (2002) Coupling of GTP hydrolysis by elongation factor G to translocation and factor recycling on the ribosome. *Biochemistry*, 41, 12806–12.

Kim, H. D., Puglisi, J. D. & Chu, S. (2007) Fluctuations of transfer RNAs between classical and hybrid states. *Biophys J*, 93, 3575–82.

Korostelev, A. & Noller, H. F. (2007) The ribosome in focus: new structures bring new insights. *Trends Biochem Sci*, 32, 434–41.

Laursen, B. S., Sorensen, H. P., Mortensen, K. K. & Sperling-Petersen, H. U. (2005) Initiation of protein synthesis in bacteria. *Microbiol Mol Biol Rev*, 69, 101–23.

Lee, T. H., Blanchard, S. C., Kim, H. D., Puglisi, J. D. & Chu, S. (2007) The role of fluctuations in tRNA selection by the ribosome. *Proc Natl Acad Sci USA*, 104, 13661–5.

Levene, M. J., Korlach, J., Turner, S. W., Foquet, M., Craighead, H. G. & Webb, W. W. (2003) Zero-mode waveguides for single-molecule analysis at high concentrations. *Science*, 299, 682–6.

Marshall, R. A., Aitken, C. E. & Puglisi, J. D. (2009) GTP hydrolysis by IF2 guides progression of the ribosome into elongation. *Mol Cell*, 35, 37–47.

Marshall, R. A., Dorywalska, M. & Puglisi, J. D. (2008) Irreversible chemical steps control intersubunit dynamics during translation. *Proc Natl Acad Sci USA*, 105, 15364–9.

Moazed, D. & Noller, H. F. (1989a) Interaction of tRNA with 23S rRNA in the ribosomal A, P, and E sites. *Cell*, 57, 585–97.

Moazed, D. & Noller, H. F. (1989b) Intermediate states in the movement of transfer RNA in the ribosome. *Nature*, 342, 142–8.

Mohr, D., Wintermeyer, W. & Rodnina, M. V. (2002) GTPase activation of elongation factors Tu and G on the ribosome. *Biochemistry*, 41, 12520–8.

Munro, J. B., Altman, R. B., O'Connor, N. & Blanchard, S. C. (2007) Identification of two distinct hybrid state intermediates on the ribosome. *Mol Cell*, 25, 505–17.

Munro, J. B., Altman, R. B., Tung, C. S., Cate, J. H., Sanbonmatsu, K. Y. & Blanchard, S. C. (2009a) Spontaneous formation of the unlocked state of the ribosome is a multistep process. *Proc Natl Acad Sci USA*, 107, 709–14.

Munro, J. B., Altman, R. B., Tung, C. S., Sanbonmatsu, K. Y. & Blanchard, S. C. (2009b) A fast dynamic mode of the EF-G-bound ribosome. *Embo J*, 29, 770–81.

Munro, J. B., Sanbonmatsu, K. Y., Spahn, C. M. & Blanchard, S. C. (2009c) Navigating the ribosome's metastable energy landscape. *Trends Biochem Sci*, 34, 390–400.

Nikulin, A., Eliseikina, I., Tishchenko, S., Nevskaya, N., Davydova, N., Platonova, O., Piendl, W., Selmer, M., Liljas, A., Drygin, D., Zimmermann, R., Garber, M. & Nikonov, S. (2003) Structure of the L1 protuberance in the ribosome. *Nat Struct Biol*, 10, 104–8.

Odom, O. W., Picking, W. D. & Hardesty, B. (1990) Movement of tRNA but not the nascent peptide during peptide bond formation on ribosomes. *Biochemistry*, 29, 10734–44.

Ogle, J. M., Murphy, F. V., Tarry, M. J. & Ramakrishnan, V. (2002) Selection of tRNA by the ribosome requires a transition from an open to a closed form. *Cell*, 111, 721–32.

Parker, J. (1989) Errors and alternatives in reading the universal genetic code. *Microbiol Rev*, 53, 273–98.

Pestka, S. (1969) Studies on the formation of transfer ribonucleic acid-ribosome complexes. VI. Oligopeptide synthesis and translocation on ribosomes in the presence and absence of soluble transfer factors. *J Biol Chem*, 244, 1533–9.

Petry, S., Weixlbaumer, A. & Ramakrishnan, V. (2008) The termination of translation. *Curr Opin Struct Biol*, 18, 70–7.

Pioletti, M., Schlunzen, F., Harms, J., Zarivach, R., Gluhmann, M., Avila, H., Bashan, A., Bartels, H., Auerbach, T., Jacobi, C., Hartsch, T., Yonath, A. & Franceschi, F. (2001) Crystal structures of complexes of the small ribosomal subunit with tetracycline, edeine and IF3. *Embo J*, 20, 1829–39.

Ratje, A. H., Loerke, J., Mikolajka, A., Brunner, M., Hildebrand, P. W., Starosta, A. L., Donhofer, A., Connell, S. R., Fucini, P., Mielke, T., Whitford, P. C., Onuchic, J. N., Yu, Y., Sanbonmatsu, K. Y., Hartmann, R. K., Penczek, P. A., Wilson, D. N. & Spahn, C. M. (2010) Head swivel on the ribosome facilitates translocation by means of intra-subunit tRNA hybrid sites. *Nature*, 468, 713–6.

Rodnina, M. V., Gromadski, K. B., Kothe, U. & Wieden, H. J. (2005) Recognition and selection of tRNA in translation. *FEBS Lett*, 579, 938–42.

Rodnina, M. V., Savelsbergh, A., Katunin, V. I. & Wintermeyer, W. (1997) Hydrolysis of GTP by elongation factor G drives tRNA movement on the ribosome. *Nature*, 385, 37–41.

Rodnina, M. V. & Wintermeyer, W. (2001) Ribosome fidelity: tRNA discrimination, proofreading and induced fit. *Trends Biochem Sci*, 26, 124–30.

Savelsbergh, A., Katunin, V. I., Mohr, D., Peske, F., Rodnina, M. V. & Wintermeyer, W. (2003) An elongation factor G-induced ribosome rearrangement precedes tRNA-mRNA translocation. *Mol Cell*, 11, 1517–23.

Savelsbergh, A., Mohr, D., Kothe, U., Wintermeyer, W. & Rodnina, M. V. (2005) Control of phosphate release from elongation factor G by ribosomal protein L7/12. *Embo J*, 24, 4316–23.

Schmeing, T. M. & Ramakrishnan, V. (2009) What recent ribosome structures have revealed about the mechanism of translation. *Nature*, 461, 1234–42.

Schmeing, T. M., Voorhees, R. M., Kelley, A. C., Gao, Y. G., Murphy, F. V. T., Weir, J. R. & Ramakrishnan, V. (2009) The crystal structure of the ribosome bound to EF-Tu and aminoacyl-tRNA. *Science*, 326, 688–94.

Spahn, C. M., Gomez-Lorenzo, M. G., Grassucci, R. A., Jorgensen, R., Andersen, G. R., Beckmann, R., Penczek, P. A., Ballesta, J. P. & Frank, J. (2004) Domain movements of elongation factor eEF2 and the eukaryotic 80S ribosome facilitate tRNA translocation. *Embo J*, 23, 1008–19.

Spirin, A. S. (1968) How does the ribosome work? A hypothesis based on the two subunit construction of the ribosome. *Curr Mod Biol*, 2, 115–27.

Spirin, A. S. (1985) Ribosomal translocation: facts and models. *Prog Nucleic Acid Res Mol Biol*, 32, 75–114.

Stanley, R. E., Blaha, G., Grodzicki, R. L., Strickler, M. D. & Steitz, T. A. (2010) The structures of the anti-tuberculosis antibiotics viomycin and capreomycin bound to the 70S ribosome. *Nat Struct Mol Biol*, 17, 289–93.

Stapulionis, R., Wang, Y., Dempsey, G. T., Khudaravalli, R., Nielsen, K. M., Cooperman, B. S., Goldman, Y. E. & Knudsen, C. R. (2008) Fast in vitro translation system immobilized on a surface via specific biotinylation of the ribosome. *Biol Chem*, 389, 1239–49.

Steitz, T. A. (2008) A structural understanding of the dynamic ribosome machine. *Nat Rev Mol Cell Biol*, 9, 242–53.

Sternberg, S. H., Fei, J., Prywes, N., Mcgrath, K. A. & Gonzalez, R. L., Jr. (2009) Translation factors direct intrinsic ribosome dynamics during translation termination and ribosome recycling. *Nat Struct Mol Biol*, 16, 861–8.

Subramanian, A. R. & Dabbs, E. R. (1980) Functional studies on ribosomes lacking protein L1 from mutant Escherichia coli. *Eur J Biochem*, 112, 425–30.

Sytnik, A., Vladimirov, S., Jia, Y., Li, L., Cooperman, B. S. & Hochstrasser, R. M. (1999) Peptidyl transferase center activity observed in single ribosomes. *J Mol Biol*, 285, 49–54.

Taylor, D. J., Nilsson, J., Merrill, A. R., Andersen, G. R., Nissen, P., & Frank, J. (2007) Structures of modified eEF2 80S ribosome complexes reveal the role of GTP hydrolysis in translocation. *Embo J*, 26, 2421–31.

Thompson, R. C. & Stone, P. J. (1977) Proofreading of the codon-anticodon interaction on ribosomes. *Proc Natl Acad Sci USA*, 74, 198–202.

Traut, R. R. & Monro, R. E. (1964) The puromycin reaction and its relation to protein synthesis. *J Mol Biol*, 10, 63–72.

Uemura, S., Aitken, C. E., Korlach, J., Flusberg, B. A., Turner, S. W. & Puglisi, J. D. (2010) Real-time tRNA transit on single translating ribosomes at codon resolution. *Nature*, 464, 1012–7.

Valle, M., Zavialov, A., Sengupta, J., Rawat, U., Ehrenberg, M. & Frank, J. (2003) Locking and unlocking of ribosomal motions. *Cell*, 114, 123–34.

Villa, E., Sengupta, J., Trabuco, L. G., Lebarron, J., Baxter, W. T., Shaikh, T. R., Grassucci, R. A., Nissen, P., Ehrenberg, M., Schulten, K. & Frank, J. (2009) Ribosome-induced changes in elongation factor Tu conformation control GTP hydrolysis. *Proc Natl Acad Sci USA*, 106, 1063–8.

Wang, Y., Qin, H., Kudaravalli, R. D., Kirillov, S. V., Dempsey, G. T., Pan, D., Cooperman, B. S. & Goldman, Y. E. (2007) Single-molecule structural dynamics of EF-G–ribosome interaction during translocation. *Biochemistry*, 46, 10767–75.

Yamada, T., Mizugichi, Y., Nierhaus, K. H. & Wittmann, H. G. (1978) Resistance to viomycin conferred by RNA of either ribosomal subunit. *Nature*, 275, 460–1.

Zavialov, A. V., Buckingham, R. H. & Ehrenberg, M. (2001) A posttermination ribosomal complex is the guanine nucleotide exchange factor for peptide release factor RF3. *Cell*, 107, 115–24.

CHAPTER 7

Structure and Dynamics of the Ribosome as Revealed by Cryo-Electron Microscopy

Xabier Agirrezabala

Mikel Valle

I. INTRODUCTION

Protein synthesis is a central process for living entities. In all kingdoms of life, the key component of this process is the ribosome, a large macromolecular assembly composed of two distinctly sized subunits, whose cores are largely composed of ribosomal RNA, or rRNA. This molecular machine mediates the sequential incorporation of amino acids carried by the transfer RNAs (tRNAs) into the nascent protein chain. This process is also known as translation, because the genetic message encoded in the messenger RNA (mRNA) is deciphered into the language of proteins (Figure 7.1).

The functional complexity of molecular machines such as the ribosome is coupled to their structural intricacy. Whereas initial imaging by electron microscopy showed nothing but dense granules ~200Å in diameter (Palade, 1955), decades later the development of cryo-electron microscopy (cryo-EM), combined with image processing methodology, brought evidence for the extreme complexity of the ribosome with its multiple mobile and flexible parts. Recently, as the quality of electron density maps has greatly improved thanks to methodological advances in this field (see Chapter 2 in this volume by Joachim Frank), new intermediate states have been observed and the dynamic behavior of the ribosome and its ligands has been further characterized in conjunction with other biophysical techniques such as X-ray crystallography, kinetic analysis, and single-molecule fluorescent resonance energy transfer (smFRET) (see Figure 7.2). Considering that the 3D reconstructions obtained by cryo-EM are snapshots representing functional states along the translation pathway, it is now widely acknowledged that this experimental approach has been indispensable in unraveling essential processes that cooperate in translation. Furthermore, many structural insights that have come from cryo-EM work currently drive experimental design in a wide range of studies of the ribosome, from biochemical to crystallographic studies and single-molecule FRET.

In this chapter, we will focus on the analysis of the structure and dynamics of ribosomal complexes by means of cryo-EM. We will first cover the contributions of this technique to advances in the understanding of the four major steps of protein translation (initiation, elongation, termination, and recycling) by relating the emerging X-ray structures to models inferred from cryo-EM by flexible fitting and discussing their functional meanings. We will then turn to exciting discoveries related to the rotation of the small subunit with respect to the large subunit (i.e., the so-called ratchet-like movement). After analyzing the implications of this reconfiguration in the study of the protein biosynthesis, the review will conclude with an account of recent cryo-EM data related to eukaryotic ribosomes.

II. AN OVERVIEW OF PROTEIN TRANSLATION AS SEEN BY CRYO-ELECTRON MICROSCOPY

II.1 The Initiation Process

During translation initiation, the small subunit binds directly to the translation initiation region (TIR) of the mRNA, which contains the Shine-Dalgarno (SD) sequence (5–9 nucleotides situated upstream of the start codon). This binding is thought to be a two-step process: The first step, a preliminary binding of the small subunit to the mRNA, requires single-stranded regions of the mRNA, whereas the second step, slower than the first, further stabilizes the mRNA and requires the presence of both the initiator tRNA (fMet-tRNA^{fMet} in prokaryotes) and IF2 in its GTP form. This second stage involves the unfolding of structured mRNAs, which facilitates the interaction between the SD and the anti-SD sequence present at the 3′ end of the 16S rRNA (Studer and Joseph, 2006).

The small, 30S subunit is roughly divided into a body and a head. A single RNA duplex, formed by helix 28, constitutes the neck of the 30S subunit, connecting the head and the body. Correct placement of the mRNA transcript places it into the cleft at the neck of the 30S subunit. Subsequently, a 30S initiation complex containing the three initiation factors (IF1, IF2, and IF3), mRNA, and initiator tRNA is formed, in which the anticodon of the tRNA base is

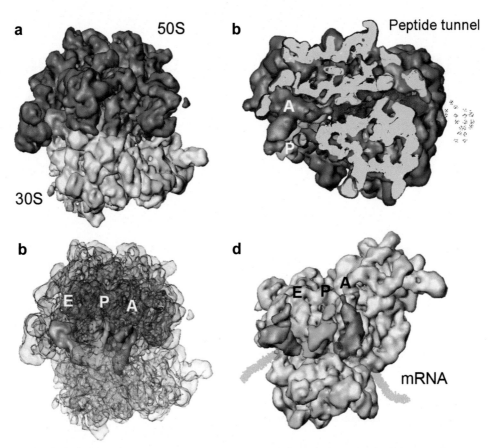

FIGURE 7.1: The architecture of the ribosome. (a) The structure of the ribosome is defined by the architecture of its two subunits, with ribosomal RNA (rRNA) being the central component in both cases. Whereas in prokaryotes the small and large subunits are composed of 16SrRNA and 5S and 23S rRNA, respectively, in eukaryotes the small subunit contains the 18S rRNA molecule and the large subunit 5S, 5.8S and 28S rRNAs (termed according to their sedimentation rates). (b) The ribosome has three tRNA binding sites: A (for amino-acyl), P (for peptidyl), and E (for exit). (c) The nascent polypeptides transit through the exit tunnel of the large subunit during translation. (d) Correct placement of the mRNA transcript during the initiation step places it in the cleft of the neck of the small subunit. The large and small subunits are shown in blue and yellow, respectively._

paired with the initiation AUG codon of the mRNA, ensuring that the initiator tRNA is correctly positioned in the P site. This specific tRNA has unique structural features when compared to the rest of the tRNAs, which enter the elongation phase of translation. It shows conformational variations in the anticodon, acceptor, and D stems (Gualerzi and Pon, 1990; Varshney et al., 1993), which are structural features specifically recognized by the initiation factors.

The initiation factors IF1, IF2 (which exhibits GTPase activity), and IF3 participate in recruitment and positioning of mRNA and initiator tRNA in different ways. In addition, IF3 (stimulated by IF1) keeps the 50S subunit dissociated from the initiation complex being formed. The last step of the initiation process is the association of the large subunit via recognition of a particular conformation of the small subunit (Milon et al., 2008). This event – that is, the binding of the large subunit – activates the hydrolysis of GTP by IF2, which in turn leads to dissociation of the initia-

tion factors and marks the irreversibility of the initiation process (Figure 7.3). Overall, the initiation factors ensure the accuracy of the initiation process by kinetically controlling the formation of the 30S and 70S initiation complexes (Gualerzi and Pon, 1990; La Teana et al., 1995; Boelens and Gualerzi, 2002; Antoun et al., 2003; Laursen et al., 2005; Antoun et al., 2006) and eventually lead to the correct configuration of the initiator tRNA.

An initial low-resolution cryo-EM image of IF3-bound 30S subunit complexes showed evidence of the ability of IF3 to dissociate 70S ribosomes into the small and large subunits, as its C-terminal domain was visualized on the platform side of the head where an essential inter-subunit bridge is situated (McCutcheon et al., 1999). Later on, higher-resolution cryo-EM studies visualized the structure of 70S initiation complexes that were thought to represent the conformations before (Allen et al., 2005) and after GTP hydrolysis (Myasnikov et al., 2005). In both studies,

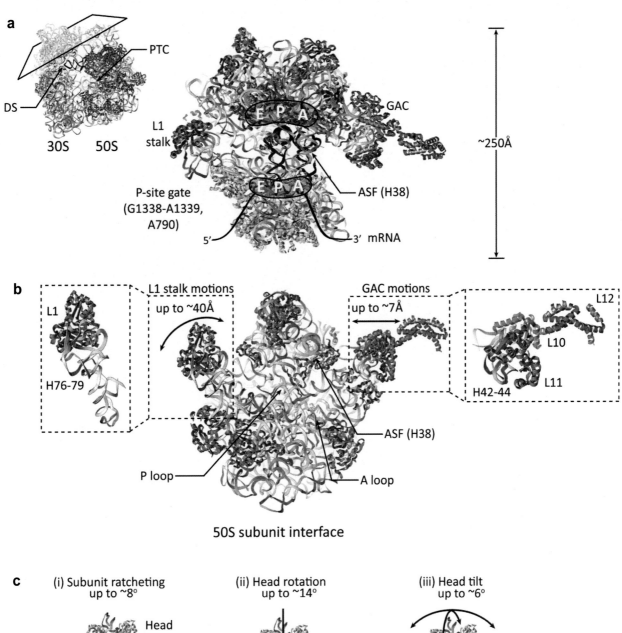

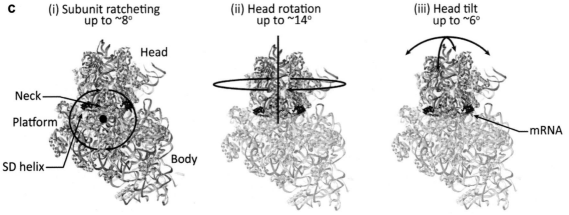

FIGURE 7.2: Overview of the ribosomal structure and dynamics. *(a) Schematic illustration of the dynamic components of the protein translation machinery within the 70S ribosome. (b) Large subunit and the estimated range of motion of the L1 stalk and GAC. (c) Small subunit and its estimated range of motion: (i) inter-subunit ratchet-like motion, (ii) head rotation, and (iii) head tilt. Details regarding each motion are provided in the main text. Ribosomal proteins are shown in tan (small subunit) and blue (large subunit), and the rRNA in gray. The A- and P-site tRNAs and the single-stranded mRNA in (a) and (c), respectively, are shown in red. Conserved elements in contact with the tRNAs such as the P and A loops or the A-site finger (ASF) are shown in green. Labels and landmarks: decoding site (DS), peptidyl-transfer center (PTC), L1 stalk (L1), GTPase-associated center (GAC), A-site finger (ASF; H38 of 23S rRNA), P-site gate (G1338–A1339, A790 of 16S rRNA). Data reproduced with permission from (Munro et al., 2009).*

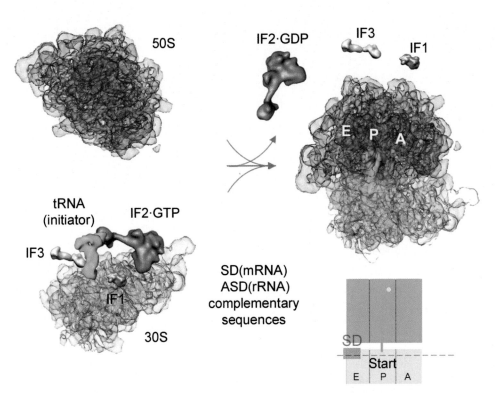

FIGURE 7.3: The process of initiation. *The formation of the 70S translation initiation complexes, containing the 30S (transparent yellow) and 50S (transparent blue) subunits, initiator fMet-tRNA^{fMet} (green), mRNA (not shown for clarity) and initiation factors IF1 (blue), IF2 (gray) and IF3 (yellow), proceeds from the initial formation of 30S initiation complexes. The association between the large subunit and subsequent GTP hydrolysis promotes release of the factors. The schematic drawing illustrates the position of fMet-tRNA^{fMet} in the 70S initiation complex. More details regarding each sub-step are provided in the main text. Some intermediate stages in the overview are omitted for simplicity.*

the initiator tRNA was observed in a transition state, off the PTC (the so-called P/I state), and IF2 was localized at the entrance to the inter-subunit space; however, it is noteworthy that these studies showed discrepancies regarding tRNA positioning and interactions with the initiation factor, attributed by Allen and coworkers to the absence of IF1 in the post-hydrolysis structure (Allen and Frank, 2007). A very relevant difference between the pre-hydrolysis and post-hydrolysis state was the rotation of the small subunit with respect to the large subunit (ratchet-like inter-subunit motion), shown by the pre-hydrolysis initiation ribosome.

At the same time, very interesting cryo-EM data regarding the docking of structured mRNA onto the platform of the 30S subunit were reported (Marzi et al., 2007). Once unfolded, this mRNA enters the channel through the neck of the small subunit. The next stage was described by the reconstructions presented by Simonetti and coworkers (Simonetti et al., 2008). These structures, obtained from a single sample after image classification techniques and in the absence of IF3, described the interaction between the small subunit and the initiator tRNA, the mRNA and initiation factors 1 and 2 (bound to GTP). These data, in which no interaction between the two initiation factors was discerned, showed the initiator tRNA in a conformation

different from that of the structures shown in the aforementioned 70S initiation ribosomes (Figure 7.4). In addition, based on a comparison of the pre- and post-hydrolysis structures, the authors suggested that discrete conformational changes of IF2 had occurred along the initiation pathway, which may be relevant for the transition of the initiation complexes from 30S to 70S.

Recent structural data include high-resolution crystal structures of ribosomal complexes related to the initiation process (Carter et al., 2001; Pioletti et al., 2001; Yusupova et al., 2001; Myasnikov et al., 2005; Jenner et al., 2005; Yusupova et al., 2006; Jenner et al., 2007; Korostelev et al., 2007). These data further include the X-ray structure of the ribosome-bound elongation factor P (EF-P), a protein that stimulates the formation of the first peptide bond by facilitating proper positioning of the initiator tRNA in the P site (Blaha et al., 2009). Cryo-electron microscopic reconstructions have also provided important clues to the mechanism of translation initiation by visualizing biologically active 30S and 70S initiation complexes. The binding sites of the different factors involved in translation initiation have been delineated throughout these studies, as well as the successive functional conformations associated with each stage of the initiation process (Myasnikov et al., 2009; Simonetti

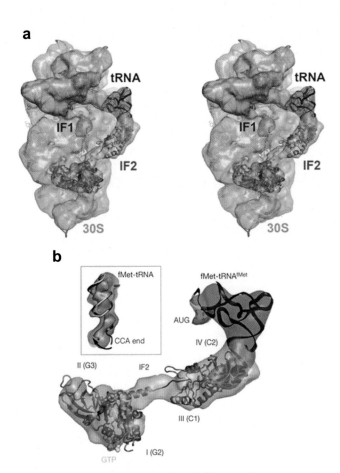

FIGURE 7.4: The 30S-IC complex. *(a) Stereo-view representation of the cryo-EM reconstruction of the 30S initiation complex and corresponding atomic model: 30S subunit (yellow), IF2 (green), IF1 (blue), initiator tRNA (green). (b) Close-up view of IF2 and fMet-tRNA^{fMet} structures. The inset shows the deviation of the 3′ CCA end of the initiator tRNA when compared with the classical tRNA (yellow). Labels and landmarks: I-IV (domains I-IV of IF2), AUG (initiation codon). Data reproduced with permission from (Simonetti et al., 2008).*

et al., 2009). However, this process is still far from being understood and several questions regarding the interplay between initiation factors remain unaddressed, as complete 30S initiation complexes containing the three factors have not yet been characterized.

II.2 Incorporation of the Ternary Complex into the Ribosome: The Decoding Process

In the first step of the elongation cycle – the incorporation of aa-tRNA – the ribosome has to recognize cognate ternary complexes (formed by aa-tRNA•EF-Tu•GTP) and reject the incorrect ones based on the codon string of the mRNA presented at the decoding center (Figure 7.5). This decoding process, which is divided into two stages separated by the hydrolysis of GTP (initial selection and proofreading), is kinetically controlled (Gromadski and Rodnina,

2004; Rodnina and Wintermeyer, 2001). Structurally, the process is based on the Watson-Crick complementarity of the mRNA codon and the tRNA anticodon. X-ray studies of anticodon stem loops (ASL) bound to isolated 30S subunits have shown that the nucleotides G530, A1492, and A1493 from 16S rRNA monitor the correct geometry of the codon-anticodon match (Ogle et al., 2001). A reconfiguration of the head and shoulder domains was also detected upon cognate codon-anticodon pairing (Ogle et al., 2002), a rearrangement that, along with local changes in the decoding center, accounts for the accuracy of the process (Ogle and Ramakrishnan, 2005; Agirrezabala and Frank, 2009).

The stepwise movement of the aa-tRNA into the ribosomal A site (i.e., accommodation) involves the formation of several discrete structural intermediates (Rodnina et al., 1994; Rodnina et al., 1995; Rodnina et al., 1996; Blanchard et al., 2004). Crystallography of the aforementioned 30S subunits (Ogle et al., 2001; Ogle et al., 2002) has provided valuable information on the decoding site interactions; however, essential insights into the structural dynamics of the aa-tRNA incorporation process have been obtained by means of cryo-EM through the analysis of an intermediate: the A/T state of tRNA binding. This state was observed in 70S ribosomes bound with cognate aa-tRNA•EF-Tu•GDP ternary complexes stalled by the antibiotic kirromycin (Stark et al., 2002; Valle et al., 2002; Valle et al., 2003a; Li et al., 2008; Villa et al., 2009; Schuette et al., 2009). In this state, the elongation factor Tu is stalled in the GTPase-activated configuration (Rodnina et al., 1995). In these cryo-EM maps, the aa-tRNA molecule adopts a distorted configuration. The distortion, located between the anticodon- and D-stem loop regions, could be essential during decoding as it probably sets the threshold for triggering the GTPase activity of EF-Tu (Yarus et al., 2003; Frank et al., 2005). Moreover, the structural data from cryo-EM have shown that this deformation in the aa-tRNA structure and the spatial relationship with EF-Tu allows for sampling and recognition of the cognate codon, which in turn leads to the activation of the elongation factor and the hydrolysis of GTP.

In a recent paper, Ramakrishnan and coworkers (Schmeing et al., 2009) confirmed the existence of the aforementioned pronounced conformational distortion of the tRNA bound to the ribosome in the A/T state by X-ray crystallography. This study corroborates at the molecular level the previous observations made by cryo-EM studies of ribosomes from different organisms and cognate aa-tRNAs carried out by independent groups. The X-ray study thus confirms the validity of the majority of conclusions drawn by several single-particle cryo-EM works, some of which were even carried out at moderate resolution. Looking back through the history of cryo-EM reconstructions of ribosomes bound with ternary complexes, we find structures at 16.8Å (Valle et al., 2002), 12–14Å (Stark et al., 2002), 11Å (Valle et al., 2003a), 9–11Å (Li et al., 2008), 6.7Å (Villa

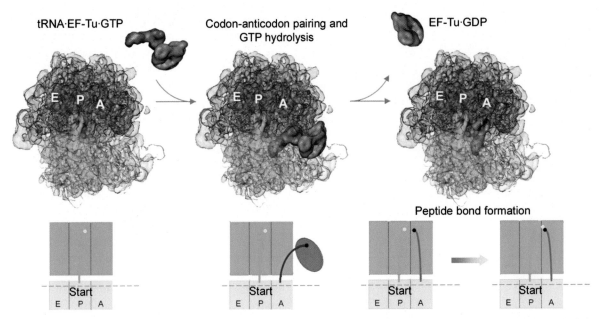

FIGURE 7.5: Incorporation of the ternary complex into the ribosome. *The process of cognate tRNA acceptance is initiated by the delivery of aa-tRNA (shown in purple) to the ribosome as part of a ternary complex with EF-Tu (shown in red) and GTP. The codon sampling and recognition is enabled by the characteristic A/T conformation adopted by the incoming aa-tRNA. A cognate codon–anticodon pairing leads to the activation of the GTPase and subsequent hydrolysis of GTP. The resulting conformational change and release of the factor facilitates the accommodation of the acceptor end into the PTC of the 50S subunit. The schematic drawings illustrate the position and movement of EF-Tu and tRNAs during each step of the process. More details regarding each step are provided in the main text. Some intermediate stages in the overview are omitted for simplicity.*

et al., 2009) and 6.4Å (Schuette et al., 2009). In Figure 7.6, we show the progress of cryo-EM of this particular complex within the last decade. Obviously, the gain in the definition of details with the increase in resolution and quality (following the advances in electron microscopy as well as data processing and analysis) has allowed differences due to different aa-tRNAs to be addressed (Li et al., 2008), as well as key fundamental mechanisms to be discovered, which were not discernible at lower resolutions (Villa et al., 2009) (Schuette et al., 2009). In essence, however, the basic structural features reported early on still remain completely valid, namely that the flexibility of the anticodon arm of the incoming aa-tRNA allows the codon-anticodon pairing while the tRNA is in the A/T state, i.e., while being delivered by EF-Tu.

II.3 Translocation of tRNAs by EF-G

Peptide bond formation is a key step during the protein elongation cycle. The reaction, facilitated by the architectural framework of the large subunit, occurs after the GTP hydrolysis-induced EF-Tu conformational change and dissociation, and subsequent accommodation of the incoming tRNA in the A site. The ribosome-promoted peptide transfer relies on accurate substrate positioning. The ribosome, via the peptidyl-transferase center, assists in the reaction by orienting and placing in reaction distance two groups:

the α-amino group of the amino acid attached to the A-site tRNA, and the terminal carboxyl group of the aa/peptide-chain annexed to the P-site tRNA. Kinetic and structural X-ray studies have characterized this reaction and its mechanisms with increasing detail (Beringer and Rodnina, 2007; Simonovic and Steitz, 2009).

Following the accommodation of aa-tRNA and rapid peptide transfer, the A site is emptied for the next round of aa-tRNA incorporation. First, the aminoacyl ends of the tRNAs move with respect to the large subunit, leaving the pretranslocational ribosome complex in the so-called hybrid A/P and P/E state (Moazed and Noller, 1989). This configuration is coupled to the ratchet-like rotation of the small subunit relative to the large subunit, which was first observed by Frank and Agrawal in the presence of EF-G (Frank and Agrawal, 2000), and is now known to be spontaneously formed due to the inherent architectural properties of the ribosome itself as it is exposed to thermal fluctuations (Ermolenko et al., 2007a; Ermolenko et al., 2007b; Kim et al., 2007; Munro et al., 2007; Agirrezabala et al., 2008; Cornish et al., 2008; Fei et al., 2008; Julian et al., 2008; Marshall et al., 2008b). This rearrangement, which is essential for advancement of the process, will be covered in more detail in section III.1. Translocation, the last step in the elongation cycle, requires the action of the GTPase EF-G to advance the tRNA-mRNA complex by the span of one codon. Kinetic data have shown that the hydrolysis of

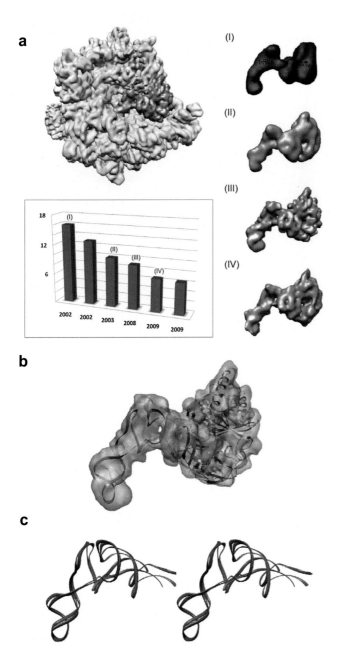

FIGURE 7.6: Cryo-EM of ternary complexes. *(a) 3D reconstruction of a ternary complex bound 70S ribosome (EMD-1849). EF-Tu is shown in red, the A/T tRNA in purple, and the P-site tRNA in green. The large and small subunits are shown in blue and yellow, respectively. The plot shown below illustrates the progress of cryo-EM work on ribosomes bearing ternary complexes. While the vertical axis shows the resolution in Angstroms (FSC = 0.5 criterion), the horizontal axes shows the year of publication. The reference to the original article in which each reconstruction was presented is mentioned in the main text. (i-iv). A gallery of close-up views of isolated densities corresponding to reconstructed ternary complexes from (Valle et al., 2002, Valle et al., 2003a, Li et al., 2008, Villa et al., 2009), respectively. (b) Atomic model of the ternary complex generated by flexible fitting into the cryo-EM densities. (c) Stereo-view representation of the the first cryo-EM derived A/T tRNA model ((PDB code: 1QZA (Valle et al., 2003a)) superimposed with the X-ray structure at 3.6Å resolution (PDB code: 2WRN; Schmeing et al. (2009)).*

GTP generates this rearrangement of the ribosomal complex and precedes translocation (Peske et al., 2000; Savelsbergh et al., 2003; Wilden et al., 2006). The outcome of this event is a ribosome bearing a peptidyl-tRNA in the P site and a deacylated tRNA in the E site, which is ready to exit the translating complex. The post-translocational 70S complex thus formed is prepared to enter a new round of aa-tRNA incorporation and continue the protein elongation cycle (Figure 7.7).

The biggest contribution to the knowledge we currently have regarding the structural basis for the action of EF-G on the pretranslocational ribosome comes from cryo-EM (see Figure 7.8). These cryo-EM studies included different prokaryotic (Agrawal et al., 1998; Agrawal et al., 1999; Frank and Agrawal, 2000; Stark et al., 2000; Valle et al., 2003b; Datta et al., 2005; Connell et al., 2007) and eukaryotic systems (Gomez-Lorenzo et al., 2000; Spahn et al., 2004a, Taylor et al., 2007; Sengupta et al., 2008). In all of these structures, which were obtained both in GDP and GTP states (in the latter case either using non-hydrolyzable GTP analogs, a transition-state analog, or different translocation-blocking drugs), the A site was always empty and the ribosome was in the inter-subunit-rotated configuration. The L1 stalk was seen in the "closed" position, displaced toward the body of the large subunit, and the P-site tRNA was in the hybrid P/E configuration. The apparent flexibility of the L1 stalk most likely accounts for its absence in many of the crystallographic structures.

While X-ray atomic structures of EF-G and its eukaryotic counterpart eEF2 show almost no variability in the structural disposition of its domains (domains I-V and G') (Aevarsson et al., 1994; Czworkowski et al., 1994; al-Karadaghi et al., 1996; Jorgensen et al., 2003; Jorgensen et al., 2004; Hansson et al., 2005a; Hansson et al., 2005b; Jorgensen et al., 2005), the cryo-EM maps show that the factor, once bound to the ribosome, undergoes a drastic rearrangement of its domain IV, which is placed in the vicinity of the decoding site (top part of helix h44 of the small subunit) in the new configuration. Cryo-EM also showed that h44 is displaced by 8Å in the mRNA-tRNA translocation direction (VanLoock et al., 2000). Recently, the first ribosome-bound EF-G structure at atomic resolution was reported from X-ray data (Gao et al., 2009). This structure, arrested in the post-translocational state by fusidic acid, where it shows no ratchet-like rotation of the small subunit, presents EF-G in virtually the same conformation as shown by the cryo-EM studies (Valle et al., 2003b). Other findings previously reported by cryo-EM studies, such as the interaction of the G' domain with the C-terminal domain of L12, were also now confirmed and shown in atomic detail.

To conclude this section, we will mention the EF4 (LepA)-bound ribosomal structure solved by cryo-EM, which shows a novel A-site tRNA configuration (Connell et al., 2008). This structure offers clues on how this

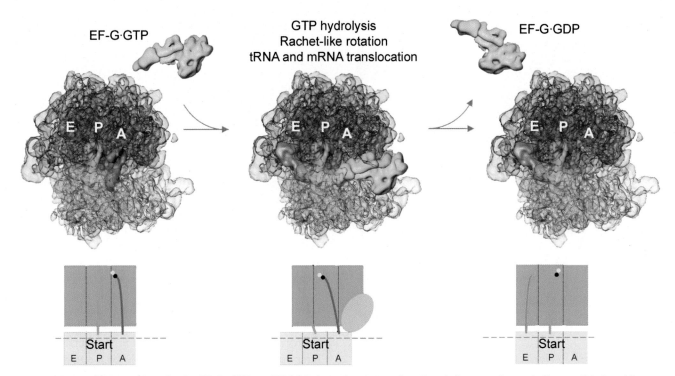

FIGURE 7.7: tRNA translocation by EF-G. *GTPase EF-G binds to the pretranslocational ribosome formed after peptide bond formation which bears a deacylated tRNA in the P site and a dipeptidyl-tRNA in the A site. The binding of the factor stabilizes the hybrid configuration of the tRNAs along with the rotated position of the small subunit. The hydrolysis of GTP by EF-G promotes the translocation of the mRNA-tRNA complex from the small subunit's A and P sites to the P and E sites, respectively. The schematic drawings illustrate the position and movement of the EF-G and tRNAs during each step of the process. More details regarding each sub-step are provided in the main text. Some intermediate stages are omitted in the overview for simplicity.*

highly conserved bacterial GTPase mediates the back-translocation reaction of the tRNAs, reversing the action of EF-G. The functional role of this elongation factor seems to be related to the efficiency of protein synthesis control via interactions with ribosomes after a defective translocation (Qin et al., 2006).

II.4 Termination of Protein Synthesis: The Role of Release Factors

Three of the 64 triplet combinations, UAG, UAA, and UGA, do not specify any amino acid. These so-called stop codons signal the termination of protein synthesis (as opposed to the "sense" codons, including the initiation AUG codon, which are recognized by a tRNA and thus code for an amino acid). The synthesized polypeptide chain is released from the ribosome when such a codon is encountered on the mRNA. This event is of critical importance in translation, as faulty misreading of the stop codon as a sense codon will allow read-through of the sequence. The class I release factors (RF1 and RF2) effect the release of the peptide chain through specific interactions between the highly conserved GGQ motif and the ribosomal PTC,

and trigger the hydrolysis of the ester bond between the P-site tRNA CCA end and the nascent chain (Kisselev et al., 2003; Nakamura and Ito, 2003). Whereas RF1 recognizes codons UAA and UAG, RF2 recognizes codons UAA and UGA. In addition, kinetic studies have shown that stop codon recognition promotes specific conformational changes in the decoding center, which also help promote the release of the nascent protein (Youngman et al., 2007). In the last step of termination, class II RF3, a small GTPase, releases the class I factor from the ribosome (Figure 7.9).

Comparison of X-ray structures of the factors (Vestergaard et al., 2001; Shin et al., 2004) with cryo-EM reconstructions shows (see Figure 7.10) that both RF1 and RF2 undergo large conformational changes as they bind to the ribosome, attaining a conformation that allows them to simultaneously reach the decoding center with their codon recognition motif and the PTC with their GGQ motif (Rawat et al., 2003; Klaholz et al., 2003; Rawat et al., 2006). Initial low-resolution crystal structure images of ribosome-bound RF1 and RF2 complexes (Petry et al., 2005), as well as small-angle X-ray scattering data of RF1 in solution (Vestergaard et al., 2005) confirmed this finding. More

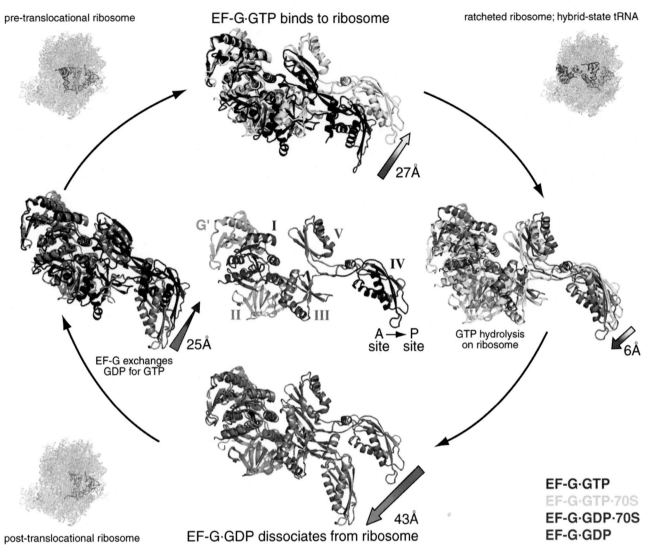

pre-translocational ribosome

EF-G·GTP binds to ribosome

ratcheted ribosome; hybrid-state tRNA

27Å

G' I V

IV

II III

A → P
site site

EF-G exchanges
GDP for GTP

25Å

GTP hydrolysis
on ribosome

6Å

post-translocational ribosome

EF-G·GDP dissociates from ribosome

43Å

EF-G·GTP
EF-G·GTP·70S
EF-G·GDP·70S
EF-G·GDP

FIGURE 7.8: The conformational changes observed in EF-G. *In the ribosome-bound form (yellow), the tip of domain IV of EF-G•GTP shifts by ~27Å with respect to the conformation that is shown in solution (red). A comparison of the ribosome-bound structures before (yellow) and after (blue) GTP hydrolysis shows a ~6Å shift toward the decoding center. Changes of EF-G•GDP from the ribosome-bound conformation (blue) with respect to the GDP-bound structure in solution (pink) include a shift of ~43Å. A comparison of the GDP-bound (pink) and GTP-bound (red) structures obtained in solution reveal a shift of ~25Å at the tip of domain IV. The ribosome-bound forms were obtained by cryo-EM, whereas the structures in solution were obtained by X-ray crystallography. Data reproduced with permission from (Frank et al., 2007). "Copyright (2007) National Academy of Sciences, U.S.A."*

recently, atomic data from X-ray crystallography showed the stop codon recognition mechanisms by RF1 (Laurberg et al., 2008) and RF2 (Korostelev et al., 2008; Weixlbaumer et al., 2008) in greater detail.

The only structural data so far regarding the mechanism of action of RF3 have come from cryo-EM studies (Klaholz et al., 2004; Gao et al., 2007). The study by Gao and coworkers (see Figure 7.11) showed that binding of RF3 (in the GDPNP form) to the ribosomal release complex stabilizes the inter-subunit-rotated conformation of the ribosome in conjunction with the "closed" position of

L1 and the P/E hybrid tRNA disposition, changes similar to those observed with EF-G binding in the GDPNP form (Valle et al., 2003b). The implications of the ratchet-related rearrangements in translation termination will be discussed in Section III.3.

II.5 Ribosome Recycling

Once protein synthesis is terminated, the remaining mRNA and deacylated P-site tRNA must be released and the ribosome must be split into its constituent subunits. In

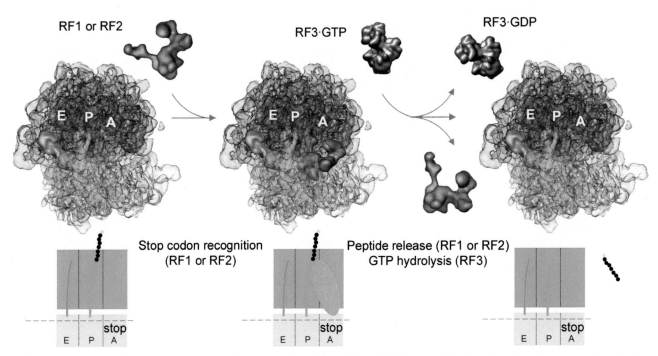

FIGURE 7.9: Model of termination process. *The stop codon (UAA, UAG or UGA) recognition by class I release factors leads to the end of the translation of the coding region, liberating the nascent polypeptide chain. Class II release factor RF3 uses the energy of GTP hydrolysis to trigger the dissociation of class I factors from the ribosomal release complex. The schematic drawings illustrate the position and movement of RFs and tRNAs during the process. Details regarding each step are provided in the main text. Some intermediate stages are omitted in the overview for simplicity.*

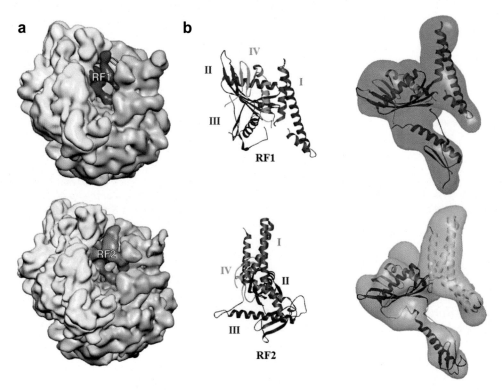

FIGURE 7.10: Comparison between cryo-EM and X-ray structures of the class I release factors. *(a) Binding positions of RF1 (purple) and RF2 (magenta) on the ribosome. (b) Comparison of RF1 and RF2 from X-ray structures in solution (left) and ribosome-bound conformations (right) as obtained by cryo-EM and flexible fitting. Color coding (for RF1 and RF2): domain I, gold; domain II, blue; domain III, red; domain IV, green. Data reproduced with permission from (Rawat et al., 2006).*

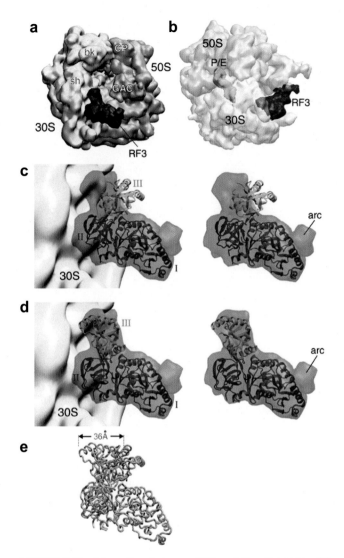

FIGURE 7.11: Interaction between RF3 and the ribosome. *(a)
Side and (b) top views of RF3•GDPNP-bound ribosomal release
complexes. RF3 is shown in red, the small and large subunits
in yellow and blue, respectively, and the hybrid P/E tRNA in
green. (c) Stereo-view representation of the crystal structure of
RF3•GDP fitted into the cryo-EM density. (d) Stereo-view representation of the model generated by flexible fitting. (e) Superim-
position of crystal (green) and cryo-EM-derived (orange) models. Color coding (for RF3): domain I, red; domain II, blue; and
domain III, green. Labels and landmarks: bk (beak), sh (shoulder), CP (central protuberance), GAC (GTPase-associated center). Data reproduced with permission from (Gao et al., 2007).*

prokaryotes, the recycling of this post-termination complex requires the action of the recycling factor (RRF), as
well as of EF-G (Figure 7.12). Kinetic studies showed that
the release of mRNA and tRNA occurs after dissociation
of the subunits, once IF3 is bound to the small subunit.
Notably, these studies showed that there is no translocation of the mRNA-tRNA, complex even though the

recycling reaction requires the hydrolysis of GTP by EF-G
(Karimi et al., 1999; Peske et al., 2005).

Crystal structures of isolated RRF from several species
(Selmer et al., 1999; Kim et al., 2000) have shown that this
L-shaped factor is composed of two domains connected
by a flexible link which is known to be essential for its
function (Toyoda et al., 2000). Cryo-electron microscopy
studies have visualized this factor as positioned in the inter-
subunit space, overlapping the A and P sites (Agrawal et al.,
2004; Gao et al., 2005b). In addition, these studies showed
that RRF stably binds to ribosomes in the inter-subunit-
rotated form, with the deacylated tRNA in the P/E hybrid
configuration, which is in agreement with previous observations (Lancaster et al., 2002; Peske et al., 2005). Changes
in the conformation of helix H69, an element involved in
the formation of inter-subunit bridges, were also reported
in the cryo-EM structures. Moreover, the reconstruction of
RRF-bound subunits with EF-G and GDPNP, as reported
by Gao and coworkers, showed that the binding of EF-
G triggers an interdomain rotation of RRF (Figure 7.13).
According to the proposed model, the hydrolysis of GTP
on EF-G would drive the head rotation of RRF to a position that leads to the destabilization or weakening of the
bridges connecting the subunits. More recently, a cryo-
EM study reported the existence of a second binding site
for RRF (Barat et al., 2007). This novel binding position, which overlaps with the P-site tRNA binding site,
was only found in a small population of ribosomes, suggesting it is a short-lived intermediate. These findings led
the authors to suggest an alternative model for recycling
and subunit disassembly which involves several sequential
steps.

The crystal structures of RRF bound to different ribosomal complexes such as the 50S subunit (Wilson et al., 2005),
the empty 70S ribosome (Borovinskaya et al., 2007), and a
ribosome containing mRNA, ASL in the P site, and a tRNA
in the E site (Weixlbaumer et al., 2007) confirmed most
of the observations made previously by cryo-EM. However, they showed differences regarding the conformation
of H69; therefore, the exact atomic details on which the
ribosomal subunit dissociation mechanism relies on are still
unclear (see Petry et al. [2008] for further insights).

III. THE RATCHET-LIKE INTER-SUBUNIT MOTION AS A CENTRAL REQUIREMENT FOR PROTEIN BIOSYNTHESIS BY THE RIBOSOME

A few years after the first EM images of the ribosome had
been obtained, in the 1960s, a working model was proposed based on its two-subunit construction that implied an
inter-subunit motion during tRNA translocation in which
the ribosome oscillates between "unlocked" (open) and
"locked" (closed) states (Bretscher, 1968; Spirin, 1968).
Decades later, cryo-EM visualized such a rearrangement

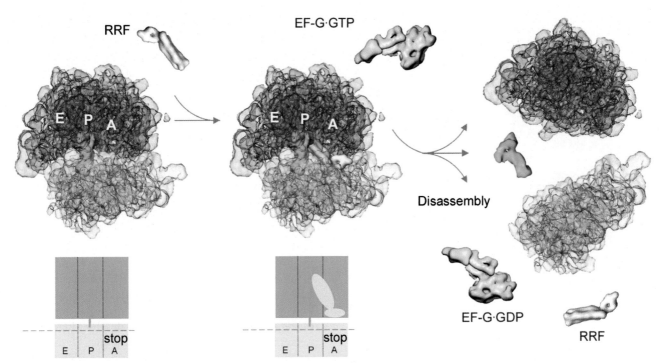

FIGURE 7.12: Ribosome recycling. *The disassembly of the post-termination complex (which bears a deacylated tRNA in the P site) resulting from the action of the release factors is catalyzed by RRF (shown in gray) and EF-G (shown in light blue). Thereafter, IF3 is postulated to bind to the small subunit to prevent the re-association of the subunits and to lead a new round of initiation (not shown). The schematic drawings illustrate the position and movement of RRF and tRNAs during this process. More details regarding each step are provided in the main text. Some intermediate stages are omitted in the overview for simplicity.*

coupled to the EF-G-mediated translocation (Frank and Agrawal, 2000). The observed subunit rotation (or ratchet-like motion), is a complex structural rearrangement that involves the remodeling of inter-subunit contacts at the

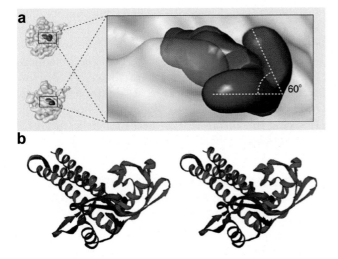

FIGURE 7.13: The mechanisms of RRF actions during recycling. *(a) Superimposed RRF structures in its 50S (dark blue) and 70S (purple) bound conformations. The orientation of the subunits is shown as successive thumbnails on the left. (b) Stereo-view representation of superimposed RRF conformations derived by flexible fitting. Data reproduced with permission from (Gao et al., 2005b).*

interface of the small and large subunits (see Figure 7.14a). In the new configuration, which would be equivalent to the unlocked state proposed earlier, the small subunit is rotated with respect to the large subunit by an overall rotation angle of ~6°. This rotation was initially thought to be exclusively related to translocation, but subsequent research has demonstrated that it is also coupled to other steps. As already mentioned in previous sections, cryo-EM studies have shown that rearrangements involving ribosomal elements, equivalent to those observed in translocation-related complexes, also occur during initiation (Allen et al., 2005; Myasnikov et al., 2005), termination (Klaholz et al., 2004; Gao et al., 2007), and recycling (Agrawal et al., 2004; Gao et al., 2005b). More recently, single-molecule FRET studies characterized the motion during the course of initiation (Marshall et al., 2009) and termination and recycling (Sternberg et al., 2009).

It is well established that inter-subunit rotation is only detected after aa-tRNA accommodation and peptide transfer, that is, when the ribosome is in the unlocked state (Valle et al., 2003b; Zavialov and Ehrenberg, 2003). However, there are some contradictory FRET data regarding the process of back-rotation. Some groups reported spontaneous back-rotation activity for the ribosome (Cornish et al., 2008), whereas one group has reported that back-rotation only occurs after EF-G-stimulated translocation of the tRNA-mRNA complex from the A to the P site

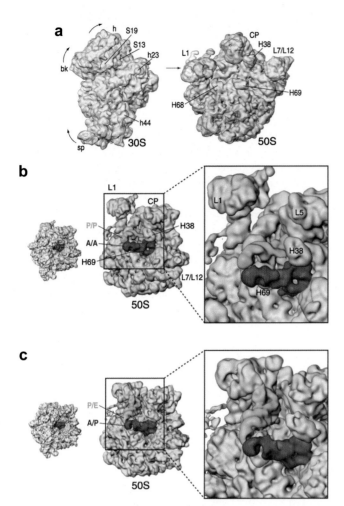

FIGURE 7.14: *The hybrid tRNA configuration. (a) Superimposed 30S (left) and 50S (right) subunits. Small and large subunits in the hybrid, rotated configuration are shown in yellow and blue, respectively, and the classic-state subunits in transparent gray. (b) Close-up view of the classic 50S configuration. (c) Close-up view of the hybrid 50S configuration. The orientations of the small subunits are shown in thumbnails on the left. Labels and landmarks: S19 (small subunit protein S19), S13 (SSU protein S13), h23 (SSU helix 23), h44 (SSU helix 44), H38 (large subunit/LSU helix 38), H68 (LSU helix 68), H69 (LSU helix 69), H76 (LSU helix 76), L5 (LSU protein L5). Data reproduced with permission from (Agirrezabala et al., 2008).*

III.1 Spontaneous Inter-Subunit Rotation and the Formation of Hybrid States

The existence of intermediate states for tRNA binding has been long postulated (Bretscher, 1968; Spirin, 1968). However, attempts to provide experimental support for the existence of these intermediates failed until, decades later, chemical protection experiments demonstrated its validity. These experiments, conducted by Moazed and Noller, found that the acceptor ends of the A- and P-site tRNAs spontaneously relocate to form the so-called hybrid state of tRNA binding (Moazed and Noller, 1989). In this configuration, the ASL of tRNAs reside in the small subunit's A and P sites, whereas the acceptor ends are placed in the large subunit's P and E sites. Although this configuration was initially elusive to structural characterization, and was later thought to be exclusively promoted by the action of EF-G since hybrid P/E tRNA was only visualized in its presence (Valle et al., 2003b), more recent cryo-EM data showed that the tRNA reconfiguration along with the inter-subunit rotation is spontaneously acquired, apparently as a consequence of the ribosome's architecture and its immersion in the thermal bath (Agirrezabala et al., 2008; Julian et al., 2008). FRET studies also showed the reversible nature of transitions between classic and hybrid forms, giving us an appreciation of the ribosome as a stochastic molecular machine constantly in motion (Frank and Gonzalez Jr., 2010).

Along these lines, research studies found a magnesium ion-dependence of the fluctuations of tRNAs between the classic and hybrid state in the pretranslocational ribosome (Kim et al., 2007). As previously stated, two independent groups analyzed the pretranslocational ribosome by cryo-EM (Agirrezabala et al., 2008; Julian et al., 2008). As a result of the application of image classification procedures, these studies showed the existence of two major sub-populations of ribosomal complexes: one a classical, unrotated form with tRNAs in the classic configuration, and the other a rotated ribosome configuration where both the A- and P-site tRNAs are in the hybrid configuration. The L1 stalk, displaced toward the inter-subunit space, was found to be in contact with the P/E-tRNA in the hybrid state (see Figure 7.14). Following the trend established by the smFRET study of Kim and coworkers, the different experimental settings (i.e., Mg^{2+} concentration) used in the two cryo-EM analyses led to different relative populations of classic and hybrid states. As the ratchet-like motion implies disruption and reforming of the interactions between the two subunits, this observation provides evidence of the Mg^{2+} ion-mediated stabilization of the tertiary structure of the unrotated configuration.

Further characterization of the dynamic nature of the pretranslocation state included the smFRET analysis of the coupling between the L1 stalk and tRNA (Fei et al., 2008). This analysis detected spontaneous and reversible changes

(Marshall et al., 2008b). What is clear is that the acylation state of the P-site tRNA controls the locking and unlocking of the ribosome: Only a deacylated tRNA at the P site allows conformational changes related to inter-subunit rotation and formation of the hybrid tRNA configuration necessary for the tRNA translocation step. Based on these observations, we now understand the rotation as a Brownian effect that takes place when the temperature is high enough and when the subunits are free to move – that is, when the presence of a charged tRNA in the P site does not provide an additional bridge between ribosomal subunits that impedes the conformational change.

in the L1-tRNA interaction, in line with studies by Cornish and coworkers (Cornish et al., 2008) and cryo-EM studies (Agirrezabala et al., 2008; Julian et al., 2008). The importance of the identity of the tRNA occupying the A site in the stability of the hybrid state configuration was also revealed (Munro et al., 2007; Fei et al., 2008).

III.2 L1 Stalk Movement, Inter-Subunit Rotation, and Formation of the Hybrid tRNA Configuration: Concerted Motions?

As previously mentioned, independent studies established the dynamic relationships between the hybrid tRNA configuration, inter-subunit rotation, and movement of the L1 stalk. However, it seems futile to look for causality or an ordered sequence of events – smFRET studies show that most changes are reversible, and the only mark of time (giving it direction) is made by irreversible steps: GTP hydrolysis and diffusion of Pi.

For example, recent studies showed that the L1 stalk presents at least three conformations, dependent on the acylation state of the tRNA at the P site (Cornish et al., 2009). Two of these states were observed in the pretranslocational ribosome (i.e., open state with vacant E site and fully closed state with P/E hybrid tRNA configuration) and the third was associated with post-translocational ribosomes with a half-closed L1 stalk and a deacylated tRNA in the classical E/E state. X-ray crystallography also offers clues about the meta-stable nature of pretranslocational ribosomes after peptide transfer since the analysis of different crystal structures (albeit in the presence of ASLs instead of complete tRNAs) found evidence of the existence of intermediate degrees of inter-subunit rotation (Zhang et al., 2009). These structures suggested that the ribosome remodels the interface of the subunits in discrete steps. Based on these structures, Cate and coworkers proposed a model where the inter-subunit rotation begins with a rotation of the 30S subunit body, continues with motions of the 30S platform and head domains, and finishes with the remodeling of the central inter-subunit bridges.

Evidence for the existence of intermediate states also comes from smFRET, in the form of an intermediate A/A-P/E tRNA configuration (Munro et al., 2007). Moreover, this study also reported distinct rates for the hybrid P/E state formation compared to the rates for inter-subunit rotation or back-rotation. The temporal coupling of L1 movements and subunit rotation was also analyzed by a comparison of forward rates by Cornish and coworkers (Cornish et al., 2009). This analysis showed that whereas classical-to-hybrid transition (L1 closing and inter-subunit rotation) occurs at similar rates (suggesting that they may be coupled), the rates of the hybrid-to-classical transition do not match, with L1 opening at a faster rate than back-rotation of the 30S subunit.

More recently, smFRET studies suggested that the L1 stalk movement and formation of the tRNA hybrid configuration might be independent (Munro et al., 2010). This study reported that while inter-subunit rotation and L1 stalk closure take place at similar rates, the classical-to-hybrid P/E reconfiguration of the tRNAs is 10–20 times faster. Based on the different time-scales of the rearrangements involved, the authors suggested that formation of the hybrid configuration of the pretranslocational ribosome (i.e. the unlocked state of the ribosome, the state in which all of the rearrangements converge) is achieved throughout a multi-step mechanism in which the different motions are linked, but in a loosely coupled way.

As stated previously, the cryo-EM structures from different research groups depicted ribosomes with inter-subunit rotation and hybrid A/P and P/E tRNAs in synchrony with the closed position of the L1 stalk. These studies characterized the two major populations existing in pretranslocational ribosome preparations and did not report on any additional intermediate stages of these components. This result is indeed consistent with findings reported by other groups (Fei et al., 2008). However, the first point to be discussed in relation to these cryo-EM findings is that of conformational averaging, because it is possible that the hybrid state observed via cryo-EM might actually be a mixture of states whose structural variabilities could not be differentiated due to the limited resolution of the image classification approaches employed. This would be difficult to determine because only indirect support could be used (e.g. smFRET data).

On the other hand, it is also noteworthy that these cryo-EM studies were carried out using wild-type ribosomes. In contrast, intermediates such as A-A/P-E (Munro et al., 2007) were obtained using mutants that directly affect the A/P configuration (Dorner et al., 2006) and that might stabilize certain conformations and/or increase their lifetimes. The structures with intermediate rotations depicted by X-ray crystallography (Zhang et al., 2009) may also have reduced significance due to the absence of complete tRNA structures in the solved structures.

Single-molecule FRET provides data on the ongoing dynamic process itself and allows the characterization of transitions between different states with great accuracy as it avoids the averaging of subpopulations of bulk preparations (Marshall et al., 2008a; Munro et al., 2009), which is a source of problems when visualizing complexes with cryo-EM (see (Spahn and Penczek, 2009) for further insights). However, smFRET offers a limited structural perspective, that is, the fluorescence signals can only report on the relative movement of elements that are part of a more complex system. Thus, dynamic information from cryo-EM is central to complement and/or to offer alternative views of the processes analyzed by smFRET in real time, as it can provide snapshot-like structural information in 3D. The progress in automated cryo-EM data collection, in combination with the use of sophisticated classification strategies and the availability of vast computational resources (i.e., supercomputers) are mitigating the problem of conformational averaging, allowing large data sets to be subdivided into smaller and more homogenous subsets. In addition, future experiments aiming to reconcile results not only from FRET and cryo-EM, but also from other experimental approaches will require the use of identical experimental settings, because it is evident that the conformations adopted by the ribosome are sensitive to multiple variables, as previously noted by Frank and Gonzalez (Frank and Gonzalez Jr., 2010). The emergence of such collaborative efforts has great promise for the convergence of results from different sources, promising to shed light on how subunit ratcheting, L1 movements and hybrid tRNA configurations are truly related.

III.3 Ratchet-Like Inter-Subunit Rotation During Initiation, Termination, and Recycling

As is the case after peptide transfer and before translocation, a deacylated tRNA also occupies the P site during translation termination and ribosome recycling, processes in which hybrid P/E tRNAs are known to occur, as observed by cryo-EM (e.g., Gao et al., 2005b; Gao et al., 2007). Single-molecule FRET data have shown that the binding of RF1 blocks the ratchet-like motion (Sternberg et al., 2009), which is otherwise an inherent characteristic of ribosomes bound with a deacylated tRNA in the P site. This is line with cryo-EM reconstructions of ribosomes bearing RF1/RF2, which show no rotation of the small subunits (Klaholz et al., 2003; Rawat et al., 2003; Rawat et al., 2006). These same data also suggest that stabilization of the rotated state occurs after RF1 dissociation and the binding of RF3 in the GTP form, as shown by cryo-EM studies (Gao et al., 2007). It is not possible to determine from these results whether the primary role of RF3 is in the active promotion of the rotated state (as proposed by Gao and coworkers) or the dissociation of RF1, or both.

This same study also corroborates findings made by cryo-EM related to RRF-mediated stabilization of the rotated conformation during recycling (Gao et al., 2005b). These findings led the authors to suggest a very interesting idea regarding the control of the inter-subunit rotation in both directions for organizing the binding and action of different translation factors (with otherwise overlapping binding sites) during the various phases of protein synthesis. In view of the fact that inter-subunit rotation is not observed during aa-tRNA incorporation, this may be especially relevant during the protein elongation cycle. According to this view, the unrotated ribosome would be ready to interact with EF-Tu, whereas in the rotated state it would favor the binding of EF-G. The fact that the

inter-subunit-rotated, hybrid state of the ribosome is kinetically more efficient than the classic configuration during EF-G-mediated translocation would support this suggestion (Dorner et al., 2006).

The scenario is different during initiation, because an initiator tRNA is present in the P site instead of a deacylated tRNA, and the ribosome seems to progress, in an inverse manner, from the rotated to the unrotated state. The ribosome shows two rotational states during initiation, one being the unrotated form in which the 70S initiation ribosome is ready to accept incoming aa-tRNAs in the A site (Marshall et al., 2008b; Marshall et al., 2009). The IF2-catalyzed GTP hydrolysis has been linked to the back-rotation of the small subunit, controlling the orientation of subunits as visualized by cryo-EM (Allen et al., 2005; Myasnikov et al., 2005). Cryo-electron microscopy results suggest that conformational changes of IF2 triggered by the hydrolysis of GTP interfere with the normal configuration of certain inter-subunit bridges, promoting the back-rotation of the small subunit, a process thought to be essential for the incorporation of fMet-tRNA into the 70S initiation complex (Myasnikov et al., 2005; Simonetti et al., 2008). According to the smFRET-derived model proposed by Marshall and coworkers, the rotated-to-unrotated state transition would be coupled to the displacement of the acceptor end of initiator tRNA, originally placed in the novel P/I configuration (Allen et al., 2005) by the action of IF2 (as shown by cryo-EM maps), to the PTC region in the P site.

IV. CHARACTERIZATION OF EUKARYOTIC RIBOSOMES: STRUCTURAL INSIGHTS FROM CRYO-EM

The eukaryotic extra rRNA expansion segments and proteins (the roles of which still remain largely undetermined) are predominantly located in the external part of the resulting 80S ribosomes, distant from the structurally conserved decoding and peptide transfer centers (Figure 7.15). As an example, the *Saccharomyces cerevisiae* ribosomal 60S large subunit is composed of 25S rRNA, 5S rRNA, and 5.8S rRNA molecules (*E. coli*: 23S rRNA and 5S rRNA) and 46 proteins (vs. 34 in *E. coli*), and the small 40S subunit of 18S rRNA (16S rRNA in *E. coli*) and 32 proteins (vs. 21 in *E. coli*). This greater structural complexity is reflected in the increased complexity of the biogenesis process. In prokaryotes, for example, a small number of factors (modification enzymes, RNA helicases, and chaperones, among others) are required for ribosome biogenesis, whereas in the highly compartmentalized eukaryotic ribosome biogenesis, more than 170 non-ribosomal factors and 70 small nucleolar rRNAs are required, most of which are essential (see Dinman, 2009; Kressler et al., 2010; Panse and Johnson, 2010 for further insights).

a

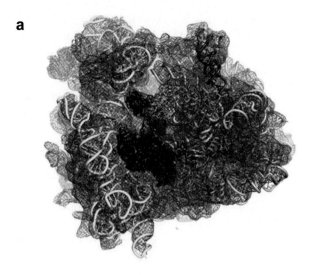

b

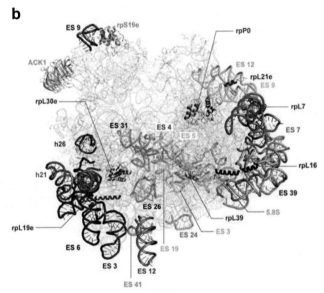

Table 7.1: Ribosomal Factors in Protein Synthesis

Translation Step	Bacteria	Archaea	Eukarya
Initiation	IF1	aIF1A	eIF1A
	IF2	aIF5B	eIF5B
	IF3	aIF1	eIF1
		aIF2α	eIF2α
		aIF2β	eIF2β
		aIF2γ	eIF2γ
		aIF2Bα	eIF2Bα
			eIF2Bβ
			eIF2Bγ
		aIF2Bδ	eIF2Bδ
			eIF2Bε
			eIF3 (13 subunit)
		aIF4A	eIF4A
			eIF4B
			eIF4E
			eIF4G
			eIF4H
		aIF5	eIF5
		aIF6	eIF6
			PABP
Elongation	EF-Tu	aEF1α	eEF1A
	EF-Ts	aEF1B	eEF1 B (2 or 3 subunits)
	SelB	SelB	eEFSec
			SBP2
	EF-G	aEF2	eEF2
Termination	RF1	aRF1	eRF1
	RF2		
	RF3		eRF3
Recycling	RRF		
	EFG		
			eIF3
			eIF3j
			eIF1A
			eIF1

Note: The orthologous or functionally homologous factors are aligned. Data reproduced with permission from Rodnina and Wintermeyer (2009).

FIGURE 7.15: The eukaryotic ribosome. *(a) Cryo-EM reconstruction of yeast 80S ribosome (transparent mesh) and the derived atomic model (for experimental and modeling details, see Taylor et al. (2007) and Taylor et al. (2009), respectively). (b) Highlight of expansion segments (ES) and additional ribosomal eukaryotic proteins as localized in the small (rpS) and large (rpL) subunits. For clarity, eEF2 is omitted in (b). Data reproduced with permission from (Agirrezabala and Frank, 2010).*

Due to the higher degree of complexity, which is also reflected in the number of ribosomal factors involved throughout the whole protein synthesis process (see Table 7.1 for a comparison), our knowledge of the molecular mechanisms of eukaryotic protein translation lags far behind those of the prokaryotic (and archaeal) counterparts. In addition, until recently, all of the attempts to crystallize eukaryotic ribosomes have failed, and all of the structural information regarding eukaryotic systems has come from cryo-EM studies. The cryo-EM data include comparative analyses between ribosomes from different sources (Gao et al., 2005a; Nilsson et al., 2007), analysis of the structural conservation of expansion segments by the generation of quasi-atomic models (Spahn et al., 2001a; Chandramouli et al., 2008; Taylor et al., 2009), characterization of species-specific translation factors (Andersen et al., 2006), and the study of structurally divergent mitochondrial and chloroplast ribosomes (Sharma et al., 2003; Sharma et al., 2007; Sharma et al., 2009), which are structurally closer to prokaryotic ribosomes than to 80S eukaryotic ribosomes. In the following paragraphs, we will summarize the main contributions of cryo-EM to our knowledge of eukaryotic protein synthesis. Rather than detailing each stage, we will focus exclusively on the areas in which cryo-EM studies have mainly contributed: initiation and tRNA translocation. For more general

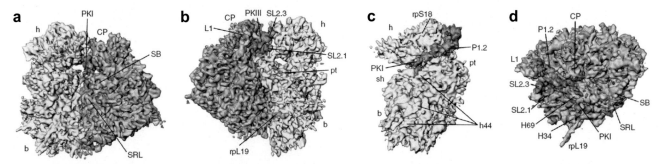

FIGURE 7.16: **IRES-mediated initiation of protein synthesis.** *(a), (b) Cryo-EM reconstruction of the CrPV (cricket paralysis virus) IRES-bound yeast 80S ribosome shown from distinct views: IRES (magenta), 40S subunit (yellow), 60S subunit (blue) (c), (d) Distinct views of the small and large subunit, with the apposing subunit computationally removed. Labels and landmarks: b (body) h (head), pt (platform), sh (shoulder), CP (central protuberance), L1 (L1 stalk), SB (stalk base), SRL (sarcin-ricin loop), h44 (SSU helix 44), H34 (LSU helix 34), H69 (LSU helix 69), rpS18 (SSU protein 18), rpL19 (LSU protein 19), P1.2, SL2.1, SL2.3 and PKI (secondary structure elements of the CrPV IRES). Data reproduced with permission from (Schuler et al., 2006).*

information in eukaryotic protein synthesis, we direct readers to more specific and comprehensive review articles (e.g., Kapp and Lorsch, 2004; Rodnina and Wintermeyer, 2009 and references therein).

The complexity of eukaryotic systems is evident in translation initiation, which is an extremely complex and highly regulated process (Acker and Lorsch, 2008; Myasnikov et al., 2009; Sonenberg and Hinnebusch, 2009). This process involves several sequential steps: Briefly, a ternary complex formed by eIF2, GTP and initiator tRNA binds the small subunit to form the 43S pre-initiation complex, a reaction guided by the factors eIF1, eIF1A and eIF3. The cap-dependent interaction of the 5′ end of the mRNA with the 43S complex is mediated by eIF4E, and also involves eIF4Ga and PABP (polyA binding protein). After initiation codon recognition (a process in which the eIF4A, eIF2, and eIF5 factors are implicated), the 48S complex previously formed binds to the large 60S subunit. After hydrolysis of GTP by eIF5B, and subsequent release of the factors and correct positioning of initiator tRNA, the 80S initiation ribosome complex is formed.

Cryo-electron microscopy has contributed to insights in the first stages of this process. Three-dimensional reconstructions of 40S small subunit-bound eIF1-eIF1A complexes – factors known to be sufficient for promoting assembly of the 43S complex under certain experimental conditions (Algire et al., 2002) – have shown that the factor binding process leads to an open, scanning-competent complex that facilitates access to the mRNA, the next step in the process (Passmore et al., 2007). Similarly, structures of the factors eIF1, eIF2, eIF3, and eIF5 (which form the so-called multi-factor complex in budding yeast) bound to the small subunit show how factor binding leads to the rearrangement of the head and platform regions of the small subunit, conformational changes that facilitate access of the mRNA to the binding channel (Gilbert et al., 2007).

In addition, several cryo-EM studies have analyzed the non-canonical pathway of mRNA binding to the small subunit that involves the binding of IRES (internal ribosomal entry sites) elements. These structured RNA sequences, located in mRNA un-translated regions (UTRs), bypass the requirement of having a modified cap on the 5′ end in the mRNA. They are recruited via distinct mechanisms. In the hepatitis C virus, for example, the uncapped mRNA IRES element binds eIF3 and the small subunit, leading to initiation codon recognition and positioning (Pestova et al., 1998; Kieft et al., 1999; Kieft et al., 2001). These IRES elements, bound to 40S subunit and 40S-eIF3 complexes, have been analyzed by cryo-EM (Spahn et al., 2001b; Siridechadilok et al., 2005). More strikingly, the IRES elements found in the cricket paralysis virus-like (CrPV) particle are able to assemble elongation-competent ribosomes "from scratch," without requiring any initiation factor or even initiator tRNA (Pestova and Hellen, 2003). Cryo-EM reconstructions of such IRES elements bound to 40S subunits and 80S ribosomes have shown that the IRES elements contact the three tRNA binding sites (see Figure 7.16 for more details), generating conformational changes in the ribosome, which are thought to be an integral part of the general mechanism of translation initiation, such as mRNA positioning in the entry channel via conformational changes in the small subunit (Spahn et al., 2004b; Schuler et al., 2006).

In contrast to translation initiation, elongation relies on universally conserved mechanisms. This conservation is reflected in the tRNA translocation process, which also implicates the ratchet-like rotation of the small 40S subunit. In fact, much of what we know about the underlying general mechanism of translocation comes from cryo-EM studies on yeast ribosomes (Gomez-Lorenzo et al., 2000; Spahn et al., 2004a; Taylor et al., 2007; Sengupta et al., 2008). As previously discussed in Section III.1, the binding of EF-G stabilizes the small subunit in its rotated state as well as the bending of h44 toward the P site by ~8Å. A mechanism by which the remaining 10–12Å required to fully translocate tRNAs from the A site to the P site has been proposed

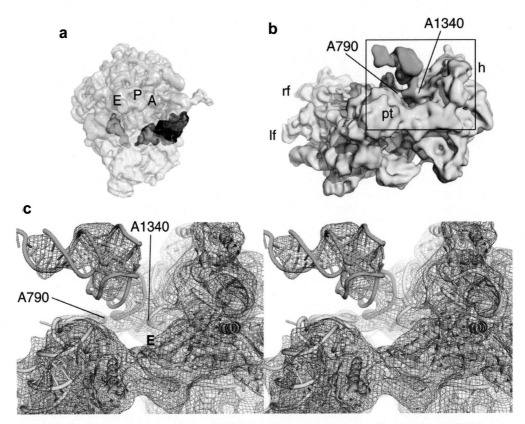

FIGURE 7.17: The passage gate for the P/E tRNA. *(a) Cryo-EM reconstruction of eEF2• GDPNP-bound 80S ribosomes: hybrid P/E tRNA (green), eEF2 (red), 40S subunit (yellow), 60S subunit (blue). (b) Close-up view of the 40S subunit (c) Stereo-view representation of the interactions between the small subunit and the hybrid tRNA. The view in (c) corresponds to a close-up of that shown in (b). Labels and landmarks: lf (left foot), rf (right foot), pt (platform), h (head), E (E-site). Data reproduced with permission from (Frank et al., 2007). "Copyright (2007) National Academy of Sciences, U.S.A."*

based on the analysis of eEF2-bound 80S ribosomes reconstructed by cryo-EM. The changes observed in domain IV upon the hydrolysis of GTP (see Figure 7.8) form the structural basis for the decoupling of the mRNA-tRNA complex in the A site from the decoding center (Taylor et al., 2007). This event leads to the rotation of the head with respect to the body of the small subunit, which goes in the direction of the E site, a rearrangement that was not only observed in the aforementioned cryo-EM reconstructions but also in the crystal structures of vacant *E. coli* 70S ribosomes (Schuwirth et al., 2005). These crystallographic structures further suggest that certain nucleotides, positioned in the P to E site path, may form a "revolving door" that alters its position depending on the status of the rotated head (Figure 7.17). If this hypothesis turns out to be correct, then it may be possible to cover the remaining distance from the A and P sites, to the P and E sites, respectively, with this head rotation.

The proposed mechanism by which the interaction between the decoding center and mRNA-tRNA complex is decoupled has recently received indirect support from X-ray structures of the post-translocational ribosomes, which

were obtained by trapping EF-G with fusidic acid (Gao et al., 2009). In this atomic structure, as expected in a post-translocational ribosome, no contacts between domain IV and the A site codon are discernible. Instead, the structure shows contacts between the tip of domain IV and the P site tRNA. Thus, it is likely that these interactions, primarily established in the A site, are kept moving along the translocation pathway of the tRNA from the A site to the P site.

V. CONCLUDING REMARKS

In this chapter we covered the advances in the exploration of structural dynamics of the protein synthesis cycle from cryo-EM studies, most of which have only come to light in the past few years. As we have seen, cryo-EM has provided evidence that the tRNA shows a distorted conformation during incorporation and that after peptidyl transfer, the ribosome constantly fluctuates between different configurations related to the inter-subunit rotation. Other examples are the conformational changes associated with the binding of elongation factors during the course of the

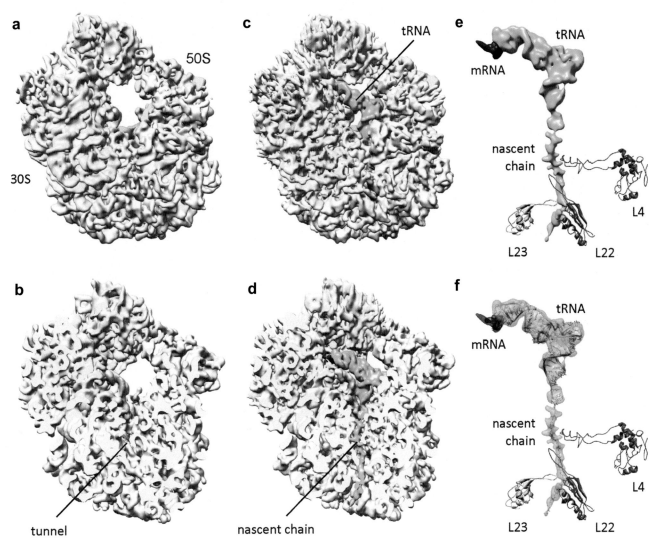

FIGURE 7.18: Nascent polypeptide chain–mediated translational stalling. *(a), (b) Cryo-EM reconstruction of vacant 70S ribosomes at 6.6Å resolution (c), (d) Cryo-EM reconstruction of TnaC•70S complexes at 5.8Å resolution. (e), (f) Close-up view of the TnaC-tRNA structure. The TnaC-tRNA is shown in green and the mRNA in red. The small and large subunits are shown in yellow and blue, respectively. For orientation purposes, the relative positions of L4 (purple), L22 (blue), and L23 (yellow) are shown. Labels and landmarks: L4 (LSU protein 4), L22 (LSU protein 22), L23 (LSU protein 23). Data reproduced with permission from (Seidelt et al., 2009).*

elongation cycle and the interplay between different factors during translation initiation and termination.

Although not mentioned throughout the text, cryo-EM-derived findings of enormous biological relevance are those related to ribosomal complexes containing nascent chains as they emerge from the ribosomal exit tunnel (Gilbert et al., 2004; Seidelt et al., 2009; Bhushan et al., 2010) (Figure 7.18), as well as ribosomes tagged with signal recognition particles (SRP) that enable the emerging proteins to target cell membranes by docking with SRP receptors (Halic et al., 2004; Halic et al., 2006a; Halic et al., 2006b; Schaffitzel et al., 2006). These secreted membrane proteins, which contain N-term hydrophobic signal sequences recognized by the SRP, are translocated across or into cell membranes (cytoplasmic membranes in prokaryotes and endo-

plasmic reticulum membranes in eukaryotes) via protein-conducting channels, also known as translocons. In fact, several of such integral membrane protein devices, either prokaryotic or eukaryotic, bound to ribosomal complexes, have also been depicted by cryo-EM studies and their workings analyzed (Beckmann et al., 1997; Beckmann et al., 2001; Menetret et al., 2000; Menetret et al., 2005; Mitra et al., 2005; Menetret et al., 2007; Menetret et al., 2008; Becker et al., 2009).

It is noteworthy that the current limitation of cryo-EM as a general approach is the limited resolution of the resulting reconstructions: Whereas a resolution in the range of 9–12Å is routinely reached, reconstructions showing a resolution in the range of 5–7Å, displaying near-atomic details, are still the exception (Schuette et al., 2009, Seidelt et al.,

2009; Villa et al., 2009) (Figure 7.18). One of the crucial advances needed to improve the quality of 3D structures, apart from more massive data collection and processing capabilities, is the means to efficiently sort out the different populations that coexist in the sample. Notwithstanding, as we have seen, cryo-EM contributions toward the understanding of the workings of protein synthesizing machinery have already been immense. In addition, until quite recently (Rabl et al., 2010; Ben-Shem et al., 2010) cryo-EM has been the only source of structural information about eukaryotic systems, the knowledge of which will be essential in the future for understanding the various ribosomal dysfunctions associated with human disease. Needless to say, kinetic studies, smFRET data, and X-ray structures have been of pivotal importance in unraveling the workings of the ribosome. Moreover, atomic data have been indispensable when interpreting cryo-EM results by fitting coordinates into the maps. In this chapter, we mainly focused on summarizing the contributions made by cryo-EM and image processing to deciphering the dynamics of this complex molecular machine. However, enormous challenges are still ahead, because the function of the ribosome is far from fully understood, and ultimately the key for understanding the workings and functions of ribosomes will come from merging and interpreting the results from a wide range of experimental approaches.

ACKNOWLEDGMENTS

The authors thank Derek Taylor for constructive criticisms of the manuscript. X.A. is a recipient of a "Ramon y Cajal" Fellowship from the Spanish Government (RYC-2009–04885). Support from the Department of Education, Universities and Research of the Basque Country Government to X.A., the Etortek Research Programmes 2008/2010 (Department of Industry, Tourism and Trade of the Basque Country Government) and the Innovation Technology Department of the Bizkaia County (Basque Country) to M.V. are also acknowledged.

REFERENCES

Acker, M. G. & Lorsch, J. R. (2008) Mechanism of ribosomal subunit joining during eukaryotic translation initiation. *Biochem Soc Trans*, 36, 653–7.

Aevarsson, A., Brazhnikov, E., Garber, M., Zheltonosova, J., Chirgadze, Y., Al-Karadaghi, S., Svensson, L. A. & Liljas, A. (1994) Three-dimensional structure of the ribosomal translocase: elongation factor G from Thermus thermophilus. *EMBO J*, 13, 3669–77.

Agirrezabala, X. & Frank, J. (2009) Elongation in translation as a dynamic interaction among the ribosome, tRNA, and elongation factors EF-G and EF-Tu. *Q Rev Biophys*, 42, 159–200.

Agirrezabala, X. & Frank, J. (2010) From DNA to proteins via the ribosome: structural insights into the workings of the translation machinery. *Hum Genomics*, 4, 226–37.

Agirrezabala, X., Lei, J., Brunelle, J. L., Ortiz-Meoz, R. F., Green, R. & Frank, J. (2008) Visualization of the hybrid state of tRNA binding promoted by spontaneous ratcheting of the ribosome. *Mol Cell*, 32, 190–7.

Agrawal, R. K., Heagle, A. B., Penczek, P., Grassucci, R. A. & Frank, J. (1999) EF-G-dependent GTP hydrolysis induces translocation accompanied by large conformational changes in the 70S ribosome. *Nat Struct Biol*, 6, 643–7.

Agrawal, R. K., Penczek, P., Grassucci, R. A. & Frank, J. (1998) Visualization of elongation factor G on the Escherichia coli 70S ribosome: the mechanism of translocation. *Proc Natl Acad Sci USA*, 95, 6134–8.

Agrawal, R. K., Sharma, M. R., Kiel, M. C., Hirokawa, G., Booth, T. M., Spahn, C. M., Grassucci, R. A., Kaji, A. & Frank, J. (2004) Visualization of ribosome-recycling factor on the Escherichia coli 70S ribosome: functional implications. *Proc Natl Acad Sci USA*, 101, 8900–5.

Al-Karadaghi, S., Aevarsson, A., Garber, M., Zheltonosova, J. & Liljas, A. (1996) The structure of elongation factor G in complex with GDP: conformational flexibility and nucleotide exchange. *Structure*, 4, 555–65.

Algire, M. A., Maag, D., Savio, P., Acker, M. G., Tarun, S. Z., Jr., Sachs, A. B., Asano, K., Nielsen, K. H., Olsen, D. S., Phan, L., Hinnebusch, A. G. & Lorsch, J. R. (2002) Development and characterization of a reconstituted yeast translation initiation system. *RNA*, 8, 382–97.

Allen, G. S. & Frank, J. (2007) Structural insights on the translation initiation complex: ghosts of a universal initiation complex. *Mol Microbiol*, 63, 941–50.

Allen, G. S., Zavialov, A., Gursky, R., Ehrenberg, M. & Frank, J. (2005) The cryo-EM structure of a translation initiation complex from Escherichia coli. *Cell*, 121, 703–12.

Andersen, C. B., Becker, T., Blau, M., Anand, M., Halic, M., Balar, B., Mielke, T., Boesen, T., Pedersen, J. S., Spahn, C. M., Kinzy, T. G., Andersen, G. R. & Beckmann, R. (2006) Structure of eEF3 and the mechanism of transfer RNA release from the E-site. *Nature*, 443, 663–8.

Antoun, A., Pavlov, M. Y., Andersson, K., Tenson, T. & Ehrenberg, M. (2003) The roles of initiation factor 2 and guanosine triphosphate in initiation of protein synthesis. *EMBO J*, 22, 5593–601.

Antoun, A., Pavlov, M. Y., Lovmar, M. & Ehrenberg, M. (2006) How initiation factors tune the rate of initiation of protein synthesis in bacteria. *EMBO J*, 25, 2539–50.

Barat, C., Datta, P. P., Raj, V. S., Sharma, M. R., Kaji, H., Kaji, A. & Agrawal, R. K. (2007) Progression of the ribosome recycling factor through the ribosome dissociates the two ribosomal subunits. *Mol Cell*, 27, 250–61.

Becker, T., Bhushan, S., Jarasch, A., Armache, J. P., Funes, S., Jossinet, F., Gumbart, J., Mielke, T., Berninghausen, O., Schulten, K., Westhof, E., Gilmore, R., Mandon, E. C. & Beckmann, R. (2009) Structure of monomeric yeast and mammalian Sec61 complexes interacting with the translating ribosome. *Science*, 326, 1369–73.

Beckmann, R., Bubeck, D., Grassucci, R., Penczek, P., Verschoor, A., Blobel, G. & Frank, J. (1997) Alignment of conduits for the nascent polypeptide chain in the ribosome-Sec61 complex. *Science*, 278, 2123–6.

Beckmann, R., Spahn, C. M., Eswar, N., Helmers, J., Penczek, P. A., Sali, A., Frank, J. & Blobel, G. (2001) Architecture of the protein-conducting channel associated with the translating 80S ribosome. *Cell*, 107, 361–72.

Ben-Shem, A., Jenner, L., Yusupova, G. & Yusupov, M. (2010) Crystal structure of the eukaryotic ribosome. Science, 26, 1203–9.

Beringer, M. & Rodnina, M. V. (2007) The ribosomal peptidyl transferase. *Mol Cell*, 26, 311–21.

Bhushan, S., Gartmann, M., Halic, M., Armache, J. P., Jarasch, A., Mielke, T., Berninghausen, O., Wilson, D. N. & Beckmann, R. (2010) alpha-Helical nascent polypeptide chains visualized within distinct regions of the ribosomal exit tunnel. *Nat Struct Mol Biol*, 17, 313–7.

Blaha, G., Stanley, R. E. & Steitz, T. A. (2009) Formation of the first peptide bond: the structure of EF-P bound to the 70S ribosome. *Science*, 325, 966–70.

Blanchard, S. C., Gonzalez, R. L., Kim, H. D., Chu, S. & Puglisi, J. D. (2004) tRNA selection and kinetic proofreading in translation. *Nat Struct Mol Biol*, 11, 1008–14.

Boelens, R. & Gualerzi, C. O. (2002) Structure and function of bacterial initiation factors. *Curr Protein Pept Sci*, 3, 107–19.

Borovinskaya, M. A., Pai, R. D., Zhang, W., Schuwirth, B. S., Holton, J. M., Hirokawa, G., Kaji, H., Kaji, A. & Cate, J. H. (2007) Structural basis for aminoglycoside inhibition of bacterial ribosome recycling. *Nat Struct Mol Biol*, 14, 727–32.

Bretscher, M. S. (1968) Translocation in protein synthesis: a hybrid structure model. *Nature*, 218, 675–7.

Carter, A. P., Clemons, W. M., Jr., Brodersen, D. E., Morgan-Warren, R. J., Hartsch, T., Wimberly, B. T. & Ramakrishnan, V. (2001) Crystal structure of an initiation factor bound to the 30S ribosomal subunit. *Science*, 291, 498–501.

Chandramouli, P., Topf, M., Menetret, J. F., Eswar, N., Cannone, J. J., Gutell, R. R., Sali, A. & Akey, C. W. (2008) Structure of the mammalian 80S ribosome at 8.7 Å resolution. *Structure*, 16, 535–48.

Connell, S. R., Takemoto, C., Wilson, D. N., Wang, H., Murayama, K., Terada, T., Shirouzu, M., Rost, M., Schuler, M., Giesebrecht, J., Dabrowski, M., Mielke, T., Fucini, P., Yokoyama, S. & Spahn, C. M. (2007) Structural basis for interaction of the ribosome with the switch regions of GTP-bound elongation factors. *Mol Cell*, 25, 751–64.

Connell, S. R., Topf, M., Qin, Y., Wilson, D. N., Mielke, T., Fucini, P., Nierhaus, K. H. & Spahn, C. M. (2008) A new tRNA intermediate revealed on the ribosome during EF4-mediated back-translocation. *Nat Struct Mol Biol*, 15, 910–5.

Cornish, P. V., Ermolenko, D. N., Noller, H. F. & Ha, T. (2008) Spontaneous inter-subunit rotation in single ribosomes. *Mol Cell*, 30, 578–88.

Cornish, P. V., Ermolenko, D. N., Staple, D. W., Hoang, L., Hickerson, R. P., Noller, H. F. & Ha, T. (2009) Following movement of the L1 stalk between three functional states in single ribosomes. *Proc Natl Acad Sci USA*, 106, 2571–6.

Czworkowski, J., Wang, J., Steitz, T. A. & Moore, P. B. (1994) The crystal structure of elongation factor G complexed with GDP, at 2.7 Å resolution. *EMBO J*, 13, 3661–8.

Datta, P. P., Sharma, M. R., Qi, L., Frank, J. & Agrawal, R. K. (2005) Interaction of the G′ domain of elongation factor G and the C-terminal domain of ribosomal protein L7/L12 during translocation as revealed by cryo-EM. *Mol Cell*, 20, 723–31.

Dinman, J. D. (2009) The eukaryotic ribosome: current status and challenges. *J Biol Chem*, 284, 11761–5.

Dorner, S., Brunelle, J. L., Sharma, D. & Green, R. (2006) The hybrid state of tRNA binding is an authentic translation elongation intermediate. *Nat Struct Mol Biol*, 13, 234–41.

Ermolenko, D. N., Majumdar, Z. K., Hickerson, R. P., Spiegel, P. C., Clegg, R. M. & Noller, H. F. (2007a) Observation of intersubunit movement of the ribosome in solution using FRET. *J Mol Biol*, 370, 530–40.

Ermolenko, D. N., Spiegel, P. C., Majumdar, Z. K., Hickerson, R. P., Clegg, R. M. & Noller, H. F. (2007b) The antibiotic viomycin traps the ribosome in an intermediate state of translocation. *Nat Struct Mol Biol*, 14, 493–7.

Fei, J., Kosuri, P., Macdougall, D. D. & Gonzalez, R. L., Jr. (2008) Coupling of ribosomal L1 stalk and tRNA dynamics during translation elongation. *Mol Cell*, 30, 348–59.

Frank, J. & Agrawal, R. K. (2000) A ratchet-like inter-subunit reorganization of the ribosome during translocation. *Nature*, 406, 318–22.

Frank, J., Gao, H., Sengupta, J., Gao, N. & Taylor, D. J. (2007) The process of mRNA-tRNA translocation. *Proc Natl Acad Sci USA*, 104, 19671–8.

Frank, J. & Gonzalez Jr, R. L. (2010) Structure and Dynamics of a Processive Brownian Motor: The Translating Ribosome. *Annu Rev Biochem*, 79, 381–412.

Frank, J., Sengupta, J., Gao, H., Li, W., Valle, M., Zavialov, A. & Ehrenberg, M. (2005) The role of tRNA as a molecular spring in decoding, accommodation, and peptidyl transfer. *FEBS Lett*, 579, 959–62.

Gao, H., Ayub, M. J., Levin, M. J. & Frank, J. (2005a) The structure of the 80S ribosome from Trypanosoma cruzi reveals unique rRNA components. *Proc Natl Acad Sci USA*, 102, 10206–11.

Gao, H., Zhou, Z., Rawat, U., Huang, C., Bouakaz, L., Wang, C., Cheng, Z., Liu, Y., Zavialov, A., Gursky, R., Sanyal, S., Ehrenberg, M., Frank, J. & Song, H. (2007) RF3 induces ribosomal conformational changes responsible for dissociation of class I release factors. *Cell*, 129, 929–41.

Gao, N., Zavialov, A. V., Li, W., Sengupta, J., Valle, M., Gursky, R. P., Ehrenberg, M. & Frank, J. (2005b) Mechanism for the disassembly of the posttermination complex inferred from cryo-EM studies. *Mol Cell*, 18, 663–74.

Gao, Y. G., Selmer, M., Dunham, C. M., Weixlbaumer, A., Kelley, A. C. & Ramakrishnan, V. (2009) The structure of the ribosome with elongation factor G trapped in the posttranslocational state. *Science*, 326, 694–9.

Gilbert, R. J., Fucini, P., Connell, S., Fuller, S. D., Nierhaus, K. H., Robinson, C. V., Dobson, C. M. & Stuart, D. I. (2004) Three-dimensional structures of translating ribosomes by Cryo-EM. *Mol Cell*, 14, 57–66.

Gilbert, R. J., Gordiyenko, Y., Von Der Haar, T., Sonnen, A. F., Hofmann, G., Nardelli, M., Stuart, D. I. & Mccarthy, J. E. (2007) Reconfiguration of yeast 40S ribosomal subunit domains by the translation initiation multifactor complex. *Proc Natl Acad Sci USA*, 104, 5788–93.

Gomez-Lorenzo, M. G., Spahn, C. M., Agrawal, R. K., Grassucci, R. A., Penczek, P., Chakraburtty, K., Ballesta, J. P., Lavandera, J. L., Garcia-Bustos, J. F. & Frank, J. (2000) Three-dimensional cryo-electron microscopy localization of EF2 in the Saccharomyces cerevisiae 80S ribosome at 17.5 Å resolution. *EMBO J*, 19, 2710–8.

Gromadski, K. B. & Rodnina, M. V. (2004) Kinetic determinants of high-fidelity tRNA discrimination on the ribosome. *Mol Cell*, 13, 191–200.

Gualerzi, C. O. & Pon, C. L. (1990) Initiation of mRNA translation in prokaryotes. *Biochemistry*, 29, 5881–9.

Halic, M., Becker, T., Pool, M. R., Spahn, C. M., Grassucci, R. A., Frank, J. & Beckmann, R. (2004) Structure of the signal recognition particle interacting with the elongation-arrested ribosome. *Nature*, 427, 808–14.

Halic, M., Blau, M., Becker, T., Mielke, T., Pool, M. R., Wild, K., Sinning, I. & Beckmann, R. (2006a) Following the signal sequence from ribosomal tunnel exit to signal recognition particle. *Nature*, 444, 507–11.

Halic, M., Gartmann, M., Schlenker, O., Mielke, T., Pool, M. R., Sinning, I. & Beckmann, R. (2006b) Signal recognition particle receptor exposes the ribosomal translocon binding site. *Science*, 312, 745–7.

Hansson, S., Singh, R., Gudkov, A. T., Liljas, A. & Logan, D. T. (2005a) Crystal structure of a mutant elongation factor G trapped with a GTP analogue. *FEBS Lett*, 579, 4492–7.

Hansson, S., Singh, R., Gudkov, A. T., Liljas, A. & Logan, D. T. (2005b) Structural insights into fusidic acid resistance and sensitivity in EF-G. *J Mol Biol*, 348, 939–49.

Jenner, L., Rees, B., Yusupov, M. & Yusupova, G. (2007) Messenger RNA conformations in the ribosomal E site revealed by X-ray crystallography. *EMBO Rep*, 8, 846–50.

Jenner, L., Romby, P., Rees, B., Schulze-Briese, C., Springer, M., Ehresmann, C., Ehresmann, B., Moras, D., Yusupova, G. & Yusupov, M. (2005) Translational operator of mRNA on the ribosome: how repressor proteins exclude ribosome binding. *Science*, 308, 120–3.

Jorgensen, R., Merrill, A. R., Yates, S. P., Marquez, V. E., Schwan, A. L., Boesen, T. & Andersen, G. R. (2005) Exotoxin A-eEF2 complex structure indicates ADP ribosylation by ribosome mimicry. *Nature*, 436, 979–84.

Jorgensen, R., Ortiz, P. A., Carr-Schmid, A., Nissen, P., Kinzy, T. G. & Andersen, G. R. (2003) Two crystal structures demonstrate large conformational changes in the eukaryotic ribosomal translocase. *Nat Struct Biol*, 10, 379–85.

Jorgensen, R., Yates, S. P., Teal, D. J., Nilsson, J., Prentice, G. A., Merrill, A. R. & Andersen, G. R. (2004) Crystal structure of ADP-ribosylated ribosomal translocase from Saccharomyces cerevisiae. *J Biol Chem*, 279, 45919–25.

Julian, P., Konevega, A. L., Scheres, S. H., Lazaro, M., Gil, D., Wintermeyer, W., Rodnina, M. V. & Valle, M. (2008) Structure of ratcheted ribosomes with tRNAs in hybrid states. *Proc Natl Acad Sci USA*, 105, 16924–7.

Kapp, L. D. & Lorsch, J. R. (2004) The molecular mechanics of eukaryotic translation. *Annu Rev Biochem*, 73, 657–704.

Karimi, R., Pavlov, M. Y., Buckingham, R. H. & Ehrenberg, M. (1999) Novel roles for classical factors at the interface between translation termination and initiation. *Mol Cell*, 3, 601–9.

Kieft, J. S., Zhou, K., Jubin, R. & Doudna, J. A. (2001) Mechanism of ribosome recruitment by hepatitis C IRES RNA. *RNA*, 7, 194–206.

Kieft, J. S., Zhou, K., Jubin, R., Murray, M. G., Lau, J. Y. & Doudna, J. A. (1999) The hepatitis C virus internal ribosome entry site adopts an ion-dependent tertiary fold. *J Mol Biol*, 292, 513–29.

Kim, H. D., Puglisi, J. D. & Chu, S. (2007) Fluctuations of transfer RNAs between classical and hybrid states. *Biophys J*, 93, 3575–82.

Kim, K. K., Min, K. & Suh, S. W. (2000) Crystal structure of the ribosome recycling factor from Escherichia coli. *EMBO J*, 19, 2362–70.

Kisselev, L., Ehrenberg, M. & Frolova, L. (2003) Termination of translation: interplay of mRNA, rRNAs and release factors? *EMBO J*, 22, 175–82.

Klaholz, B. P., Myasnikov, A. G. & Van Heel, M. (2004) Visualization of release factor 3 on the ribosome during termination of protein synthesis. *Nature*, 427, 862–5.

Klaholz, B. P., Pape, T., Zavialov, A. V., Myasnikov, A. G., Orlova, E. V., Vestergaard, B., Ehrenberg, M. & Van Heel, M. (2003) Structure of the Escherichia coli ribosomal termination complex with release factor 2. *Nature*, 421, 90–4.

Korostelev, A., Asahara, H., Lancaster, L., Laurberg, M., Hirschi, A., Zhu, J., Trakhanov, S., Scott, W. G. & Noller, H. F. (2008) Crystal structure of a translation termination complex formed with release factor RF2. *Proc Natl Acad Sci USA*, 105, 19684–9.

Korostelev, A., Trakhanov, S., Asahara, H., Laurberg, M., Lancaster, L. & Noller, H. F. (2007) Interactions and dynamics of the Shine Dalgarno helix in the 70S ribosome. *Proc Natl Acad Sci USA*, 104, 16840–3.

Kressler, D., Hurt, E. & Babetaler, J. (2010) Driving ribosome assembly. *Biochim Biophys Acta*, 1803, 673–83.

La Teana, A., Gualerzi, C. O. & Brimacombe, R. (1995) From stand-by to decoding site. Adjustment of the mRNA on the 30S ribosomal subunit under the influence of the initiation factors. *RNA*, 1, 772–82.

Lancaster, L., Kiel, M. C., Kaji, A. & Noller, H. F. (2002) Orientation of ribosome recycling factor in the ribosome from directed hydroxyl radical probing. *Cell*, 111, 129–40.

Laurberg, M., Asahara, H., Korostelev, A., Zhu, J., Trakhanov, S. & Noller, H. F. (2008) Structural basis for translation termination on the 70S ribosome. *Nature*, 454, 852–7.

Laursen, B. S., Sorensen, H. P., Mortensen, K. K. & Sperling-Petersen, H. U. (2005) Initiation of protein synthesis in bacteria. *Microbiol Mol Biol Rev*, 69, 101–23.

Li, W., Agirrezabala, X., Lei, J., Bouakaz, L., Brunelle, J. L., Ortiz-Meoz, R. F., Green, R., Sanyal, S., Ehrenberg, M. & Frank, J. (2008) Recognition of aminoacyl-tRNA: a common molecular mechanism revealed by cryo-EM. *EMBO J*, 27, 3322–31.

Marshall, R. A., Aitken, C. E., Dorywalska, M. & Puglisi, J. D. (2008a) Translation at the single-molecule level. *Annu Rev Biochem*, 77, 177–203.

Marshall, R. A., Aitken, C. E. & Puglisi, J. D. (2009) GTP hydrolysis by IF2 guides progression of the ribosome into elongation. *Mol Cell*, 35, 37–47.

Marshall, R. A., Dorywalska, M. & Puglisi, J. D. (2008b) Irreversible chemical steps control intersubunit dynamics during translation. *Proc Natl Acad Sci USA*, 105, 15364–9.

Marzi, S., Myasnikov, A. G., Serganov, A., Ehresmann, C., Romby, P., Yusupov, M. & Klaholz, B. P. (2007) Structured mRNAs regulate translation initiation by binding to the platform of the ribosome. *Cell*, 130, 1019–31.

Mccutcheon, J. P., Agrawal, R. K., Philips, S. M., Grassucci, R. A., Gerchman, S. E., Clemons, W. M., Jr., Ramakrishnan, V. &

Frank, J. (1999) Location of translational initiation factor IF3 on the small ribosomal subunit. *Proc Natl Acad Sci USA*, 96, 4301–6.

Menetret, J. F., Hegde, R. S., Aguiar, M., Gygi, S. P., Park, E., Rapoport, T. A. & Akey, C. W. (2008) Single copies of Sec61 and TRAP associate with a nontranslating mammalian ribosome. *Structure*, 16, 1126–37.

Menetret, J. F., Hegde, R. S., Heinrich, S. U., Chandramouli, P., Ludtke, S. J., Rapoport, T. A. & Akey, C. W. (2005) Architecture of the ribosome-channel complex derived from native membranes. *J Mol Biol*, 348, 445–57.

Menetret, J. F., Neuhof, A., Morgan, D. G., Plath, K., Radermacher, M., Rapoport, T. A. & Akey, C. W. (2000) The structure of ribosome-channel complexes engaged in protein translocation. *Mol Cell*, 6, 1219–32.

Menetret, J. F., Schaletzky, J., Clemons, W. M., Jr., Osborne, A. R., Skanland, S. S., Denison, C., Gygi, S. P., Kirkpatrick, D. S., Park, E., Ludtke, S. J., Rapoport, T. A. & Akey, C. W. (2007) Ribosome binding of a single copy of the SecY complex: implications for protein translocation. *Mol Cell*, 28, 1083–92.

Milon, P., Konevega, A. L., Gualerzi, C. O. & Rodnina, M. V. (2008) Kinetic checkpoint at a late step in translation initiation. *Mol Cell*, 30, 712–20.

Mitra, K., Schaffitzel, C., Shaikh, T., Tama, F., Jenni, S., Brooks, C. L., III, Ban, N. & Frank, J. (2005) Structure of the E. coli protein-conducting channel bound to a translating ribosome. *Nature*, 438, 318–24.

Moazed, D. & Noller, H. F. (1989) Intermediate states in the movement of transfer RNA in the ribosome. *Nature*, 342, 142–8.

Munro, J. B., Altman, R. B., O'Connor, N. & Blanchard, S. C. (2007) Identification of two distinct hybrid state intermediates on the ribosome. *Mol Cell*, 25, 505–17.

Munro, J. B., Altman, R. B., Tung, C. S., Cate, J. H., Sanbonmatsu, K. Y. & Blanchard, S. C. (2010) Spontaneous formation of the unlocked state of the ribosome is a multistep process. *Proc Natl Acad Sci USA*, 107, 709–14.

Munro, J. B., Sanbonmatsu, K. Y., Spahn, C. M. & Blanchard, S. C. (2009) Navigating the ribosome's metastable energy landscape. *Trends Biochem Sci*, 34, 390–400.

Myasnikov, A. G., Marzi, S., Simonetti, A., Giuliodori, A. M., Gualerzi, C. O., Yusupova, G., Yusupov, M. & Klaholz, B. P. (2005) Conformational transition of initiation factor 2 from the GTP- to GDP-bound state visualized on the ribosome. *Nat Struct Mol Biol*, 12, 1145–9.

Myasnikov, A. G., Simonetti, A., Marzi, S. & Klaholz, B. P. (2009) Structure-function insights into prokaryotic and eukaryotic translation initiation. *Curr Opin Struct Biol*, 19, 300–9.

Nakamura, Y. & Ito, K. (2003) Making sense of mimic in translation termination. *Trends Biochem Sci*, 28, 99–105.

Nilsson, J., Sengupta, J., Gursky, R., Nissen, P. & Frank, J. (2007) Comparison of fungal 80 S ribosomes by cryo-EM reveals diversity in structure and conformation of rRNA expansion segments. *J Mol Biol*, 369, 429–38.

Ogle, J. M., Brodersen, D. E., Clemons, W. M., Jr., Tarry, M. J., Carter, A. P. & Ramakrishnan, V. (2001) Recognition of cognate transfer RNA by the 30S ribosomal subunit. *Science*, 292, 897–902.

Ogle, J. M., Murphy, F. V., Tarry, M. J. & Ramakrishnan, V. (2002) Selection of tRNA by the ribosome requires a transition from an open to a closed form. *Cell*, 111, 721–32.

Ogle, J. M. & Ramakrishnan, V. (2005) Structural insights into translational fidelity. *Annu Rev Biochem*, 74, 129–77.

Palade, G. E. (1955) A small particulate component of the cytoplasm. *J Biophys Biochem Cytol*, 1, 59–68.

Panse, V. G. & Johnson, A. W. (2010) Maturation of eukaryotic ribosomes: acquisition of functionality. *Trends Biochem Sci*, 35, 260–6.

Passmore, L. A., Schmeing, T. M., Maag, D., Applefield, D. J., Acker, M. G., Algire, M. A., Lorsch, J. R. & Ramakrishnan, V. (2007) The eukaryotic translation initiation factors eIF1 and eIF1A induce an open conformation of the 40S ribosome. *Mol Cell*, 26, 41–50.

Peske, F., Matassova, N. B., Savelsbergh, A., Rodnina, M. V. & Wintermeyer, W. (2000) Conformationally restricted elongation factor G retains GTPase activity but is inactive in translocation on the ribosome. *Mol Cell*, 6, 501–5.

Peske, F., Rodnina, M. V. & Wintermeyer, W. (2005) Sequence of steps in ribosome recycling as defined by kinetic analysis. *Mol Cell*, 18, 403–12.

Pestova, T. V. & Hellen, C. U. (2003) Translation elongation after assembly of ribosomes on the Cricket paralysis virus internal ribosomal entry site without initiation factors or initiator tRNA. *Genes Dev*, 17, 181–6.

Pestova, T. V., Shatsky, I. N., Fletcher, S. P., Jackson, R. J. & Hellen, C. U. (1998) A prokaryotic-like mode of cytoplasmic eukaryotic ribosome binding to the initiation codon during internal translation initiation of hepatitis C and classical swine fever virus RNAs. *Genes Dev*, 12, 67–83.

Petry, S., Brodersen, D. E., Murphy, F. V. T., Dunham, C. M., Selmer, M., Tarry, M. J., Kelley, A. C. & Ramakrishnan, V. (2005) Crystal structures of the ribosome in complex with release factors RF1 and RF2 bound to a cognate stop codon. *Cell*, 123, 1255–66.

Petry, S., Weixlbaumer, A. & Ramakrishnan, V. (2008) The termination of translation. *Curr Opin Struct Biol*, 18, 70–7.

Pioletti, M., Schlunzen, F., Harms, J., Zarivach, R., Gluhmann, M., Avila, H., Bashan, A., Bartels, H., Auerbach, T., Jacobi, C., Hartsch, T., Yonath, A. & Franceschi, F. (2001) Crystal structures of complexes of the small ribosomal subunit with tetracycline, edeine and IF3. *EMBO J*, 20, 1829–39.

Qin, Y., Polacek, N., Vesper, O., Staub, E., Einfeldt, E., Wilson, D. N. & Nierhaus, K. H. (2006) The highly conserved LepA is a ribosomal elongation factor that back-translocates the ribosome. *Cell*, 127, 721–33.

Rabl, J., Leibundgut, M., Ataide, S. F., Haag, A. & Ban, N. (2010) Crystal structure of the eukaryotic 40S ribosomal subunit in complex with initiation factor 1. *Science*, 331, 730–6.

Rawat, U., Gao, H., Zavialov, A., Gursky, R., Ehrenberg, M. & Frank, J. (2006) Interactions of the release factor RF1 with the ribosome as revealed by cryo-EM. *J Mol Biol*, 357, 1144–53.

Rawat, U. B., Zavialov, A. V., Sengupta, J., Valle, M., Grassucci, R. A., Linde, J., Vestergaard, B., Ehrenberg, M. & Frank, J. (2003) A cryo-electron microscopic study of ribosome-bound termination factor RF2. *Nature*, 421, 87–90.

Rodnina, M. V., Fricke, R., Kuhn, L. & Wintermeyer, W. (1995) Codon-dependent conformational change of elongation factor

Tu preceding GTP hydrolysis on the ribosome. *EMBO J*, 14, 2613–9.

Rodnina, M. V., Fricke, R. & Wintermeyer, W. (1994) Transient conformational states of aminoacyl-tRNA during ribosome binding catalyzed by elongation factor Tu. *Biochemistry*, 33, 12267–75.

Rodnina, M. V., Pape, T., Fricke, R., Kuhn, L. & Wintermeyer, W. (1996) Initial binding of the elongation factor Tu.*GTP.aminoacyl-tRNA complex preceding codon recognition on the ribosome. J Biol Chem*, 271, 646–52.

Rodnina, M. V. & Wintermeyer, W. (2001) Fidelity of aminoacyl-tRNA selection on the ribosome: kinetic and structural mechanisms. *Annu Rev Biochem*, 70, 415–35.

Rodnina, M. V. & Wintermeyer, W. (2009) Recent mechanistic insights into eukaryotic ribosomes. *Curr Opin Cell Biol*, 21, 435–43.

Savelsbergh, A., Katunin, V. I., Mohr, D., Peske, F., Rodnina, M. V. & Wintermeyer, W. (2003) An elongation factor G-induced ribosome rearrangement precedes tRNA-mRNA translocation. *Mol Cell*, 11, 1517–23.

Schaffitzel, C., Oswald, M., Berger, I., Ishikawa, T., Abrahams, J. P., Koerten, H. K., Koning, R. I. & Ban, N. (2006) Structure of the E. coli signal recognition particle bound to a translating ribosome. *Nature*, 444, 503–6.

Schmeing, T. M., Voorhees, R. M., Kelley, A. C., Gao, Y. G., Murphy, F. V. T., Weir, J. R. & Ramakrishnan, V. (2009) The crystal structure of the ribosome bound to EF-Tu and aminoacyl-tRNA. *Science*, 326, 688–94.

Schuette, J. C., Murphy, F. V. T., Kelley, A. C., Weir, J. R., Giesebrecht, J., Connell, S. R., Loerke, J., Mielke, T., Zhang, W., Penczek, P. A., Ramakrishnan, V. & Spahn, C. M. (2009) GTPase activation of elongation factor EF-Tu by the ribosome during decoding. *EMBO J*, 28, 755–65.

Schuler, M., Connell, S. R., Lescoute, A., Giesebrecht, J., Dabrowski, M., Schroeer, B., Mielke, T., Penczek, P. A., Westhof, E. & Spahn, C. M. (2006) Structure of the ribosome-bound cricket paralysis virus IRES RNA. *Nat Struct Mol Biol*, 13, 1092–6.

Schuwirth, B. S., Borovinskaya, M. A., Hau, C. W., Zhang, W., Vila-Sanjurjo, A., Holton, J. M. & Cate, J. H. (2005) Structures of the bacterial ribosome at 3.5 A resolution. *Science*, 310, 827–34.

Seidelt, B., Innis, C. A., Wilson, D. N., Gartmann, M., Armache, J. P., Villa, E., Trabuco, L. G., Becker, T., Mielke, T., Schulten, K., Steitz, T. A. & Beckmann, R. (2009) Structural insight into nascent polypeptide chain-mediated translational stalling. *Science*, 326, 1412–5.

Selmer, M., Al-Karadaghi, S., Hirokawa, G., Kaji, A. & Liljas, A. (1999) Crystal structure of Thermotoga maritima ribosome recycling factor: a tRNA mimic. *Science*, 286, 2349–52.

Sengupta, J., Nilsson, J., Gursky, R., Kjeldgaard, M., Nissen, P. & Frank, J. (2008) Visualization of the eEF2–80S ribosome transition-state complex by cryo-electron microscopy. *J Mol Biol*, 382, 179–87.

Sharma, M. R., Booth, T. M., Simpson, L., Maslov, D. A. & Agrawal, R. K. (2009) Structure of a mitochondrial ribosome with minimal RNA. *Proc Natl Acad Sci USA*, 106, 9637–42.

Sharma, M. R., Koc, E. C., Datta, P. P., Booth, T. M., Spremulli, L. L. & Agrawal, R. K. (2003) Structure of the mammalian mitochondrial ribosome reveals an expanded functional role for its component proteins. *Cell*, 115, 97–108.

Sharma, M. R., Wilson, D. N., Datta, P. P., Barat, C., Schluenzen, F., Fucini, P. & Agrawal, R. K. (2007) Cryo-EM study of the spinach chloroplast ribosome reveals the structural and functional roles of plastid-specific ribosomal proteins. *Proc Natl Acad Sci USA*, 104, 19315–20.

Shin, D. H., Brandsen, J., Jancarik, J., Yokota, H., Kim, R. & Kim, S. H. (2004) Structural analyses of peptide release factor 1 from Thermotoga maritima reveal domain flexibility required for its interaction with the ribosome. *J Mol Biol*, 341, 227–39.

Simonetti, A., Marzi, S., Jenner, L., Myasnikov, A., Romby, P., Yusupova, G., Klaholz, B. P. & Yusupov, M. (2009) A structural view of translation initiation in bacteria. *Cell Mol Life Sci*, 66, 423–36.

Simonetti, A., Marzi, S., Myasnikov, A. G., Fabbretti, A., Yusupov, M., Gualerzi, C. O. & Klaholz, B. P. (2008) Structure of the 30S translation initiation complex. *Nature*, 455, 416–20.

Simonovic, M. & Steitz, T. A. (2009) A structural view on the mechanism of the ribosome-catalyzed peptide bond formation. *Biochim Biophys Acta*, 1789, 612–23.

Siridechadilok, B., Fraser, C. S., Hall, R. J., Doudna, J. A. & Nogales, E. (2005) Structural roles for human translation factor eIF3 in initiation of protein synthesis. *Science*, 310, 1513–5.

Sonenberg, N. & Hinnebusch, A. G. (2009) Regulation of translation initiation in eukaryotes: mechanisms and biological targets. *Cell*, 136, 731–45.

Spahn, C. M., Beckmann, R., Eswar, N., Penczek, P. A., Sali, A., Blobel, G. & Frank, J. (2001a) Structure of the 80S ribosome from Saccharomyces cerevisiae–tRNA-ribosome and subunit-subunit interactions. *Cell*, 107, 373–86.

Spahn, C. M., Gomez-Lorenzo, M. G., Grassucci, R. A., Jorgensen, R., Andersen, G. R., Beckmann, R., Penczek, P. A., Ballesta, J. P. & Frank, J. (2004a) Domain movements of elongation factor eEF2 and the eukaryotic 80S ribosome facilitate tRNA translocation. *EMBO J*, 23, 1008–19.

Spahn, C. M., Jan, E., Mulder, A., Grassucci, R. A., Sarnow, P. & Frank, J. (2004b) Cryo-EM visualization of a viral internal ribosome entry site bound to human ribosomes: the IRES functions as an RNA-based translation factor. *Cell*, 118, 465–75.

Spahn, C. M., Kieft, J. S., Grassucci, R. A., Penczek, P. A., Zhou, K., Doudna, J. A. & Frank, J. (2001b) Hepatitis C virus IRES RNA-induced changes in the conformation of the 40s ribosomal subunit. *Science*, 291, 1959–62.

Spahn, C. M. & Penczek, P. A. (2009) Exploring conformational modes of macromolecular assemblies by multiparticle cryo-EM. *Curr Opin Struct Biol*, 19, 623–31.

Spirin, A. S. (1968) How does the ribosome work? A hypothesis based on the two subunit construction of the ribosome. *Curr Mod Biol*, 2, 115–27.

Stark, H., Rodnina, M. V., Wieden, H. J., Van Heel, M. & Wintermeyer, W. (2000) Large-scale movement of elongation factor G and extensive conformational change of the ribosome during translocation. *Cell*, 100, 301–9.

Stark, H., Rodnina, M. V., Wieden, H. J., Zemlin, F., Wintermeyer, W. & Van Heel, M. (2002) Ribosome interactions of aminoacyl-tRNA and elongation factor Tu in the codon-recognition complex. *Nat Struct Biol*, 9, 849–54.

Sternberg, S. H., Fei, J., Prywes, N., Mcgrath, K. A. & Gonzalez, R. L., Jr. (2009) Translation factors direct intrinsic ribosome

0

dynamics during translation termination and ribosome recycling. *Nat Struct Mol Biol*, 16, 861–8.

Studer, S. M. & Joseph, S. (2006) Unfolding of mRNA secondary structure by the bacterial translation initiation complex. *Mol Cell*, 22, 105–15.

Taylor, D. J., Devkota, B., Huang, A. D., Topf, M., Narayanan, E., Sali, A., Harvey, S. C. & Frank, J. (2009) Comprehensive molecular structure of the eukaryotic ribosome. *Structure*, 17, 1591–604.

Taylor, D. J., Nilsson, J., Merrill, A. R., Andersen, G. R., Nissen, P. & Frank, J. (2007) Structures of modified eEF2 80S ribosome complexes reveal the role of GTP hydrolysis in translocation. *EMBO J*, 26, 2421–31.

Toyoda, T., Tin, O. F., Ito, K., Fujiwara, T., Kumasaka, T., Yamamoto, M., Garber, M. B. & Nakamura, Y. (2000) Crystal structure combined with genetic analysis of the Thermus thermophilus ribosome recycling factor shows that a flexible hinge may act as a functional switch. *RNA*, 6, 1432–44.

Valle, M., Sengupta, J., Swami, N. K., Grassucci, R. A., Burkhardt, N., Nierhaus, K. H., Agrawal, R. K. & Frank, J. (2002) Cryo-EM reveals an active role for aminoacyl-tRNA in the accommodation process. *EMBO J*, 21, 3557–67.

Valle, M., Zavialov, A., Li, W., Stagg, S. M., Sengupta, J., Nielsen, R. C., Nissen, P., Harvey, S. C., Ehrenberg, M. & Frank, J. (2003a) Incorporation of aminoacyl-tRNA into the ribosome as seen by cryo-electron microscopy. *Nat Struct Biol*, 10, 899–906.

Valle, M., Zavialov, A., Sengupta, J., Rawat, U., Ehrenberg, M. & Frank, J. (2003b) Locking and unlocking of ribosomal motions. *Cell*, 114, 123–34.

Vanloock, M. S., Agrawal, R. K., Gabashvili, I. S., Qi, L., Frank, J. & Harvey, S. C. (2000) Movement of the decoding region of the 16 S ribosomal RNA accompanies tRNA translocation. *J Mol Biol*, 304, 507–15.

Varshney, U., Lee, C. P. & Rajbhandary, U. L. (1993) From elongator tRNA to initiator tRNA. *Proc Natl Acad Sci USA*, 90, 2305–9.

Vestergaard, B., Sanyal, S., Roessle, M., Mora, L., Buckingham, R. H., Kastrup, J. S., Gajhede, M., Svergun, D. I. & Ehrenberg, M. (2005) The SAXS solution structure of RF1 differs from its crystal structure and is similar to its ribosome bound cryo-EM structure. *Mol Cell*, 20, 929–38.

Vestergaard, B., Van, L. B., Andersen, G. R., Nyborg, J., Buckingham, R. H. & Kjeldgaard, M. (2001) Bacterial polypeptide release factor RF2 is structurally distinct from eukaryotic eRF1. *Mol Cell*, 8, 1375–82.

Villa, E., Sengupta, J., Trabuco, L. G., Lebarron, J., Baxter, W. T., Shaikh, T. R., Grassucci, R. A., Nissen, P., Ehrenberg, M., Schulten, K. & Frank, J. (2009) Ribosome-induced changes in elongation factor Tu conformation control GTP hydrolysis. *Proc Natl Acad Sci USA*, 106, 1063–8.

Weixlbaumer, A., Jin, H., Neubauer, C., Voorhees, R. M., Petry, S., Kelley, A. C. & Ramakrishnan, V. (2008) Insights into translational termination from the structure of RF2 bound to the ribosome. *Science*, 322, 953–6.

Weixlbaumer, A., Petry, S., Dunham, C. M., Selmer, M., Kelley, A. C. & Ramakrishnan, V. (2007) Crystal structure of the ribosome recycling factor bound to the ribosome. *Nat Struct Mol Biol*, 14, 733–7.

Wilden, B., Savelsbergh, A., Rodnina, M. V. & Wintermeyer, W. (2006) Role and timing of GTP binding and hydrolysis during EF-G-dependent tRNA translocation on the ribosome. *Proc Natl Acad Sci USA*, 103, 13670–5.

Wilson, D. N., Schluenzen, F., Harms, J. M., Yoshida, T., Ohkubo, T., Albrecht, R., Buerger, J., Kobayashi, Y. & Fucini, P. (2005) X-ray crystallography study on ribosome recycling: the mechanism of binding and action of RRF on the 50S ribosomal subunit. *EMBO J*, 24, 251–60.

Yarus, M., Valle, M. & Frank, J. (2003) A twisted tRNA intermediate sets the threshold for decoding. *RNA*, 9, 384–5.

Youngman, E. M., He, S. L., Nikstad, L. J. & Green, R. (2007) Stop codon recognition by release factors induces structural rearrangement of the ribosomal decoding center that is productive for peptide release. *Mol Cell*, 28, 533–43.

Yusupova, G., Jenner, L., Rees, B., Moras, D. & Yusupov, M. (2006) Structural basis for messenger RNA movement on the ribosome. *Nature*, 444, 391–4.

Yusupova, G. Z., Yusupov, M. M., Cate, J. H. & Noller, H. F. (2001) The path of messenger RNA through the ribosome. *Cell*, 106, 233–41.

Zavialov, A. V. & Ehrenberg, M. (2003) Peptidyl-tRNA regulates the GTPase activity of translation factors. *Cell*, 114, 113–22.

Zhang, W., Dunkle, J. A. & Cate, J. H. (2009) Structures of the ribosome in intermediate states of ratcheting. *Science*, 325, 1014–7.

CHAPTER 8

Viewing the Mechanisms of Translation through the Computational Microscope

James Gumbart

Eduard Schreiner

Leonardo G. Trabuco

Kwok-Yan Chan

Klaus Schulten

I. INTRODUCTION

Biological molecular machines span a range of sizes, from the 80-Å-sized helicases (e.g., PcrA) (Dittrich and Schulten, 2006; Dittrich et al., 2006), to the 250-Å-sized ATP synthase (Aksimentiev et al., 2004), to the 10-μm-long Bacterial flagellum (Arkhipov et al., 2006; Kitao et al., 2006). The machines all share certain traits, particularly the ability to utilize energy to perform useful work. Like macroscopic machines, those on the molecular scale are typically comprised of different components that carry out a cycle of well-regulated steps. Unlike macroscopic machines, however, molecular machines must contend with, and even take advantage of, thermal fluctuations that are omnipresent at their scale.

A quintessential example of a large molecular machine, the ribosome, is found in all organisms and in all cells. It is a large (2.5–4.5 MDa) nucleo-protein complex responsible for translating a cell's genetic information into proteins (Korostelev et al., 2008; Steitz, 2008; Schmeing and Ramakrishnan, 2009; Frank and Gonzalez, Jr., 2010). The ribosome is composed of a multitude of interacting components (more than fifty) that assemble into two subunits, denoted large and small. Translation can be broken down into four fundamental stages, initiation, elongation, termination and recycling, each composed of multiple steps and requiring the involvement of additional specialized components. In the first stage (step 1), the two ribosomal subunits join together with a messenger RNA (mRNA) strand to initiate its translation. Initiation is followed by elongation (step 2) of the nascent protein, enabled via the delivery of each amino acid by a transfer RNA (tRNA) in complex with elongation factor Tu (EF-Tu) (Agirrezabala and Frank, 2009). The translocation of tRNAs through the ribosome also occurs in discrete steps, brought about by a large-scale ratchet-like motion of the two ribosomal subunits (Frank

and Agrawal, 2000; Dunke and Cate, 2010). The nascent protein leaves the ribosome through an exit tunnel, which is not merely a passive conduit but can play a regulatory role. Some nascent proteins control their own translation through specific protein-tunnel interactions that halt translation or recruit other factors to the ribosome. For example, proteins not destined for immediate extrusion into the cytoplasm can direct the ribosome to a protein-conducting translocon, the SecY/Sec61 complex, which then aids the proper localization of the nascent protein (Rapoport, 2007). After elongation is completed, translation is terminated (step 3) and the ribosomal components are all recycled (step 4), making them available for the next mRNA.

Many techniques have been brought to bear on the problem of extracting the structural and functional details of molecular machines such as the ribosome, for example, X-ray crystallography, cryo-electron microscopy (cryo-EM), and Förster resonance energy transfer (FRET). At first glance, it might appear that molecular dynamics (MD) simulations, due to their limited time and size scales, cannot contribute significantly to an understanding of highly dynamic and complex processes occurring during translation by the ribosome. However, a number of advances over the years have put simulations in a unique position to approach these processes. MD simulations can now tackle systems involving millions of atoms, a scale ideal for studying ribosome function (see, e.g., Sanbonmatsu et al., 2005; Freddolino et al., 2006; Kitao et al., 2006; Sanbonmatsu and Tung, 2007; Ishida and Hayward, 2008; Klein and Shinoda, 2008; Chandler et al., 2009; Yin et al., 2009; Zink and Grubmüller, 2009). Additionally, the high temporal resolution of MD simulations (typically on the order of 1–2 fs) permits the division of composite functional stages that occur on timescales longer than a normal simulation, into their myriad sub-stages, each of which can be studied independently. For example, elongation by the ribosome requires first the accommodation of a tRNA into its initial binding site, a sub-process already examined using MD (Sanbonmatsu et al., 2005). Simulations also provide a means to bridge data from multiple sources, such as from crystallography and cryo-EM using molecular dynamics flexible fitting (MDFF, detailed later in the chapter), thus providing atomic-detailed structures of the ribosome in a variety of functional states.

To date, MD simulations have been utilized to elucidate a number of mechanisms involved in translation and protein synthesis. In this chapter, we focus on three mechanisms in particular, namely how the ribosome induces GTP hydrolysis by EF-Tu, how the peptide TnaC regulates its own synthesis, and how the translocon is regulated by bound channel partners such as the ribosome. Because a unifying aspect of all three studies is the application of MDFF to determine the structure of a relevant ribosome-protein complex, the MDFF method is first briefly explained. Finally, other studies that have examined mechanisms of ribosomal function utilizing MD simulations are reviewed.

II. MOLECULAR DYNAMICS METHODS

At its most basic level, molecular dynamics (MD) is a computational method in which one solves Newton's second law for N particles in a potential $U(\vec{r}_1, \vec{r}_2, \ldots, \vec{r}_N)$ describing their mutual interactions (Phillips et al., 2005); in other words, one solves the set of equations

$$m_i \ddot{r}_i = -\frac{\partial}{\partial r_i} U(\vec{r}_1, \vec{r}_2, \ldots, \vec{r}_N), i = 1 \cdots N. \quad (1)$$

Equation 1 presents two obvious challenges for obtaining an accurate, numerical solution. The first is how to appropriately define the potential energy function, U, whereas the second challenge is how to discretize the equation in such a way that it can be solved quickly for a large number of atoms while still being representative of the "real" underlying dynamics of the biological system being simulated. Both challenges have been met with great success over the last three decades (Phillips et al., 2005; Schulten et al., 2008).

As noted in the Introduction, the disconnect between the accessible time scale of MD simulations and that of many biological problems of interest is a common limitation, but also a motivation for the development of novel approaches. For example, significant efforts to accelerate MD simulations are being constantly undertaken, including the utilization of specially designed hardware (Shaw et al., 2009) and graphics processing units (Stone et al., 2010). Today, MD simulations covering the ms time scale are becoming feasible. Additionally, MD can serve purposes other than pure simulation of dynamic processes. MD can also aid the combination and interpretation of disjointed experimental data, a purpose that has no inherent time scale, through, for example, molecular dynamics flexible fitting (MDFF).

The MDFF method (Trabuco et al., 2008; Trabuco et al., 2009) can be used to flexibly fit high-resolution structures, typically from X-ray crystallography, into low-resolution density maps, such as those provided by cryo-EM experiments. The MDFF method is based on standard MD simulation techniques, with the addition of two new potential energy terms. The first term is derived directly from the 3D density map, resulting in forces that drive atoms into high-density regions. Because all $3N$ degrees of freedom are employed in an MD simulation, where N is the number of atoms, a natural concern for MDFF is over-fitting, that is, unphysical structural deformations resulting from the under-determined nature of the fitting problem. To address this concern, a second term is added to the potential energy function, corresponding to harmonic restraints designed to preserve secondary-structure elements. The role of this term is discussed extensively in Trabuco et al. (2008). Setup and analysis of MDFF simulations can be performed using VMD (Humphrey et al., 1996), a molecular visualization and analysis package, whereas NAMD (Phillips et al., 2005) is used to conduct the required molecular dynamics simulations. For a recent overview of other flexible fitting methods, the reader is referred to Trabuco et al. (2009).

Structure determination of ribosome complexes was the main driving force for MDFF development and the main system to which the method has been applied. Three examples where MDFF furnished atomic models of ribosome complexes in different functional states are presented in this chapter. The first application of MDFF was the study of a ribosome complex corresponding to an intermediate state of the elongation cycle where the amino acid is being delivered to the ribosome (Villa et al., 2009). The second example is a ribosome in a complex with a protein-conducting channel through which certain nascent proteins are secreted or delivered to a cellular membrane (Becker et al., 2009; Trabuco et al., 2010a). The last example presented in this chapter involves a regulatory nascent peptide that is able to control its own translation while still residing inside the ribosome (Seidelt et al., 2009; Trabuco et al., 2010a).

III. RIBOSOME-INDUCED CONFORMATIONAL CHANGES IN ELONGATION FACTOR TU

The ribosome plays a central role in every living cell by translating the genetic information stored as a sequence of nucleic acids into proteins, which are sequences of amino acids. One of the fundamental steps in ribosome function is the decoding of the gene transcript, the messenger RNA (mRNA), according to the genetic code. Each amino acid is encoded by a codon, a triplet of nucleotides; for example, UUC codes for the amino acid phenylalanine. A pivotal role in the decoding step is played by transfer RNAs (tRNAs),

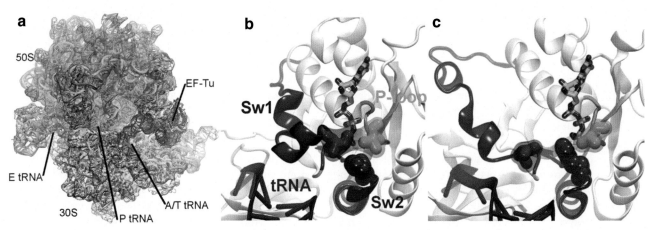

FIGURE 8.1: *Ribosome-induced GTPase activation of EF-Tu. (a) MDFF model derived from a 6.7-Å cryo-EM reconstruction of a 70S:EF-Tu:aminoacyl-tRNA:GDP complex stalled with kirromycin (Reproduced from Villa et al., 2009 with permission). (b) The hydrophobic gate within EF-Tu is formed by Val20 (P-loop, orange) and Ile60 (switch 1, blue) and prevents the catalytic residue His84 (switch 2, red) from accessing the GTP molecule. (c) The swing of switch 1 opens the hydrophobic gate; the catalytic His84 can then reorient toward GTP.*

which are each loaded with an amino acid and recognize the corresponding codon of the mRNA with their own complementary anticodon triplet. In the preceding example, the tRNA carrying the amino acid phenylalanine, tRNAPhe, has the anticodon AAG, the complement to the mRNA codon UUC.

Transfer RNAs are always delivered to the ribosome in a ternary complex (TC) with elongation factor Tu (EF-Tu) and GTP. When the anticodon binds to the matching codon, the low intrinsic GTPase activity of EF-Tu (Fasano et al., 1982) is greatly enhanced, which leads to rapid GTP hydrolysis. In the GDP-bound state, EF-Tu loses its affinity for the tRNA and dissociates from the ribosome. To hydrolyze GTP, EF-Tu has to undergo a conformational change: The hydrophobic gate (Berchtold et al., 1993; Kjeldgaard et al., 1993), which in its closed form prevents the access of the catalytically active histidine to the GTP binding site, must open. How the codon-anticodon binding sends a signal to the 70-Å-distant active site of EF-Tu and how this signal initiates gate opening is still not clear. However, what is clear is that the ribosome holds the gate open (Villa et al., 2009).

Because crystals of the ribosome in complex with additional factors are extremely difficult to obtain, until recently (Schmeing et al., 2009), all structural information on the complex formed between the ribosome and TC stemmed from cryo-EM. To be able to better interpret the densities resulting from cryo-EM, a number of methods were developed to flexibly fit molecules into the cryo-EM density, including MDFF. By fitting the individually known crystal structures of all the components, MDFF allowed the interpretation of a relatively high resolution, 6.7-Å cryo-EM map obtained from an *E. coli* 70S ribosome in complex with an aminoacyl-tRNA:EF-Tu:GDP TC, which was stalled by the antibiotic kirromycin preventing dissociation

(Villa et al., 2009) (see Figure 8.1a). The fitted models of the ribosome-TC complex clearly show how, in the context of a cognate tRNA, the ribosome is able to hold open the hydrophobic gate formed by switch I and the P-loop, allowing the catalytic residue His84 (Daviter et al., 2003) to access GTP (see Figure 8.1b). One wing of the gate, the P-loop, is kept in place relative to the rest of EF-Tu by its interaction with the sarcin-ricin loop, a feature of the 23S rRNA (ribosomal RNA). The other wing, switch I, is kept open by helices h8 and h14 of the 16S rRNA. When the gate is open, the catalytic His84 is able to reach the GTP binding site to induce hydrolysis.

The enormous advances in X-ray crystallography have recently provided model structures of the Bacterial ribosome in complex with elongation factors derived from X-ray diffraction (Gao et al., 2009; Schmeing et al., 2009). The structure of the complex with EF-Tu (Schmeing et al., 2009) is in excellent agreement with the previous cryo-EM-derived model and confirms the conformation of the opened hydrophobic gate and its mechanism of stabilization (Villa et al., 2009).

An interesting observation made from the X-ray structures is a very distinct "swing" of the D loop of the tRNA with respect to the structure of the isolated TC (see Figure 8.2a). This swing is not observed, however, in any of the MDFF-derived models. Unfortunately, straightforward comparison of the structures is prevented by the use of different P-site tRNAs, namely tRNAPhe and tRNAThr, in the cryo-EM and X-ray preparations, respectively. Additionally, the structure of the isolated TC, which was used as reference for comparisons with the crystal structure, contained tRNAPhe. Thus, there is the possibility that the observed swing is a peculiarity of tRNAThr. The second possibility is that the method used for structure determination based on cryo-EM data, namely MDFF, was not able

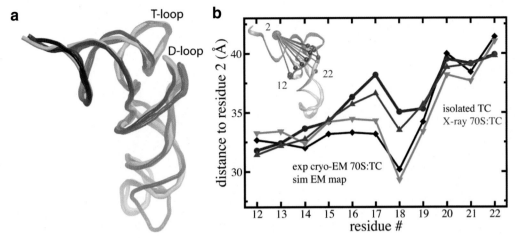

FIGURE 8.2: *A/T tRNA structure as seen in cryo-EM and X-ray crystallography. (a) Swing of the D loop relative to the T loop in tRNA in ternary complexes. (b) Distances between the phosphorus atoms of residues in the D loop relative to residue 2 in the acceptor stem (see inset). The structure and distances of the tRNA in a free ternary complex (TC) (PDB 1OB2, black, diamonds) are compared to a tRNA in a ribosome-bound TC obtained by MDFF interpretation of experimental cryo-EM data (Villa et al., 2009) (green, triangles down) and one obtained by X-ray crystallography (Schmeing et al., 2009) (red, circles). Distances and structure for the MDFF fit into a simulated cryo-EM map generated from the tRNA from the crystal structure (Schmeing et al., 2009) are shown in blue (triangles up).*

to capture this particular structural change in the tRNA. To distinguish between these two possibilities, a simulated map was generated from the crystal structure (Schmeing et al., 2009) at the same resolution as the experimental map used in Villa et al. (2009) (6.7Å). Subsequently, a system containing a TC with tRNA[Phe] was fitted into the simulated density using MDFF in explicit solvent (Trabuco et al., 2010b). The results of this fitting clearly show that MDFF does capture the swing of the D loop (see Figure 8.2a). To avoid any bias due to the alignment and to quantify the distortion, distances were calculated between residue 2 in the acceptor stem and several residues in the D loop, confirming the motion (see Figure 8.2b). The fact that MDFF was able to reproduce the swing suggests that the original experimental cryo-EM density containing tRNA[Phe] did not feature any swung conformation of the D loop. This in turn implies that the structural difference observed for the specific tRNAs in cryo-EM reconstructions and in X-ray structures reflects the specific natures of these tRNAs and may not be a general feature of the mechanism enhancing the GTPase activity of EF-Tu.

We would like to note that during the proof-reading phase of this book, new X-ray structures were reported (Voorhees et al., 2010), obtained without kirromycin. The new crystal structures represent a different, earlier, intermediate than the ones described above and call the described model of EF-Tu activation partially into question. In particular, these structures show specific interactions between the catalytic histidine and the 23S rRNA. Additionally, the reported complex features an ordered switch I. These two observations suggest, in turn, that a large opening of the hydrophobic gate is not as important as the precise positioning of the catalytic histidine.

IV. REGULATION OF THE PROTEIN-CONDUCTING CHANNEL BY THE RIBOSOME

IV.1 Role of the Translocon in Protein Development

Nascent polypeptides often control to what region of the cell they are targeted. For example, the hormone insulin, a 51-amino-acid long protein, leaves the cell via the secretory pathway. Targeting to this pathway is brought about through an N-terminal signal sequence at the beginning of the nascent chain; this sequence, comprised of 24 amino acids in the case of insulin, is cleaved after synthesis of the protein. For many secretory proteins, as well as almost all membrane proteins, the signal sequence emerging from the ribosome's exit tunnel recruits the signal recognition particle, a nucleo-protein complex that arrests translation momentarily (Halic and Beckmann, 2005). In this moment, the stalled ribosome is brought to a membrane, either the plasma membrane in Prokaryotes or the endoplasmic reticulum membrane in Eukaryotes, where the ribosome is docked to a protein-conducting translocon, the SecY/Sec61 complex. In this complex, the ribosome's exit tunnel is aligned with the channel's opening, providing a nearly uninterrupted path for the nascent chain.

After docking of the ribosome to the translocon, translation resumes and the nascent protein is directly inserted into the channel, concomitantly with its synthesis (Osborne et al., 2005; Rapoport, 2007; Papanikou et al., 2007; Driessen and Nouwen, 2008; Mandon et al., 2009). This process, known as cotranslational translocation, is present in all organisms, although it is not the only means of nascent-protein translocation. In Bacteria, a post-translational pathway, the SecA pathway, is more

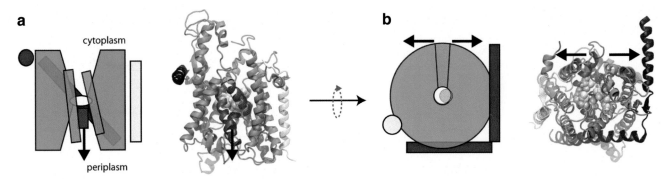

FIGURE 8.3: Schematic (left) and crystal structure (right) of a closed translocon, SecYEβ (PDB code 1RHZ; van den Berg et al.,
2004). SecY, SecE, and Secβ are shown in green, red, and yellow, respectively. The lateral gate helices 2b and 7 are colored blue,
the plug purple, and the hydrophobic pore ring is rendered in a white, space-filling representation. The translocon is shown (a) from
the membrane plane, with the opening motion of the plug denoted by an arrow, and (b) from the cytoplasmic side, with the opening
motion of the lateral gate indicated.

commonly utilized by translocated proteins (Driessen and
Nouwen, 2008), whereas in Eukaryotes, post-translational
translocation can occur through use of the ER-lumenal
protein, BiP (Rapoport, 2007). Both SecA and BiP are also
molecular machines, which, although less complex than
the ribosome, also utilize chemical energy (ATP or GTP)
to redirect newly synthesized proteins to specific locations
in the cell.

Until recently, all structural information on the nature of
complexes formed between the translocon and its partners
came from indirect sources, for example, from crosslinking data (Osborne and Rapoport, 2007) or from low-
resolution cryo-EM maps (Beckmann et al., 1997; Ménétret
et al., 2000; Beckmann et al., 2001; Morgan et al., 2002;
Ménétret et al., 2005; Mitra et al., 2005). However,
significant advances in the last few years have provided
a wealth of new structures. The first crystal structure of
a translocon-partner complex, that of a SecY-SecA complex, revealed a possible mechanism of nascent-chain insertion (Erlandson et al., 2008; Zimmer et al., 2008), although
debate about the oligomeric state of SecA remains (Sardis
and Economou, 2010). In addition, cryo-EM maps of
ribosome-translocon complexes have reached a resolution
sufficient to discern major structural features in the channel. Recent maps indicate that a monomer of the channel may be competent for translocation. In particular,
maps of an inactive ribosome-SecY complex (Ménétret
et al., 2007) and of a mammalian ribosome-Sec61 complex (Ménétret et al., 2008) provide definitive evidence
that a ribosome can bind a single channel copy, although
not that it forms a functional complex. Evidence of the
functionality of a ribosome-translocon-monomer complex
came from a high-resolution map of an actively translating ribosome bound to a monomer of Sec61 (Becker et al.,
2009), described in detail later in the chapter.

Whereas channel partners such as the ribosome provide
the driving force for protein translocation, the translocon
must determine where to place the nascent protein, either

in the membrane or across the channel. This placement is
effected through the modulation of two gates in the channel, a transverse gate controlling access to the periplas-
mic/lumenal space and a unique lateral gate controlling
access to the membrane. The structural nature of these
gates is revealed by the crystal structure of an Archaeal
SecY (van den Berg et al., 2004). The transverse gate is
formed by two elements, a small helix termed the plug
on the channel's periplasmic side and a hydrophobic constriction at the center of the hourglass-shaped channel, the
pore ring, both being necessary to close the channel (van
den Berg et al., 2004; Gumbart and Schulten, 2006; Li
et al., 2007; Saparov et al., 2007; Gumbart and Schulten,
2008). The lateral gate is formed at the junction of two
clamshell-like halves of SecY, between helices 2b and 7
(see Figure 8.3b) (van den Berg et al., 2004; Gumbart and
Schulten, 2007; du Plessis et al., 2009). How these gates
are opened and closed, discriminating between soluble and
transmembrane protein segments, is a topic of significant
importance for, for example, structure prediction of membrane proteins (Bernsel et al., 2008). Recent experiments
suggest that a given nascent chain segment samples both
the water-filled channel and, passing the lateral gate, the
hydrophobic membrane core, moving to where it is energetically more favorable (Hessa et al., 2005; Hessa et al.,
2007; White, 2007).

Although recent data have illuminated many structural
aspects of the translocon, they do not reveal much about
the dynamic mechanisms involved in its function. MD simulations, on the other hand, can animate the structures,
illustrating how function and dynamics of the channel are
intertwined. A number of MD studies of the translocon,
including those of the translocon in complex with a ribosome, have provided a greater understanding of the roles of
the previously identified structural elements. Simulations
utilizing MDFF have also permitted the determination of
novel structures of ribosome-translocon complexes from
cryo-EM data.

IV.2 Simulations of Isolated Translocons

Even though the translocon always works in concert with a channel partner such as the ribosome, and thus does not qualify as a machine on its own, to appreciate how the translocon and partner come together to form a unified molecular machine requires also to understand the translocon's behavior in isolation. To this end, numerous simulation studies have been carried out on the translocon, generally focusing on one of its two primary functions: translocation and lateral gating. These studies, in addition to providing insight into the translocon itself, aid understanding of secretion and membrane-insertion processes, which take place on a much larger scale and require the cooperation of multiple components.

Translocation of a nascent peptide through the translocon was the first target of MD studies, taking multiple approaches. In one study, artificial tests of the channel's elasticity were undertaken through forced opening by virtual spheres, revealing a channel capable of expanding to a large degree without irreversibly distorting its structure (Tian and Andricioaei, 2006). The ability to dynamically expand and contract is necessary for the channel to accommodate the wide variety of substrates – mostly unfolded peptide chains – it encounters. This ability was examined more directly by forced translocation of nascent peptide helices through the channel (Gumbart and Schulten, 2006). During translocation, the plug was observed to leave the channel center intact, returning afterward (Gumbart and Schulten, 2006); such a motion of the plug was also predicted based on a 15-ns equilibrium simulation of SecY (Haider et al., 2006). The plug can also be displaced from its closed state through mutations to the channel, which disrupt a hydrogen-bonding network involving conserved residues (Bondar et al., 2010). In addition, the pore ring formed a gasket-like seal around the polypeptide during its translocation, opening wide enough to allow it to pass but not to permit a significant number of water molecules or ions through (Gumbart and Schulten, 2006). That the pore ring is apparently sufficient to maintain the channel's seal brings into question the function of the plug. Crystallographic structures of plug-deletion mutants of the translocon reveal that new plugs form from the remaining residues (Li et al., 2007), but simulations of these mutants demonstrate that the new plugs are unstable (Gumbart and Schulten, 2008). This instability makes the pore ring unstable as well, illustrating how the two elements – plug and pore ring – must work together to maintain the channel's closed state (Gumbart and Schulten, 2008).

The translocon forms a channel across the lipid bilayer and also a pathway into the bilayer, the latter function being served by the translocon's lateral gate. The crystallographic structure of SecY alone displays a closed lateral gate (van den Berg et al., 2004), although more recent structures of SecY in complex with SecA (Zimmer et al.,

2008) and a Fab fragment (Tsukazaki et al., 2008) show that the gate is slightly opened in these complexes; full opening requires the insertion of the nascent protein's signal sequence, known to bind at the lateral gate (Plath et al., 1998). Simulations of gate opening and closing in the absence of a nascent chain, forced in a direction perpendicular to the lateral exit, illuminated the structural elements that resist gate opening, for example, SecY's plug (Gumbart and Schulten, 2007). Simulations of SecY with an open lateral gate on the 1-μs timescale also demonstrated that lipids are excluded from the exposed channel through hydrophobic effects (Gumbart and Schulten, 2007). This exclusion may be largely unnecessary, however, as free-energy calculations based on extended simulations suggest that the gate may only open briefly for hydrophobic regions of the nascent chain (Zhang and Miller, 2010). Whether this opening would be sufficient for sampling of the membrane environment and subsequent exit of the nascent chain segment is uncertain. Simulations have indicated that SecY may aid the insertion of membrane-protein segments in an additional way, by reducing the free-energy barrier of charged groups to move from water to the membrane core (Johansson and Lindahl, 2009).

IV.3 Simulations of Translocon-Partner Complexes

Before questions of overall protein translocation could be addressed by simulations, the requisite structures of complexes between the translocon and its channel partners needed to be solved. Crystallography has played some role here, with structures of SecY in complex with SecA (Zimmer et al., 2008) as well as with an artificial Fab fragment (Tsukazaki et al., 2008) now available. For ribosome-translocon complexes, structural data have been limited to cryo-EM maps. However, by combining those maps with individual crystallographic structures of ribosome and channel through MDFF, atomic-scale structures of complexes can be generated (Trabuco et al., 2008, Trabuco et al., 2009). MDFF has been used to determine structures of two ribosome-translocon complexes, that of a Bacterial ribosome-SecY-monomer complex (Gumbart et al., 2009) and that of a mammalian ribosome-Sec61-monomer complex (Becker et al., 2009). The two structures, which present the first high-resolution visualizations of the cotranslational translocation system, display significant similarities in the ribosome-translocon interface. For example, in both structures, the two cytoplasmic loops of the translocon insert into the ribosome's exit tunnel. In addition, SecE/Sec61γ and the C-terminus of the translocon interact with the ribosome (see Figure 8.4). All connections between channel and ribosome, except for that with SecE, have analogs in the SecY-SecA structure (Zimmer et al., 2008). These similarities suggest that the mode of binding is conserved across all species and even across different channel partners.

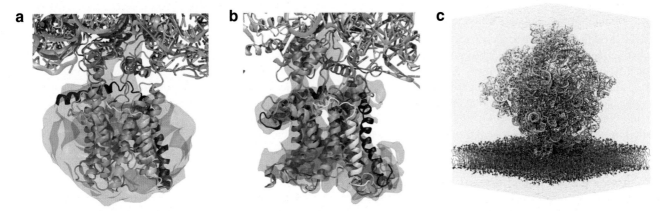

FIGURE 8.4: *Ribosome-translocon complexes. In all panels, the translocon is colored as in Figure 8.3, and the large and small subunits of the ribosome are colored in cyan and yellow, respectively. (a) Bacterial ribosome-SecY complex (map from Ménétret et al., 2007; PDB code 3KC4 from Gumbart et al., 2009). (b) Mammalian ribosome-Sec61 complex (PDB code 2WWB; Becker et al., 2009). In both (a) and (b), the cryo-EM map used to determine the structure is indicated as a transparent gray surface. (c) Simulated ribosome-SecY system from Gumbart et al. (2009). The membrane lipids are indicated as gray sticks, with specific headgroup atoms shown as red and blue spheres. The water box is shown as a transparent blue surface.*

MD simulations of two translocon-partner complexes have been carried out to date. In the first case, the structure of the ribosome-SecY-monomer was used to build a complete system including membrane and water, requiring nearly 3 million atoms in total (Gumbart et al., 2009). Equilibrium simulations of this system established the persistence of connections between the ribosome and SecY as well as the stability of the complex. To determine if the resolved structure, which was based on a map of a non-translating complex, is functional, translocation of a polypeptide from ribosome's exit tunnel to channel was performed. During the simulated translocation, the connection between ribosome and channel was stable, and the ribosome's exit tunnel accommodated both the inserted loops of SecY and a nascent-peptide helix (Gumbart et al., 2009). This result, combined with the structure of an actively translating ribosome-Sec61-monomer complex (Becker et al., 2009), demonstrates that a translocon monomer is sufficient for translocation, although higher-order oligomers may yet be relevant under certain conditions (Schaletzky and Rapoport, 2006; Gumbart et al., 2009). The second set of translocon-partner simulations examined the structure of SecY bound to a Fab fragment (Tsukazaki et al., 2008; Mori et al., 2010). Although not a traditional binding partner for SecY, the Fab fragment interacts with SecY's cytoplasmic loops, just as the ribosome and SecA do, leading to a slight opening in SecY's lateral gate (Tsukazaki et al., 2008). Long simulations (~100 ns) of the Fab-bound SecY show that this slight opening is stably maintained, whereas it closes upon removal of the Fab fragment (Mori et al., 2010). Additionally, the opening remains stable when the Fab fragment is replaced by SecA in simulation (Mori et al., 2010). The effects of the channel partner on SecY were also examined in simulations of the ribosome-SecY-monomer complex.

During the simulation, SecY, and particularly its plug, was less stable compared to simulation of SecY alone; calculation of the dominant motions of the plug in simulation (see Figure 8.5c) also reveals a tendency of the plug to leave the channel when the ribosome is bound (Gumbart et al., 2009). On longer time scales, transient exit of the plug may occur; in experiments, the translocon does permit the passage of ions and small molecules when the ribosome is bound (Simon and Blobel, 1991; Heritage and Wonderlin, 2001; Wonderlin, 2009). The destabilization of the plug in simulation could also be caused solely by restraining SecY's cytoplasmic loops, without the ribosome present (Gumbart et al., 2009). This result explains how a variety of disparate channel partners, which all bind to and restrict the motion of SecY's two exposed loops, can induce the opening of the channel.

IV.4 Future Developments

The atomic-scale details of protein translocation and membrane insertion are beginning to come into focus, thanks in large part to the growing amount of structural data emanating from crystallography and cryo-EM. However, of all the data currently available, none have permitted resolution of a nascent protein caught inside the channel. A major goal of future studies will be to determine structures of so-called insertion intermediates, allowing the observation of the nascent chain at different stages of its translocation through the channel or insertion into the membrane; recent results indicate the latter has already been achieved (Frauenfeld et al., 2011). Structures of insertion intermediates also provide opportunities for further simulation studies, which can develop trajectories of the nascent chain between the intermediate states. Ultimately, these experimental and computational studies will lead to visualization

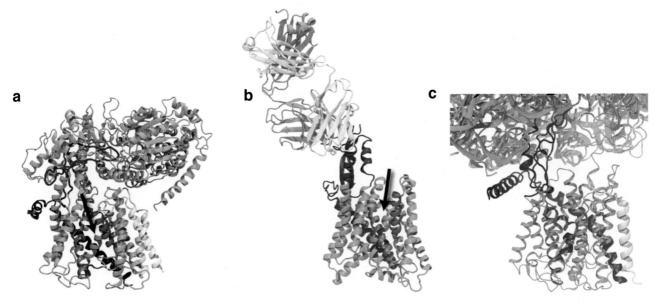

FIGURE 8.5: *Structurally characterized translocon-partner complexes. In all panels, the translocon is colored as in Figure 8.3 and the channel partner is in (a, c) cyan or (b) cyan and yellow. (a) SecA-translocon (PDB code 3DIN; Zimmer et al., 2008) and (b) Fab-translocon (PDB code 2ZJS; Tsukazaki et al., 2008) complexes. The opening of the lateral gate in each structure is indicated by an arrow. (c) Ribosome-translocon complex. The orange arrows represent an exaggeration of the dominant motion of the plug calculated from simulation (Gumbart et al., 2009).*

of the entire cycle of the translocation machinery, including docking of the channel partner with the translocon, energy-driven nascent-chain insertion, and subsequent undocking.

V. THE REGULATORY NASCENT CHAIN TNaC

V.1 Translational Stalling by Regulatory Nascent Chains

The process of protein synthesis by the ribosome begins with the formation of an initiation complex, in which two ribosomal subunits bind to the mRNA to be translated, together with an initiator tRNA which provides the first amino acid of the peptide chain to be synthesized. This process then moves to the subsequent phase, elongation, in which amino acids are repeatedly incorporated into the nascent protein. The addition of an amino acid to the nascent chain involves the formation of a peptide bond, a chemical reaction that takes place in the "active site" of the ribosome termed peptidyl transferase center (PTC). As it is extended, the nascent peptide chain travels through the exit tunnel, which connects the PTC with the ribosome surface, corresponding to a distance of ~100 Å.

The exit tunnel was originally thought to serve as an inert passage for nascent proteins. However, an increasing number of peptide chains are found to be able to regulate their own translation while they are still being synthesized (Tenson and Ehrenberg, 2002). Such regulatory nascent chains interact with the ribosomal exit tunnel, eliciting responses such as stalling of protein elongation (Nakatogawa and Ito, 2002) or termination (Gong and Yanofsky,

2002). The mere presence of certain regulatory nascent chains in the exit tunnel is sufficient to induce translational arrest (Nakatogawa and Ito, 2002). Others require a small molecule to also bind to the ribosome, such as regulatory peptides that control the expression of antibiotic resistance genes in Bacteria (Ramu et al., 2009).

A classical example of a regulatory nascent chain is TnaC, the leader peptide of the tryptophanase (*tna*) operon in *E. coli*, which is responsible for tryptophan degradation (Yanofsky, 2007). At normal tryptophan concentrations, synthesis of the mRNA that codes for the genes in the *tna* operon is terminated before the structural genes are transcribed. At high concentrations of free tryptophan, however, the presence of TnaC inside the exit tunnel leads to translational arrest, preventing termination. Synthesis of the *tna* mRNA then proceeds, leading to the expression of the downstream structural genes and thus activating the *tna* operon. Extensive biochemical investigations by Yanofsky and colleagues provided detailed information on which TnaC and ribosome residues are critical for stalling (summarized in Trabuco et al., 2010a).

V.2 Structure of 70S:TnaC and the Mechanism of Stalling

An *E. coli* 70S ribosome stalled by TnaC was recently visualized by cryo-EM at 5.8-Å resolution (Figure 8.6) (Seidelt et al., 2009). This was the first time a nascent chain was visualized inside the ribosome at such a level of detail. The cryo-EM map revealed that TnaC adopts a distinct conformation in the exit tunnel with several points of

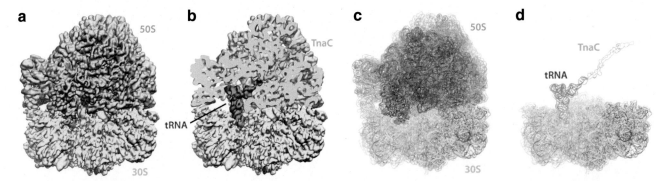

FIGURE 8.6: *Cryo-EM single-particle reconstruction of an E. coli TnaC:70S complex (Seidelt et al., 2009; EMDB-1657). Panel (a) shows the entire density, and in panel (b) the 50S subunit is sliced for clarity. (c) Atomic model obtained with the MDFF method (PDB 2WWL/2WWQ), with the cryo-EM map shown in transparent surface. (d) Same as in (c), but with the 50S subunit is sliced for clarity to reveal the tunnel. Cyan: 50S subunit; yellow: 30S subunit; red: P-site tRNA; green: TnaC.*

contact with the ribosome. It became clear that the TnaC structure inside the ribosome does not feature any secondary-structure elements. The sample used in the cryo-EM experiment contained ribosomes stalled during translation of the tnaC gene in the presence of free tryptophan. Comparison with a control density map lacking TnaC revealed clear density for the TnaC nascent chain attached to tRNA^Pro; Pro24 is the last amino acid residue in TnaC before the stop codon, where stalling occurs.

The MDFF method was employed to obtain an atomic model of the 70S:TnaC complex, aiding the interpretation of the cryo-EM data (Seidelt et al., 2009). Of particular interest is the analysis of conformations of highly conserved residues in the PTC, the active site of the ribosome

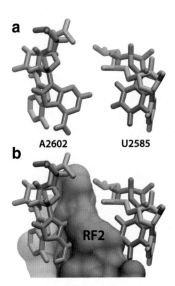

FIGURE 8.7: *Mechanism of translational stalling. (a) Conformations adopted by critical residues of the 23S rRNA (A2602 and U2585) in the TnaC:70S (orange; PDB 2WWL/2WWQ) and RF2:70S (blue; PDB 2WH1/2WH2 [Weixlbaumer et al., 2008]) complexes. (b) Same as in (a), but with part of the RF2 structure represented as a gray surface, showing that in TnaC:70S these 23S residues adopt a conformation that would clash with RF2, thus explaining how termination of protein synthesis is inhibited by TnaC.*

where peptide bond formation and termination of protein synthesis take place. TnaC-mediated stalling comes about through the inhibition of termination at the PTC. Analysis of the atomic model together with the cryo-EM data showed that the highly conserved 23S rRNA residue A2602 adopts a distinct conformation in the 70S:TnaC complex, in contrast with the control map where this residue is disordered. Interestingly, the conformation of A2602 in the 70S:TnaC complex resembles the conformation seen in a crystal structure containing sparsomycin, an antibiotic that inhibits the PTC (Schmeing et al., 2005). Furthermore, the flexible base U2585 also adopts a distinct conformation in the 70S:TnaC complex, shifting to interact with TnaC's Pro24. Release factor 2 (RF2) is the protein factor that usually terminates TnaC synthesis at low tryptophan concentrations – in other words, when there is no stalling. The conformations of A2602 and U2585 seen in the 70S:TnaC complex are incompatible with cohabitation by release factors (Figure 8.7), explaining the mechanism of TnaC-mediated translational arrest.

V.3 Recognition of TnaC by the Ribosomal Exit Tunnel

The MDFF-derived 70S:TnaC atomic model provided structural insight into the mechanism of PTC silencing by TnaC, as described earlier (Seidelt et al., 2009). Even though the model was derived from one of the highest-resolution cryo-EM reconstructions of the ribosome to date, the positions of TnaC amino acid residues became only approximately known. MDFF results suggest that an ensemble of structures is consistent with the cryo-EM density, similar to results from structural analysis by nuclear magnetic resonance for other proteins. To investigate in more detail the precise interactions between critical TnaC and ribosomal residues, extensive MD simulations were performed. Starting from ten independent, MDFF-derived atomic models of TnaC inside the exit tunnel (Figure 8.8), a reduced system containing TnaC and the exit tunnel was simulated in explicit solvent for a total aggregate time of 2.1 μs (Trabuco et al., 2010a). Analysis of the simulations

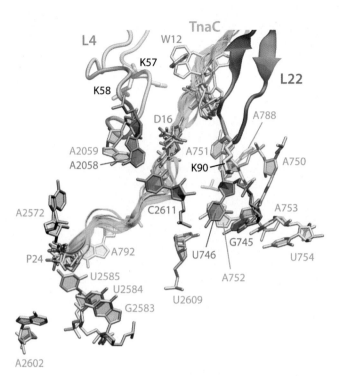

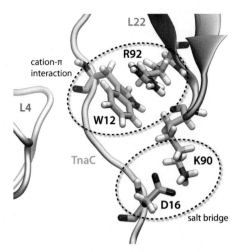

FIGURE 8.8: *Structure of TnaC inside the exit tunnel. Ten models of TnaC obtained by applying MDFF are shown in green, with the three critical residues highlighted (Trp12, Asp16, and Pro24). Also shown are part of ribosomal proteins L4 and L22, which correspond to the constriction site of the exit tunnel, and 23S rRNA residues for which biochemical data are available (orange), with sequence signatures characteristic of Bacteria (gray).*

FIGURE 8.9: *Recognition of critical TnaC residues by the ribosomal exit tunnel. A snapshot from one of the MD simulations is shown that highlights a cation-π interaction between critical residue W12 (TnaC) and R92 (L22), as well as a salt bridge between critical residues D16 (TnaC) and K90 (L22).*

revealed how critical recognition of TnaC residues by the ribosome is for function as well as suggested mutations predicted to reduce TnaC-mediated stalling.

According to several mutational studies, the highly conserved TnaC residues W12, D16, and P24 are essential for stalling (see Trabuco et al., 2010a for a comprehensive review of mutational data). W12 is located near the constriction site of the exit tunnel formed by ribosomal proteins L4 and L22, as seen in the MDFF-derived 70S:TnaC model (Seidelt et al., 2009). In the equilibrium MD simulations, W12 is seen to form cation-π interactions with positively charged residues from these ribosomal proteins, in particular with highly conserved R92 of L22 (Figure 8.9) (Trabuco et al., 2010a). Such cation-π interactions between aromatic and positively charged residues are strong and specific, and are seen in various cases of molecular recognition in biology (Gallivan and Dougherty, 1999). Another observation from the equilibrium MD simulations is that TnaC's D16 is recognized by ribosomal proteins via salt bridges, in particular one involving K90 of L22, a conserved residue that is also critical for TnaC-mediated stalling (Figure 8.9). Moreover, the MD simulations show that TnaC residues I19 and V20 interact with conserved residues from the ribosome known to be important for translational arrest. Bioinformatic analysis combined with further simulations suggests mutations in TnaC at positions 19 and

20 that are predicted to reduce stalling (Trabuco et al., 2010a).

V.4 Open Questions

The recent studies described in the preceding subsection provided a structural view of the 70S:TnaC complex, revealing the mechanisms of translational stalling (Seidelt et al., 2009) and recognition of TnaC by the ribosome (Trabuco et al., 2010a). One question that remains is the precise location where tryptophan – a required cofactor for TnaC-mediated stalling – binds in the PTC. Analysis of the 70S:TnaC structure suggested communication pathways between critical TnaC residues and the PTC (Seidelt et al., 2009). The possible communication pathways remain to be tested both computationally and experimentally. There are presumably many hidden communication pathways within the ribosome linking various functional sites. Identification and understanding of such "communication highways" is one of the main open questions in ribosome research, as well as in the study of molecular machines in general.

VI. REVIEW OF COMPUTATIONAL STUDIES OF RIBOSOME FUNCTION

VI.1 Probing Ribosome Function through Computational Experiments

In the previous sections, three particular functional aspects of the ribosome addressed by simulations were described, namely activation of EF-Tu during decoding, regulation of the protein-conducting channel, and translational control by TnaC. However, almost every step of elongation has been studied computationally. In this section, examples of these computational studies and the insights they yielded are presented.

VI.2 Decoding the Message

A fundamental function of the ribosome function is the decoding of the genetic information. This decoding involves the assignment of the correct amino acid to a given mRNA codon, but also the recognition of stop codons that specify the end of the protein-coding region. For the former type of decoding, the ribosome uses tRNAs, whereas for the latter, proteins called release factors are recruited. Although crystal structures of the ribosome bound to tRNAs and release factors exist (Schmeing and Ramakrishnan, 2009), the aspects governing the energetics of the decoding process at atomic detail are not well understood.

The ribosome itself plays a decisive role in the decoding of mRNA. Apart from large-scale rearrangements within the small subunit, the decoding process carried out by tRNAs involves structural changes in two specific nucleotides of the ribosome: A1492 and A1493. These two bases flip out of helix 44 in the small subunit in order to monitor the geometry of the decoding helix formed by the mRNA codon and the anticodon of the tRNA. Simulations of the ribosomal decoding center addressed the role of steric fit, hydrogen bonding, and base pair stability at different positions of the decoding helix in the discrimination mechanism of the ribosome (Sanbonmatsu and Joseph, 2003). Further simulations of the ribosomal decoding center using enhanced sampling techniques revealed the free-energy landscape of the "conformational switch", that is, the nucleotides A1492 and A1493 in the small subunit. The results suggest that a low energy barrier separating the two states of the switch, flipped into helix 44 or flipped out, is relevant for easy shifting of the equilibrium between the two states brought about by binding of ligands such as aminoglycoside antibiotics or tRNAs (Sanbonmatsu, 2006).

Decoding involves not only the correct recognition of amino acid–specifying codons by tRNAs, but also the recognition of the stop codons by release factors. Recently, free-energy perturbation calculations of the interactions between release factors and stop codons were able to quantitatively describe how stop codons are recognized (Sund et al., 2010). Moreover, these calculations explained the specificity of the two Bacterial release factors for their corresponding stop codons. The simulations also suggest that the decoding process carried out by the two release factors goes far beyond simple mimicry of tRNA molecules. For example, no monitoring rRNA bases a required; instead, the recognition of the codon involves a wide range of interactions with side chains and the backbone of the proteins.

VI.3 Accommodation of the Incoming tRNA

Successful decoding of the mRNA leads to dissociation of EF-Tu, and the incoming tRNA has to be accommodated into the ribosome where it brings the new amino acid into the ribosomal reaction center. However, there are no structural data available describing the path(s) the tRNA takes as it moves into the ribosome.

In the first simulations of the accommodation process, the tRNA was biased to go from the state it is in immediately after EF-Tu dissociation into a fully accommodated state (Sanbonmatsu et al., 2005). The simulations revealed specific interactions and rearrangements that are necessary to allow the tRNA to be accommodated and demonstrated the importance of tRNA flexibility during this process. More recently, a study employing unrestrained structure-based as well as explicit-solvent simulations (Whitford et al., 2010) confirmed the "accommodation corridor" described earlier (Sanbonmatsu et al., 2005), but also uncovered parallel paths for accommodation. Supported by FRET experiments presented in the same work, the identified mechanism of accommodation involves stochastic rather than concerted conformational changes in the tRNA and the ribosome.

VI.4 The Ribosome is a Chemical Machine

The main purpose of the complex ribosome machinery is to catalyze two reactions. The first is the formation of a peptide bond between the nascent protein chain and the new amino acid brought by the accommodated tRNA. The second reaction catalyzed by the ribosome together with release factors, once the nascent protein is complete, is the hydrolysis of the bond connecting the nascent chain to the tRNA during termination. This reaction releases the synthesized protein from the ribosome. Although X-ray crystallography yielded important insights into the structure of the ribosomal reaction center in complex with tRNAs and release factors (Schmeing and Ramakrishnan, 2009), insights into the reaction mechanism at atomic detail are very limited for both the formation of the peptide bond and for the termination reaction.

To explore possible mechanisms of the peptide-bond formation, molecular dynamics and free-energy perturbation simulations were combined with an empirical valence bond description (Trobro and Åqvist, 2005). Apart from revealing a structured hydrogen-bond network in the active site, the simulation results suggested that the most favorable mechanism for peptide-bond formation involves proton-shuttling via the O2′ oxygen of the terminal P-site tRNA adenine rather than acid-base catalysis by ribosomal residues. The ribosome, however, preorganizes reactants and solvent in the reaction site, leading to a large entropic contribution to the catalytic effect. A more recent study of the transition states along the reaction path using density functional theory modified the proton shuttle mechanism to a "double proton shuttle" (Wallin and Åqvist, 2010). In this mechanism, a water molecule directly participates in the proton transduction within a transition state, while a second water molecule stabilizes the negative charge that arises during the reaction.

Interestingly, a proton-shuttle mechanism involving the $2'$-OH group of the terminal P-site tRNA adenine was also suggested by computational investigations of the hydrolysis reaction during termination executed by release factors (Trobro and Åqvist, 2009). Also in this case, additional water molecules were found to be essential for organizing the reactants in favorable orientations. Additionally, the simulations showed that the role of the methylated Gln in the universally conserved GGQ motif of the release factors is to orient the water molecule attacking the ester bond, thus stabilizing the transition state.

VI.5 Crosstalk with the Nascent Chain

As described earlier, translation can be controlled by nascent proteins. One example that was given is TnaC, which regulates the expression of genes for tryptophan metabolism. For this particular example, it was shown how the nascent chain interacts with the ribosome.

Earlier studies have also addressed the question of how nascent chains interact with the ribosomal exit tunnel on a more general basis. MD simulations were used to obtain free-energy maps describing the physicochemical environment within the tunnel as seen by different amino acids (Petrone et al., 2008). These maps provide a unique view of the interior of the tunnel, revealing binding pockets and repelling regions specific to each amino acid. Moreover, these free-energy maps offered for the first time a context for an interpretation of the sequence-dependent interplay between a nascent chain and the ribosome, which proved useful in later investigations of TnaC (Trabuco et al., 2010a). MD simulations were also used to obtain directly a picture of a nascent peptide within the exit tunnel (Ishida and Hayward, 2008). The authors modeled a nascent chain within the ribosome, finding two possible paths at the constriction site formed by the ribosomal proteins L4 and L22. Comparison to simulations of an empty exit tunnel also suggested that the nascent chain itself opens the constriction site. Subsequently, MD simulations were used to study the motion of a modeled peptide as it is being pulled out of the tunnel. The simulations showed that the nascent protein could be moved only along one path but not the other.

VI.6 The Ribosome is a Large Dynamic Machine

The studies described up to now addressed almost every aspect of elongation: decoding, accommodation, peptide bond formation, termination, and the interaction of the nascent protein with the exit tunnel. Apart from accommodation (Whitford et al., 2010), these investigations did not involve large structural rearrangements of the ribosome. However, cryo-EM, FRET, and X-ray crystallography all indicate that the ribosome is a dynamic molecular machine that undergoes large conformational changes like the inter-subunit rotation (ratcheting) and large-scale movement of the L1 stalk (Frank and Agrawal, 2000; Korostelev et al., 2008; Steitz, 2008; Agirrezabala and Frank, 2009; Schmeing and Ramakrishnan, 2009; Dunke and Cate, 2010; Frank and Gonzalez, Jr., 2010). It is not clear, however, how the movements of different parts of the ribosome arise and how they are coordinated.

To gain an understanding of the intrinsic dynamics of the ribosome, two approaches have been taken. One approach is to explore the dynamics of specific architectural elements of the ribosome in isolation. For example, simulations of kink-turns (Rázga et al., 2006; Réblová et al., 2009) have directly demonstrated that kink-turns and analogous elements (Réblová et al., 2009) can act as joints mediating and directing large-scale motions. More recently, it was shown that another type of architectural element, three-way junctions, also features hinge-like flexibility (Beššeová et al., 2010). Owing to their positions in the ribosome, these flexible elements can facilitate accommodation, large-scale motion of the L7/L12 stalk, and the motion of the A-site finger.

In the second approach to studying ribosome dynamics, full ribosome structures were used at different levels of spatial and temporal resolution. For example, simulations using a low-resolution anharmonic network model were carried out to investigate the large-scale cooperative motions of the ribosome (Trylska et al., 2005). The subsequent analysis revealed correlations between the motions of distant parts of the ribosome, for example, between the L1 and L7/L12 stalks, which are located on opposite sides of the large subunit. A similar method for exploring large motions of the ribosome is the use of normal mode analysis with different kinds of network models, such as an anisotropic network model (Wang et al., 2004) or an elastic network model (Tama et al., 2003; Kurkcuoglu et al., 2008). These simulations show the ribosome as a highly dynamic machine with correlations between distant parts. More importantly, these studies revealed the spontaneous nature of large-scale rearrangements like the inter-subunit rotation, suggesting that the ribosome is a stochastic molecular machine rather than a clock-like automaton.

VII. OUTLOOK

MD simulations have made notable contributions to the illumination of the mechanistic aspects of translation by the ribosome. However, practically all aspects addressed so far, including the function of EF-Tu, regulation by the nascent chain TnaC, and interaction of the ribosome with the translocon, relate to only two out of four stages of translation: elongation and termination. Even within these stages, numerous intermediate steps remain at least somewhat uncharacterized, such as the movement of tRNAs through the ribosome induced by ratchet-like motion of the two subunits (Agirrezabala and Frank, 2009). The other two stages – initiation and recycling – are almost completely untouched by MD simulations so far, although they provide a wealth of additional open questions. Clearly, despite the progress already made, a full systematic understanding of the ribosome's machinery remains elusive.

Advancement in understanding of how the ribosome functions will continue to be attained through traditionally applied methods, such as the determination of new structures (Schmeing et al., 2009) and pure MD simulations (Ishida and Hayward, 2008). However, it appears likely that it will be at the interface of different methods where an increasing number of advances will be made. Computational modeling, and simulation in particular, is well poised to provide a means to integrate data from multiple sources. Such integration is already taking place for crystallographic structures and cryo-EM data, merged via MDFF simulations. This integration is providing views of functional states of the ribosome previously unseen at atomic resolution. Visualization of these states, and knowledge of the dynamics connecting them, can reveal how even the smallest components of the ribosome serve their part in the operation of the entire molecular machine. The computational microscope, not built from brass tubes and glass lenses, but based on general chemical knowledge, principles from physics, well-selected mathematical algorithms, efficient and easy-to-use software, ever-improving computer hardware, and – last but not least – intuitive computer graphics, will play an increasingly important role in future studies.

ACKNOWLEDGMENTS

This work was supported by grants from the National Institutes of Health (P41-RR005969 and R01-GM067887) and the National Science Foundation (PHY0822613). We would like to thank Charles Brooks for useful comments on the manuscript. All-atom MD simulations discussed here were performed using the package NAMD (Phillips et al., 2005). Molecular images in this article were rendered using the molecular visualization software VMD (Humphrey et al., 1996).

REFERENCES

Agirrezabala, X., and Frank, J. (2009). Elongation in translation as a dynamic interaction among the ribosome, tRNA, and elongation factors EF-G and EF-Tu. *Quart. Rev. Biophys.* 43, 159–200.

Aksimentiev, A., Balabin, I.A., Fillingame, R.H., and Schulten, K. (2004). Insights into the molecular mechanism of rotation in the Fo sector of ATP synthase. *Biophys. J.* 86, 1332–1344.

Arkhipov, A., Freddolino, P.L., Imada, K., Namba, K., and Schulten, K. (2006). Coarse-grained molecular dynamics simulations of a rotating bacterial flagellum. *Biophys. J.* 91, 4589–4597.

Becker, T., Bhushan, S., Jarasch, A., Armache, J.P., Funes, S., Jossinet, F., Gumbart, J., Mielke, T., Berninghausen, O., Schulten, K., et al. (2009). Structure of monomeric yeast and mammalian Sec61 complexes interacting with the translating ribosome. *Science* 326, 1369–1373.

Beckmann, R., Bubeck, D., Grassucci, R., Penczek, P., Verschoor, A., Blobel, G., and Frank, J. (1997). Alignment of conduits for the nascent polypeptide chain in the ribosome-Sec61 complex. *Science* 278, 2123–2126.

Beckmann, R., Spahn, C.M.T., Eswar, N., Helmers, J., Penczek, P.A., Sali, A., Frank, J., and Blobel, G. (2001). Architecture of the protein-conducting channel associated with the translating 80S ribosome. *Cell* 107, 361–372.

Berchtold, H., Reshetnikova, L., Reiser, C.O., Schirmer, N.K., Sprinzl, M., and Hilgenfeld, R. (1993). Crystal structure of active elongation factor Tu reveals major domain rearrangements. *Nature* 365, 126–132.

Bernsel, A., Viklund, H., Falk, J., Lindahl, E., von Heijne, G., and Elofsson, A. (2008). Prediction of membrane-protein topology from first principles. *Proc. Natl. Acad. Sci. USA* 105, 7177–7181.

Beššeová, I., Réblová, K., Leontis, N.B., and Šponer, J. (2010). Molecular dynamics simulations suggest that RNA three-way junctions can act as flexible RNA structural elements in the ribosome. *Nucleic Acids Res.* 38, 6247–6264.

Bondar, A.N., del Val, C., Freites, J.A., Tobias, D.J., and White, S.H. (2010). Dynamics of SecY translocons with translocation-defective mutations. *Structure* 18, 847–857.

Chandler, D.E., Gumbart, J., Stack, J.D., Chipot, C., and Schulten, K. (2009). Membrane curvature induced by aggregates of LH2s and monomeric LH1s. *Biophys. J.* 97, 2978–2984.

Daviter, T., Wieden, H.J., and Rodnina, M.V. (2003). Essential role of histidine 84 in elongation factor Tu for the chemical step of GTP hydrolysis on the ribosome. *J. Mol. Biol.* 332, 689–699.

Dittrich, M., and Schulten, K. (2006). PcrA helicase, a prototype ATP-driven molecular motor. *Structure* 14, 1345–1353.

Dittrich, M., Yu, J., and Schulten, K. (2006). PcrA helicase, a molecular motor studied from the electronic to the functional level. *Topics in Current Chemistry* 268, 319–347.

Driessen, A.J.M., and Nouwen, N. (2008). Protein translocation across the bacterial cytoplasmic membrane. *Annu. Rev. Biochem.* 77, 643–667.

du Plessis, D.J.F., Berrelkamp, G., Nouwen, N., and Driessen, A.J.M. (2009). The lateral gate of SecYEG opens during proteins translocation. *J. Biol. Chem.* 284, 15805–15814.

Dunke, J.A., and Cate, J.H.D. (2010). Ribosome structure and dynamics during translocation and termination. *Annu. Rev. Biophys.* 39, 227–244.

Erlandson, K.J., Miller, S.B.M., Nam, Y., Osborne, A.R., Zimmer, J., and Rapoport, T.A. (2008). A role for the two-helix finger of the SecA ATPase in protein translocation. *Nature* 455, 984–987.

Fasano, O., Vendittis, E.D., and Parmeggiani, A. (1982). Hydrolysis of GTP by elongation factor Tu can be induced by monovalent cations in the absence of other effectors. *J. Biol. Chem.* 257, 3145–3150.

Fischer, N., Konevega, A.L., Wintermeyer, W., Rodnina, M.V., and Stark, H. (2010). Ribosome dynamics and tRNA movement by time-resolved electron cryomicroscopy. *Nature* 466, 329–333.

Frank, J., and Agrawal, R.K. (2000). A ratchet-like inter-subunit reorganization of the ribosome during translocation. *Nature* 406, 318–322.

Frank, J., and Gonzalez, Jr., R.L. (2010). Structure and dynamics of a processive brownian motor: the translating ribosome. *Annu. Rev. Biochem.* 79, 1–32.

Frauenfeld, J., Gumbart, J., van der Sluis, E.O., Funes, S., Gartmann, M., Beatrix, B., Mielke, T., Berninghausen, O., Becker, T., Schulten, K., and Beckmann, R. (2011). Cryo-EM structure of the ribosome-SecYE complex in the membrane environment. *Nat. Struct. Mol. Biol.* 18, 614–621.

Freddolino, P.L., Arkhipov, A.S., Larson, S.B., McPherson, A., and Schulten, K. (2006). Molecular dynamics simulations of the complete satellite tobacco mosaic virus. *Structure* 14, 437–449.

Gallivan, J.P., and Dougherty, D.A. (1999). Cation-π interactions in structural biology. *Proc. Natl. Acad. Sci. USA* 96, 9459–9464.

Gao, Y.G., Selmer, M., Dunham, C.M., Weixlbaumer, A., Kelley, A.C., and Ramakrishnan, V. (2009). The structure of the ribosome with elongation factor G trapped in the posttranslocational state. *Science* 326, 694–699.

Gong, F., and Yanofsky, C. (2002). Instruction of translating ribosome by nascent peptide. *Science* 297, 1864–1867.

Gumbart, J., and Schulten, K. (2006). Molecular dynamics studies of the archaeal translocon. *Biophys. J.* 90, 2356–2367.

Gumbart, J., and Schulten, K. (2007). Structural determinants of lateral gate opening in the protein translocon. *Biochemistry* 46, 11147–11157.

Gumbart, J., and Schulten, K. (2008). The roles of pore ring and plug in the SecY protein- conducting channel. *J. Gen. Physiol.* 132, 709–719.

Gumbart, J., Trabuco, L.G., Schreiner, E., Villa, E., and Schulten, K. (2009). Regulation of the protein-conducting channel by a bound ribosome. *Structure* 17, 1453–1464.

Haider, S., Hall, B.A., and Sansom, M.S.P. (2006). Simulations of a protein translocation pore: SecY. *Biochemistry* 45, 13018–13024.

Halic, M., and Beckmann, R. (2005). The signal recognition particle and its interactions during protein targeting. *Curr. Opin. Struct. Biol.* 15, 116–125.

Heritage, D., and Wonderlin, W.F. (2001). Translocon pores in the endoplasmic reticulum are permeable to a neutral, polar molecule. *J. Biol. Chem.* 276, 22655–22662.

Hessa, T., Kim, H., Bihlmaier, K., Lundin, C., Boekel, J., Andersson, H., Nilsson, I., White, S.H., and von Heijne, G. (2005). Recognition of transmembrane helices by the endoplasmic reticulum translocon. *Nature* 433, 377–381.

Hessa, T., Meindl-Beinker, N.M., Bernsel, A., Kim, H., Sato, Y., Lerch-Bader, M., Nilsson, I., White, S.H., and von Heijne, G. (2007). Molecular code for transmembrane-helix recognition by the Sec61 translocon. *Nature* 450, 1026–1030.

Humphrey, W., Dalke, A., and Schulten, K. (1996). VMD – Visual Molecular Dynamics. *J. Mol. Graphics* 14, 33–38.

Ishida, H., and Hayward, S. (2008). Path of nascent polypeptide in exit tunnel revealed by molecular dynamics simulation of ribosome. *Biophys. J.* 95, 5962–5973.

Johansson, A.C.V., and Lindahl, E. (2009). Protein contents in biological membranes can explain abnormal solvation of charged and polar residues. *Proc. Natl. Acad. Sci. USA* 106, 15684–15689.

Kitao, A., Yonekura, K., Maki-Yonekura, S., Samatey, F.A., Imada, K., Namba, K., and Go, N. (2006). Switch interactions control energy frustration and multiple flagellar filament structures. *Proc. Natl. Acad. Sci. USA* 103, 4894–4899.

Kjeldgaard, M., Nissen, P., Thirup, S., and Nyborg, J. (1993). The crystal structure of elongation factor EF-Tu from *Thermus aquaticus* in the GTP conformation. *Structure* 1, 35–50.

Klein, M.L., and Shinoda, W. (2008). Large-scale molecular dynamics simulations of self- assembling systems. *Science* 321, 798–800.

Korostelev, A., Ermolenko, D.N., and Noller, H.F. (2008). Structural dynamics of the ribosome. *Curr. Opin. Chem. Biol.* 12, 674–683.

Kurkcuoglu, O., Doruker, P., Sen, T.Z., Kloczkowski, A., and Jernigan, R.L. (2008). The ribosome structure controls and directs mRNA entry, translocation and exit dynamics. *Phys. Biol.* 5, 46005-1–14.

Li, W., Schulman, S., Boyd, D., Erlandson, K., Beckwith, J., and Rapoport, T.A. (2007). The plug domain of the SecY protein stabilizes the closed state of the translocation channel and maintains a membrane seal. *Mol. Cell* 26, 511–521.

Mandon, E.C., Trueman, S.F., and Gilmore, R. (2009). Translocation of proteins through the Sec61 and SecYEG channels. *Curr. Opin. Cell Biol.* 21, 501–507.

Ménétret, J.F., Hegde, R.S., Agular, M., Gygi, S.P., Park, E., Rapoport, T.A., and Akey, C.W. (2008). Single copies of Sec61 and TRAP associate with a nontranslating mammalian ribosome. *Structure* 16, 1126–1137.

Ménétret, J.F., Hegde, R.S., Heinrich, S.U., Chandramouli, P., Ludtke, S.J., Rapoport, T.A., and Akey, C.W. (2005). Architecture of the ribosome-channel complex derived from native membranes. *J. Mol. Biol.* 348, 445–457.

Ménétret, J.F., Neuhof, A., Morgan, D.G., Plath, K., Radermacher, M., Rapoport, T.A., and Akey, C.W. (2000). The structure of ribosome-channel complexes engaged in protein translocation. *Mol. Cell* 6, 1219–1232.

Ménétret, J.F., Schaletzky, J., Clemons, W.M., Jr., Osborne, A.R., Skånland, S.S., Denison, C., Gygi, S.P., Kirkpatrick, D.S., Park, E., Ludtke, S.J., et al. (2007). Ribosome binding of a single copy of the SecY complex: implications for protein translocation. *Mol. Cell* 28, 1083–1092.

Mitra, K., Schaffitzel, C., Shaikh, T., Tama, F., Jenni, S., Brooks, C.L., Ban, N., and Frank, J. (2005). Structure of the *E. coli* protein-conducting channel bound to a translating ribosome. *Nature* 438, 318–324.

Morgan, D.G., Ménétret, J.F., Neuhof, A., Rapoport, T.A., and Akey, C.W. (2002). Struc- ture of the mammalian ribosome-channel complex at 17 Å resolution. *J. Mol. Biol.* 324, 871–886.

Mori, T., Ishitani, R., Tsukazaki, T., Nureki, O., and Sugita, Y. (2010). Molecular mechanisms underlying the early stage of protein translocation through the Sec translocon. *Biochemistry* 49, 945–950.

Nakatogawa, H., and Ito, K. (2002). The ribosomal exit tunnel functions as a discriminating gate. *Cell* 108, 629–636.

Osborne, A.R., and Rapoport, T.A. (2007). Protein translocation is mediated by oligomers of the SecY complex with one SecY copy forming the channel. *Cell* 129, 97–110.

Osborne, A.R., Rapoport, T.A., and Van Den Berg, B. (2005). Protein translocation by the Sec61/SecY channel. *Annu. Rev. Cell. Dev. Biol.* 21, 529–550.

Papanikou, E., Karamanou, S., and Economou, A. (2007). Bacterial protein secretion through the translocase nanomachine. *Nat. Rev. Microbiol.* 5, 839–851.

Petrone, P.M., Snow, C.D., Lucent, D., and Pande, V.S. (2008). Side-chain recognition and gating in the ribosome exit tunnel. *Proc. Natl. Acad. Sci. USA* 105, 16549–16554.

Phillips, J.C., Braun, R., Wang, W., Gumbart, J., Tajkhorshid, E., Villa, E., Chipot, C., Skeel, R.D., Kale, L., and Schulten, K. (2005). Scalable molecular dynamics with NAMD. *J. Comp. Chem.* 26, 1781–1802.

Plath, K., Mothes, W., Wilkinson, B.M., Stirling, C.J., and Rapoport, T.A. (1998). Signal sequence recognition in post-translational protein transport across the yeast ER membrane. *Cell* 94, 795–807.

Ramu, H., Mankin, A., and Vazquez-Laslop, N. (2009). Programmed drug-dependent ribosome stalling. *Mol. Microbiol.* 71, 811–824.

Rapoport, T.A. (2007). Protein translocation across the eukaryotic endoplasmic reticulum and bacterial plasma membranes. *Nature* 450, 663–669.

Rázga, F., Zacharias, M., Réblová, K., Koča, J., and Šponer, J. (2006). RNA kink-turns as molecular elbows: hydration, cation binding, and large-scale dynamics. *Structure* 14, 825–835.

Réblová, K., Rázga, F., Li, W., Gao, H., Frank, J., and Šponer, J. (2009). Dynamics of the base of ribosomal A-site finger revealed by molecular dynamics simulations and Cryo-EM. *Nucleic Acids Res.* 38, 1325–1340.

Sanbonmatsu, K.Y. (2006). Energy landscape of the ribosomal decoding center. *Biochimie* 88, 1053–1059.

Sanbonmatsu, K.Y., and Joseph, S. (2003). Understanding discrimination by the ribosome: stability testing and groove measurement of codon-anticodon pairs. *J. Mol. Biol.* 328, 33–47.

Sanbonmatsu, K.Y., Joseph, S., and Tung, C.S. (2005). Simulating movement of tRNA into the ribosome during decoding. *Proc. Natl. Acad. Sci. USA* 102, 15854–15859.

Sanbonmatsu, K.Y., and Tung, C.S. (2007). High performance computing in biology: Multimillion atom simulations of nanoscale systems. *J. Struct. Biol.* 157, 470–480.

Saparov, S.M., Erlandson, K., Cannon, K., Schaletzky, J., Schulman, S., Rapoport, T.A., and Pohl, P. (2007). Determining the conductance of the SecY protein translocation channel for small molecules. *Mol. Cell* 26, 501–509.

Sardis, M.F., and Economou, A. (2010). SecA: a tale of two protomers. *Mol. Microbiol.* 76, 1070–1081.

Schaletzky, J., and Rapoport, T.A. (2006). Ribosome binding to and dissociation from translocation sites of the endoplasmic reticulum membrane. *Mol. Biol. Cell* 17, 3860–3869.

Schmeing, T.M., Huang, K.S., Strobel, S.A., and Steitz, T.A. (2005). An induced-fit mechanism to promote peptide bond formation and exclude hydrolysis of peptidyl-tRNA. *Nature* 438, 520–524.

Schmeing, T.M., and Ramakrishnan, V. (2009). What recent ribosome structures have revealed about the mechanism of translation. *Nature* 461, 1234–1242.

Schmeing, T.M., Voorhees, R.M., Kelley, A.C., Gao, Y.G., Murphy, F.V., Weir, J.R., and Ramakrishnan, V. (2009). The crystal structure of the ribosome bound to EF-Tu and aminoacyl-tRNA. *Science* 326, 688–694.

Schulten, K., Phillips, J.C., Kalé, L.V., and Bhatele, A. (2008). Biomolecular modeling in the era of petascale computing. In D. Bader, editor, *Petascale Computing: Algorithms and Applications*. Chapman and Hall/CRC Press, Taylor and Francis Group, New York, 165–181.

Seidelt, B., Innis, C.A., Wilson, D.N., Gartmann, M., Armache, J.P., Villa, E., Trabuco, L.G., Becker, T., Mielke, T., Schulten, K., et al. (2009). Structural insight into nascent polypeptide chain-mediated translational stalling. *Science* 326, 1412–1415.

Shaw, D.E., Dror, R.O., Salmon, J.K., Grossman, J., Mackenzie, K.M., Bank, J.A., Young, C., Deneroff, M.M., Batson, B., Bowers, K.J., et al. (2009). Millisecond-scale molecular dynamics simulations on Anton. In *SC'09: Proceedings of the Conference on High Performance Computing Networking, Storage and Analysis*, pp. 39:1–39:11, New York, NY, USA, ACM.

Simon, S., and Blobel, G. (1991). A protein-conducting channel in the endoplasmic- reticulum. *Cell* 65, 371–380.

Steitz, T.A. (2008). A structural understanding of the dynamic ribosome machine. *Nat. Rev. Mol. Cell Biol.* 9, 242–253.

Stone, J.E., Hardy, D.J., Ufimtsev, I.S., and Schulten, K. (2010). GPU-accelerated molecular modeling coming of age. *J. Mol. Graph. Model.* 29, 116–125.

Sund, J., Ander, M., and Aqvist, J. (2010). Principles of stop-codon reading on the ribosome. *Nature* 465, 947–951.

Tama, F., Valle, M., and Brooks III, C.L. (2003). Dynamic reorganization of the functionally active ribosome explored by normal mode analysis and cryo-electron microscopy. *Proc. Natl. Acad. Sci. USA* 100, 9319–9323.

Tenson, T., and Ehrenberg, M. (2002). Regulatory nascent peptides in the ribosomal tunnel. *Cell* 108, 591–594.

Tian, P., and Andricioaei, I. (2006). Size, motion, and function of the SecY translocon revealed by molecular dynamics simulations with virtual probes. *Biophys. J.* 90, 2718–2730.

Trabuco, L.G., Harrison, C.B., Schreiner, E., and Schulten, K. (2010a). Recognition of the regulatory nascent chain TnaC by the ribosome. *Structure* 18, 627–637.

Trabuco, L.G., Schreiner, E., Gumbart, J., Hsin, J., Villa, E., and Schulten, K. (2010b). Applications of the molecular dynamics flexible fitting method. *J. Struct. Biol.* 173, 420–427.

Trabuco, L.G., Villa, E., Mitra, K., Frank, J., and Schulten, K. (2008). Flexible fitting of atomic structures into electron microscopy maps using molecular dynamics. *Structure* 16, 673–683.

Trabuco, L.G., Villa, E., Schreiner, E., Harrison, C.B., and Schulten, K. (2009). Molecular Dynamics Flexible Fitting: A practical guide to combine cryo-electron microscopy and X-ray crystallography. *Methods* 49, 174–180.

Trobro, S., and Åqvist, J. (2005). Mechanism of peptide bond synthesis on the ribosome. *Proc. Natl. Acad. Sci. USA* 102, 12395–12400.

Trobro, S., and Åqvist, J. (2009). Mechanism of the translation termination reaction on the ribosome. *Biochemistry* 48, 11296–11303.

Trylska, J., Tozzini, V., and McCammon, J.A. (2005). Exploring global motions and correlations in the ribosome. *Biophys. J.* 89, 1455–1463.

Tsukazaki, T., Mori, H., Fukai, S., Ishitani, R., Mori, T., Dohmae, N., Perederina, A., Sugita, Y., Vassylyev, D.G., Ito, K., et al. (2008). Conformational transition of Sec machinery inferred from bacterial SecYE structures. *Nature* 455, 988–991.

van den Berg, B., Clemons, W.M., Jr., Collinson, I., Modis, Y., Hartmann, E., Harrison, S.C., and Rapoport, T.A. (2004).

X-ray structure of a protein-conducting channel. *Nature* 427, 36–44.

Villa, E., Sengupta, J., Trabuco, L.G., LeBarron, J., Baxter, W.T., Shaikh, T.R., Grassucci, R.A., Nissen, P., Ehrenberg, M., Schulten, K., et al. (2009). Ribosome-induced changes in elongation factor Tu conformation control GTP hydrolysis. *Proc. Natl. Acad. Sci. USA* 106, 1063–1068.

Voorhees, R.M., Schmeing, T.M., Kelley, A.C., and Ramakrishnan, V. (2010). The mechanism for activation of GTP hydrolysis on the ribosome. *Science* 330, 835–838.

Wallin, G., and Åqvist, J. (2010). The transition state for peptide bond formation reveals the ribosome as a water trap. *Proc. Natl. Acad. Sci. USA* 107, 1888–1893.

Wang, Y., Rader, A.J., Bahar, I., and Jernigan, R.L. (2004). Global ribosome motions revealed with elastic network model. *J. Struct. Biol.* 147, 302–314.

Weixlbaumer, A., Jin, H., Neubauer, C., Voorhees, R.M., Petry, S., Kelley, A.C., and Ramakrishnan, V. (2008). Insights into translational termination from the structure of RF2 bound to the ribosome. *Science* 322, 953–956.

White, S.H. (2007). Membrane protein integration: The biology-physics nexus. *J. Gen. Physiol.* 129, 363–369.

Whitford, P.C., Geggier, P., Altman, R.B., Blanchard, S.C., Onuchic, J.N., and Sanbonmatsu, K.Y. (2010). Accommodation of aminoacyl-tRNA into the ribosome involves reversible excursions along multiple pathways. *RNA* 16, 1196–1204.

Wonderlin, W.F. (2009). Constitutive, translation-independent opening of the protein-conducting channel in the endoplasmic reticulum. *Pflug. Arch. Eur. J. Physiol.* 457, 917–930.

Yanofsky, C. (2007). RNA-based regulation of genes of tryptophan synthesis and degradation in bacteria. *RNA* 13, 1141–1154.

Yin, Y., Arkhipov, A., and Schulten, K. (2009). Simulations of membrane tubulation by lattices of amphiphysin N-BAR domains. *Structure* 17, 882–892.

Zhang, B., and Miller, T.F. (2010). Hydrophobically stabilized open state for the lateral gate of the Sec translocon. *Proc. Natl. Acad. Sci. USA* 107, 5399–5404.

Zimmer, J., Nam, Y., and Rapoport, T.A. (2008). Structure of a complex of the ATPase SecA and the protein-translocation channel. *Nature* 455, 936–943.

Zink, M., and Grubmüller, H. (2009). Mechanical properties of the icosahedral shell of southern bean mosaic virus: A molecular dynamics study. *Biophys. J.* 96, 1350–1363.

CHAPTER 9

The Ribosome as a Brownian Ratchet Machine

Alexander S. Spirin

Alexei V. Finkelstein

Ribosomes are ribonucleoprotein nanoparticles responsible for the synthesis of all proteins in living cells. The function of a ribosome is to translate the genetic information encoded in the nucleotide sequence of mRNA into the amino acid sequence of a protein. During this process, the ribosome performs the unidirectional driving of a single mRNA chain and numerous mRNA-bound tRNA macromolecules through itself. In this process, the free energies of the transpeptidation reaction and GTP hydrolysis are consumed. Thus, the translating ribosome can be considered as a conveying protein-synthesizing molecular machine (Spirin 2002, 2004, 2009a; Frank & Gonzalez 2010). The purpose of this chapter is to analyze the conveying mechanism of this machine.

I. GENERAL PRINCIPLES OF STRUCTURE AND FUNCTION OF THE RIBOSOME

I.1 Subdivision of the Ribosome into Two Unequal Subunits

At first approximation, the ribosome can be described as a compact, almost spherical body with linear dimensions of 25 to 35 nm (reviewed in Spirin 1999, pp. 50–51). Detailed structural analyses reveal the highly complex, asymmetric quaternary structure of the ribosome (Ban et al. 2000; Wimberly et al. 2000; Harms et al. 2001; Yusupov et al. 2001; Ramakrishnan 2002; Gao et al. 2003; Schuwirth et al. 2005; Selmer et al. 2006).

One of the most remarkable features of the ribosome is its universal subdivision into two structurally and functionally different sub-particles, called the small and large ribosomal subunits, either the 30S and 50S in the case of the Bacterial (prokaryotic) 70S ribosomes, or the 40S and 60S in the case of the cytoplasmic 80S ribosomes of Eukaryotes – animals, plants, and fungi. The association between the ribosomal subunits is relatively labile, maintained mostly by several Mg^{2+}-dependent RNA-RNA contacts, and also by a few protein-RNA and protein-protein interactions ("intersubunit bridges"). The

non-functioning ribosome reversibly dissociates into its subunits upon depletion of Mg^{2+}, or at high concentrations of monovalent cations. (It should be mentioned, however, that aminoacyl-tRNA and peptidyl-tRNA bound within the functioning ribosome form additional inter-subunit bridges strongly stabilizing the association of the subunits; see Belitsina & Spirin 1970). Only the entire ribosome is capable of performing mRNA-dependent protein synthesis, termed translation. After finishing the translation, the ribosome dissociates into its subunits, under the effect of special protein factors. A new round of translation begins with the dissociated ribosomal subunits; the initiation of translation is coupled with the re-association of the subunits into the entire ribosome. The inter-subunit contacts are labile enough to allow some movement of the subunits relative to each other during translation (see Section II).

The principal morphological features of the two ribosomal subunits (reviewed in Spirin 1999, pp. 63–73) are designated on contours presented in Figure 9.1. The following features are important for subsequent consideration of functional centers of the ribosome. The small ribosomal subunit is sub-divided into three main lobes – a body, a side lobe or platform, and a head. The head is covalently connected to the rest of the subunit (specifically, the side lobe, or platform) by a thin neck. The prominent part of the head is called the beak. The part of the body under the beak is called the shoulder. The large subunit has three protuberances forming a crown-like image over the rest of the subunit body. The central protuberance (sometimes designated as the head of the large subunit) is rather massive and separated from the body by a wide "valley" that traverses the contacting (facing the other subunit) side of the large subunit from one lateral protuberance to the other. The lateral protuberance situated on the right-hand side when the subunit is viewed from its contacting surface is called the L10/L12 stalk (by name of the ribosomal proteins involved in its formation). The other lateral protuberance is designated as the L1 protuberance, or the L1 stalk (again by the name of the ribosomal protein localized there).

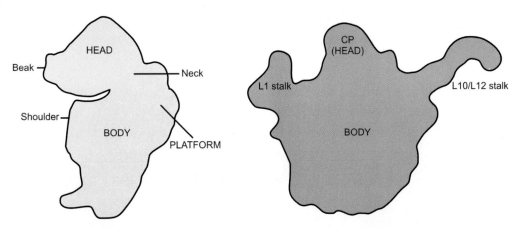

FIGURE 9.1: *Contours of the two ribosomal subunits viewed from their contacting (facing each other) surfaces. Left, small subunit; right, large subunit. Drawn from a compilation of X-ray structural analysis data on* Thermus thermophilus *and* Escherichia coli *ribosomes (mainly from Schuwirth et al. 2005, and Schmeing et al. 2009).*

I.2 Self-Folding of Ribosomal RNAs

Each ribosomal subunit is organized largely by one molecule of a compactly folded high-polymer ribosomal RNA (rRNA), which comprises two-thirds (in Bacterial ribosomes) or about half (in Eukaryotic ribosomes) of the total mass of a ribosomal subunit (reviewed in Spirin 1999, pp. 75–95). The rRNA chain of the large ribosomal subunit (50S or 60S) is approximately twice as long as the rRNA of the small subunit (30S or 40S, respectively). Thus, the rRNA chain of the Bacterial 30S subunit (the 16S rRNA) typically consists of approximately 1,500 nucleotide residues, whereas the rRNA of the 50S subunit (the 23S rRNA) comprises approximately 3,000 nucleotides. The chains of two high-polymer rRNAs of mammalian cells (18S rRNA and 28S rRNA) have lengths of approximately 1,900 nucleotides and 4,800 nucleotides, respectively.

Under conditions of suppressed electrostatic repulsion of the phosphate groups (i.e., at moderate ionic strengths), in the presence of sufficient Mg^{2+} concentrations, the chains of the isolated rRNAs are capable of self-folding into compact particles of specific shapes. The dimensions and shapes of the compactly folded 16S rRNA and 23S rRNA are similar to the dimensions and shapes of the corresponding ribosomal subunits, 30S and 50S, respectively (Vasiliev et al. 1978; Vasiliev & Zalite 1980; Vasiliev et al. 1986). From this it can be concluded that the specific self-folding of rRNAs determines the formation of the ribosomal particles with their characteristic quaternary structures, including assembly of numerous ribosomal proteins on the RNA cores.

The two high-polymer rRNAs are folded so that their sections form individualized compact domains. The 16S

(18S) rRNA is folded into three large compact domains forming the body (5′-proximal part of the RNA, domain I), the platform (the middle section of the RNA, domain II), and the head (the 3′-proximal part of the RNA, domain III) of the small ribosomal subunit (Wimberly et al. 2000). In addition, the forth minor domain representing the 3′-terminal section of the 16S rRNA chain is folded into two double helixes (h44 and h45) located on the contacting surface of the small subunit. The covalent link between the head and the rest of the small subunit is constituted by helix h28, connecting domain III to the side lobe domain II; it is the neck of the subunit.

The 23S (28S) rRNA is folded into six compact domains (from I to VI), which are not so distinctly separated in three-dimensional space, but rather tightly interact with and mutually penetrate each other (Ban et al. 2000), so that the large ribosomal subunit is more rigid, compared to the small subunit. The tertiary structures of the folded Bacterial 16S and 23S rRNAs with differently colored domains and numeral designations of some double helical regions are shown in Figure 9.2.

In addition to the high-polymer rRNA, the large ribosomal subunit (both 50S and 60S) contains an rRNA of relatively small mass – the 5S rRNA of 120 nucleotides in length. It is localized on the central protuberance (the head) of the large subunit, looking like a hat on one side of the head (see Figure 9.2). Also, the Eukaryotic 60S ribosomal subunits have another relatively low-polymer rRNA species called the 5.8S rRNA, which is a structural homolog of the 5′-terminal 160-nucleotide section of the Prokaryotic 23S rRNA. These small rRNAs are comparable in size with ribosomal proteins and together with the

a

b

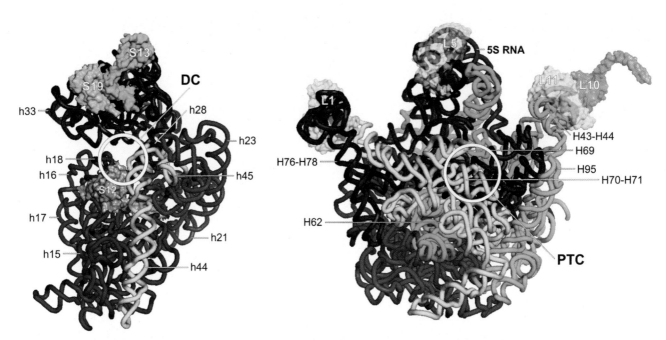

FIGURE 9.2: *Folding of Bacterial* (E. coli) *rRNAs, 16S rRNA (a) and 23S rRNA (+ 5S rRNA) (b), into secondary and tertiary structures viewed from the contacting surfaces of the ribosomal subunits, 30S and 50S, respectively; shown in the* tube *format. Some interface ribosomal proteins are also shown, being presented in the* surface *format. The drawing of ribosomes, here and below, is done using a program PyMOL. RNA domains are differently colored (by analogy with the presentation of Yusupov et al. 2001, figure 3 C, D, for* T. thermophilus *rRNAs). Compiled from X-ray structural analysis data on* E. coli *ribosomes (Berk et al. 2006, PDB codes 2I2V and 2I2U; Zhang et al. 2009, PDB code 3I1M); position and structure of E. coli protein L1 are imaged according to Trabuco et al. 2008, and Gumbart et al. 2009. Position and structure of protein L10 are taken from the data on* T. thermophilus *ribosomes (Gao et al. 2009, PDB code 2WRJ). (a) (16S rRNA): domain I, blue; domain II, magenta; domain III, red; minor domain IV, yellow. Ribosomal protein S12 is located near the decoding center (DC), and the heterodimer S13-S19 is positioned at the top of the head, and is in contact with protein L5 on the 50S subunit. (b) (23S rRNA and 5S rRNA): domain I, blue; domain II, green; domain III, magenta; domain IV, yellow; domain V, red; domain VI, cyan; 5S rRNA, blue. Ribosomal protein L5 (half-transparent surface* format*) sits at the top of the central protuberance, and is bound with the 5S rRNA. Protein L1 is attached to the block of helixes H76-H78, proteins L11 and L10 are bound to the block of helixes H43-H44; and not shown here, the two highly movable dimers of protein L12 are attached to the elongated α-helical domain of protein L10.*

proteins are accommodated on the high-polymer rRNA core.

The contacts, or bridges, between the two ribosomal subunits in the entire ribosome are organized mostly by rRNAs. The contacts are relatively labile and re-arrangeable, thus allowing the ribosomal subunits to be mutually shifted during translation. The main inter-subunit contacts are grouped in the middle of the bodies of the two subunits between the longitudinal helix h44 of the small subunit and the regions of transverse helixes H69, H71, and H62 of domain IV of the large subunit. Helix h44 may be considered as a principal pivot or hinge in the labile association of two ribosomal subunits.

I.3 Assembly of Ribosomal Proteins on Compact RNA Cores

The small (30S or 40S) and the large (50S or 60S) ribosomal subunits have their own assortments of ribosomal proteins,

each being present in one copy per particle (reviewed in Spirin 1999, pp. 97–129). The important exception is the protein L12, which, together with its acetylated derivative L7, is present in four copies. This tetramer will be further termed as (L12)₄. The majority of ribosomal proteins are basic polypeptides owing to the high content of lysine and arginine residues. The molecular masses of ribosomal proteins vary mostly between 10 and 30 kDa, although several strongly basic proteins of the large subunit are even smaller, whereas the largest protein of the Bacterial 30S subunit is about 60 kDa. The nomenclature of the proteins is based on their electrophoretic mobility in a highly cross-linked polyacrylamide gel, under denaturing conditions, where polypeptides are separated mainly according to their sizes; the proteins are designated by numbers from 1 for the largest (or rather the slowest migrating on the gel) polypeptide to 2, 3, 4, and so forth, for progressively smaller proteins, with prefixes either S (i.e., S1, S2, S3, etc.) in the case of the small ribosomal subunit proteins, or L (i.e., L1,

L2, L3, etc.) in the case of the large subunit (Kaltschmidt & Wittmann 1970).

Most ribosomal proteins are distributed on the surfaces of the compact RNA cores in patches, rather than covering the surfaces evenly, and no symmetry in their arrangements has been revealed. Parts of some ribosomal proteins, especially their unfolded tails, can penetrate deep into the rRNA cores. Unlike the situation of spherical viral particles, the numerous different proteins are assembled into the ribosomal particles according the principle of recognition of their own sites on the compactly folded rRNAs. Thus, each protein recognizes and attaches to a specific local structure of the corresponding rRNA; the attachment is often accompanied by some alterations of local rRNA conformations (Vasiliev et al. 1986; Ban et al. 2000; Wimberly et al. 2000). Rather extensive areas of the rRNA surfaces remain free of proteins and are directly exposed to the medium. Most proteins are located on the external (facing the surroundings) surfaces of the subunits, whereas the contacting surfaces contain few proteins, preferentially located on the periphery of the surfaces. In Figure 9.2a, three functionally important proteins of the small subunit are indicated on the contacting surface: S12, S13, and S19. In the 70S ribosome, the protein heterodimer S13/S19 of the 30S subunit head contacts protein L5 of the 50S subunit head (central protuberance, shown in Figure 9.2b). As also shown in Figure 9.2b, one lateral protuberance is formed by the block of helixes H76–H78 with protein L1, and the other lateral protuberance includes the block of helixes H42–H44 with attached protein L11 and the pentameric protein complex L10•(L12)$_4$.

Three roles of the ribosomal proteins in the ribosome can be distinguished. First, they can simply stabilize or modify local structures of the rRNA (see Vasiliev et al. 1986; Ban et al. 2000; Wimberly et al. 2000). Most ribosomal proteins seem to perform this auxiliary structural role. Second, some of the proteins can participate in the function of ribosomal active centers, formed mainly by rRNA. In particular, the catalytic activity of the ribosome – its peptidyl transferase activity – is fully provided by local rRNA structures within the large ribosomal subunit (particularly, by part of domain V) (Nissen et al. 2000), but some ribosomal proteins (e.g., L27; not shown in Figure 9.2b) may be required for maintaining the optimal conformations of these structures (Voorhees et al. 2009). Third, ribosomal proteins can fulfill their own specific functions. An example of this is the pentameric protein complex L10•(L12)$_4$ of the large ribosomal subunit. The complex L10•(L12)$_4$ plays an important role in the translating ribosome by providing the recruitment of two external (non-ribosomal) proteins – elongation factors EF1 (EF-Tu) and EF2 (EF-G) – required for the catalysis of aminoacyl-tRNA binding and translocation, respectively (see section V for references). The other lateral protuberance – the L1 stalk – is involved in the exit of the deacylated tRNA

from the translating ribosome (see Sections IV and V for references).

I.4 "Division of Labor" between the Ribosomal Subunits

To synthesize proteins, the ribosome uses aminoacyl-tRNAs as substrates. The synthesis proceeds by the sequential addition of amino acid residues to the carboxyl terminus (C-terminus) of the growing polypeptide chain (reviewed in Spirin 1999, pp. 133–161). The addition is accomplished via a transpeptidation reaction between the ribosome-retained peptidyl-tRNA and the newly arrived ribosome-bound aminoacyl-tRNA, whose anticodon has been correctly recognized by the ribosome-mRNA complex. As a result, the C-terminus of the peptidyl residue is transferred from its tRNA residue to the amino group of the newly bound aminoacyl-tRNA, and the tRNA from the substrate peptidyl-tRNA is released:

$$\text{Pept}(n)\text{-tRNA}' + \text{Aa-tRNA}'' \rightarrow \text{Pept}(n+1)\text{-tRNA}'' + \text{tRNA}'$$

(Pept is the peptidyl residue of peptidyl-tRNA, n is the number of amino acid residues in it, and Aa is the amino acid residue).

The *peptidyl-transferase center* (PTC, see Figure 9.2b) is an integral part of the large ribosomal subunit (50S or 60S). The PTC is organized by an rRNA component (part of domain V) of the large subunit at the subunit interface side, in the "valley" below the central protuberance, under helix H71 of domain IV (see Figure 9.2b). It contains two substrate-binding sites: the *a* site,[1] responsible for binding of the 3'-aminoacyl acceptor end of the incoming aminoacyl-tRNA as an acceptor substrate; and the *p* site,

---

[1] In this work, for the sake of avoiding confusion, we designate the part of the aminoacyl-tRNA-binding site located at the DC of the small ribosomal subunit as the *A* site, whereas the part bound by the PTC at the large subunit, the acceptor substrate-binding site of PTC, as the *a* site. Analogously, the part of the site retaining the peptidyl-tRNA on the small subunit prior to the peptidyl transferase reaction is designated here as the *P* site, and the part bound in the PTC at the large subunit, the donor substrate-binding site of PTC, is designated as the *p* site. *A* and *P* in capital letters are proposed for the sites on the small subunit, because the size of each site is large (nucleotide triplet, molecular scale, ~10 Å), whereas *a* and *p* designate small sites at PTC of the large subunit (atomic scale, ~3 Å). The binding sites of the deacylated tRNA on the small and the large ribosomal subunits are designated as *E* and *e*, respectively. Correspondingly, the positions of the tRNAs in the ribosome are denoted as *A/a, A/p, P/p, P/e, E/e*, as well as *A/T* where *T* is the binding site of the aminoacylated acceptor arm of the aminoacyl-tRNA at the ribosome-bound elongation factor EF-Tu. The sequential functional states of the translating ribosome are signified as (*P/p*), (*A/T, P/p*), (*A/a, P/p*), (*A/p, P/e*), and (*P/p, E/e*). Note that in the natural translation stages, no more than two tRNA residues are simultaneously present in the ribosome (Spirin 1984; Robertson & Wintermeyer 1987; Semenkov et al. 1996; Spiegel et al. 2007; Uemura et al. 2010).

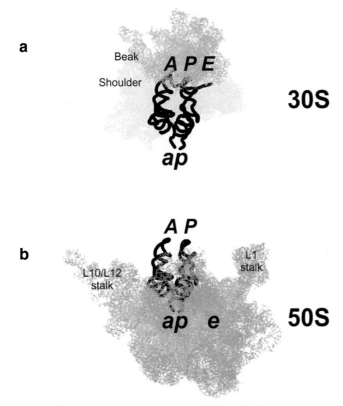

FIGURE 9.3: *Functional sites of the ribosomal subunits. The subunits are represented on the basis of recent X-ray structural analysis data on* T. thermophilus *ribosomes (Schmeing et al. 2009, PDB code 2WRN; Gao et al. 2009, PDB codes 2WRI and 2WRJ; Voorhees et al. 2009, PDB code 2WDG) in the line format, with the two bound tRNAs in the tube format. The aminoacyl-tRNA (left) and peptidyl-tRNA (right) positions are those of the tRNA residues in the translating 70S ribosome prior to the transpeptidation reaction. See footnote 1 for the nomenclature of tRNA binding sites used in this paper. (a) Small (30S) ribosomal subunit oriented with its head to the viewer and the interface side downward. The head is colored in a darker yellow, as compared with the rest of the subunit. The anticodons of the tRNAs are paired with two adjacent cognate codons of the mRNA shown in red. The elbows (external angles) of the L-shaped tRNAs are facing in the same direction as the head is (towards the viewer). The acceptor ends of the tRNAs are close to each other in the positions equivalent to those in a and p sites of the peptidyl transferase center (PTC) on the 50S subunit. (b) Large (50S) ribosomal subunit oriented with its central protuberance (head) toward the viewer and the interface side upward. The elbows (external angles) of the L-shaped tRNAs face toward the central protuberance (i.e., to the viewer). The lateral protuberances are on the left (L10/L12 stalk) and on the right (L1 stalk). The anticodon arms of the tRNAs are in the positions equivalent to those in the A and P sites on the 30S subunit.*

which binds the 3'-aminoacyl acceptor end of the peptidyl-tRNA (Figure 9.3). In the course of the peptidyl transferase reaction, the nitrogen atom of the amino group of the aminoacyl-tRNA at the *a* site attacks the carbon of the carbonyl of the ester group of the peptidyl-tRNA at the *p* site, resulting in the formation of the tetrahedral intermediate (Lim & Spirin 1986; Nissen at al. 2000; Hansen

et al. 2002). The latter is spontaneously cleaved into the products – the elongated peptidyl-tRNA and the deacylated tRNA (ibid).

On the other hand, the order of amino acid residues in the synthesized polypeptide chain is determined by the sequence of the nucleotide residues in the messenger RNA (mRNA) chain, which binds to the ribosome and thus programs it. The sequence of nucleotide triplets (codons) determines the amino acid sequence of the polypeptide chain by setting the order of entering of various aminoacyl-tRNAs into the mRNA-programmed ribosome. Thus, the translating ribosome continuously retains elongating peptidyl-tRNAs and periodically accepts an aminoacyl-tRNA. To enter the ribosome, a molecule of aminoacyl-tRNA must be recognized by a nucleotide triplet (codon) of the mRNA in the *decoding center* (DC, see Figure 9.2a). The DC is organized by the rRNA of the small ribosomal subunit and is localized on the contacting side in the region of the neck (helix h28) at the upper end of the longitudinal helix h44 and the loop of helix h18 (see Figure 9.2a). The DC is a part of the *A* site, where the anticodon arm of the incoming aminoacyl-tRNA binds and specifically interacts with a complementary (cognate) codon of mRNA. In the immediate vicinity is the *P* site, which retains the anticodon arm of the peptidyl-tRNA bound with the preceding codon of mRNA (also via the complementary codon-anticodon interaction) (Figure 9.3). After transpeptidation and moving from the *P* site, the deacylated tRNA is temporarily found in the *E* site (exit site).

The large and small ribosomal subunits associate with each other in such a way that a significant inter-subunit space is formed. The two tRNA substrates – the aminoacyl-tRNA and peptidyl-tRNA – are accommodated in this inter-subunit space. Thus, the two substrate tRNAs prior to the transpeptidation reaction are bound to two adjacent codons of the mRNA by their anticodons at the *A* and *P* sites on the small ribosomal subunit, and at the same time interact by their 3' ends with the *a* and *p* sites of the PTC on the large subunit.

Entrance of the incoming aminoacyl-tRNA into the inter-subunit space is localized between the beak and the shoulder of the small subunit and the base of the L10/L12 stalk of the large subunit (Figures 9.1 and 9.2). The area at the base of the L10/L12 stalk of the large subunit together with the shoulder of the small subunit form a pocket for transient binding of special proteins called elongation factors, EF-Tu (EF1) and EF-G (EF2). These factors catalyze the entry of the incoming aminoacyl-tRNA into the *A* site before transpeptidation and the intra-ribosomal translocation of the newly elongated peptidyl-tRNA after transpeptidation, respectively (see Section V).

The exit of the deacylated tRNA from the *E* site occurs near the subunit interface also, but on the opposite side, at the other lateral protuberance, the L1 stalk (see Figures 9.1b and 9.2b). There is evidence that the L1 stalk is involved in the temporary retention of the deacylated tRNA in the *E* site after transpeptidation and its subsequent release from the translating ribosome (see Sections IV and V).

Thus, this overview of the functional centers of the entire ribosome shows the three aspects of its function. First, the ribosome is a polypeptide-synthesizing device, a ribozyme that catalyzes the peptidyl transferase reaction. This is the *biochemical function* of the ribosome. Another biochemical function of the ribosome is inducing the hydrolytic GTPase activities of several proteins – translation factors – that transiently interact with the ribosome during translation. Second, the ribosome is an apparatus that decodes the genetic information recorded in the nucleotide sequence of mRNA. This is the *genetic function*. It is remarkable that the biochemical and genetic functions are distinctly divided between the two ribosomal subunits: the large subunit is only catalytic, whereas the small subunit is solely decoding. Third, the translating ribosome is a molecular machine, which accomplishes the passing of compact tRNAs through itself and simultaneously drives the mRNA chain. This function of the ribosome can be considered as the *conveying function*. The conveying function requires the involvement of both ribosomal subunits.

I.5 Translation of mRNA as a Conveying Process

The function of the ribosome is to translate the genetic information encoded in the nucleotide sequences of mRNA into amino acid sequences of the polypeptide chains of proteins. The reading of the nucleotide sequence of the mRNA chain proceeds by its unidirectional running through the DC of the ribosome, which works as a tape-driving mechanism. The role of movers for the mRNA is played by the tRNAs passed through the ribosome. During the process of translation, the ribosome performs the unidirectional drawing of tRNA molecules through itself, and the mRNA chain moves due to the coupling of its codons with the complementary anticodons of the tRNAs. The most direct evidence of the complementary coupling between tRNA and mRNA during translocation came from experiments where the mutant tRNA with a quadruplet instead of a triplet as an anticodon suppressed the (+1) frame-shift mutation in mRNA (Riddle & Carbon 1973). In other words, the four-nucleotide anticodon tRNA provided the translocational shift of mRNA in the ribosome by four (not by three) nucleotide residues. The discovery of ribosomal synthesis of a polypeptide from aminoacyl-tRNA in the absence of any template polynucleotide demonstrated the lack of dependence of the translocation mechanism on the template polyribonucleotide (Belitsina et al. 1981, 1982; Spirin et al. 1988). The authentic elongation cycle, including the EF-G-catalyzed translocation stage, was demonstrated in the case of the template-free elongation (ibid). It follows that the tRNA's entry into the ribosome and successively

stepping from the *A* site to the *P* site, and then to the *E* site, carries the attached mRNA through the ribosome. Thus, the translocation of tRNAs with their anticodons within the ribosome is an active process, whereas the coupled movement of the mRNA chain is a passive one. The direction of tRNA translocation is from the L10/L12 stalk to the L1 stalk of the ribosome (which implies the tRNA movement from *A* to *P* and then to *E* site), and the mRNA chain is driven through the ribosome from its 5′ end to the 3′ end.

In the course of translation, the free energies of the transpeptidation reaction and the GTP hydrolysis reactions are consumed (Figure 9.4). Hence, the translating ribosome can be considered as a conveying molecular machine (and simultaneously a technological protein-synthesizing machine). The conveying function has been suggested to be based on the anisotropic motions generated by thermal Brownian movements of large blocks of the ribosome and the ribosomal subunits and rectified by energy-dependent mechanisms (Spirin 1985, 2009a; Frank & Gonzalez 2010).

II. CONFORMATIONAL MOBILITY OF THE RIBOSOME

II.1 Large-Block Mobility in the Ribosomal Subunits

The unique conformations of the two specifically folded high-polymer rRNAs (Figure 9.2) allow specific anisotropic motility of their structural blocks. This underlies the structural basis for functional intra-subunit movements. In accordance with the dense mutual packing and interpenetration of the rRNA domains in the ribosomal 23S rRNA of the large ribosomal subunit (Ban et al. 2000), as compared with the loosely arranged and weakly interacting rRNA domains of the 16S rRNA of the small subunit (Wimberly et al. 2000), the two subunits were found to manifest very different levels of thermal mobility of their blocks (Korostelev & Noller 2007). The main body of the large subunit proved to be almost monolithic, with low levels of structural fluctuations inside. At the same time, the two peripheral protuberances of the large subunit – the block of helixes H43-H44 with proteins L11-L10-(L12)$_4$ called the L10/L12 stalk, and the block of helixes H76-H77-H78 with protein L1 called L1 stalk (Figures 9.1b and 9.2b) – demonstrated extremely high levels of thermal mobility (Korostelev & Noller 2007). The small ribosomal subunit has high levels of thermal mobility in all domains of the 16S rRNA (ibid).

The movable L10/L12 stalk (Gudkov et al. 1982) was found to be involved in binding of elongation factors to the ribosome (Girshovich et al. 1981; Girshovich et al. 1986; Moazed et al. 1988), leading to the immobilization of the stalk (Gongadze et al. 1984). Cryo-electron microscopy observations demonstrated that GTP-dependent binding of EF-G, as well as EF-Tu, to the ribosome is accompanied by positioning of the L10/L12 stalk closer to the central

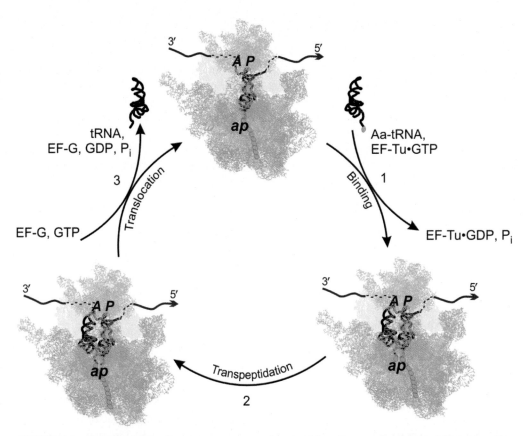

FIGURE 9.4: *The translating ribosome as a conveying machine: three main stages of the elongation cycle of the translating ribosome resulting in passing one tRNA molecule and one mRNA nucleotide triplet through the inter-subunit channel of the ribosome. In this and all the following figures with ribosome images, the ribosomes are oriented in such a way that the small (30S, yellow) subunit is on top of the large (50S, pink) subunit, the heads of both being directed to the viewer. The ribosomal subunits and tRNA residues are presented in the same way as in Figure 9.3. The tRNA residues bound to the A and P sites are paired with their anticodons with two adjacent codons of the mRNA (red). The arrowhead at the 5′ end of mRNA chain indicates the direction of mRNA movement through the ribosome during translocation (the conveying direction). Small gray ovals at the acceptor ends of tRNAs symbolize aminoacyl residues; correspondingly, the growing polypeptide chain is symbolized by the ovals in the chain. A codon-cognate aminoacyl-tRNA enters into the translating ribosome from the left side (at the L10/L12 stalk), whereas a deacylated tRNA exits from the right side (at the L1 stalk). The stages of aminoacyl-tRNA binding and translocation are catalyzed by elongation factors EF-Tu and EF-G, respectively (correspondingly marked on the scheme, but not drawn), with coupled hydrolysis of GTP, whereas transpeptidation is catalyzed by the ribosome itself.*

protuberance of the 50S ribosomal subunit; this movement was proposed to be part of a general mechanism of loading translation factors into the ribosome's factor-binding site (Valle et al. 2003a; Valle et al. 2003b; Spahn et al. 2004).

The variations in the position of the other-side protuberance of the large ribosomal subunit – the L1-H76-H77-H78 block (the L1 stalk) – were revealed by cryo-electron microscopy reconstructions and X-ray crystallographic studies of the ribosomes in different states, and its mobility was proposed to be involved in the displacement and exit of the deacylated tRNA during the final stage of translocation (Gomez-Lorenzo et al. 2000; Harms et al. 2001; Yusupov

et al. 2001; Valle et al. 2003a; Spahn et al. 2004; Schuwirth et al. 2005). In experiments using single-molecule Förster resonance energy transfer (FRET), the real-time dynamics of the L1 stalk was followed and its movement relative to the body of the large ribosomal subunit was demonstrated (Cornish et al. 2009; Fei et al. 2009); three distinct conformational states – open, half-closed, and fully closed – were observed.

The first experimental evidence for an intra-subunit conformational change in the small (30S) ribosomal subunit – the movement of the head of the small ribosomal subunit relative to its body – came from neutron-scattering experiments (Serdyuk & Spirin 1986; Serdyuk et al. 1992).

It was proposed that this movement may have a direct relation to the mechanical act of translocation. More recently, cryo-electron microscopy reconstructions of Eukaryotic 80S ribosomes demonstrated a small rotational shift of the head relative to the body of the small ribosomal subunit upon the binding of elongation factor eEF2 to the ribosome (Spahn et al. 2004; Taylor et al. 2007). Independently, two different conformations of Bacterial 70S ribosomes (designated as I and II) were revealed by X-ray crystallographic analysis; the main difference between the two was that the head of the small subunit in the ribosome of type II, when compared with the ribosome of type I, was rotated as a rigid block around the neck in the direction of the mRNA and tRNA conveying path during translocation, meaning counter-clockwise if viewed from top of the head (Schuwirth et al. 2005). The rotational shift was estimated to be up to 12°, or about 20 Å at the subunit interface.

Analysis of the dynamics of thermal structural movements in the small ribosomal subunit by the translation-libration-screw (TLS) crystallographic refinement method (Korostelev & Noller 2007) showed that the most movable region of this subunit is the block formed by helixes h30-h34 of domain III of the 16S RNA, which constitutes the beak – the prominent part of the head at the entrance into the inter-subunit channel (see h33 in Figure 9.2a). X-ray crystallographic analysis demonstrated that the anisotropic displacement of this structural block is realized in the process of aminoacyl-tRNA binding to the A site of the ribosome: Upon binding of the anticodon hairpin of tRNA in the A site, this block moves toward the shoulder – the part of the body at the other side of the entrance into the inter-subunit channel (Ogle et al. 2001; Ogle et al. 2002). As a result, the anticodon hairpin is found to be occluded (locked) in the A site.

Another highly movable part of the small ribosomal subunit is the minor 3′-terminal domain of the 16S RNA (helixes h44-h45). Helix h44, the longest hairpin of the subunit that ranges from the head to the end of the body and forms a number of important contacts with the large subunit, displays a tendency for rotational motions around an axis approximately parallel to the small subunit's long axis and located at the subunit interface (Korostelev & Noller 2007). This rotational movement may play a pivotal role in the mutual mobility of the ribosomal subunits and in translocation (Yusupov et al. 2001; Valle et al. 2003a; Schuwirth et al. 2005; Korostelev et al. 2006).

The side lobe formed by domain II of the 16S RNA, the so-called platform, showed significant mobility relative to the rest of the small subunit (Gabashvili et al. 2000; Korostelev & Noller 2007). The functional role of this motion may be associated with the processes of relative movements of the ribosomal particle and mRNA during initiation of translation (Yusupova et al. 2001; Yusupova et al. 2006).

II.2 Locked and Unlocked Conformations[2]

The intra-ribosomal movements are based on thermal anisotropic fluctuations where the anisotropy is determined by the structure of a movable body and its environment. It is evident that the binding of a ligand can induce fixation of one of the alternative conformations. This can be called *locking*, induced fit, or maximization of non-covalent bonds between a ligand and surrounding groups; it is usually accompanied by the closing of a binding pocket around a ligand. The destruction (e.g., hydrolysis) or modification (e.g., group transfer) of a bound ligand leading to a decrease of the non-covalent interactions with its binding site can induce *unlocking*, or the opening of the binding pocket.

Considering the full elongation cycle of the ribosome from this position, it was proposed that during translation the ribosomal parts move relative to each other, and a series of various locking and unlocking events underlie the intra-ribosomal movements. Mutual mobility of the two ribosomal subunits was postulated to provide the main type of locking-unlocking dynamics of the translating ribosome (Spirin 1968a, 1968b, 1969). The idea was that the presence of two substrate ligands, aminoacyl-tRNA and peptidyl-tRNA, in the A and P sites of the translating ribosome induces a tight association between the ribosomal subunits (locking), and that this locking is required for the intraribosomal transpeptidation reaction between the substrates. Transpeptidation changes the non-covalent bonding landscape in the binding pockets of the ligands, thus resulting in loosening the subunits association (unlocking), this being a prerequisite for inter-subunit movements of the ligands during the subsequent intra-ribosomal translocation. Recently the inter-subunit locking-unlocking model for the translating ribosome has been supported by the experiments using single-molecule FRET technique during elongation process (Aitken & Puglisi 2010). In more detail, the inter-subunit mobility problem is discussed in the following section.

---

[2] The terms "locked" and "unlocked" are used here for description of two alternative states of the conformation of a macromolecule or a macromolecular complex, one of which is more or less rigid or fixed, and the other is relaxed, flexible, and fluctuating, respectively. In most cases (although not always), the locked conformation is equivalent to a closed form of a macromolecule or a macromolecular complex, whereas the unlocked implies an open form. The closing or locking is usually a result of ligand binding that is accompanied by an induced fit or maximization of non-covalent interactions between a macromolecular partner and a ligand. As to the entire ribosome (70S or 80S) specifically, a locked state means a tight association between the ribosomal subunits, whereas an unlocked state implies a looser association of the mutually oscillating subunits with possibilities for ligands to move within the inter-subunit space. In this case the terms "inter-subunit unlocking" and "inter-subunit locking" are used. Another case is "entrance unlocking" and "entrance locking" when a more local conformation at the side of the entry of aminoacyl-tRNA into ribosome is correspondingly changed.

Generally, locking and unlocking events in translating ribosome are not confined to the inter-subunit locking and unlocking. The binding of the cognate codon-anticodon helix (anticodon stem-loop) to the *A* site on the 30S ribosomal subunit described by Ogle et al. (2001, 2002) in terms of the transition from an open to a closed form is a remarkable example of the locking of the aminoacyl-tRNA-binding pocket of the small ribosomal subunit. Moreover, as the closing of the aminoacyl-tRNA-binding pocket also involves the shifts of the shoulder and the beak of the small subunit (see Figures 9.1 and 9.2) toward the subunit interface (Ogle et al. 2002), this conformational movement can be considered to be a locking of the entrance into the inter-subunit space of the ribosome. (To distinguish this process from the previously described inter-subunit locking-unlocking, the terms "entrance locking" and "entrance unlocking" will be used in the following text and figures). The induced fit allowing maximization of contacts in the tRNA-binding pocket is contemplated as a physical mechanism of the selectivity of codon-directed binding of a cognate tRNA (Ogle et al. 2001). The open (unlocked) form of the 30S subunit should be more relaxed (fluctuating), whereas the closed (locked) form seems to be more rigid due to addition of non-covalent contacts. The situation may also be considered as an oscillation between two (or more) alternative local conformations, with the equilibrium shifted toward the open form when a ligand is absent, whereas the presence of the proper ligand (cognate anticodon stem-loop) shifts the equilibrium to the closed form and reduces the rate of the reverse rearrangement.

The attachment of a GTP-bound translation factor to the ribosome leads to immobilization of the L10/L12 stalk and formation of a closed pocket around the factor with the participation of the adjacent tRNA entrance region, in particular the shoulder of the small subunit (Stark et al. 2002; Valle et al. 2003a; Valle et al. 2003b; Gao et al. 2009; Schmeing et al. 2009). Next, hydrolytic cleavage of the factor-bound GTP and release of orthophosphate would lead to relaxation of the closed conformation of the factor (unlocking of its domains) and the ribosomal pocket. The conformational changes result in the loss of the high affinity of the factor for the L10/L12 pocket and thus its local unlocking (relaxation). This can also allow movements at the subunit interface.

It is logical to infer that the binding of an aminoacyl-tRNA to the *A* site of the post-translation ribosome should occur with unlocked conformations of the *A* and *a* sites on both subunits and lead to fixation of locked conformations. This may involve both the entrance locking upon binding of a codon-cognate aminoacyl-tRNA and the locked state of inter-subunit contacts. In any case, the subsequent transpeptidation reaction requires the locked conformation in the region of the PTC as it firmly positions the aminoacyl residue in the *a* site, in the immediate vicinity of the peptidyl group of the *p*-site peptidyl-tRNA.

As a result of the transpeptidation, two strong inter-subunit bridges, formed by bound aminoacyl-tRNA and peptidyl-tRNA (Belitsina & Spirin 1970), become disrupted: The *P*-site tRNA is now deacylated and cannot be retained further in the *p* site of PTC on the large ribosomal subunit, and the *A*-site tRNA has lost affinity for the *a* site of the PTC. Thus the ribosome in this post-transpeptidation state is allowed to be unlocked at the subunit interface (Spirin 1968a, 1969; Valle et al. 2003a; Aitken & Puglisi 2010).

Generally, according to the physical logics used here, the binding of a ligand to the ribosome must induce locking of a certain type, either local or more generalized, whereas a chemical reaction destroying the original structure of a bound ligand leads to unlocking. Movements of a bound ligand within the ribosome, as well as its entry into and release from the ribosome, require an unlocked state of the ribosome or its corresponding binding pocket. Hence, both factor-free and EF-Tu-promoted entry of an aminoacyl-tRNA into the ribosome should demand an open, or unlocked, conformation at the entrance into the ribosome, that is, in the region of its inter-subunit space adjacent to the beak/shoulder side of the small subunit and L10/L12 stalk of the large subunit (see Figures 9.1 and 9.3). As mentioned earlier, binding of a codon-cognate aminoacyl-tRNA simultaneously to the *A* site at the small subunit and the *a* site of PTC at the large subunit, side-by-side with peptidyl-tRNA, locks the ribosome entrance, and may stabilize the inter-subunit locking. Transfer of peptide group from the *P*-site tRNA to amino group of the *A*-site aminoacyl-tRNA destabilizes the chemical landscape between the ribosomal subunits (see above) and leads to inter-subunit unlocking; this unlocking has been recently shown by the FRET technique (Aitken & Puglisi 2010). The movements of the ligands and their parts during subsequent translocation phase should be accomplished in the unlocked ribosome. The completion of translocational shifts may result in intersubunit locking, as peptidyl-tRNA and deacylated tRNA now occupy their favorable *P/p* and *E/e* positions in the inter-subunit space. We will follow this logic in the descriptions of our models for factor-free translation, EF-Tu-promoted binding of aminoacyl-tRNA and EF-G-promoted translocation (Figures 9.7, 9.10, and 9.11, respectively), with the reservation that not all locking and unlocking events are proven yet experimentally.

II.3 Inter-Subunit Mobility

The fact that ribosomes are universally built from two loosely associated and easily separable subunits in all living beings is one of the most fascinating properties of the translation machinery. The two subunits – the small one (30S in Prokaryotes or 40S in Eukaryotes) and the large one (50S in Prokaryotes or 60S in Eukaryotes) – have different functions: The small subunit is responsible for the genetic functions of the ribosome, whereas the large subunit acts as its catalytic partner. However, neither of the subunits alone is capable of performing the coupled unidirectional movement of mRNA and tRNA, that is, the conveying function called translocation. Earlier it was proposed that (1) the main functional purpose of the two-subunit construction of the ribosome is the organization of the translocation mechanism of the ribosome; (2) translocation requires the mutual mobility of the ribosomal subunits; and (3) translocation proceeds through an intermediate state, when the 3′-ends of the products of the transpeptidation reaction – peptidyl-tRNA and deacylated tRNA – occupy shifted positions on the large subunit, but their codon-anticodon duplexes are as yet unshifted on the small subunit (Bretscher 1968; Spirin 1968a, 1968b, 1969). The mechanistic principle of these models was based on the idea that the associated subunits of the translating ribosome pass through an intermediate stage when they are allowed to move relative to each other.

The first experimental evidence supporting intra-ribosomal conformational mobility coupled with translocation came from comparison of the compactness of the particles, before and after translocation, using the methods of sedimentation (Baranov et al. 1979) and neutron scattering in various mixtures of H_2O and D_2O (Serdyuk & Spirin 1986; Spirin et al. 1987). The sedimentation coefficient of the post-translocation ribosome was shown to be somewhat less than that of the pretranslocation ribosome (the difference was about 1S), and the radius of gyration (R_g) of the protein component of the post-translocation ribosome, irrespective of the number of bound tRNA molecules, is somewhat greater than that of the pretranslocation ribosome (ΔR_g of 1 to 3 Å). In other words, translocation made the whole ribosome slightly less compact. However, these data did not answer the question of whether the slight decrease in ribosome compactness upon translocation reflects an inter-subunit change or a conformational alteration within one of the subunits. Later neutron scattering experiments with contrast-matched large subunits showed that the effect was mainly due to conformational changes in the small ribosomal subunit (Serdyuk et al. 1992). (By now it seems likely that the effect reflected the unlocking of the ribosome entrance as a result of A site vacation upon translocation).

Recent developments in the cryo-electron microscopy technique allowed J. Frank and colleagues to demonstrate an inter-subunit movement coupled with translocation: they detected a rotational shift of one ribosomal subunit relative to the other around an axis perpendicular to the subunit interface (Frank & Agrawal 2000; Valle et al. 2003a). This rotational shift of the small subunit relative to the large subunit was estimated to be approximately 6° counter-clockwise, if viewed from the small subunit. The observation of such a shift was confirmed in studies by H. Noller's group, using a cross-linking technique, FRET methodology, and TLS crystallographic refinement (Ermolenko

et al. 2007a; Horan & Noller 2007; Korostelev & Noller 2007). The ribosome was shown to be fixed in the "rotated" form[3] upon binding of EF-G, the catalyst of translocation, until EF-G and deacylated tRNA were released from the ribosome (Frank & Agrawal 2000; Spiegel et al. 2007). More recently, using single-molecule FRET methodology, it was found that ribosomes undergo spontaneous inter-subunit movements, oscillating between the original and "rotated" forms, with the equilibrium shifted toward either the original or "rotated" forms depending on the functional state of the ribosome (Cornish et al. 2008). The inter-subunit unlocking accompanying rapid spontaneous counter-clockwise inter-subunit rotation was observed in FRET experiments as an immediate result of transpeptidation reaction (Marshall et al. 2008; Aitken & Puglisi 2010). Spontaneous inter-subunit movement and associated moves of the tRNAs to the hybrid state were also observed by cryo-EM under factor-free conditions (Agirrezabala et al. 2008). The binding of EF-G to the ribosome in the presence of uncleavable GTP analogue results in fixation of the "rotated" form (Cornish et al. 2008; Fei et al. 2008). The hydrolysis of EF-G-bound GTP and the resultant EF-G release leads to restoration of the original (non-rotated) mutual positions of the ribosomal subunits in the temporarily locked state (Marshall et al. 2008; Aitken & Puglisi 2010).

The following conclusions can be made from the FRET data (Cornish et al. 2008; Fei et al. 2008; Fei et al. 2009; Aitken & Puglisi 2010; see also Chapter 6 chapter by R. Gonzalez in this book).

(1) Vacant ribosomes thermally oscillate between the original and the "rotated" forms with relatively low forward and reverse rotation rates; the equilibrium is somewhat shifted toward the original form (the proportion of the two forms in the equilibrium mixture is about 3:2).

(2) The binding of N-acylated aminoacyl-tRNA to the P site reduces the forward rotation rate and correspondingly shifts the equilibrium toward the original form.

(3) Occupancy of the A site and the P site, respectively, by N-acylated aminoacyl-tRNA and by deacylated tRNA, which models the translating ribosome after transpeptidation, induces rapid oscillation of ribosomes between

---

[3] After Frank and colleagues (Frank & Agrawal 2000; Valle et al. 2003a) discovered a new form of the 70S ribosome, where the small subunit is somewhat rotated relative to the large subunit, as compared with the standard classical orientation of the subunits, the term "ratcheted" for designation of the new "rotated" form became widely used. The authors of this chapter decided to avoid this term and to use only the term "rotated," as the term "ratchet" has been already engaged by both classic physicists (see Feynman et al. 1963) and contemporary workers in the field of molecular machines (see, e.g., Vale & Oosawa 1990; Cordova et al. 1992; Ait-Haddou & Herzog 2003) for description of devices capable of rectifying random Brownian motion into directional movements.

the original and "rotated" forms, with the equilibrium shifted to the "rotated" form; this state corresponds to the pretranslocation ribosome.

(4) The binding of the translocation catalyst EF-G with a non-hydrolyzable GTP analog, with deacylated tRNA remaining in the *P* site (situating the first step of translocation), fixes the "rotated" form of the ribosome.

(5) The transition to the final post-translocation state, after GTP hydrolysis, and the release of EF-G and deacylated tRNA, when peptidyl-tRNA occupies the *P* site but the *A* site becomes vacant, principally restores the situation mentioned in (2).

(6) It is the inter-subunit unlocking that allows both the oscillation between the original and the "rotated" forms of the translating ribosome and the migration of acceptor ends of tRNAs in the inter-subunit space (see the next section).

Thus, the pretranslocation-state ribosomes in the absence of elongation factors oscillate between the original and the "rotated" forms. The presence of deacylated tRNA in the *P* site after transpeptidation strongly stimulates the rates of both forward and, to a lesser extent, reverse rotational shifts of the ribosomal subunits, shifting the equilibrium toward the "rotated" form. The properties of the ribosome when it is allowed to oscillate between the alternative conformations imply its unlocked state. The binding of EF-G with an uncleavable GTP analog fixes the "rotated" form of the pretranslocation state ribosome, temporarily locking the ribosome in the intermediate position.

II.4 Conformational Intermediates

In both models of translocation proposed four decades ago (Bretscher 1968; Spirin 1968a, 1968b, 1969) inter-subunit movement was considered as a mechanism required for translocation of peptidyl-tRNA and deacylated tRNA, and an intermediate state in the process of moving of the peptidyl-tRNA from the *A/a* to *P/p* position was postulated. The intermediate state was assumed to be the result of the spontaneous transition of products of the transpeptidation reaction from the *A/a* to *A/p* and from the *P/p* to *P/e* positions. The point is that transpeptidation radically changes the chemical situation in the region of the PTC. First, in place of the peptidyl-tRNA (having a high affinity of its N-blocked CCA-aminoacyl group for the *p* site), a deacylated tRNA (lacking such an affinity for the *p* site) appears. Second, instead of the aminoacylated 3'-adenosyl terminus with a specific affinity for the *a*-site, the newly formed peptide group appears in the place of the free amino group in the *a* site. This new situation induces a spontaneous transition from a weak to a strong binding state in the unlocked ribosome: The acceptor ends of the newly formed peptidyl-tRNA and deacylated tRNA move from *a* to *p* and from

p to *e* sites, respectively. As a result, the newly formed peptidyl-tRNA and deacylated tRNA are found in the intermediate hybrid positions *A/p* and *P/e*; this is a prerequisite for subsequent EF-G•GTP-driven translocation of the rest of the tRNAs, with their bound codons, from the *A* to *P* and from the *P* to *E* sites, respectively, on the small ribosomal subunit. The fact that translocation of tRNA molecules proceeds stepwise through a discrete intermediate state was first established in a series of chemical foot-printing studies (Moazed & Noller 1986, 1989a,b). More recent cryo-electron microscopy studies allowed the direct visualization of the hybrid states as intermediates preceding translocation (Agirrezabala et al. 2008; Julián et al. 2008). The hybrid situation was found to be unstable, and the acceptor ends of both tRNAs spontaneously oscillate between the *a* and *p*, and the *p* and *e* sites (Blanchard et al. 2004a; Kim et al. 2007; Munro et al. 2007; Fei et al. 2008).

In addition to the affinity changes, transpeptidation results in the disruption of the chemical groups responsible for firm retention of the ribosomal subunits in the locked pretranspeptidation state. Such destabilization of some inter-subunit contacts leads to restoration of the equilibrium between the "non-rotated" and "rotated" conformations of the ribosome. Thus, the spontaneous shift of the acceptor ends of the product tRNA residues and the appearance of the hybrid state after transpeptidation is allowed due to unlocking the translating ribosome. Indeed, cryo-electron microscopy studies, chemical foot-printing and FRET analyses suggested that the hybrid situation correlates with establishing the equilibrium between the classical and "rotated" states of the ribosome (Valle et al. 2003a; Ermolenko et al. 2007a; Kim et al. 2007; Spiegel et al. 2007; Cornish et al. 2008). It is remarkable that binding of EF-G with a non-cleavable GTP analog (simulating EF-G with GTP prior to GTP hydrolysis) was shown to lock the "rotated" state of the ribosome (Cornish et al. 2008; Fei et al. 2008). By analogy with EF-Tu, the ribosome-bound EF-G in GTP-induced form should be in a rigid (locked) conformation, because the presence of the effector ligand (GTP) provides additional non-covalent bonds within the complex. The ligand destruction – that is, the hydrolysis of GTP (with subsequent release of orthophosphate) – should result in relaxation of the EF-G structure. Indeed, extensive inter-domain rearrangements in EF-G (and its Eukaryotic homolog eEF2) were reported when mutual domain positions were compared in the factor complexed with the pre-translocation or post-translocation ribosomes, and in the presence of uncleavable GTP analog, or GDP and fusidic acid (Agrawal et al. 1998; Agrawal et al. 1999; Stark et al. 2000; Valle et al. 2003a; Spahn et al. 2004; Connell et al. 2007; Taylor et al. 2007). It may be suggested that the ribosome-bound EF-G with GTP (or its analog) has affinity for a transition state of the translocation reaction, and the GTP hydrolysis results in relaxation of the rigid conformations of the EF-G•ribosome complex. This relaxed complex allows slippage of the two codon-anticodon duplexes from the *A* and *P* sites to the *P* and *E* sites on the small ribosomal subunit and thus establishes the *P/p*, *E/e* situation in the "non-rotated" form of the ribosome.

Passing through intermediate states also occurs during EF-Tu-promoted entry of the aminoacyl-tRNA into the empty *A* site of the post-translocation state ribosome (Moazed & Noller 1989a, 1989b; Valle et al. 2003b). An unlocked form of the ribosomal decoding center required for the entry of the anticodon arm of an aminoacyl-tRNA is realized in the open state of the empty *A* site of the small ribosomal subunit, when the beak and the shoulder of the subunit (see Figures 9.1a and 9.2a) are drawn slightly apart (Ogle et al. 2001; Ogle et al. 2002). When a codon-cognate Aa-tRNA•EF-Tu•GTP complex is selected, its anticodon arm interacts with the mRNA in the *A* site of the small ribosomal subunit and the protein moiety binds in the factor-binding pocket at the base of L10/L12 stalk of the large subunit. This leads to closing of the pocket around EF-Tu and setting of the aminoacyl-tRNA in an intermediate, hybrid *A/T* state ("*T*" designating a site on EF-Tu, Figure 9.10).

II.5 GTP-Dependent Catalysis of Conformational Rearrangements

The discovery of factor-free translocation together with the previous knowledge of factor-free binding of aminoacyl-tRNA to the mRNA-programmed ribosome (Section IV) led to the following principal conclusions concerning the mechanism and energetics of translocation (Spirin 1978, 1985, 1988; Chetverin & Spirin 1982). First, translocation in the absence of EF-G clearly showed that the molecular mechanism of translocation is intrinsic to the ribosome itself and not introduced by EF-G. Second, translocation without GTP as an energy substrate shows that translocation is a thermodynamically spontaneous (downhill[4]) process and, hence, principally does not require input energy. Third, the realization of the full elongation cycle without both elongation factors and any other energy source except aminoacyl-tRNA proved that the transpeptidation reaction in the ribosome is the source of energy driving the elongation cycle. This implies that the free energy of the transpeptidation reaction is accumulated in the pretranslocation-state ribosome (the products are not released yet), which makes this state thermodynamically unstable.

---

[4] The term "downhill" is used only to say that the process is spontaneous, that is, accompanied by the overall decrease of thermodynamic potential and not requiring additional free energy for its performing. This term does not imply, here and later in the chapter, that the process does not need overcoming free-energy barriers. A limited rate of translocation, ~10 codons per second (Lodish & Jacobsen 1972), implies the existence of kinetic barriers, because diffusion of a ribosome-size particles at ~10 nm distance in an aqueous medium needs a few microseconds only – the time that is 10^5 times shorter (Finkelstein & Ptitsyn 2002, pp. 100–102). [To page 25].

At the same time, the elongation factors with GTP strongly increase the elongation rate (approximately by two orders of magnitude; see Kakhniashvili & Spirin 1977; Spirin 1978; Chetverin & Spirin 1982), and EF-G with GTP specifically accelerates translocation (at least by an order of magnitude; see ibid). Also, EF-Tu makes codon-dependent binding of aminoacyl-tRNA much faster, as compared with the slow, factor-free binding of the substrate. Thus, both elongation factors can be considered as enzyme-like catalysts of thermodynamically allowed, spontaneous processes. However, there are two important peculiarities of the catalytic action of the elongation factors (as well as other GTP-dependent translation factors): (1) the catalysis is coupled with GTP hydrolysis, and (2) the processes catalyzed are not chemical reactions of covalent transformations but are acts of conformational transitions.

Conformational flexibility and large-block mobility can provide conditions for association-dissociation (attachment-detachment) processes to pass through intermediate states with, respectively, partially formed or disrupted contacts between partner macromolecules. It is likely that intermediate states are required to avoid a kinetic blockade in the case of extensive multi-center interactions between macromolecules. From this assertion, the possibility of catalysis of conformational rearrangements and transitions can be directly deduced. Similarly to the enzymatic catalysis of covalent chemical reactions, which is based on the affinity of an enzyme for the *transition state* of the reaction, the catalysis of conformational rearrangements is made possible by virtue of the affinity of a protein (a catalyst of the rearrangement) for *conformational transition state*. Elongation factors EF-Tu and EF-G, as well as other GTP-dependent translation factors (IF2 and RF3), are such catalysts of conformational rearrangements of the ribosome.

The requirement for nucleoside triphosphates and their hydrolysis in the processes of conformational catalysis could be inferred from the following consideration. In the enzymatic catalysis of a covalent reaction, formation of a complex between an enzyme and a transition state intermediate is followed by decay of the intermediate and formation of the reaction products spontaneously released from the enzyme (due either to low affinity for the enzyme or low concentration of the products in the medium). Hence, liberation of the enzyme upon completion of the reaction is paid for by the change in free energy of the catalyzed covalent reaction itself. In the case of catalysis of a conformational transition, the catalyzed process is usually not a reaction accompanied by a significant decrease in free energy of the system. This circumstance can be overcome by coupling the catalysis with an exergonic chemical reaction, such as hydrolysis of a nucleoside triphosphate. Thus, when an elongation factor with GTP has an affinity for a conformational intermediate and binds to it, the detachment of the factor or its displacement will be required in order to complete the conformational transition. This will demand energy compensation at the expense of an exergonic process that would be capable of sufficiently lowering the free energy of the system. It is the coupled chemical reaction of GTP hydrolysis that can provide energy for the factor detachment from a conformational intermediate through the change in its affinity.

In light of the preceding statement, it is meaningful that translocation can be catalyzed in vitro by EF-G without GTP cleavage. It was demonstrated that EF-G with a non-cleavable GTP analog can interact with pretranslocation state ribosomes, and the subsequent removal of EF-G from the ribosomes by a physical washing-off procedure results in the appearance of post-translocation state ribosomes capable of continuing the elongation cycle (Belitsina et al. 1975, 1976). Hence, translocation can be promoted by attachment and subsequent detachment of EF-G, without coupled GTP hydrolysis. This observation suggests that the main role of GTP cleavage in the process of translocation is the destruction of a ligand (GTP) that imparted the conformation of EF-G having affinity for an intermediate translocational state of the ribosome. Hence, the GTP hydrolysis leads to the removal of the factor from its complex with the translocational intermediate and thus allows the completion of translocation.

Coupling conformational catalysis with hydrolysis of nucleoside triphosphates is not a phenomenon uniquely characteristic of translation processes. Similar GTP or ATP-dependent conformational catalysis occurs in various processes where acceleration of non-covalent macromolecular rearrangements by proteins with GTPase or ATPase activities, such as chaperonins, DNA topoisomerases, RNA helicases, transmembrane transporters, G-protein interactions, and so forth, is observed. These catalytic proteins can be considered as a special class of enzymes called "energases" (see review by Purich 2001). "Although most enzymes employ noncovalent interactions to facilitate covalent bond scission, the 'energases' use $\Delta G_{covalent\text{-}bond\text{-}scission}$ (the change in the Gibbs free energy during covalent-bond-scission) to modify noncovalent interactions" (cited from Purich 2001, p. 420). Also, ATP- or GTP-dependent catalysis of directional molecular displacements through alternating attachment and detachment of a moving part of a substrate takes place in molecular movement systems, such as myosin locomotion on actin filaments and kinesin or dynein locomotion on microtubules (see, e.g., Cordova et al. 1992; Nishiyama et al. 2003; Vale 2003).

III. PRINCIPLES OF BROWNIAN RATCHET-AND-PAWL MECHANISMS

III.1 Peculiarities of Nanomachines

The manifestations of physical laws in a microworld can strongly differ from those in the macroworld. In particular, a number of principles that underlie the work of power-stroke macromachines, such as the internal-combustion engine, cannot be realized at the molecular level. Five main features of molecular machines should be mentioned:

(1) *Macromolecules and their complexes have small masses and move in viscous molecular medium rather than in vacuum or air.* It follows that the structural blocks of molecular machines are practically inertialess: A particle of 3 to 30 nm in diameter is incapable of conserving its momentum in aqueous environments for longer than a picosecond or a small fraction of a nanosecond (Finkelstein & Ptitsyn 2002, pp. 100–102). During this time, such a particle cannot cover more than a fraction of an Ångstrom unit. Therefore, flywheels, pendulums, and other kinetic energy storage systems cannot be used in molecular machines made of biological macromolecules, such as proteins and nucleic acids.

(2) *Molecular machines have small volumes that do not allow them to store heat* in molecular media for a time longer than a small fraction of a nanosecond (for reasons similar to those given above).

(3) *Structural blocks and joints have conformational flexibility.* Molecular bodies, such as proteins, nucleic acids, and their complexes, are made of flexible polymers forming semi-rigid modules of secondary and tertiary structures with movable side groups, so that they can hardly satisfy the requirements of mechanical accuracy. That is why it is unlikely that force transmission in molecular machines can be provided by rigid levers, cranks, hooks, and other mechanical constructions. (It should be mentioned, however, that in the case of force transmission at atomic distances, i.e., at the distance of a covalent bond or van der Waals radius, short-range pushers or pullers cannot be excluded, as this takes place when oxidized Fe (Fe^{3+}) moves the F-helix in hemoglobin; see Perutz 1970). At the same time, the semi-rigid structural blocks of biological macromolecules can serve as molecular ratchets-and-pawls that block motions in certain directions (see Sections IV and V).

(4) *All parts of molecular machines experience Brownian motions and thermal oscillations.* As a result, the structural elements of molecular machines are not strictly fixed in space relative to each other, but rather undergo permanent conformational fluctuations, or Brownian motions. These Brownian motions are fast (a particle of 10 nm in diameter can cover 10 nm in an aqueous medium within a fraction of a microsecond; see Finkelstein & Ptitsyn 2002, pp. 100–102); they must not be imagined as completely random: rather, they are tamed within the limited space formed by surrounding structures and stochastically directed by the free-energy landscape (transitions from a weak to a strong binding site). Therefore, the work of molecular machines should have a stochastic rather than a strictly mechanically determined character.

(5) *Conformational changes occur that create or destroy catalytic sites and modify free-energy landscapes for the moving parts of molecular complexes.* This is the reason

why certain conformational transitions or chemical reactions take place only at certain configurations of a molecular complex and are prohibited at others, thus creating a definite order of operations of the molecular machine.

Hence, because neither mechanical kinetic energy nor high-precision mechanics can be realized at the molecular level, molecular machines, including machines of the conveying type, must be considered as constructions moving without heat or mechanical engines, as well as without precise long-range transmission of mechanical forces. They are based on the quite different principles that follow from the five features of molecular systems, where the driving force for molecular directional movements is essentially a Brownian motion but biased in a certain direction by energy-supplied "Maxwell demons" (see Feynman et al. 1963), that is, special mechanisms of random motion rectification (Cordova et al. 1992; Spirin 2009a; Frank & Gonzalez 2010).

III.2 Maxwell's Demon and the Rectification of Random Motions

Following its concept, the Maxwell demon should be capable of rectifying Brownian motions and thus converting them into a directional motion. However, to work in such a way, any actual mechanism of such a type must be supplied with free energy; in other words, the demon has to consume some fuel. The engines of most molecular machines are fueled by high-energy compounds, usually ATP or GTP, or the products of high-energy group transfers from ATP, such as aminoacyl-tRNA. As a rule, binding of a high-energy substrate to a specific site of the engine induces a transition from a fluctuating, loose (unlocked, open) conformation of the site into a fixed (locked, closed) conformation, due to formation of non-covalent bonds between the site and the substrate (induced fit). This internal conformational change, which can be considered as the *first stroke* of an engine, increases its affinity for some binding sites and decreases the affinity for others. Next, the bound high-energy substrate is catalytically hydrolyzed (or chemically transformed in another way) at the binding pocket; as a result, the affinity of the reaction products for the pocket decreases, and the conformation again becomes mobilized (unlocked, open). This is the *second stroke* of the engine. The conformational change that restores the initial affinities of the engine for the binding sites is followed by the release of the reaction products or their displacements to neighboring sites. All displacements that exceed the distances at which the atomic forces act require no special mechanical motive forces, such as directional pulling or pushing, but are generated by Brownian motions and are properly fixed by binding affinities.

The conversion of a substrate (i.e., GTP) to products (i.e., GDP + P_i) is always unidirectional (substrate binding → conversion → product release → ...) when P_i is

released (i.e., the direction of the overall reaction is determined by the total free-energy decrease). The problem of realizing a machine is how to couple this free-energy decrease with a unidirectional molecular movement. This is the problem of molecular transmissions and movers.

Unidirectional molecular movement is always the result of a transition from a weak- to a strong-binding state. The induced fit is the most evident example of a downhill step of the transition of this kind. A free-energy rise is not excluded at certain steps of movement, if it ultimately leads to the free-energy decrease, but this free-energy barrier must not exceed ~10 kT units: otherwise, it will slow down the transition too much.

III.3 Ratchet-and-Pawl Mechanisms for Rectification of Random Motions[5]

The following considerations are based on the thermal ratchet-and-pawl model (Figure 9.5). As molecular machines are quite different from the simple mechanical device described in the famous Feynman's lectures, it is not out of place to describe the ratchet-and-pawl mechanism in general terms applicable for both molecular and mechanical devices.

The classicall Feynman (or rather, Feynman-Smoluchowski-Lippmann; see http://en.wikipedia.org/wiki/Brownian_ratchet#cite_note-Harmor-2; Smoluchowski 1912; Feynman et al. 1963) ratchet-and-pawl mechanism consists of the impeller, ratchet, and pawl. The function of the Feynman impeller is to collect random Brownian motions and to pass them to the ratchet. In molecular machines, Brownian motions are collected by all parts, including ratchets and pawls, so that the molecular machines have no separate impeller.

The function of the Feynman pawl is to rectify the Brownian motions of the ratchet, that is, to select the motions in the required direction and block the backward motions (it is similar to the work of the Maxwell demon). The Feynman pawl can perform this role only if the energy profile created by the ratchet dent is reduced, in the forward direction, by an external force (cf. Figure 9.5c and 9.5d), or if the pawl temperature is different from the temperature of the Brownian motions of the ratchet (Feynman et al. 1963). In isothermal molecular machines, such as ribosomes, the pawl can perform its role at the cost of chemical fuel (consumed by either the pawl or the ratchet), which reduces the free energy for motion in the required direction and does not reduce it for the backward motion.

The function of the conventional ratchet is twofold. The first is to create an indented energy profile to provide conditions for blocking the ratchet by the pawl. The ratchet

---

[5] The term "ratchet" is used here only in the classical sense (Feynman et al. 1963), namely for designation of devices working in combination with a pawl and thus allowing rotation or linear movement in one direction and preventing the backward motions.

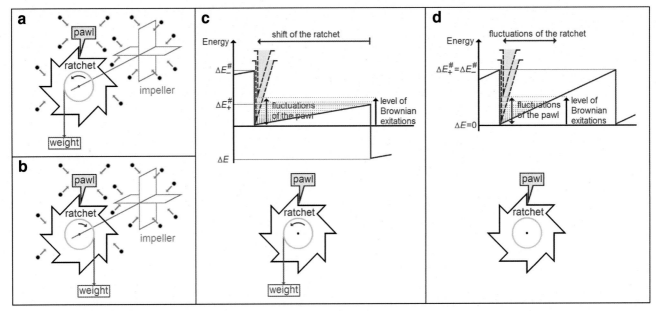

FIGURE 9.5: *The ratchet-and-pawl mechanism. (a, b) The classical isothermal Feynman ratchet-and-pawl mechanism with the impeller; all parts of the mechanism are attacked by randomly moving Brownian particles. (a) The "weight" tends to rotate the ratchet in the direction, in which the pawl slides along the flat slope of the ratchet dent. (b) The "weight" tends to rotate the ratchet in the direction in which the pawl runs against the steep slope of the ratchet dent. The panel (b) may make an impression that the Brownian particles attacking the impeller can raise the weight (because just this "weight-raising" direction of rotation makes the pawl sliding along the flat slope of the ratchet dent). However, this direction is not preferable, because the pawl also experiences random fluctuations, being attacked by the particles having the same temperature of Brownian movements as the particles attacking the impeller; as a result, it is the energy spent to raise the pawl and to turn the loaded ratchet rather than the steepness of the ratchet dent that dictates the preferred direction of rotation. Therefore, in this case the weight also goes down, though slower than in the case shown in panel (a) (Feynman et al. 1963). (c) The energy profile created by the ratchet dent for the pawl in the presence of external force (the hanging weight) acting in the direction of required movement of the ratchet (to the left). The probability of the force-required forward movement of the ratchet is proportional to $exp(-\Delta E_+^{\#}/kT)$, and the probability of the backward (to the right) movement of the ratchet is proportional to $exp(-\Delta E_-^{\#}/kT)$, where T is the temperature, K is the Boltzmann constant, and $\Delta E_+^{\#}$, $\Delta E_-^{\#}$ are the heights of the corresponding the right and left energy barriers. These barriers differ due to the action of the weight, which makes the energy of the machine higher when the pawl is in the current ratchet's indent, and lower when the pawl is in the next indent (to the right). Thus, the shift of the next ratchet's indent is a "transition from a weak to strong binding site." ΔE is the gain of energy achieved at one step of the ratchet in the direction of the external force. (d) The energy profile for the pawl in the absence of external force.*

of a molecular machine also creates an indented energy, or rather free-energy profile for the pawl. Importantly, this profile can be modified by chemical reactions and conformational changes (Astumian & Derényi 1998). The second function specific of the classical Feynman ratchet is to turn on its axis, back or forth, under the action of Brownian excitations or an auxiliary external force (like a suspended weight). In molecular machines, the move along the reaction coordinate can be done both by the molecular ratchets and molecular pawls.

It is worthwhile to add the following. The classical Feynman ratchet-and-pawl mechanism assumes that Brownian motions are strong enough to force a pawl to jump sometimes over the ratchet dent (Feynman et al. 1963) and to allow the backward motion, which was blocked by the down pawl. Molecular machines, at least those considered in this paper, seem to avoid this possibility because their pawls are fed by the binding energies and energies of chemical reactions, which are much greater than the energies of Brownian motions acting on the pawl.

III.4 Rectification of Random Motions Illustrated by the 40S-5′UTR Scanning Model

Initiation of translation of Eukaryotic mRNA is a complex process (for review, see Pestova et al. 2007). It includes a stage of unidirectional movement along the mRNA chain of the initiation 43S ribosomal complex, consisting of the 40S ribosomal particle, several initiation factors (eIF1, eIF1A, eIF2, eIF3, and eIF4F), and initiator methionyl-tRNA. This complex moves along the 5′ untranslated region (5′ UTR) of an mRNA from its cap-structure to the initiation codon in the 5′ to 3′ direction, with accompanying hydrolysis of ATP into ADP and orthophosphate (ATP-dependent scanning of 5′ UTR). A recently proposed model of this ATP-dependent unidirectional movement (Spirin 2009b) illustrates a simple Brownian ratchet-and-pawl mechanism at the molecular level. According to the model, the random one-dimensional diffusion of the 43S complex along the mRNA chain is rectified into the unidirectional 5′-to-3′ movement by the ratchet-and-pawl mechanism.

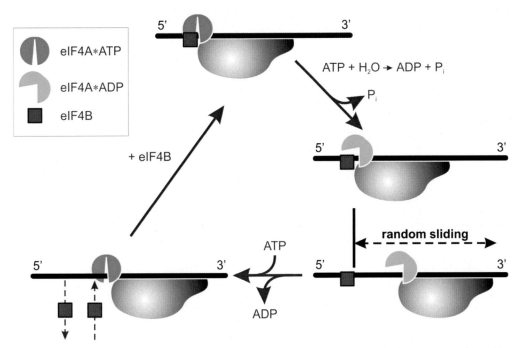

FIGURE 9.6: *Hypothetical scheme of the working cycle of a scanning 43S ribosomal complex. The Eukaryotic 40S ribosomal particle together with bound initiation factors and initiator methionyl-tRNA (the initiation 43S complex) is symbolized by the upside-down "mouse" contour. The ATP-binding initiation factor eIF4A (represented as two filled semicircles, locked with bound ATP and unlocked after ATP hydrolysis) is a subunit of the ribosome-bound initiation factor eIF4F (not shown). eIF4A is localized at the rear (trailing side) of the particle. The mRNA-binding initiation factor eIF4B is shown by grey-filled square. (Adapted from Spirin 2009b, with small modifications).*

The proposed mechanism is organized by one of the subunits of the 43S complex-bound heterotrimeric initiation factor eIF4F, designated eIF4A, and a free mRNA-binding protein called initiation factor eIF4B (Figure 9.6). The subunit eIF4A, a two-domain protein, is an RNA-dependent ATPase considered the engine of the 43S ribosomal complex, with ATP as the fuel. The hydrolysis of the eIF4A-bound ATP is induced when eIF4A binds to mRNA and eIF4B. The hydrolysis of the eIF4A-bound ATP leads to the unlocking of the protein: its two domains move somewhat apart, and the affinities of eIF4A for mRNA and eIF4B drop. In the model shown in Figure 9.6, it is the binding of eIF4B that plays the pivotal role of the pawl for the moving 43S complex, which plays the role of a moving dent of the molecular ratchet.

The sequence of events in the course of one cycle of the ATP-dependent scanning can be described as follows. State I (the mRNA-fixed complex): The 43S ribosomal complex with eIF4A in the ATP form (locked) and the initiation factor eIF4B are bound to the mRNA (shown as the straight 5′—3′ line) and to each other. The entire ribosomal complex is anchored on the mRNA chain by the eIF4B•eIF4A•ATP complex. The mRNA-induced hydrolysis of the eIF4A-bound ATP leads to state II, in which eIF4A•ADP has lost its interactions with the mRNA and

eIF4B, resulting in unanchoring the ribosomal complex from the mRNA chain. In this state, the complex can diffusionally slide along the mRNA chain, while eIF4B remains attached to the mRNA for some time due to its low rate of dissociation from the mRNA. Thus, as the mRNA-attached eIF4B prevents backward diffusional movement, a stochastic net-directed displacement (rather than a step of a definite length) occurs; the forward diffusional shift is realized if eIF4B remains attached to mRNA for the time necessary to form state III. (The distance of the shift of the ribosomal complex from eIF4B must exceed some minimal length that allows eIF4A to release ADP and bind ATP from solution; this requirement for a minimal length allows the ribosomal complex to move from eIF4B to the 3′ end of the mRNA chain, even if this movement leads to the free-energy increase as the cost for melting the mRNA secondary structure). State III, the stand-by complex, results from the binding of ATP to eIF4A, which restores the affinity of eIF4A for mRNA and thus fixes the complex at a new, forward-shifted position. The complex again includes eIF4A•ATP and now attracts eIF4B from the pool of free initiation factors. Then the ribosomal complex returns to state I, but its position on the mRNA is shifted toward its 3′ end. (It should be emphasized that a directed cycle – and, therefore, a directed motion – is possible only when the number of states exceeds two; see Figure 9.6. Two states would allow only oscillation back and forth; see Finkelstein & Ptitsyn 2002, pp. 341–342).

IV. NON-CATALYZED (FACTOR-FREE) TRANSLATION CYCLE

To analyze the conveying function of the ribosomal molecular machine, it is worthwhile to start with the simplest case, namely factor-free translation that has no additional free-energy support.

IV.1 Discovery of Spontaneous Translocation

The translating ribosome is a molecular conveying machine that directionally passes compact tRNA molecules and the tRNA-bound mRNA chain through itself (Figure 9.4). The main driving act in the conveying process is realized at the stage of translocation. A milestone discovery at the end of 1960s–beginning of the 1970s was the finding that the ribosome is principally capable of performing the function of translocation in the absence of both the translocation factor EF-G and additional free energy in the form of GTP (Pestka 1968, 1969; Gavrilova & Spirin 1971, 1972; Gavrilova et al. 1976). It was shown that the rate of the factor-free translocation, usually low with intact ribosomes, could be significantly increased in cell-free translation systems by modification or removal of interface ribosomal proteins S12 and S13 (see Figure 9.2a), which were believed to restrain spontaneous translocation in the absence of EF-G with GTP (Gavrilova & Spirin 1974; Gavrilova et al. 1974; Cukras et al. 2003). These discoveries implied that in the intact ribosome, these interface proteins prevent spontaneous translocation by retention of the particle in a high-energy pretranslocation state. At the same time, in vitro experiments using non-modified (intact) ribosomes demonstrated that low Mg^{2+} concentrations (e.g., 3 mM Mg^{2+} at 100 mM NH_4Cl) can also strongly stimulate spontaneous (factor-free) translocation, the high Mg^{2+} concentrations being inhibitory both for factor-free and EF-G-promoted translocation (Belitsina & Spirin 1979). The above-mentioned observations suggested that translocation is somehow connected to inter-subunit association and inter-subunit mobility (see Section II). In any case, these findings led to the following conclusions: (1) the translocation mechanism is principally inherent to the ribosome itself and not introduced by EF-G; (2) the translocation process per se is thermodynamically spontaneous – in other words, this is a downhill process accompanied by decrease of thermodynamic potential and not requiring additional free energy for its performance; (3) EF-G must be considered as a GTP-dependent catalyzer of conformational transitions that contributes mainly to the kinetics of the process (Spirin 1978). More recently, it was demonstrated that the antibiotic sparsomycin added to the pretransloca-

tion state ribosome induces accurately coupled movement of mRNA with tRNA, thus also providing evidence that the translocation mechanism is inherent to the ribosome (Fredrick & Noller 2003).

IV.2 Factor-Free Translation

On the other hand, it has been known for a long time that in a cell-free system, a cognate aminoacyl-tRNA can specifically bind to the peptidyl-tRNA-carrying ribosome in the absence of the elongation factor EF-Tu (the factor was discovered later) and then successfully participate in the transpeptidation reaction (Kaji & Kaji 1963; Nakamoto et al. 1963; Gottesman 1967). Based on this knowledge and the aforementioned discovery of spontaneous translocation, factor-free translation systems were made where both EF-Tu and EF-G, as well as GTP, were absent, and aminoacylated tRNA was the sole high-energy substrate present in the medium (Gavrilova et al. 1976; Belitsina & Spirin 1979). The effective factor-free translation systems could include either SH-modified or S12-depleted ribosomes, or factor-free translation could be performed with intact ribosomes at a lowered (sub-optimal) Mg^{2+} concentration. Another alternative was proposed when factor-free translation was performed in a stepwise way by alternating high Mg^{2+} (to provide effective factor-independent binding of aminoacyl-tRNA) and low Mg^{2+} (to stimulate factor-free translocation) using the matrix-bound polyribonucleotide column technique (Belitsina & Spirin 1979).

It should be emphasized that the factor-free process is sensitive – even more sensitive than the factor-catalyzed process – to all specific inhibitors of ribosomal mechanisms of aminoacyl-tRNA binding, transpeptidation, and translocation, such as tetracycline, streptomycin, chloramphenicol, erythromycin, spectinomycin, thiostrepton, and viomycin (Spirin et al. 1976; Southworth et al. 2002). As expected, it is insensitive to fusidic acid, an inhibitor of EF-G-promoted translocation (ibid). Factor-free translation was found to be authentic by other criteria, including the accuracy of codon-anticodon recognition, the formation of peptide bonds, and the processivity of mRNA movement through the ribosome (Gavrilova & Spirin 1974; Gavrilova et al. 1976; Gavrilova et al. 1981; Southworth et al. 2002), as well as the value of energy of activation of the overall process (Kakhniashvili & Spirin 1977).

IV.3 Model of Factor-Free Elongation Cycle

Hence, the factor-free elongation cycle consisting of three main stages – codon-dependent binding of aminoacyl-tRNA, ribosome-catalyzed transpeptidation, and spontaneous translocation – can be considered as the basic route of the normal translational process, but without auxiliary catalytic mechanisms and accompanying additional free-energy support. Certainly, it is much slower than the full,

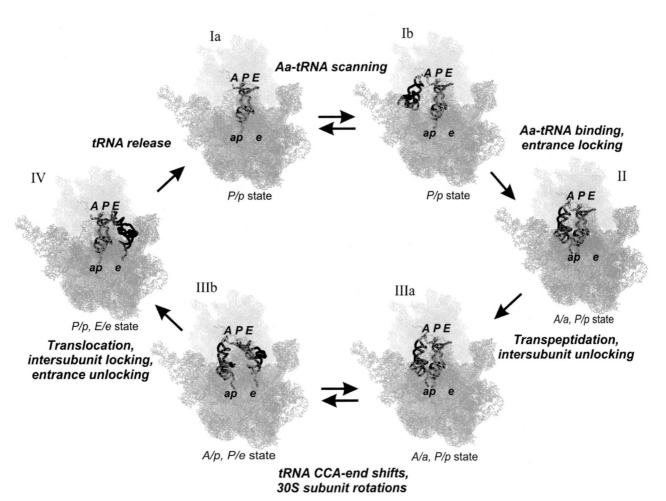

FIGURE 9.7: *Factor-free (non-enzymatic) elongation cycle. The ribosomes and tRNAs are oriented and presented in the same way as in Figures 9.3 and 9.4. Positions of two lateral protuberances are conditionally shown unchangeable, although in reality they are movable and depend on a functional state of the ribosome. All explanations are in the text.*

factor-promoted cycle. Nevertheless, the principal possibility of translating a template polyribonucleotide and elongating a polypeptide using a basic ribosomal mechanism without elongation factors and GTP allows analysis of the factor-free translating ribosome as a self-sufficient molecular machine.

It follows that translation can proceed at the expense of aminoacyl-tRNA molecules as the sole source of free energy, and the transpeptidation reaction can be the only exergonic reaction of the translation cycle. Thus, considering the factor-free elongation cycle in terms of a molecular machine, the fuel for the translating ribosome is the molecules of aminoacylated tRNA, and the engine is the PTC catalyzing the exergonic reaction of transpeptidation. As discussed in Section I, the PTC is organized by domain V of the compactly folded ribosomal RNA of the large ribosomal subunit (see Figure 9.2b) and is localized in the middle of the large subunit at the subunit interface (Ban et al. 2000; Nissen et al. 2000). The PTC can be sub-divided into an acceptor substrate site, the *a* site, and a donor substrate site,

the *p* site (Figure 3b). In the following discussion, the cyclic process of factor-free translation where the PTC moves the entire working cycle is considered. The factor-free elongation cycle is schematically represented in Figure 9.7.

The following is a plausible model of the sequence of events in factor-free ribosome cycling. It is convenient to begin the consideration with the post-translocation state ribosome (Figure 9.7, state Ia). In this state, the ribosome has peptidyl-tRNA accommodated in the *P/p* position and is competent for binding a new aminoacyl-tRNA to the *A* site, into the *A/a* position. The capability to bind the large substrate implies that the ribosome in this state must be somehow open or unlocked. The open character of the empty *A* site at the 30S subunit has been convincingly demonstrated by the X-ray structural analysis (Ogle et al. 2001; Ogle et al. 2002). Earlier, NMR analysis of a fragment of the DC suggested that the state of the *A* site in the absence of ligands is not rigid, but rather fluctuates between more open and more closed conformations (Fourmy et al. 1998). According to X-ray analysis data supported by the results of the TLS crystallographic refinement (Korostelev & Noller 2007), the fluctuations seem to involve the mobility of the beak (helixes h30–h34) and the shoulder (helixes h15–h18) of the small ribosomal subunit (see Figures 9.1 and 9.2). Generally, the beak and the shoulder of the small subunit with the empty *A* site are somewhat away from the subunit interface and probably fluctuate, thus leaving open the entrance into the ribosome. This unlocked state of the ribosome entrance and the *A*-site pocket is favorable for the primary scanning of aminoacyl-tRNA molecules (see Figure 9.7, state Ia ↔ Ib), during which the anticodons of aminoacyl-tRNA molecules are browsed and eventually the codon-cognate aminoacyl-tRNA becomes selected. In any case, aminoacyl-tRNA molecules are allowed to enter the unlocked *A* site of the ribosome and try on the codon set there. Studies of transient kinetics of factor-free tRNA binding to the ribosome suggest the formation of a short-lived intermediate complex prior to the codon-anticodon recognition step (Wintermeyer & Robertson 1982). The possibility cannot be excluded that some groups of the L10/L12 stalk base (possibly protein L11) at the 50S subunit may transiently interact with the elbow of aminoacyl-tRNA; this interaction may be analogous to the L11-tRNA interaction in the case of EF-Tu-promoted binding of the aminoacyl-tRNA (see Schmeing et al. 2009, and Section V.1).

When the anticodon of the aminoacyl-tRNA is recognized by the cognate codon of the mRNA in the *A* site, it binds first at the *A* site of the small subunit and then with the PTC *a* site that has an affinity for the aminoacyl-adenosyl residue (Figure 9.7, state II). The shift of the acceptor arm of the aminoacyl-tRNA between the ribosomal subunits (the Ib→II transition) must require an unlocked state of the entrance into the ribosome, with its following locking. Upon the first step of the binding (the recognition step), the beak and the shoulder close around the anticodon stem-loop of the aminoacyl-tRNA and lock the *A* site-surrounding structure (Ogle et al. 2001; Ogle et al. 2002). The correct codon-anticodon recognition induces conformational changes in the DC, the upper part of helix h44, and the loop of helix h18 of 16S rRNA (see Figure 9.2a); these changes recognize the geometry of the codon-anticodon helix (Ogle et al. 2001) and allow the rearrangement of the *A*-site complex. This rearrangement is the second step of the aminoacyl-tRNA binding, which is completed by the accommodation of the acceptor end of aminoacyl-tRNA in the *a* site of the large ribosomal subunit. The rearrangement step in the factor-free binding, however, is significantly slower as compared to the EF-Tu•GTP-promoted binding (see Wintermeyer & Robertson 1982). It should be mentioned that the binding of a near-cognate aminoacyl-tRNA does not cause such induced-fit changes at the first step of the binding (Ogle et al. 2002), explaining why tRNA selection by the ribosome is more accurate than can be expected from energy

differences between matched and mismatched anticodons only (see, e.g., Pape et al. 1999; Ogle et al. 2001; Ramakrishnan 2002; Rodnina et al. 2005).

Thus, the resulting conformational fit allows the acceptor end of the cognate aminoacyl-tRNA (its aminoacyl-adenosyl residue) to be accommodated into the PTC *a* site. Now the peptidyl-tRNA (donor substrate) and aminoacyl-tRNA (acceptor substrate) are in the *P/p* and *A/a* positions, respectively. Because both substrates of the transpeptidation reaction are side by side in the PTC of the locked ribosome, their reacting groups are tightly drawn together and locked in the PTC to provide the accurate nucleophilic attack: The reactive carbonyl group at the peptide C-end of the peptidyl-tRNA is now positioned against the amino group of the aminoacyl-tRNA for an in-line nucleophilic attack that transfers the nascent polypeptide from the *P*-site bound peptidyl-tRNA to the *A*-site bound aminoacyl-tRNA (Nissen et al. 2000; Hansen et al. 2002; Schmeing et al. 2005a; Schmeing et al. 2005b; Trobro & Aqvist 2005; Bieling et al. 2006; Voorhees et al. 2009).

The subsequent transpeptidation reaction results in replacement of the free amino group of the aminoacyl-tRNA in the *a* site of the PTC by the amide group connecting the aminoacyl-tRNA residue with the peptidyl residue, as well as in the appearance of deacylated tRNA in the *p site* of the PTC:

$$\text{Aa-tRNA}'' + \text{Pept-tRNA}' \rightarrow \text{Pept-Aa-tRNA}'' + \text{tRNA}'.$$

As a result, the chemical situation in the PTC is strongly changed. Now the *a* and *p* sites of the PTC accommodate reaction products that no longer have strong affinities for these sites (Figure 9.7, state IIIa). This implies that the previous interactions have disappeared and the inter-subunit contacts are destabilized. In such a situation, the ribosome becomes unlocked (Aitken & Puglisi 2010) and starts to oscillate between "non-rotated" and "rotated" forms (see Section II). Simultaneously, the products tend to leave their previous sites in the PTC. Under conditions of Brownian motions, the weakly bound groups can dissociate from their previous biding sites and be caught by other sites with higher affinities for them (Figure 9.7, IIIa→IIIb transition). As a result, the newly formed peptide group with the ester group at the CCA-terminus of the tRNA starts to fluctuate between the *a* site and the *p* site, while the deacylated terminus of the tRNA is oscillating between the *p* site and the *e* site outside the PTC. Thus, when the ribosome is in the unlocked state, the hybrid positions *A/p* and *P/e* appear (state IIIb) and exist in equilibrium with the original *A/a* and *P/p* positions (state IIIa) (Blanchard et al. 2004a; Kim et al. 2007; Munro et al. 2007). At the same time, this post-transpeptidation state (IIIa ↔ IIIb) displays equilibrium between the "non-rotated" and "rotated" forms (Cornish et al. 2008). The peptidyl-tRNA and the deacylated tRNA were reported to fluctuate between the *A/a* and *A/p* positions and the *P/p* and *P/e* positions, respectively, at frequencies of about one order of magnitude higher than the frequency of intersubunit rotations (Blanchard et al. 2004a; Kim et al. 2007; Munro et al. 2007; Cornish et al. 2008). (It should be noted, however, that tRNA and inter-subunit rotation dynamics have not yet been measured on identically prepared samples under identical conditions).

The resultant unlocked state is thermodynamically unstable, in other words, the transition of two tRNAs from the *A/p* and *P/e* positions to the *P/p* and *E/e* positions (Figure 9.7, state IIIb to state IV transition) will reduce the free energy. The free energy of the transpeptidation reaction is stored in the pretranslocation state ribosome (states III). In state IIIb, this free energy is probably reserved in sterically strained (unfavorable) conformations of the acceptor arms of the tRNA residues and distorted inter-subunit contacts. In the absence of the translocation catalyst, EF-G, a high kinetic barrier prevents a fast downhill transition from the pretranslocation hybrid state IIIb to the post-translocation state IV which has a lower free energy. A plausible model of the spontaneous transition from the pretranslocation state IIIb to the post-translocation state IV that involves two rotational inter-subunit movements (forward and backward) is presented in Figure 9.8.

After the transpeptidation reaction, the translating ribosome becomes unlocked and displays two kinds of thermal motions: One is a forward and backward migration of the acceptor parts of the tRNAs within the unlocked ribosome (Figure 9.7, states IIIa ↔ IIIb), and the other is a rotational shift of the small subunit relative to the large subunit and its reversal (Figure 9.8). When the *A/p*, *P/e*, or the hybrid state is attained (Figure 9.8, the upper state) and the small subunit is "rotated" (Figure 9.8, the middle transient state), the *A* and *P* sites at the small subunit together with the pair of codon-anticodon duplexes bound to them are found to be shifted relative to the large subunit in the direction of the mRNA 5'-end. Occasional dissociation of tRNA residues together with their mRNA codons from the *A* and *P* sites on the small subunit (i.e., uncoupling of the DC from codon-anticodon duplexes) can take place in this position. In this case, during the reverse rotational movement, the two shifted codon-anticodon duplexes can slide along the DC of the small subunit and be anchored by the *P* and *E* sites, respectively (Figure 9.8, the lower state). The sliding of the duplexes along the DC during the reverse inter-subunit rotation can be facilitated by the attraction of the elbow of the deacylated tRNA by the L1 stalk (Valle et al. 2003a; Korostelev et al. 2006; Selmer et al. 2006; Fei et al. 2008; Cornish et al. 2009; Gao et al. 2009; Fei et al. 2009). Thus, the L1 stalk can play the role of a pawl retaining the pair of codon-anticodon duplexes. This pair of duplexes performs the role of a caught ratchet dent retained at the shifted position, while the *A*, *P*, and *E* sites of the 30S subunit move back. The reversal of the small

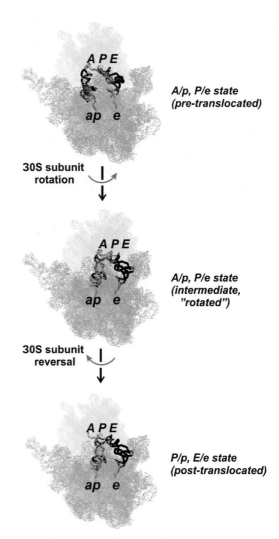

FIGURE 9.8: *Factor-free (spontaneous) translocation via an intermediate "rotated" state. Explanations are in the text.*

subunit will set the next codon of mRNA at the vacant *A* site. Hence, in this way, the forward and backward rotations of the small subunit accomplish the translocational shift of mRNA by one nucleotide triplet in the direction of the mRNA movement during translation.

In other words, spontaneous translocation occurs during the reverse motion of the small subunit, which is a thermally induced slippage of the codon-anticodon pair from the adjacent *A* and *P* sites of the DC to the *P* and *E* sites, respectively. The reverse rotational shift, that is, the transition into the original form (Figure 9.7, state IV, and Figure 9.8, the lower state) allows the strained conformations of the tRNAs and the involved binding groups of the large subunit (in Figure 9.7, state IIIb, and Figure 9.8, the upper state) to return to sterically favorable conformations, with simultaneous locking of the ribosome in the *P/p*, *E/e* state (Figure 9.7, state IV, and Figure 9.8, the lower state). Therefore, translocation is a slow transition from a thermodynamically unfavorable state to more favorable state;

in other words, it is an act of overcoming a high free-energy barrier, leading to the decrease of the free-energy level. The post-translocation state ribosome becomes locked due to favorable inter-subunit positions of two ligands: peptidyl-tRNA and deacylated tRNA.

It is worth noting that the antibiotic sparsomycin firmly fixes peptidyl-tRNA after transpeptidation in the *p* site in state IIIb (see Figure 9.7, as well as Figure 9.8, the upper state), and this fixation results in stimulation of factor-free translocation (Fredrick & Noller 2003). It seems that retention of the 3′-CCA-group of the tRNA together with the linked peptide group of the newly elongated peptidyl-tRNA in the *p* site decreases the activation barrier of the subsequent forward rotational shift of the ribosomal subunits, which is the prerequisite of the codon-anticodon translocation. Another antibiotic, viomycin, induces inter-subunit rotational movement of the ribosome and traps the ribosome in this intermediate – "rotated" – state (Figure 9.8, the middle state) (Ermolenko et al. 2007b). The same intermediate state is found to be fixed by EF-G with uncleavable GTP analog, GMPPNP (Section V.2).

Spontaneous release of the deacylated tRNA from the *E* site completes the factor-free translocation step, making the cycle virtually irreversible, and the next cycle can start again from the post-translocation state (Figure 9.7, position Ia).

V. AUXILIARY BROWNIAN RATCHET-AND-PAWL DEVICES INVOLVED IN GTP-DEPENDENT CATALYSIS OF TRANSLATION

During the natural factor-promoted elongation cycle, the additional free energy of GTP is expended in overcoming the high kinetic barriers in the processes of (1) aminoacyl-tRNA binding and (2) translocation. These GTP-dependent catalytic acts are performed by two subsidiary engines assembled for a time at the ribosome entrance from an elongation factor (either EF-Tu in the case aminoacyl-tRNA binding catalysis, or EF-G in the case of the catalysis of translocation) and the side protuberance of the large ribosomal subunit, the so-called L10/L12 stalk, with participation of its base (protein L11, L11-binding helixes H43-H44 and the loop of helix H95 of the 23S rRNA), as well as the shoulder domain of the small ribosomal subunit (Stark et al. 2002; Valle et al. 2003a; Valle et al. 2003b; Gao et al. 2009; Schmeing et al. 2009).

V.1 EF-Tu-Catalyzed Binding of Aminoacyl-tRNA

V.1.1 GTP-Dependent Shuttle Function of EF-Tu. Prior to entering into the elongation cycle, the aminoacylated tRNA forms a complex with the protein, EF-Tu, or EF1A, outside the ribosome. The protein is capable of specifically binding GDP or GTP, but the immediate phosphorylation of GDP in cells results in only GTP being bound to

EF-Tu under normal cellular conditions. The conformations of both EF-Tu•GDP and EF-Tu•GTP complexes were studied by crystallographic methods at atomic resolution, and comparisons of the structures revealed a strong conformational rearrangement depending on the bound ligand (Berchtold et al. 1993; Kjeldgaard et al. 1993; Abel et al. 1996; Polekhina et al. 1996). It was shown that the globular molecule of EF-Tu is composed of two blocks of approximately equal masses (domain I and domains II + III), connected by a flexible strand. Domain I has a nucleotide-binding site accommodating either GDP or GTP. Both in the ligand-free state and in the complex with GDP, the contact between the two halves is weak, so that the protein is in a relaxed conformation, probably oscillating between open (with the halves drawn somewhat apart) and closed forms, which are in equilibrium, although shifted toward the open form (the *unlocked* conformation). When GDP is replaced by GTP, the two halves of the globule are drawn together, and the conformation becomes fixed in the closed state, in other words, becomes *locked*. The sequence of events during the process of GTP-dependent shuttling of EF-Tu resulting in delivery of aminoacyl-tRNA molecules from the surroundings to the ribosome can be schematically presented as follows (Figure 9.9).

Binding of GTP with domain I of EF-Tu leads to conformational changes of the nucleotide-binding pocket of EF-Tu, particularly around the γ-phosphate group of GTP (induced fit), resulting in local rearrangements in domain I and the appearance of new groups on the domain interface that have an affinity for the surface of the other half of the protein (domains II + III). Thus, the two halves firmly stick together and become fixed in the closed state. Now this locked conformation (Figure 9.9, state I) is competent for binding of aminoacyl-tRNA.

The aminoacyl-tRNA binds to the EF-Tu•GTP complex by its aminoacylated acceptor arm, so that the stem helix and the aminoacyl residue are accommodated in the groove between the two stuck halves of the protein globule (Figure 9.9, state II). Various ternary Aa-tRNA•EF-Tu•GTP complexes formed with participation of aminoacyl-tRNAs of different specificities freely diffuse around the ribosome and compete to be recognized by the cognate mRNA codon on the ribosome.

Binding of the codon-cognate ternary Aa-tRNA•EF-Tu•GTP complex to the ribosome (Figure 9.9, state III) precedes the hydrolysis of GTP by EF-Tu (Blanchard et al. 2004b; Gonzalez et al. 2007; Lee et al. 2007). This binding positions EF-Tu in such a way that it triggers the opening of a hydrophobic gate within EF-Tu and allows its crucial His 84 to orient toward the GTP and activate a water molecule, subsequently leading to GTP hydrolysis (Schmeing et al. 2009; Schuette et al. 2009; Villa et al. 2009). The release of the split-off phosphate follows, leading to a reverse local rearrangement and consequent unlocking of

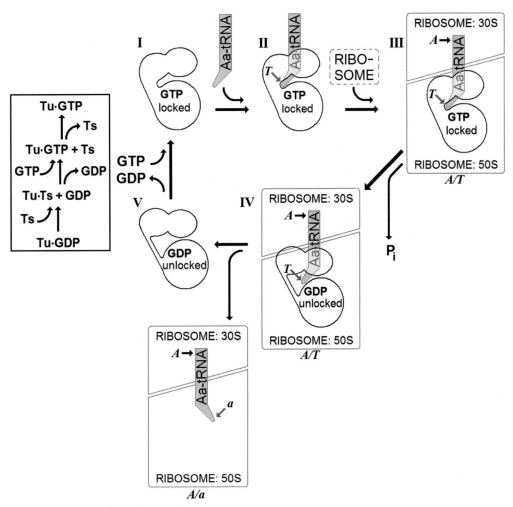

FIGURE 9.9: *Schematic representation of the shuttle cycle of the elongation factor EF-Tu. The process of a factor-promoted aminoacyl-tRNA binding is based on the locking-unlocking cycle of EF-Tu. The aminoacyl tRNA-binding sites A and a of the ribosome and the T-site on EF-Tu are shown. The GDP to GTP replacement is promoted by the elongation factor EF-Ts, as shown in the insert in the left part of the figure.*

the overall conformation of EF-Tu (state IV). This event causes a loss of EF-Tu affinity for both aminoacyl-tRNA and the ribosome; the EF-Tu•GDP complex is released from the ribosome (state V), with a subsequent exchange of GDP for GTP (state V to state I transition) catalyzed by another factor, called EF-Ts, or EF1B. Now the shuttle cycle of EF-Tu can be repeated.

In the previously described cycle, the fuel is GTP, which is combusted in the chemical reaction of hydrolysis. The engine is domain I, where the catalytic center for GTP hydrolysis is located. The transmission is the coupling between local conformational changes around the place where the energy substrate GTP binds and decays, and gross conformational movements of the locking-unlocking type. Thus, the shuttle delivery of aminoacyl-tRNA into the translating ribosome is accomplished. In this way, EF-Tu works as a molecular machine, yet its shuttle function is

relatively simple and cannot be considered a true conveying function.

V.1.2 GTP-Dependent Function of EF-Tu in Ribosomal Dynamics.

In the process of EF-Tu-catalyzed aminoacyl-tRNA binding to the translating ribosome (Figure 9.10), a ternary Aa-tRNA•EF-Tu•GTP complex interacts first with the L10/L12 stalk and its base (protein L11 and L11-binding RNA helixes) (states I). As with factor-free binding of aminoacyl-tRNA, the initial interaction of the ternary complex with the translating ribosome occurs independently of the nature of the A-site codon (Rodnina et al. 1996), being rather weak and rapidly reversible. This stage of probing involves the unlocked state of the tRNA entrance region of the ribosome. A dynamic equilibrium between the two sub-states of the ribosome, one with a vacant factor-binding site (sub-state

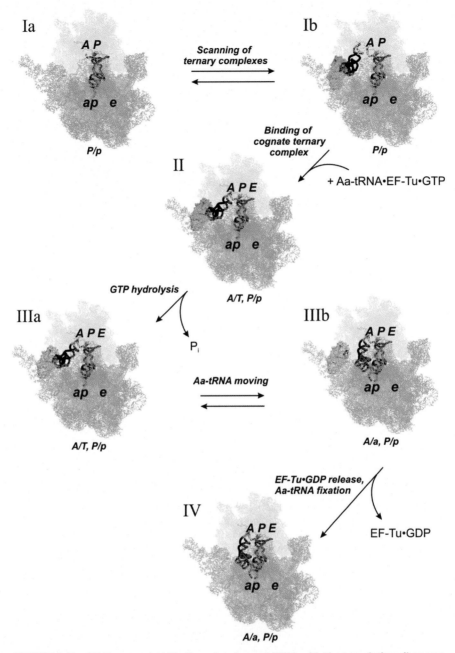

FIGURE 9.10: *EF-Tu-promoted binding of aminoacyl-tRNA with the translating ribosome. Positions of two lateral protuberances are conditionally shown unchangeable, although in reality they are movable and change during EF-Tu binding, subsequent GTP hydrolysis and EF-Tu release. Explanations are in the text.*

Ia) and the other with the transiently bound EF-Tu (substate Ib), provides browsing among anticodons of the ternary complexes with various aminoacyl-tRNAs from the surroundings.

Recognition of the cognate codon and formation of a correct codon-anticodon pair induces conformational rearrangements leading to the locking of the *A* site of the small subunit (Ogle et al. 2001; Ogle et al. 2002), and simultaneously to the formation of the closed factor-binding pocket

around EF-Tu at the large subunit, mostly due to shifting the shoulder domain of the small subunit toward both the ternary complex and the 30S subunit head (Stark et al. 2002; Valle et al. 2003b; Schmeing et al. 2009). Thus, after the entrance-unlocked ribosome admits the aminoacyl-tRNA anticodon stem-loop of the ternary complex into the *A* site of the small ribosomal subunit and the correct codon-anticodon pair is formed, the aminoacyl-tRNA is set in the intermediate hybrid *A/T* position (Figure 9.10, state II)

(Blanchard et al. 2004b; Gonzalez et al. 2007; Lee et al. 2007). In the *A/T* position, the structure of the tRNA becomes significantly distorted (bent at the anticodon stem) due to fixation of its anticodon at the DC on the small subunit and its acceptor arm at the factor-binding pocket (*T* site) of EF-Tu bound to the large subunit, this being inconsistent with the classical L-shape of tRNA (Valle et al. 2002; Valle et al. 2003b; Schmeing et al. 2009; Schuette et al. 2009; Villa et al. 2009). Evidently, it is the binding energy derived from both the codon-anticodon pairing and the ternary complex interactions with the EF-Tu-binding pocket that is used to induce and maintain the strained tRNA conformation.

Next (Figure 9.10, II → IIIa transition), the correct codon-anticodon recognition triggers fast GTP hydrolysis on EF-Tu (Rodnina et al. 1996; Schmeing et al. 2009; Schuette et al. 2009; Villa et al. 2009). The hydrolysis of GTP on the ribosome-bound EF-Tu leads to its unlocking (inter-domain mobility) and thus to the decrease of the affinity of the EF-Tu•GDP complex for the aminoacylated acceptor arm of aminoacyl-tRNA (reviewed in more detail by Abel & Jurnak 1996; Nyborg & Kjeldgaard 1996; Nyborg & Liljas 1998; Rodnina & Wintermeyer 2001; Rodnina et al. 2005). Now the tRNA may relax into its classical conformation (Valle et al. 2003b), and its acceptor arm is allowed to move, possibly fluctuating within the space between EF-Tu (sub-state IIIa) and the PTC (sub-state IIIb) (Blanchard et al. 2004b). This step is accompanied by the rearrangement of the *A*-site complex, possibly similar to the factor-free binding of aminoacyl-tRNA (see the previous section and Figure 9.7, Ib → II transition), but the participation of EF-Tu with GTP strongly accelerates this step (Wintermeyer & Robertson 1982; Pape et al. 1999). The EF-Tu•GDP complex becomes weakly retained in the pocket at the L10/L12 stalk and is now allowed to be spontaneously released (see, e.g., the reviews by Rodnina & Wintermeyer 2001; Rodnina et al. 2005).

The dissociation of the EF-Tu•GDP complex from the ribosome allows the shift of the aminoacylated CCA terminus of the cognate aminoacyl-tRNA toward the PTC (Ogle et al. 2002; Valle et al. 2003b). At the same time, when a near-cognate aminoacyl-tRNA is occasionally bound in the *A* site, it has a chance to be released from the ribosome (Pape et al. 1999; Yarus et al. 2003; Blanchard et al. 2004b; Lee et al. 2007) following the Ib → Ia path, as in the factor-free cycle (see Figure 9.7). The conformation of the PTC-bound tRNA, which was strained when the aminoacyl-tRNA was bound both to EF-Tu and to the *A* site, now changes and becomes close to the conformation of the free tRNA molecule (Schmeing et al. 2009; Valle et al. 2003b). The released acceptor arm of the codon-bound aminoacyl-tRNA has a high affinity for the *a* site of the PTC and thus inevitably should be caught there. Hence, the aminoacyl-tRNA becomes fixed in the final *A/a* position (Figure 9.10, state IV). In such a state, the two

substrates – peptidyl-tRNA in position *P/p* and aminoacyl-tRNA in position *A/a* – are in a close proximity with a strictly defined mutual orientation to provide the proper nucleophilic attack during the transpeptidation step. This implies that the immediate pretranspeptidation state ribosome should be fully and firmly locked.

V.2 EF-G-Catalyzed Translocation

Catalysis of translocation by EF-G starts with the interaction of the factor with the L10/L12 stalk and its base of the large ribosomal subunit. In this case, the target is the ribosome in the state after transpeptidation – that is, in the pretranslocation state (Figure 9.11).

As already mentioned (Sections II and IV), in this state, the ribosome is spontaneously oscillating between "nonrotated" and "rotated" forms (Cornish et al. 2008). At the same time, the acceptor arms of both the newly elongated peptidyl-tRNA and the deacylated tRNA were shown to oscillate between the *a* and *p* sites of the PTC and between the *p* and *e* sites, respectively (Blanchard et al. 2004a; Kim et al. 2007; Munro et al. 2007), the *a* → *p* transition being possible only after the *p* → *e* transition (Munro et al. 2007; Pan et al. 2007). The L10/L12 stalk of the large ribosomal subunit is free of interactions with ligand, so that it is also movable and fluctuating between allowed conformations. The L1 stalk is also found to spontaneously fluctuate between fully open (moved away from the 50S subunit bodies) and fully closed (drawn together with the 50S subunit body) positions (Fei et al. 2008; Cornish et al. 2009; Fei et al. 2009). Thus, the large-scale inter-subunit and intra-subunit movements testify to an unlocked state of the ribosome at the pretranslocation stage of the elongation cycle (Figure 9.11, Ia ⇆ Ib equilibrium state). (This state is the same as IIIa ⇆ IIIb in Figure 9.7 for factor-free translocation, and Ib is the same as the state shown at the top of Figure 9.8).

The interaction of EF-G (in the GTP form) with the L10/L12 stalk and its base on the large ribosomal subunit leads to immobilization of the flexible stalk (Gudkov et al. 1982; Gongadze et al. 1984) and formation of the factor-binding pocket around EF-G, also involving the shoulder and protein S12 of the small subunit (Connell et al. 2007; Gao et al. 2009). The elongated domain IV of the bound EF-G in the GTP-form is found to penetrate into the DC at the 30S subunit and overlap with the anticodon arm of the *A*-site tRNA (ibid). The accommodation of EF-G in the GTP-induced rigid conformation shifts the equilibrium between the two forms of the unlocked ribosome towards the "rotated" form, thus providing virtual fixation of the "rotated" state (Cornish et al. 2008; Fei et al. 2008). When the tRNA CCA-terminal sequences of the peptidyl-tRNA and the deacylated tRNA are caught by the *p* and *e* sites of the PTC, respectively, the fixation of the ribosome in the "rotated" form sets the intermediate state of the ribosome,

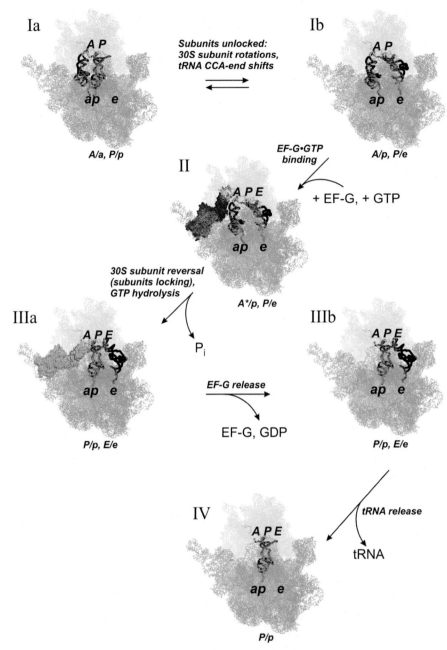

FIGURE 9.11: *EF-G-promoted translocation in the translating ribosome. Positions of two lateral protuberances are conditionally shown unchangeable, although in reality they are movable and change during EF-G binding, subsequent GTP hydrolysis, and EF-G release. Explanations are in the text.*

shown in Figure 9.11, state II. (Note that the "rotated" state of the 30S subunit and the positions of aminoacyl-tRNA and peptidyl-tRNA in this state correspond to those in the intermediate state shown in the middle of Figure 9.8 for the case of factor-free translocation).

The rotational movement of the small subunit relative to the large one has been shown to display several intermediate positions (Zhang et al. 2009). Although it is not clear whether all the mentioned structures do relate to the

translocation pathway, some transient intermediates are likely during the rotational movement of the small subunit. It is possible that the entering of domain IV of EF-G can provide a pawl effect along the stepwise path of the 30S subunit movement: It may allow only forward shifts of this subunit (which here plays the role of a ratchet dent) and block backward rotational motions. It is noteworthy that concomitantly with the rotational motion, the L1 stalk moves to interact with the elbow of the deacylated tRNA

(Fei et al. 2008; Cornish et al. 2009; Fei et al. 2009). In fact, the binding interactions, which EF-G forms with non-cleavable GTP analog, were clearly shown to make the pretranslocation ribosomal complex to significantly alter the dynamics of the L1 stalk and to shift the equilibrium between the open and closed conformations of the L1 stalk toward the closed conformation (Cornish et al. 2009; Fei et al. 2009). The formation of the multiple contacts of EF-G with both ribosomal subunits in their mutually "rotated" position represents a striking example of an induced fit, a maximization of non-covalent interactions. The situation as a whole may be described as a translocational intermediate state fixed in the "rotated"-state ribosome. However, the intermediate "rotated" state is unstable, and the ribosome may slowly reverse into the "non-rotated" state, even prior to the hydrolysis of EF-G-bound GTP (or in the presence of a non-cleavable GTP analog).

The GTP hydrolysis induced by the interactions of EF-G with the ribosome and the following release of orthophosphate result in the domains rearrangement in the bound EF-G (Connell et al. 2007; Taylor et al. 2007) and, probably, some relaxation of its structure inside the ribosome. This may lead to disruption of certain previous interactions of EF-G with the factor-binding pocket of the ribosome causing weakening of EF-G retention there. At the same time, domain IV present in the DC of the small ribosomal subunit may lose its interactions with the *A* site of the decoding center but still retain the interactions with the codon-anticodon duplex. As a result, the unlocking of the ribosome as a whole and the transition of EF-G into the relaxed GDP form must facilitate the small ribosomal subunit to rotate back into the "non-rotated" position (Figure 9.11, state IIIa). While the codon-anticodon duplexes are held by domain IV of EF-G, the DC, during the backward movement of the small subunit, can slide relative to the pair of codon-anticodon duplexes, leading to setting the duplexes in the shifted positions *P* and *E* on the small subunit, instead of the previous *A* and *P* positions, respectively. At the same time, while moving back and slipping along the mRNA in the 5′ to 3′ direction, the DC runs over the next codon of mRNA and sets it in the *A* site.

The recent intersubunit FRET measurements (Ermolenko & Noller, 2011) directly demonstrated that binding of EF-G•GTP first supports relative rotational movement (unlocking) of the ribosomal subunits, followed by a slower reversal (subunits locking), during which mRNA translocation occurs and tRNAs are set at their classical posttranslocational *P/p*, *E/e* positions.

Hence, according to the model under consideration, the slippage of the pair of codon-anticodon duplexes from the *A* and *P* sites to the *P* and *E* sites (Figure 9.11, the II → IIIa transition) is facilitated by the contact of the elongated domain IV of EF-G with the anticodon arm of the newly formed peptidyl-tRNA (see Figure 9.12): It allows the backward rotation of the small subunit and, hence, the backward

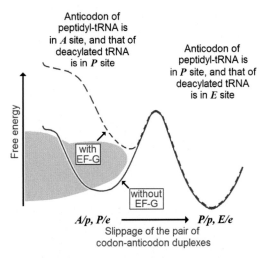

FIGURE 9.12: *Schematic representation of the change in the free-energy profile for the system (peptidyl-tRNA + deacylated tRNA) by domain IV of EF-G (shown in gray). The (A/p, P/e) → (P/p, E/e) transition can occur spontaneously (in the case of factor-free translocation), because the free energy of the (P/p, E/e) state is lower than that of the (A/p, P/e) state. The EF-G-promoted translocation is faster, as it has to overcome a lower free-energy barrier.*

shift of the decoding center, but blocks an accompanying backward shift of the codon-anticodon duplexes. This can be considered as the pawl function of EF-G during translocation of the tRNA residues pair, which plays the role of a ratchet dent. In addition to the pawl function, the interactions of domain IV with the anticodon arm of the peptidyl-tRNA may weaken the contacts of the DC with the codon-anticodon duplexes located in the *A* and *P* sites of the small subunit. (Note that both the "non-rotated" state of the 30S subunit and the positions of tRNAs in state IIIa correspond to those in the state shown at the bottom of Figure 9.8 for the case of factor-free translocation).

The role of the L1 stalk in the EF-G-promoted translocation should be the same as in factor-free translocation (see Section IV and Figures 9.7 and 9.8): The interaction of the L1 stalk with the elbow of the deacylated tRNA contributes to the sliding of the decoding center relative to codon-anticodon duplexes during the reverse rotational movement of the small subunit (Fei et al. 2008). Thus, the L1 stalk serves as a ribosome-intrinsic pawl retaining the pair of the codon-anticodon duplexes, which play the role of a ratchet dent at the shifted position while the sites of the 30S subunit return back (Figure 9.8, state *P/p*, *E/e*).

The translocational displacement corresponds to transition from a higher free-energy intermediate state to a lower free-energy post-translocation state (Figure 9.11, state IIIa). The shift of the codon-anticodon duplexes relative to the small subunit is accompanied by domain IV of EF-G in the GDP-form entering into the *A* site of the 30S subunit, and now the tip of domain IV is found in contact with the *P* site tRNA and its codon. It is this state (IIIa) that has been analyzed in the crystallographic X-ray study of the structure of the ribosome with EF-G trapped in the post-translocation state (Gao et al. 2009). The authors suppose that the contacts of domain IV with the *P* site tRNA and its codon initially occur when they are in the *A* site (state II) and are maintained throughout their translocation into the *P* site. Whereas EF-G in the GDP-form is still retained in the ribosome, the L10/L12 stalk continues to be immobilized (Connell et al. 2007), although now the factor seems to have weakened contacts and is allowed to dissociate (Figure 9.11, state IIIb). As in the case of the factor-free translocation (see Section IV and Figure 9.7), now the translating ribosome has restored the original ("non-rotated") mutual positions of its subunits and retains two tRNAs in the inter-subunit space, this being favorable for the locked state (Marshall et al. 2008; Aitken & Puglisi 2010).

The last stage is the dissociation of the deacylated tRNA from its *E/e* position. Earlier it was shown that the process of translocation is concluded by the release of deacylated tRNA from the translating ribosomes (Spirin 1984; Robertson & Wintermeyer 1987), and more recently it was confirmed (Spiegel et al. 2007). Under natural conditions, the released tRNA is immediately removed from the equilibrium mixture, either being involved in the aminoacylation reaction or strongly diluted, so that the release is practically irreversible. There is evidence that the L1 stalk mobility induced at the concluding stage of the EF-G-promoted translocation is involved in the ejection of the deacylated tRNA from the ribosome (Valle et al. 2003a; Korostelev et al. 2006; Fei et al. 2008; Cornish et al. 2009; Fei et al. 2009). Possibly, the dissociation of EF-G allows the L1 stalk as well as the L10/L12 stalk to move and thus facilitate the deacylated tRNA release. Hence, after completion of translocation and leaving of deacylated tRNA, the ribosome contains just one tRNA residue in the form of peptidyl-tRNA that occupies its classical *P/p* position (Figure 9.11, state IV). Now the ribosome is ready to admit the next aminoacyl-tRNA into the vacant codon-charged *A* site position.

VI. CONCLUSION

The manifestations of physical laws in the microworld of molecular machines is different from those in the macroworld, because of (1) small masses of macromolecular complexes moving in viscous medium, this resulting in negligible inertia; (2) small volumes of macromolecular complexes, this preventing the storage of heat; (3) conformational flexibility of structural blocks and joints, this excluding the mechanical precision; (4) Brownian motions and thermal oscillations of all parts of molecular machines; and (5) conformational changes that create or destroy catalytic sites and modify energy landscapes for moving parts of molecular complexes. As a result, a number of

principles that underlie the work of power-stroke macromachines, such as combustion engines, cannot be realized at the molecular level. The molecular machines (including machines of the conveying type, like the ribosome) can use neither storage of kinetic energy nor high-precision mechanics. They must be considered as devices moving without heat or mechanical engines, as well as without long-range mechanical transmissions and movers. The driving force for directional molecular movements is a Brownian motion biased against wrong directions by ratchet-and-pawl mechanisms, thus working as Maxwell demons fed by high-energy compounds.

The essence of the mRNA-conveying action of the ribosome can be formulated as follows (Figure 9.13). The working cycle of the ribosome consists of transitions from weak to strong binding states alternating with transitions between the states that are in equilibrium. The main energy source is the transpeptidation reaction (this is the sole energy source in the factor-free translation). The main engine is the peptidyl transferase center (PTC) catalyzing this exergonic reaction. The changes in the acceptor ends of the tRNA that occur in the PTC region (in the *a* and *p* sites of the large subunit), and the shift of these acceptor ends from the *a* and *p* to *p* and *e* positions after transpeptidation are transmitted via the tRNAs to the inter-subunit contacts. The contacts become weakened, thus allowing the inter-subunit rotational oscillations. The "rotated" position of the small subunit makes the binding of the tRNA residues pair to *A* and *P* sites less stable than to *P* and *E* sites. Translocation is realized through the rotation of the small subunit relative to the large subunit and its reversal. The forward rotation results in the shift of the *A* and *P* sites together with the bound codon-anticodon duplexes relative to the large subunit, while the positions of CCA-ends of the tRNAs at the large subunit are retained in the *p* and *e* sites. The backward rotational movement, which occurs spontaneously (in a factor-free translocation) or after GTP hydrolysis (in a EF-G-assisted translocation), leads to restoration of the original ("non-rotated") inter-subunit position, but the codon-anticodon duplexes remain at their new positions relative to the large subunit, being now caught by the *P* and *E* sites of the small subunit returned to its original position. In this process, the pair of tRNA residues serves as a driving gear, or mover of the mRNA. The next mRNA codon is now set at the vacant *A* site. This completes one round of the working cycle of the ribosome as a conveying machine.

ACKNOWLEDGEMENTS

The authors thank A. Kommer for invaluable help in preparing illustrations for this chapter, and V. A. Kolb, A. G. Ryazanov, and K. S. Vasilenko for their critical remarks, advice, and comments on the text of the manuscript. The authors are also grateful to R. L. Gonzalez and J. Frank

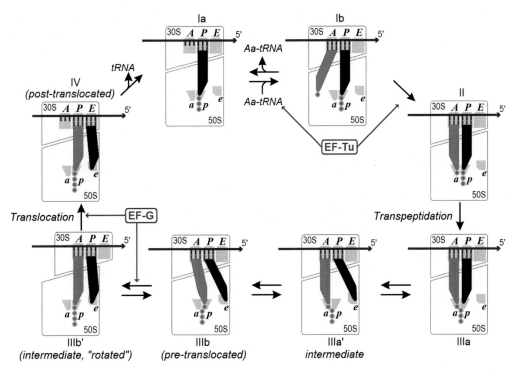

FIGURE 9.13: *A stepwise scheme of the whole working cycle of the ribosome conveying machine. The scheme is based on the factor-free elongation cycle (see Figure 9.7); its main features have been recently confirmed by electron cryo-microscopy of ribosomes involved in a factor-dependent transloca-tion (Fischer et al. 2010). The transitions catalyzed by EF-Tu and EF-G in the factor-promoted elongation cycle are pointed by long arrows. The shift of the small subunit at the IIIb→IIIb' transition is the result of its forward rotation, and the backward shift at the IIIb'→IV transition results from the reversal rotation (see Figure 9.8). All other explanations are in the text and preceding figures.*

for reviewing and editing the article. This work was sup-ported by the program on Molecular and Cell Biology of the Russian Academy of Sciences, by the Russian Program "Scientific Schools," by the Russian Foundation for Basic Research, by the Federal Agency for Science and Innova-tions, and by the International Research Scholar's Award to A.V.F. from the Howard Hughes Medical Institute.

REFERENCES

Abel, K & Jurnak, F 1996, 'A complex profile of protein elon-gation: translating chemical energy into molecular movement', *Structure*, vol. 4, no. 3, pp. 229–238.

Abel, K, Yoder, MD, Hilgenfeld, R & Jurnak, F 1996, 'An alpha to beta conformational switch in EF-Tu', *Structure*, vol. 4, no. 10, pp. 1153–1159.

Agirrezabala, X, Lei, J, Brunelle, JL, Ortiz-Meoz, RF, Green, R & Frank, J 2008, 'Visualization of the hybrid state of tRNA binding promoted by spontaneous ratcheting of the ribosome', *Molecular Cell*, vol. 32, no. 2, pp. 190–197.

Agrawal, RK, Heagle, AB, Penczek, P, Grassucci, RA & Frank, J 1999, 'EF-G-dependent GTP hydrolysis induces transloca-tion accompanied by large conformational changes in the 70S ribosome', *Nature Structural Biology*, vol. 6, no. 7, pp. 643–647.

Agrawal, RK, Penczek, P, Grassucci, RA & Frank, J 1998, 'Visu-alization of elongation factor G on the *Escherichia coli* 70S

ribosome: the mechanism of translocation', *Proceeding of the National Academy of Sciences of the USA*, vol. 95, no. 11, pp. 6134–6138.

Ait-Haddou, R & Herzog, W 2003, 'Brownian ratchet models of molecular motors', *Cell Biochemistry and Biophysics*, vol. 38, no. 2, pp. 191–214.

Aitken, CE, & Puglisi, JD 2010, 'Following the intersubunit con-formation of the ribosome during translation in real time', *Nature Structural and Molecular Biology*, vol. 17, no. 7, pp. 793–800.

Astumian, RD & Derényi, I 1998, 'Fluctuation driven transport and models of molecular motors and pumps', *European Biophysics Journal*, vol. 27, no. 5, pp. 474–489.

Ban, N, Nissen, P, Hansen, J, Moore, PB & Steitz, TA 2000, 'The complete atomic structure of the large ribosomal sub-unit at 2.4 A resolution', *Science*, vol. 289, no. 5481, pp. 905–920.

Baranov, VI, Belitsina, NV & Spirin, AS 1979, 'The use of columns with matrix-bound polyuridylic acid for isolation of translating ribosomes', *Methods in Enzymology*, vol. 59, pp. 382–397.

Belitsina, NV, Glukhova, MA & Spirin, AS 1975, 'Translocation in ribosomes by attachment-detachment of elongation factor G without GTP cleavage: Evidence from a column-bound ribo-some system', *FEBS Letters*, vol. 54, no.1, pp. 35–38.

Belitsina, NV, Glukhova, MA & Spirin, AS 1976, 'Stepwise elon-gation factor G-promoted elongation of polypeptides on the ribosome without GTP cleavage', *Journal of Molecular Biology*, vol. 108, no. 3, pp. 609–613.

Belitsina, NV & Spirin, AS 1970, 'Studies on the structure of ribosomes. IV. Participation of aminoacyl-transfer RNA and peptidyl-transfer RNA in the association of ribosomal subparticles', *Journal of Molecular Biology*, vol. 52, no. 1, pp. 45–55.

Belitsina, NV & Spirin, AS 1979, 'Ribosomal translocation assayed by the matrix-bound poly(uridylic acid) column technique', *European Journal of Biochemistry*, vol. 94, no. 1, pp. 315–320.

Belitsina, NV, Tnalina, GZ & Spirin, AS 1981, 'Template-free ribosomal synthesis of polylysine from lysyl-tRNA', *FEBS Letters*, vol. 131, no. 2, pp. 289–292.

Belitsina, NV, Tnalina, GZ & Spirin, AS 1982, 'Template-free ribosomal synthesis of polypeptides from aminoacyl-tRNAs', *BioSystems*, vol. 15, no. 3, pp. 233–241.

Berchtold, H, Reshetnikova, L, Reiser, CO, Schirmer, NK, Sprinzl, M & Hilgenfeld, R 1993, 'Crystal structure of active elongation factor Tu reveals major domain rearrangements', *Nature*, vol. 365, no. 6442, pp. 126–132.

Berk, V, Zhang, W, Pai, RD & Cate, JHD 2006, 'Structural basis for mRNA and tRNA positioning on the ribosome', *Proceeding of the National Academy of Sciences of the USA*, vol. 103, no. 43, pp. 15830–15834.

Bieling, P, Beringer, M, Adio, S & Rodnina, MV 2006, 'Peptide bond formation does not involve acid-base catalysis by ribosomal residues', *Nature Structural and Molecular Biology*, vol. 13, no. 5, pp. 423–428.

Blanchard, SC, Gonzalez, RL Jr, Kim, HD, Chu, S & Puglisi, JD 2004b, 'tRNA selection and kinetic proofreading in translation', *Nature Structural and Molecular Biology*, vol. 11, no. 10, pp. 1008–1014.

Blanchard, SC, Kim, HD, Gonzalez, RL Jr, Puglisi, JD & Chu, S 2004a, 'tRNA dynamics on the ribosome during translation", *Proceeding of the National Academy of Sciences of the USA*, vol. 101, no. 35, pp. 12893–12898.

Bretscher, MS 1968, 'Translocation in protein synthesis: a hybrid structure model', *Nature*, vol. 218, no. 5142, pp. 675–677.

Chetverin, AB & Spirin, AS 1982, 'Bioenergetics and protein synthesis', *Biochimica et Biophysica Acta*, vol. 683, no. 2, pp. 153–179.

Connell, SR, Takemoto, C, Wilson, DN, Wang, H, Murayama, K, Terada, T, Shirouzu, M, Rost, M, Schüler, M, Giesebrecht, J, Dabrowski, M, Mielke, T, Fucini, P, Yokoyama, S & Spahn, CM 2007, 'Structural basis for interaction of the ribosome with the switch regions of GTP-bound elongation factors', *Molecular Cell*, vol. 25, no. 5, pp. 751–764.

Cordova, NJ, Ermentrout, B & Oster, GF 1992, 'Dynamics of single-motor molecules: the thermal ratchet model', *Proceeding of the National Academy of Sciences of the USA*, vol. 89, no. 1, pp. 339–343.

Cornish, PV, Ermolenko, DN, Noller, HF & Ha, T 2008, 'Spontaneous intersubunit rotation in single ribosomes', *Molecular Cell*, vol. 30, no. 5, pp. 578–588.

Cornish, PV, Ermolenko, DN, Staple, DW, Hoang, L, Hickerson RP, Noller, HF & Ha, T 2009, 'Following movement of the L1 stalk between three functional states in single ribosomes', *Proceeding of the National Academy of Sciences of the USA*, vol. 106, no. 8, pp. 2571–2576.

Cukras, AR, Southworth, DR, Brunelle, JL, Culver, GM & Green, R 2003, 'Ribosomal proteins S12 and S13 function as control elements for translocation of the mRNA:tRNA complex', *Molecular Cell*, vol. 12, no. 12, pp. 321–328.

Ermolenko, DN, Majumdar, ZK, Hickerson, RP, Spiegel, PC, Clegg, RM & Noller, HF 2007a, 'Observation of intersubunit movement of the ribosome in solution using FRET', *Journal of Molecular Biology*, vol. 370, no. 3, pp. 530–540.

Ermolenko, DN, Spiegel, PC, Majumdar, ZK, Hickerson, RP, Clegg, RM & Noller, HF 2007b, 'The antibiotic viomycin traps the ribosome in an intermediate state of translocation', *Nature Structural and Molecular Biology*, vol. 14, no. 6, pp. 493–497.

Ermolenko, DN, & Noller, HF 2011, 'mRNA translocation occurs during the second step of ribosomal intersubunit rotation', *Nature Structural & Molecular Biology*, vol. 18, no. 4, pp. 457–463.

Fei, J, Bronson, JE, Hofman, JM, Srinivas, RL, Wiggins, CH & Gonzalez, RL Jr 2009, 'Allosteric collaboration between elongation factor G and the ribosomal L1 stalk directs tRNA movements during translation', *Proceeding of the National Academy of Sciences of the USA*, vol. 106, no. 37, pp. 15702–15707.

Fei, J, Kosuri, P, MacDougall, DD & Gonzalez, RL Jr 2008, 'Coupling of ribosomal L1 stalk and tRNA dynamics during translation elongation', *Molecular Cell*, vol. 30, no. 3, pp. 348–359.

Feynman, R, Leighton, R & Sands, M 1963, *The Feynman Lectures on Physics*, vol. 1, chapter 46, pp. 1–5. Addison-Wesley Publishing Company, Inc., Reading, MA.

Finkelstein, AV & Ptitsyn, OB 2002, *Protein Physics*, Academic Press, London.

Fischer, N, Konevega, AL, Wintermeyer, W, Rodnina, MV & Stark, H 2010, 'Ribosome dynamics and tRNA movement by time-resolved electron cryomicroscopy', *Nature*, vol. 466, no. 7304, pp. 329–333.

Fourmy, D, Yoshizawa, S & Puglisi, JD 1998, 'Paromomycin binding induces a local conformational change in the A-site of 16 S rRNA', *Journal of Molecular Biology*, vol. 277, no. 2, pp. 333–345.

Frank, J & Agrawal, RK 2000, 'A ratchet-like inter-subunit reorganization of the ribosome during translocation', *Nature*, vol. 406, no. 6793, pp. 318–322.

Frank, J, Gao, H, Sengupta, J, Gao, N & Taylor, DJ 2007, 'The process of mRNA-tRNA translocation', *Proceeding of the National Academy of Sciences of the USA*, vol. 104, no. 50, pp. 19671–19678.

Frank, J & Gonzalez, RL Jr 2010, 'Structure and dynamics of a processive Brownian motor: The translating ribosome', *Annual Review of Biochemistry*, vol. 79, pp. 381–412.

Fredrick, K & Noller, HF 2003, 'Catalysis of ribosomal translocation by sparsomycin', *Science*, vol. 300, no. 5622, pp. 1159–1162.

Gabashvili, IS, Agrawal, RK, Spahn, CM, Grassucci, RA, Svergun, DI, Frank, J & Penczek, P 2000, 'Solution structure of the E. coli 70S ribosome at 11.5 A resolution', *Cell*, vol. 100, no. 5, pp. 537–549.

Gao, H, Sengupta, J, Valle, M, Korostelev, A, Eswar, N, Stagg, SM, Van Roey, P, Agrawal, RK, Harvey, SC, Sali, A, Chapman, MS & Frank, J 2003, 'Study of the structural dynamics of the E coli 70S ribosome using real-space refinement', *Cell*, vol. 113, no. 6, pp. 789–801.

Gao, YG, Selmer, M, Dunham, CM, Weixlbaumer, A, Kelley, AC & Ramakrishnan, V 2009, 'The structure of the ribosome with elongation factor G trapped in the posttranslocational state', *Science*, vol. 326, no. 5953, pp. 694–699.

Gavrilova, LP, Kostiashkina, OE, Koteliansky, VE, Rutkevitch, NM & Spirin, AS 1976, 'Factor-free ("non-enzymic") and factor-dependent systems of translation of polyuridylic acid by *Escherichia coli* ribosomes', *Journal of Molecular Biology*, vol. 101, no. 4, pp. 537–552.

Gavrilova, LP, Koteliansky, VE & Spirin, AS 1974, 'Ribosomal protein S12 and "non-enzymatic" translocation', *FEBS Letters*, vol. 45, no. 1, pp. 324–328.

Gavrilova, LP, Perminova, IN & Spirin, AS 1981, 'Elongation factor Tu can reduce translation errors in poly(U)-directed cell-free systems', *Journal of Molecular Biology*, vol. 149, no. 1, pp. 69–78.

Gavrilova, LP & Spirin, AS 1971, 'Stimulation of "non-enzymic" translocation in ribosomes by ρ-chloromercuribenzoate', *FEBS Letters*, vol. 17, no. 2, pp. 324–326.

Gavrilova, LP & Spirin, AS 1972, 'A modification of the 30S ribosomal subparticle is responsible for stimulation of "non-enzymatic" translocation by *p*-chloromercuribenzoate', *FEBS Letters*, vol. 22, no. 1, pp. 91–92.

Gavrilova, LP & Spirin, AS 1974, '"Nonenzymatic" translation', *Methods in Enzymology*, vol. 30, pp. 452–462.

Girshovich, AS, Bochkareva, ES & Vasiliev, VD 1986, 'Localization of elongation factor Tu on the ribosome', *FEBS Letters*, vol. 197, no. 1–2, pp. 192–198.

Girshovich, AS, Kurtskhalia, TV, Ovchinnikov, YuA & Vasiliev, VD 1981, 'Localization of the elongation factor G on Escherichia coli ribosome', *FEBS Letters*, vol. 130, no. 1, pp. 54–59.

Gomez-Lorenzo, MG, Spahn, CM, Agrawal, RK, Grassucci, RA, Penczek, PA, Chakraburtty, K, Ballesta, JP, Lavandera, JL, Garcia-Bustos, JF, & Frank, J 2000, 'Three-dimensional cryo-electron microscopy localization of EF2 in the Saccharomyces cerevisiae 80S ribosome at 17.5 A resolution', *The EMBO Journal*, vol. 19, no. 11, pp. 2710–2718.

Gongadze, GM, Gudkov, AT, Bushuev, VN & Sepetov, NF 1984, 'The attachment of elongation factor G to the ribosome changes intramolecular mobility of protein L7/L12', *Doklady Akademii Nauk SSSR*, vol. 279, no. 1, pp. 230–232.

Gonzalez, RL, Chu, S & Puglisi, JD 2007, 'Thiostrepton inhibition of tRNA delivery to the ribosome', *RNA*, vol. 13, no. 12, pp. 2091–2097.

Gottesman, ME 1967, 'Reaction of ribosome-bound peptidyl transfer ribonucleic acid with aminoacyl transfer ribonucleic acid or puromycin', *The Journal of Biological Chemistry*, vol. 242, no. 23, pp. 5564–5571.

Gudkov, AT, Gongadze, GM, Bushuev, VN & Okon, MS 1982, 'Proton nuclear magnetic resonance study of the ribosomal protein L7/L12 in situ', *FEBS Letters*, vol. 138, no. 2, pp. 229–232.

Gumbart, J, Trabuco, LG, Schreiner, E, Villa, E & Schulten, K 2009, 'Regulation of the protein-conducting channel by a bound ribosome', *Structure*, vol. 17, no. 11, pp. 1453–1464.

Hansen, JL, Schmeing, TM, Moore, PB & Steitz, TA 2002, 'Structural insights into peptide bond formation', *Proceeding of the National Academy of Sciences of the USA*, vol. 99, no. 18, pp. 11670–11675.

Harms, J, Schluenzen, F, Zarivach, R, Bashan, A, Gat, S, Agmon, I, Bartels, H, Franceschi, F & Yonath, A 2001, 'High resolution structure of the large ribosomal subunit from a mesophilic eubacterium', *Cell*, vol. 107, no. 5, pp. 679–688.

Horan, LH & Noller, HF 2007, 'Intersubunit movement is required for ribosomal translocation', *Proceeding of the National Academy of Sciences of the USA*, vol. 104, no. 12, pp. 4881–4885.

Julián, P, Konevega, AL, Scheres, SH, Lázaro, M, Gil, D, Wintermeyer, W, Rodnina, MV & Valle, M 2008, 'Structure of ratcheted ribosomes with tRNAs in hybrid states', *Proceeding of the National Academy of Sciences of the USA*, vol. 105, no. 44, pp. 16924–16927.

Kaji, A & Kaji, H 1963, 'Specific interaction of soluble RNA and polyribonucleic acid induced polysomes', *Biochemical and Biophysical Research Communications*, vol. 13, no. 3, pp. 186–192.

Kakhniashvili, DG, & Spirin, AS 1977, 'Dependence of factor-free and factor-promoted translation systems on temperature. Absence of effects of the elongation factors and GTP on the activation energy', *Doklady Akademii Nauk SSSR*, vol. 234, no. 4, pp. 958–963.

Kaltschmidt, E, & Wittmann, HG 1970, 'Ribosomal proteins. XII. Number of proteins in small and large ribosomal subunits of *E. coli* as determined by two-dimensional gel electrophoresis', *Proceeding of the National Academy of Sciences of the USA*, vol. 67, no. 3, pp. 1276–1282.

Kim, HD, Puglisi, JD & Chu, S 2007, 'Fluctuations of transfer RNAs between classical and hybrid states', *Biophysical Journal*, vol. 93, no. 10, pp. 3575–3582.

Kjeldgaard, M, Nissen, P, Thirup, S & Nyborg, J 1993, 'The crystal structure of elongation factor EF-Tu from Thermus aquaticus in the GTP conformation', *Structure*, vol. 1, no. 1, pp. 35–50.

Korostelev, A & Noller, HF 2007, 'Analysis of structural dynamics in the ribosome by TLS crystallographic refinement', *Journal of Molecular Biology*, vol. 373, no. 4, pp. 1058–1070.

Korostelev, A, Trakhanov, S, Laurberg, M & Noller, HF 2006, 'Crystal structure of a 70S ribosome-tRNA complex reveals functional interactions and rearrangements', *Cell*, vol. 126, no. 6, pp. 1065–1077.

Lee, TH, Blanchard, SC, Kim, HD, Puglisi, JD & Chu, S 2007, 'The role of fluctuations in tRNA selection by the ribosome', *Proceeding of the National Academy of Sciences of the USA*, vol. 104, no. 34, pp. 13661–13665.

Lim, VI & Spirin, AS 1986, 'Stereochemical analysis of ribosomal transpeptidation: Conformation of nascent peptide', *Journal of Molecular Biology*, vol. 188, no. 4, pp. 565–574.

Lodish, HF & Jacobsen, M 1972, 'Regulation of hemoglobin synthesis. Equal rates of translation and termination of α- and β-globin chains', *The Journal of Biological Chemistry*, vol. 247, no. 11, pp. 3622–3629.

Marshall, RA, Dorywalska, M & Puglisi, JD 2008, 'Irreversible chemical steps control intersubunit dynamics during translation', *Proceeding of the National Academy of Sciences of the USA*, vol. 105, no. 40, pp. 15364–15369.

Moazed, D & Noller, HF 1986, 'Transfer RNA shields specific nucleotides in 16S ribosomal RNA from attack by chemical probes', *Cell*, vol. 47, no. 6, pp. 985–994.

Moazed, D & Noller, HF 1989a, 'Interaction of tRNA with 23S rRNA in the ribosomal A, P, and E sites', *Cell*, vol. 57, no. 4, pp. 585–597.

Moazed, D & Noller, HF 1989b, 'Intermediate states in the movement of transfer RNA in the ribosome', *Nature*, vol. 342, no. 6246, pp. 142–148.

Moazed, D, Robertson, JM & Noller, HF 1988, 'Interaction of elongation factors EF-G and EF-Tu with a conserved loop in 23S RNA', *Nature*, vol. 334, no. 6180, pp. 362–364.

Munro, JB, Altman, RB, O'Connor, N & Blanchard, SC 2007, 'Identification of two distinct hybrid state intermediates on the ribosome', *Molecular Cell*, vol. 25, no. 4, pp. 505–517.

Nakamoto, T, Conway, TW, Allende, JE, Spyrides, GI & Lipmann, F 1963, 'Formation of peptide bonds. I. Peptide formation from aminoacyl-sRNA', *Cold Spring Harbor Symposia on Quantitative Biology*, vol. 28, pp. 227–231.

Nishiyama, M., Higuchi, H., Ishii, Y., Taniguchi, Y, & Yanagida, T 2003, 'Single molecule processes on the stepwise movement of ATP-driven molecular motors', *BioSystems*, vol. 71, no. 1–2, pp. 145–156.

Nissen, P, Hansen, J, Ban, N, Moore, PB & Steitz, TA 2000, 'The structural basis of ribosome activity in peptide bond synthesis', *Science*, vol. 289, no. 5481, pp. 920–930.

Nyborg, J, & Kjeldgaard, M 1996, 'Elongation in bacterial protein synthesis', *Current Opinion in Biotechnology*, vol. 7, no. 4, pp. 369–375.

Nyborg, J, & Liljas, A 1998, 'Protein biosynthesis: structural studies of the elongation cycle', *FEBS Letters*, vol. 430, no. 1–2, pp. 95–99.

Ogle, JM, Brodersen, DE, Clemons, WM Jr, Tarry, MJ, Carter, AP & Ramakrishnan, V 2001, 'Recognition of cognate transfer RNA by the 30S ribosomal subunit', *Science*, vol. 292, no. 5518, pp. 897–902.

Ogle, JM, Murphy, FV, Tarry, MJ & Ramakrishnan, V 2002, 'Selection of tRNA by the ribosome requires a transition from an open to a closed form', *Cell*, vol. 111, no. 5, pp. 721–732.

Pan, D, Kirillov, SV & Cooperman, BS 2007, 'Kinetically competent intermediates in the translocation step of protein synthesis', *Molecular Cell*, vol. 25, no. 4, pp. 519–529.

Pape, T, Wintermeyer, W & Rodnina, M 1999, 'Induced fit in initial selection and proofreading of aminoacyl-tRNA on the ribosome', *The EMBO Journal*, vol. 18, no. 13, pp. 3800–3807.

Perutz, MF 1970, 'Stereochemistry of cooperative effects in haemoglobin', *Nature*, vol. 228, no. 5273, pp. 726–739.

Pestka, S 1968, 'Studies on the formation of transfer ribonucleic acid-ribosome complexes. 3. The formation of peptide bonds by ribosomes in the absence of supernatant enzymes', *The Journal of Biological Chemistry*, vol. 243, no. 10, pp. 2810–2820.

Pestka, S 1969, 'Studies on the formation of transfer ribonucleic acid-ribosome complexes. VI. Oligopeptide synthesis and translocation on ribosomes in the presence and absence of soluble transfer factors', *The Journal of Biological Chemistry*, vol. 244, no. 6, pp. 1533–1539.

Pestova, TV, Lorsh, JR & Hellen, CUT 2007, 'The mechanism of translation initiation in eukaryotes', in *Translational Control in Biology and Medicine*, eds MB Mathews, N Sonenberg & JWB Hershey, Cold Spring Harbor Laboratory Press, Plainview, NY, pp. 87–128.

Polekhina, G, Thirup, S, Kjeldgaard, M, Nissen, P, Lippmann, C & Nyborg, J 1996, 'Helix unwinding in the effector region of elongation factor EF-Tu-GDP', *Structure*, vol. 4, no. 10, pp. 1141–1151.

Purich, DL, 2001, 'Enzyme catalysis: a new definition accounting for noncovalent substrate- and product-like states', *Trends in Biochemical Sciences*, vol. 26, no. 7, pp. 417–421.

Ramakrishnan, V 2002, 'Ribosome structure and the mechanism of translation', *Cell*, vol. 108, no. 4, pp. 557–572.

Riddle, DL, & Carbon, J 1973, 'Frameshift suppression: A nucleotide addition in the anticodon of a glycine transfer RNA', *Nature New Biology*, vol. 242, no. 121, pp. 230–234.

Robertson, JM & Wintermeyer, W 1987, 'Mechanism of ribosomal translocation. tRNA binds transiently to an exit site before leaving the ribosome during translocation', *Journal of Molecular Biology*, vol. 196, no. 3, pp. 525–540.

Rodnina, MV, Gromadski, KB, Kothe, U & Wieden, H-J 2005, 'Recognition and selection of tRNA in translation', *FEBS Letters*, vol. 579, no. 4, pp. 938–942.

Rodnina, MV, Pape, T, Fricke, R, Kuhn, L & Wintermeyer, W 1996, 'Initial binding of the elongation factor Tu.GTP.aminoacyl-tRNA complex preceding codon recognition on the ribosome', *The Journal of Biological Chemistry*, vol. 271, no. 2, pp. 646–652.

Rodnina, MV, & Wintermeyer, W 2001, 'Ribosome fidelity: tRNA discrimination, proofreading and induced fit', *Trends in Biochemical Sciences*, vol. 26, no. 2, pp. 124–130.

Schmeing, TM, Huang, KS, Kitchen, DE, Strobel, SA & Steitz, TA 2005a, 'Structural insights into the roles of water and the 2′ hydroxyl of the P site tRNA in the peptidyl transferase reaction', *Molecular Cell*, vol. 20, no. 3, pp. 437–448.

Schmeing, TM, Huang, KS, Strobel, SA & Steitz, TA 2005b, 'An induced-fit mechanism to promote peptide bond formation and exclude hydrolysis of peptidyl-tRNA', *Nature*, vol. 438, no. 7067, pp. 520–524.

Schmeing, TM, Voorhees, RM, Kelley, AC, Gao, YG, Murphy, FV IV, Weir, JR & Ramakrishnan, V 2009, 'The crystal structure of the ribosome bound to EF-Tu and aminoacyl-tRNA', *Science*, vol. 326, no. 5953, pp. 688–694.

Schuette, JC, Murphy, FV IV, Kelley, AC, Weir, JR, Giesebrecht, J, Connell, SR, Loerke, J, Mielke, T, Zhang, W, Penczek, PA, Ramakrishnan, V & Spahn, CM 2009, 'GTPase activation of elongation factor EF-Tu by the ribosome during decoding', *The EMBO Journal*, vol. 28, no. 6, pp. 755–765.

Schuwirth, BS, Borovinskaya, MA, Hau, CW, Zhang, W, Vila-Sanjurjo, A, Holton, JM & Cate, JH 2005, 'Structures of the bacterial ribosome at 3.5 Å resolution', *Science*, vol. 310, no. 5749, pp. 827–834.

Selmer, M, Dunham, CM, Murphy, FV IV, Weixlbaumer, A, Petry, S, Kelley, AC, Weir, JR & Ramakrishnan, V 2006, 'Structure of the 70S ribosome complexed with mRNA and tRNA', *Science*, vol. 313, no. 5795, pp. 1935–1942.

Semenkov, YP, Rodnina, MV & Wintermeyer, W 1996, 'The "allosteric three-site model" of elongation cannot be confirmed in a well-defined ribosome system from *Escherichia coli*', *Proceeding of the National Academy of Sciences of the USA*, vol. 93, no. 22, pp. 12183–12188.

Serdyuk, IN, Baranov, VI, Tsalkova, T, Gulyamova, D, Pavlov, M, Spirin, AS & May, R 1992, 'Structural dynamics of translating ribosomes', *Biochimie*, vol. 74, pp. 299–306.

Serdyuk, IN, & Spirin, AS, 1986, 'Structural dynamics of the translating ribosomes', in *Structure, Function, and Genetics of Ribosomes*, eds B Hardesty & G Kramer, Springer-Verlag New York Inc., New York, 425–437.

Smoluchowski, M von 1912, 'Experimentell nachweisbare, der Üblichen Thermodynamik widersprechende Molekularphenomene', *Physikalische Zeitschrift*, vol. 13, pp. 1069–1080.

Southworth, DR, Brunelle, JL & Green, R 2002, 'EFG-independent translocation of the mRNA:tRNA complex is promoted by modification of the ribosome with thiol-specific reagents', *Journal of Molecular Biology*, vol. 324, no. 4, pp. 611–623.

Spahn, CM, Gomez-Lorenzo, MG, Grassucci, RA, Jorgensen, R, Andersen, GR, Beckmann, R, Penczek, PA, Ballesta, JP & Frank, J 2004, 'Domain movements of elongation factor eEF2 and the eukaryotic 80S ribosome facilitate tRNA translocation', *The EMBO Journal*, vol. 23, no. 5, pp. 1008–1019.

Spiegel, PC, Ermolenko, DN & Noller, HF 2007, 'Elongation factor G stabilizes the hybrid-state conformation of the 70S ribosome', *RNA*, vol. 13, no. 9, pp. 1473–1482.

Spirin, AS 1968a, 'On the mechanism of the ribosome working: Subunits locking-unlocking hypothesis', *Doklady Akademii Nauk SSSR*, vol. 179, no. 6, pp. 1467–1470.

Spirin, AS 1968b, 'How does the ribosome work? A hypothesis based on the two subunit construction of the ribosome', *Currents in Modern Biology*, vol. 2, no. 3, pp. 115–127.

Spirin, AS 1969, 'A model of the functioning ribosome: Locking and unlocking of the ribosome subparticles', *Cold Spring Harbor Symposia on Quantitative Biology*, vol. 34, pp. 197–207.

Spirin, AS 1978, 'Energetics of the ribosome', in *Progress in Nucleic Acid Research and Molecular Biology*, ed WE Cohn, Academic Press, Inc., New York-London, vol. 21, pp. 39–62.

Spirin, AS 1984, 'Testing the classical two-tRNA-site model for the ribosomal elongation cycle', *FEBS Letters*, vol. 165, no. 2, pp. 280–284.

Spirin, AS 1985, 'Ribosomal translocation: Facts and models', in *Progress in Nucleic Acid Research and Molecular Biology*, ed WE Cohn, Academic Press, Inc., New York-London, vol. 32, pp. 75–114.

Spirin, AS 1988, 'Energetics and dynamics of the protein-synthesizing machinery', in *The Roots of Modern Biochemistry*, eds H Kleinkauf, H von Dören & L Jaenicke, Walter de Gruyter & Co., Berlin, pp. 511–533.

Spirin, AS 1999, *Ribosomes*. Kluwer Academic Publishers /Plenum Press, New York.

Spirin, AS 2002, 'Ribosome as a molecular machine', *FEBS Letters*, vol. 514, no. 1, pp. 2–10.

Spirin, AS 2004, 'The ribosome as an RNA-based molecular machine', *RNA Biology*, vol. 1, no. 1, pp. 3–9.

Spirin, AS 2009a, 'The ribosome as a conveying thermal ratchet machine', *The Journal of Biological Chemistry*, vol. 284, no. 32, pp. 21103–21119.

Spirin. AS 2009b, 'How does a scanning ribosomal particle move along the 5′-untranslated region of eukaryotic mRNA? Brownian ratchet model', *Biochemistry*, vol. 48, no. 45, pp. 10688–10692.

Spirin, AS, Baranov, VI, Polubesov, GS, Serdyuk, IN & May, RP 1987, 'Translocation makes the ribosome less compact', *Journal of Molecular Biology*, vol. 194, no. 1, pp. 119–126.

Spirin, AS, Belitsina, NV & Yusupova, GZ 1988, 'Ribosomal synthesis of polypeptides from aminoacyl-tRNA without polynucleotide template', *Methods in Enzymology*, vol. 164, pp. 631–649.

Spirin, AS, Kostiashkina, OE & Jonak, J 1976, 'Contribution of the elongation factors to resistance of ribosomes against inhibitors: Comparison of the inhibitor effects on the factor-dependent and factor-free translation systems', *Journal of Molecular Biology*, vol. 101, no. 4, pp. 553–562.

Stark, H, Rodnina, MV, Wieden, HJ, van Heel, M, & Wintermeyer, W 2000, 'Large scale movement of elongation factor G and extensive conformational change of the ribosome during translocation', *Cell*, vol. 100, no. 3, pp. 301–309.

Stark, H, Rodnina, MV, Wieden, HJ, Zemlin, F, Wintermeyer, W & van Heel, M 2002, 'Ribosome interactions of aminoacyl-tRNA and elongation factor Tu in the codon-recognition complex', *Nature Structural Biology*, vol. 9, no. 11, pp. 849–854.

Taylor, DJ, Nilsson, J, Merrill, AR, Andersen, GR, Nissen, P & Frank, J 2007, 'Structures of modified eEF2 80S ribosome complexes reveal the role of GTP hydrolysis in translocation', *The EMBO Journal*, vol. 26, no. 9, pp. 2421–2431.

Trabuco, LG, Villa, E, Mitra, K, Frank, J & Schulten, K 2008, 'Flexible fitting of atomic structures into electron microscopy maps using molecular dynamics', *Structure*, vol. 16, no. 5, pp. 673–683.

Trobro, S & Aqvist, J 2005, 'Mechanism of peptide bond synthesis on the ribosome', *Proceeding of the National Academy of Sciences of the USA*, vol. 102, no. 35, pp. 12395–12400.

Uemura S, Aitken CE, Korlach J, Flusberg BA, Turner SW, Puglisi JD 2010, 'Real-time tRNA transit on single translating ribosomes at codon resolution', *Nature*, vol. 464, no. 7291, pp. 1012–1017.

Vale, RD 2003, 'Myosin V motor proteins: marching stepwise towards a mechanism', *The Journal of Cell Biology*, vol. 163, no. 3, pp. 445–450.

Vale, RD & Oosawa, F 1990, 'Protein motors and Maxwell's demons: Does mechanochemical transduction involve a thermal ratchet?', *Advances in Biophysics*, vol. 26, pp. 97–134.

Valle, M, Sengupta, J, Swami, NK, Grassucci, RA, Burkhardt, N, Nierhaus, KH, Agrawal, RK & Frank, J 2002, 'Cryo-EM reveals an active role for aminoacyl-tRNA in the accommodation process', *The EMBO Journal*, vol. 21, no. 13, pp. 3557–3567.

Valle, M, Zavialov, AV, Sengupta, J, Rawat, U, Ehrenberg, M & Frank, J 2003a, 'Locking and unlocking of ribosomal motions', *Cell*, vol. 114, no. 1, pp. 123–134.

Valle, M, Zavialov, A, Li, W, Stagg, SM, Sengupta, J, Nielsen, RC, Nissen, P, Harvey, SC, Ehrenberg, M & Frank, J 2003b, 'Incorporation of aminoacyl-tRNA into the ribosome as seen by cryo-electron microscopy', *Nature Structural Biology*, vol. 10, no. 11, pp. 899–906.

Vasiliev, VD, Selivanova, OM & Koteliansky, VE 1978, 'Specific self-packing of the ribosomal 16S RNA', *FEBS Letters*, vol. 95, no. 2, pp. 273–276.

Vasiliev, VD, Serdyuk, IN, Gudkov, AT & Spirin, AS 1986, 'Self-organization of ribosomal RNA', in *Structure, Function, and*

Genetics of Ribosomes, eds B Hardesty & G Kramer, Springer-Verlag, New York, pp. 128–142.

Vasiliev, VD, & Zalite, OM 1980, 'Specific compact self-packing of the ribosomal 23S RNA', *FEBS Letters*, vol. 121, no. 1, pp. 101–104.

Villa, E, Sengupta, J, Trabuco, LG, LeBarron, J, Baxter, WT, Shaikh, TR, Grassucci, RA, Nissen, P, Ehrenberg, M, Schulten, KA & Frank, J 2009, 'Ribosome-induced changes in elongation factor Tu conformation control GTP hydrolysis', *Proceeding of the National Academy of Sciences of the USA*, vol. 106, no. 4, pp. 1063–1068.

Voorhees, RM, Weixlbaumer, A, Loakes, D, Kelley, AC & Ramakrishnan, V 2009, 'Insights into substrate stabilization from snapshots of the peptidyl transferase center of the intact 70S ribosome', *Nature Structural & Molecular Biology*, vol. 16, no. 5, pp. 528–533.

Wimberly, BT, Brodersen, DE, Clemons, WM Jr, Morgan-Warren, RJ, Carter AP, Vonrhein, C, Hartsch, T & Ramakrishnan, V 2000, 'Structure of the 30S ribosomal subunit', *Nature*, vol. 407, no. 6802, pp. 327–339 (also comment pp. 306–307).

Wintermeyer, W & Robertson, JM 1982, 'Transient kinetics of transfer ribonucleic acid binding to ther ribosomal A and P sites: observation of a common intermediate complex', 1982, *Biochemistry*, vol. 21, no. 9, pp. 2246–2252.

Yarus, M, Valle, M & Frank, J 2003, 'A twisted tRNA intermediate sets the threshold for decoding', *RNA*, vol. 9, no. 4, pp. 384–385.

Yusupov, MM, Yusupova, GZ, Baucom, A, Lieberman, K, Earnest, TN, Cate, JH & Noller, HF 2001, 'Crystal structure of the ribosome at 5.5 A resolution', *Science*, vol. 292, no. 5518, pp. 883–896.

Yusupova, G, Jenner, L, Rees, B, Moras, D & Yusupov, M 2006, 'Structural basis for messenger RNA movement on the ribosome', *Nature*, vol. 444, no. 7117, pp. 391–394.

Yusupova, GZ, Yusupov, MM, Cate, JH & Noller, HF 2001, 'The path of messenger RNA through the ribosome', *Cell*, vol. 106, no. 2, pp. 233–241.

Zhang, W, Dunkle, JA & Cate, JH 2009, 'Structures of the ribosome in intermediate states of ratcheting', *Science*, vol. 325, no. 5943, pp. 1014–1017.

CHAPTER 10

The GroEL/GroES Chaperonin Machine

Arthur L. Horwich
Helen R. Saibil

I. INTRODUCTION

I.1 Chaperonins – Discovery of the Machines and their Action in Assisting Protein Folding to the Native State

Chaperonin machines are large ring assemblies that mediate ATP-dependent protein folding to the native state by binding and folding proteins in the cavities of their rings. They are present in the cytosol of organisms from all three kingdoms of life and are present also in chloroplasts and mitochondria, Eukaryotic organelles that are endosymbiotically related to Eubacteria. Their biological action in assisting protein folding is essential – deletion of these components is lethal.

The possibility of a protein "folding machine" was entertained by Anfinsen and coworkers as early as 1963 (Epstein et al., 1963) and subsequently considered by others (e.g., Rothman and Kornberg, 1986). The course of experiments that demonstrated such a component was not a linear one, however. In the early 1970s, a role was identified for a Bacterial operon known as *groE* in enabling productive phage infection of Bacteria. In particular, genetic deficiency in this locus led to an accumulation of aggregated phage head "monsters" inside infected *E. coli*, suggesting a role for this operon in phage particle assembly (Georgopoulos et al., 1972; Takano and Kakefuda, 1972). A broader role, however, in cellular metabolism was suggested by the observation that *groE* mutant cells grew poorly even in the absence of phage infection. In the late 1970s, electron microscopy studies of a purified product of the *groE* operon, the "large" component called GroEL, revealed a remarkable double-ring architecture, with rings composed of seven identical subunits surrounding a central "hole" (Hendrix, 1979; Hohn et al., 1979).

In a parallel path of study in a plant system, a role was identified for large protein complex called the "Rubisco subunit binding protein" inside chloroplasts in facilitating the assembly of the hexadecameric CO_2-fixing enzyme Rubisco (Barraclough and Ellis, 1980). In particular, the binding protein was found to associate with newly synthesized large subunits of Rubisco, translated inside the chloroplast, but was absent from the mature Rubisco assembly formed by association of large subunits with imported cytosolically synthesized small subunits, implicating the

binding protein in oligomeric assembly. In the late 1980s, the predicted sequence of the Rubisco binding protein from cloned DNA revealed a close match to the similarly analyzed primary sequence of the subunit of Bacterial GroEL (Hemmingsen et al., 1988). Thus a shared role for these components in oligomeric protein assembly was entertained.

Independently, however, studies of a mutant yeast strain, *mif4* (mitochondrial import function 4), unable to fold proteins newly imported into the mitochondrial matrix compartment, revealed a GroEL homolog called Hsp60 (composed of a modestly heat-inducible 58 kDa subunit) to be affected (Cheng et al., 1989). Hsp60 was already known to be present as a double-ring assembly of seven-membered rings, resembling GroEL (McMullin and Hallberg, 1988). Here the action on newly imported polypeptides, which were known to be translocated across the organellar membranes as non-native monomers (Eilers and Schatz, 1986), implicated Hsp60 and, by homology, its relatives in mediating protein folding to the native state. Indeed, in the mitochondrial import studies, a protein that is a monomer during its lifetime in the matrix compartment, the Rieske iron-sulfur protein, was affected in its biogenesis, failing to undergo sequential proteolytic processing steps, presumably because of misfolding and aggregation. Subsequent in vitro reconstitution studies with purified GroEL and the co-chaperonin GroES, employing several aggregation-prone proteins including homodimeric Rubisco (Goloubinoff et al., 1989) and monomeric rhodanese (Martin et al., 1991; Mendoza et al., 1991), further established the role of the chaperonins in mediating polypeptide chain folding. In these reconstitution studies, the chaperonin reaction was carried out in two steps. In the first step, the substrate protein diluted from denaturant became bound to GroEL with a stoichiometry of one substrate molecule per GroEL particle, with bound substrate exhibiting a non-native state as evidenced by lack of enzyme activity and high protease susceptibility. In the second step, ATP and GroES were added. This led to renaturation of the substrate protein over a period of minutes, associated with release from GroEL.

In the case of oligomeric proteins, it has become clear that the chaperonin mediates folding of the component subunits as monomers; released monomers then undergo subsequent steps of oligomeric assembly independently of the chaperonin. This has left open the question, however, of whether the cell also contains assembly factors that assist the process of oligomerization. Indeed, such specialized factors have been identified for a number of oligomeric proteins, including Rubisco (Liu et al., 2010). In general, it seems that such factors function as transient scaffolds, stabilizing partial assemblies prior to final assembly steps.

The hydrolysis of ATP by GroEL was recognized early by Hendrix (1979). Subsequently, Georgopoulos and coworkers detected mutations in a second gene affecting lambda phage head assembly, identified inter-genic suppression between that gene and GroEL, and showed that the gene encoded a small protein called GroES (Tilly et al., 1981). In a later study (Chandrasekhar et al., 1986), they purified overproduced GroES and observed that it was a single ring assembly that could physically associate with GroEL in the presence of MgATP, with such association slowing the rate of ATP turnover by GroEL.

A negative-stain EM reconstruction resolved the two major domains of GroEL subunits and a hinge-like interconnection between them (Saibil et al., 1993). Initial cryo-EM and X-ray structures of GroEL were presented (Braig et al., 1994; Chen et al., 1994). The X-ray structure at 2.8 Å resolution revealed the architecture of the Bacterial chaperonin in an unliganded state. This was followed shortly after by EM and X-ray structures of GroEL complexed with the "lid" structure GroES (Roseman et al., 1996; Xu et al., 1997), informing about the nature of the machine movements and the folding-active state that encapsulates a substrate protein. A host of functional and structural studies over the years, carried out by many groups, involving both the Bacterial chaperonin system (Type I, also in mitochondria and chloroplasts) and a second family of chaperonins in the Eukaryotic cytosol and Archaebacteria (Type II, lacking an independent lid component, employing instead a built in one), have given us a relatively coherent view of the structure and action of these machines. The Bacterial GroEL/GroES system is particularly well analyzed and is the focus of this review. Because we have reviewed the nature of GroEL/GroES action on polypeptide substrates in considerable detail recently (Horwich and Fenton, 2009), this review focuses on the behavior of GroEL/GroES as a molecular machine, delineating the working parts and how ATP binding allosterically drives a ring from a binding-active to folding-active state. We include discussion of an ATP/GroES directed "power stroke" that occurs during this transition, a truly mechanical step required for productive release of substrate protein. We also review how ATP hydrolysis in a GroES-bound ring functions as a timer that ends a folding cycle of that ring and directs the opposite ring to become folding-active.

I.2 Overall Action

Before considering the machine in detail, it is of value to briefly summarize its overall action. Fundamentally, as Anfinsen and his coworkers unearthed in the late 1950s (e.g., White and Anfinsen, 1959), polypeptide chains contain all of the information required for reaching the native state, typically lying at an energetic minimum (Anfinsen, 1973). However, under cellular conditions of relatively high temperature and solute concentration, there is a propensity of folding polypeptides, particularly ones with more than one domain, to lodge in kinetic traps, for example, where side chains or incipient secondary structures have wrongly docked with one another. Such kinetically trapped species are liable to multi-molecular aggregation. Hugh Pelham was first to recognize that a heat shock protein, Hsp70, could bind incipiently aggregating species, via their exposed hydrophobic surfaces, preventing or reversing aggregation under heat shock conditions (Pelham, 1986). The role, more generally, of molecular chaperones as a class, including the chaperonin ring assemblies, is to prevent or reverse the formation of kinetically trapped aggregation-prone states, in particular by binding polypeptide chains, typically in monomer form, via hydrophobic surfaces specifically exposed in non-native states through solvent-accessible hydrophobic surfaces in the chaperone itself. Such binding protects a "substrate" protein from further misfolding and from multi-molecular protein aggregation that can occur through the exposed hydrophobic surfaces or through domain swapping. Upon subsequent release of substrate protein from the chaperone, typically triggered by the action of ATP binding, a further attempt can be made at reaching a biologically productive state.

The chaperonin ring assemblies are remarkable among the various molecular chaperones insofar as they release substrate proteins into an encapsulated chamber that enables productive folding. We know now that GroEL provides kinetic assistance to the folding process essentially by using ATP binding and hydrolysis to direct alternation between two major conformational states: a binding-active one, where substrate protein binds in the open ring of the chaperonin assembly, and a folding-active one, in which substrate protein is released into a now GroES-encapsulated ring where it folds in isolation. Both binding- and folding-active states serve to provide kinetic assistance to the folding process – that is, minimizing "errors" made by folding polypeptide chains.

II. INVENTORY OF PARTS OF THE GROEL/GROES MACHINE

II.1 GroEL Architecture

The GroEL double ring is a cylindrical structure composed of 14 identical 58 kDa subunits arranged as two

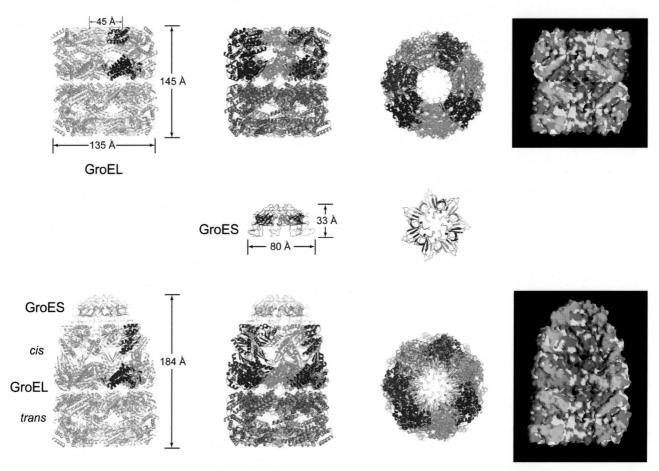

FIGURE 10.1: *Crystallographic models of GroEL (top row; PDB 1OEL), GroES (middle row; taken from PDB 1AON), and GroEL/GroES complex (bottom row; PDB 1AON). Leftmost panels of GroEL and GroEL/GroES highlight one of the identical subunits for domain arrangement: equatorial, blue; apical, red; intermediate, green. Center panels illustrate oligomeric arrangement colored by subunits in one ring, side, and end views. Right-hand panels, cutaway views removing front three subunits of GroEL and GroEL/GroES, respectively, revealing the central cavities with surfaces colored by hydrophobicity/hydrophilicity. Cut surface is in gray. Residues with hydrophobic side chains are colored yellow; residues with electrostatic or polar side chains are colored blue. Other residues are colored white. (Reproduced from Horwich et al., 2006, with permission of the American Chemical Society)*

back-to-back 7-membered rings (Figure 10.1, top row panels; Braig et al., 1994; Chen et al., 1994; Braig et al., 1995). The unliganded assembly measures ~135 Å in diameter and ~145 Å in height. Each subunit is composed of two large domains: equatorial at the waistline of the cylinder and apical at the distal end, covalently interconnected by a slender intermediate domain that is "hinged" at its top (apical) and bottom (equatorial) aspects. The collective of equatorial domains, contacting each other both side by side within a ring and bottom to bottom across the equatorial plane, serves as a "platform." The equatorial domain contains the ATP binding site (Boisvert et al., 1996; Chaudhry et al., 2003), and cooperative ATP binding in the seven equatorial domains of a ring (Yifrach and Horovitz, 1995) leads to rigid-body movements of the intermediate and apical domains of that ring (Figure 10.2a; Ranson et al., 2001), as well as small movements of the equatorial domains themselves, which program negative cooperativity

(Yifrach and Horovitz, 1995) between the rings, discussed later in this chapter. The movements within an ATP-bound ring involve downward rotation of the intermediate domains associated with small elevation and twisting movements of the apical domains (Figure 10.2; Ranson et al., 2001). These movements lead to recruitment of the co-chaperonin "lid" structure, GroES, and this is followed by a large rigid-body apical domain elevation and twisting movement that produces a domed GroES-bound ring structure, housing a chamber that is ~65 Å in diameter (Figure 10.1, bottom-row panels). These latter apical movements are attended by ejection of substrate polypeptide, which had been initially captured by the hydrophobic cavity surface of an open ring (Figure 10.1, top right panel), into a now hydrophilic encapsulated cavity where productive folding can occur (Weissman et al., 1995; Mayhew et al., 1996; Weissman et al., 1996; Rye et al., 1997; Xu et al., 1997).

a

Unliganded GroEL

D398A GroEL-ATP

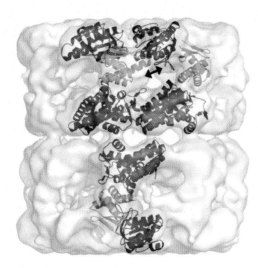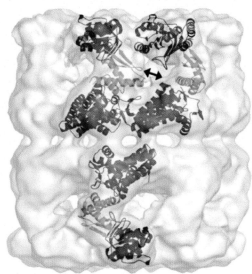

b

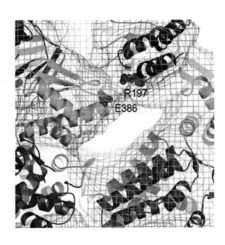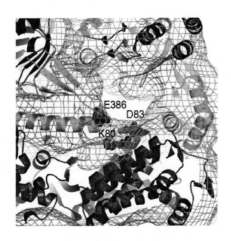

T

R

FIGURE 10.2: *Effect of cooperative ATP binding in the seven subunits of one of the two GroEL rings. Rigid-body movements reposition the apical and intermediate domains of the ATP-bound ring, determined from cryo-EM studies of unliganded GroEL and an ATP-bound hydrolysis-defective mutant D398A. The domain coloring is as in Figure 10.1: equatorial blue, intermediate green, and apical red. Panel (a), Left, GroEL in an unliganded state, electron density map from cryo-EM study with fitted crystallographic model of 3 unliganded GroEL subunits; double-headed arrow indicates contact between intermediate domain of left hand subunit and apical domain of right-hand one. Right, GroEL in the presence of ATP, map from D398A/ATP and GroEL model fit by rigid-body domain movements. Note that the intermediate domains of the ATP-bound ring (top ring) have rotated downward and that the apical domains have rotated upward. Double-headed arrow designates new contact between intermediate domain of left-hand subunit with equatorial domain of right-hand subunit. Panel (b), Left, Detail of a portion of the unliganded ring ("T" state, i.e., with low affinity for ATP), with density from cryo-EM and fitted crystallographic model showing a salt bridge between E386 from the intermediate domain of one subunit and R197 from the apical domain of its neighbor. Right, Detail of the same region of the ATP bound ring ("R" state), with cryo-EM density and rigid body fitted GroEL domains, showing that downward rotation of the intermediate domain has rearranged the salt bridge, with E386 now contacting the region around K80/D83 in the neighboring equatorial domain. Note how the long α-helix carrying 386 (left edge of the image) has closed over the occupied equatorial nucleotide pocket, positioning aa D398, distal in the same α-helix, into the nucleotide pocket to serve as a catalytic base for ATP hydrolysis. (Adapted from Ranson et al., 2001, with permission of Elsevier)*

II.2 Equatorial Domains

Considered at the level of structural detail, the equatorial domains are composed of long, roughly parallel α-helices that run tangentially with respect to the ring and are tilted, at an angle of 20–30°, relative to the equatorial plane (Figure 10.1; Braig et al., 1994). The ATP binding pocket itself has a helix-turn-helix phosphate-binding motif with the primary sequence GDGTT, differing from standard Walker (A)-type ATPases, which have GX_4GKT and usually feature a strand-turn-α helix (Ramakrishnan and Ramasarma, 2002). The chaperonin nucleotide pocket houses bound nucleotide entirely within the subunit (Boisvert et al., 1996). The aspartate residue in the GDGTT motif is present in the turn segment and coordinates a magnesium ion that contacts oxygens from all three phosphates of ATP (Figure 10.3; Boisvert et al., 1996; Chaudhry et al., 2003). Additional contacts with the phosphate oxygens are made by two other equatorial loops. These also coordinate a potassium ion that interacts with the oxygens, consistent with dependence of ATPase activity on this cation (Viitanen et al., 1990; Todd et al., 1993; Kiser et al., 2010). These various contacts stabilize the phosphate groups but are also potential mediators of allosteric signaling from the nucleotide pocket. For example, the D52-containing loop connects two antiparallel β-strands that contribute to a β-sheet with two strands from the neighboring subunit in the ring (Braig et al., 1994, Figure 10.4 therein). The loop has been observed to adjust its position upward in the presence of bound ATP (Boisvert et al., 1996; Xu et al., 1997). Such adjustments may contribute to the concerted (MWC) cooperativity reported for ATP binding within the GroEL ring, with a Hill coefficient of approximately 3 (Yifrach and Horovitz, 1995).

Hydrolysis of ATP bound in the equatorial nucleotide pocket is catalyzed by D398, the catalytic base, located in the intermediate domain (Figure 10.3; Rye et al., 1997; Xu et al., 1997). Downward rotation of the intermediate domain (25°) swings D398 into the nucleotide pocket where it can activate a water molecule to drive hydrolysis of the γ-phosphate (Chaudhry et al., 2003; see also Ditzel et al., 1998). Mutation of this residue to alanine has only small effect on affinity for ATP but reduces the rate of hydrolysis to ~2% of wild-type (Rye et al., 1997).

As mentioned, the equatorial domains make subunit-subunit contacts with each other across the ring-ring interface, which mediate negative cooperativity between the rings. Negative cooperativity programs the asymmetric behavior of the chaperonin assembly because GroES can only bind to an ATP-occupied ring (whose apical domains have been ATP-mobilized to associate with the cochaperonin; see below). The inter-ring contacts between equatorial domains present a staggered arrangement (Figure 10.4; Braig et al., 1994). That is, each GroEL subunit proffers two contacting regions at the equatorial interface. One lies at the end of so-called helix D (the second helix in the helix-turn-helix), which runs from the site of the γ-phosphate in the nucleotide pocket down to the ring-ring interface. At the interface, helix D forms a homotypic contact with a helix D coming from a subunit in the opposite ring. The contact involves a van der Waals surface as well as a hydrophobic contact. The other inter-ring contact involves a short loop segment that also forms a homotypic contact with the like structure coming from a second subunit in the opposite ring, here also involving a van der Waals surface but also a number of electrostatic contacts. The contacts between rings are adjusted in the presence of ATP, with the helix D-helix D contact becoming separated in distance (as compared with the unliganded state) and altered in angle so that it is considerably weakened while the other contact remains relatively fixed (Ranson et al., 2006). Presumably, the helix D adjustment between rings accounts for negative cooperativity, either by the repositioning of the distal helix D or via alteration of dipole-dipole interactions of the helices.

II.3 Apical Domains

A collective of seven apical domains forms the distal opening of each GroEL ring. The apical domains contain a continuous hydrophobic polypeptide-binding surface at the inside aspect of an unliganded ring (Figure 10.1, top right panel). Hydrophobic side-chains facing into solvent are lined up on a tier of three horizontally oriented secondary structures, helix H at the inlet to the cavity, helix I below it, and an underlying extended segment (Figure 10.5; Braig et al., 1994). Interestingly, all of the hydrophobics on the top two structures are aliphatics, whereas the underlying segment contains aromatics, Y203 and F204. Mutation of any of these residues to hydrophilic character abolishes substrate protein binding (Fenton et al., 1994). This, of course, involves mutation of all seven apical domains around a ring. Further studies, however, using covalently produced rings from a seven-fold tandemized coding sequence (forming double rings in the expressing cells) revealed that three consecutive wild-type apical domains are required for efficient substrate protein binding (Farr et al., 2000). This suggests that substrate binds to a continuous cavity surface of contiguous subunits. This notion was supported by covalent crosslinking studies, showing that substrates could simultaneously crosslink to three to four subunits via the apical domains. In addition, direct observation by cryo-EM of substrate in complex with GroEL revealed density from substrate docked against three or four contiguous apical domains (Elad et al., 2007). In the axial dimension, substrate density was mainly located at the level of helix I and the underlying segment (see also Clare et al., 2009).

The same cavity facing hydrophobic residues that mediate capture of non-native substrate also forms contacts with GroES, as revealed by initial mutagenesis studies showing

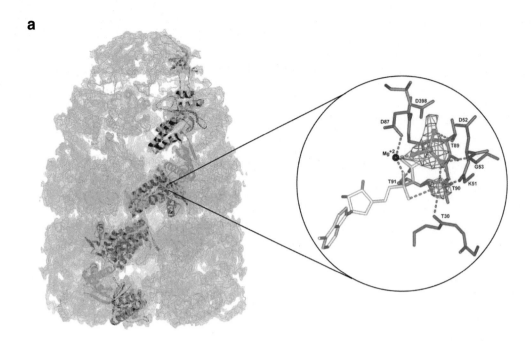

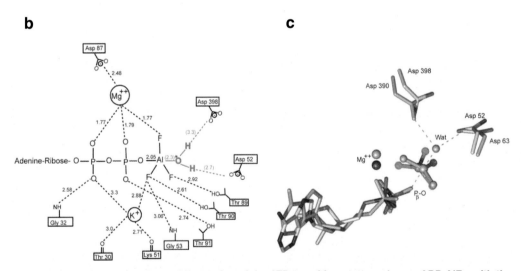

FIGURE 10.3: *Stereochemistry of interaction of the ATP transition-state analogue, ADP-AIF$_x$, with the equatorial nucleotide pocket of a GroES-bound GroEL ring (from PDB 1SVT). (a) View of the entire assembly, with single subunits in the* cis *and* trans *rings colored by domains as in Figure 10.1, and a detailed view of the nucleotide pocket in the* cis *ring, showing the nucleotide-metal complex within the equatorial pocket, with intermediate residue D398, which acts as a catalytic base, swung into the pocket by downward rotation; difference densities are shown for AIF$_3$ and a K$^+$ ion. (b) Schematic of the contacts in the nucleotide pocket. Surrounding residues form direct contacts with the phosphate oxygens (fluorines in the case of the aluminum metal complex which simulates the γ-phosphate), but also indirect ones via two cations. Also shown is an ordered water molecule positioned for in-line attack and side chains of D398 and D52 that are positioned to extract a proton for activation. The D398A mutation has a far stronger effect to reduce hydrolysis than D52A. (c) Stick diagrams of the Asp side chains in contact with the water molecule, comparing their positioning in GroEL/GroES with that of the corresponding side chains (D63 and D390) in the Archaeal thermosome (Ditzel et al., 1998; PDB 1AGE). Thermosome structures are in gray. (Reproduced from Chaudhry et al., 2003, with permission of Nature Publishing Group.)*

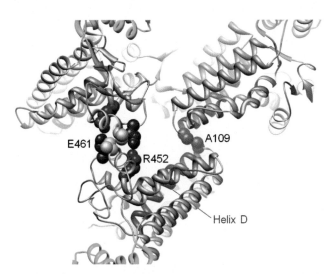

FIGURE 10.4: *Inter-ring contacts through two sites at the base of each equatorial domain, occurring in a staggered arrangement between the two GroEL rings, as viewed from the outside of a GroEL double ring, taken from unliganded GroEL (PDB 1OEL). One contact involves symmetric electrostatic interactions between E461 and R452 (side chains as colored balls) with the same residues in a subunit of the opposite ring. The other contact involves α-helices D (pink), extending from the nucleotide pocket of subunits in the respective rings to contact each other via A109. These two contacts provide the lines of allosteric communication between the two rings. Recent studies indicate that the 461 contact is a relatively fixed one, which serves as a fulcrum for movement, whereas the distances and angles between the α-helices D are altered by ATP binding and hydrolysis.*

a virtual 1:1 correspondence of loss of substrate binding with loss of ability to bind GroES (Fenton et al., 1994). X-ray structures of GroEL/GroES indeed show that the hydrophobic binding surface of each apical domain moves from facing the cavity in the unliganded state of GroEL to a 60° elevated and 90° clockwise twisted position in the nucleotide/GroES-bound state (compare apical domains top and bottom left-hand panels, Figure 10.1; Xu et al., 1997; Chaudhry et al., 2003). The mobilized hydrophobic residues form contacts with the mobile loops of GroES via helix H and form interapical contacts via helix I and the underlying segment (see Xu et al., 1997, Figure 10.5 therein). Thus, the cavity-facing hydrophobic surface of the unliganded state is effectively removed from facing the cavity during the transition to the folding-active state.

II.4 Intermediate Domains

The intermediate domains are slender interconnections formed from antiparallel segments of the GroEL polypeptide chain, ascending from the equatorial to apical domain and then descending from apical to equatorial (see Figure 10.1, left-hand panels). The domains are "hinged" at the bottom and top aspects by a pro137-gly410 and gly192-gly375 pair, respectively, allowing rigid-body movements

of the intermediate and apical domains to occur relative to each other and to the equatorial domains. The intermediate domains can make contact with nucleotide bound in the pocket in the top of the equatorial domain, thereby stabilizing their downward rotation in the ATP-liganded state (e.g. Chaudhry et al., 2003). Asp 398 in the intermediate domain, at one end of a long "descending" α-helix, makes direct contact with the γ-phosphate of ATP (Figure 10.3), but additional contacts of this α-helix (helix M; 388–408) with the equatorial domain of the same subunit are also formed, as seen in the X-ray structure of GroEL/GroES/ADP-AlF$_x$ (Chaudhry et al., 2003), including several electrostatic contacts at the proximal end.

III. THE GROEL/GROES REACTION CYCLE

III.1 Overview

The GroEL/GroES reaction cycle is summarized in Figure 10.6, with the major intermediate states illustrated. Panel [1] shows an asymmetric GroEL/GroES/ADP$_7$ complex, which is the predominant state for accepting ATP and non-native polypeptide (into the unoccupied so-called *trans* ring) during a cycling reaction (Rye et al., 1999). (Note that an unliganded GroEL complex probably never exists under physiologic conditions, where there are millimolar concentrations of ATP and an approximately two-fold molar excess of GroES over GroEL, even though such a complex has been the starting point for many in vitro biochemical experiments.) In panel [2], ATP has rapidly bound in the equatorial domains of the unoccupied *trans* ring, followed

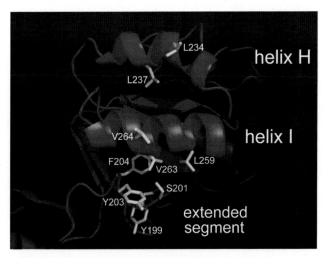

FIGURE 10.5: *Structure of the polypeptide binding apical surface of a GroEL subunit, as viewed from the inside of a GroEL ring. Ribbons diagram showing the three major secondary structures, α-helix H, α-helix I, and an underlying extended segment, each with hydrophobic side chains protruding into solvent, that have been shown by mutagenesis (Fenton et al., 1994) to be crucial for substrate polypeptide binding. (Reproduced from Horwich et al., 2007, with permission of Annual Reviews.)*

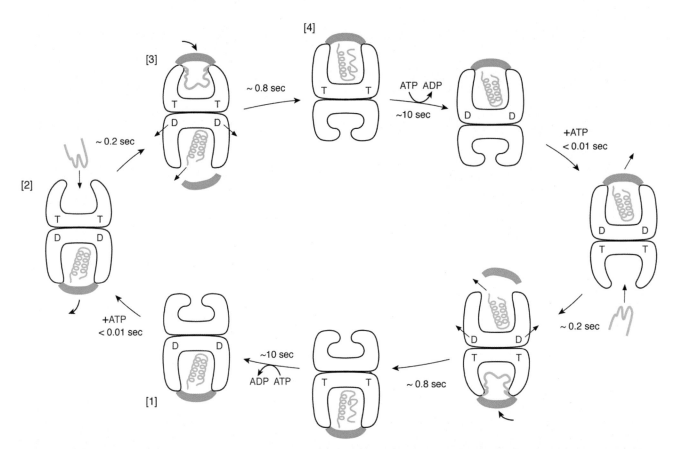

FIGURE 10.6: *GroEL/GroES reaction cycle. Trajectory of the rings of the GroEL assembly through binding-active, folding-active, and unliganded states, directed by a sequence of ATP binding and hydrolysis. GroEL shown as black outline and nucleotide-bound state of a ring is designated either T, for occupancy with ATP, or D, for occupancy with ADP. GroES illustrated as a blue "lid" structure, polypeptide as a green "squiggle." See text for detailed discussion of the cycle. Note that the cooperativity of ATP binding within a ring, anticooperativity of ATP binding between rings, and requirement for ATP occupancy to enable GroES binding dictate the asymmetric behavior of the machine. The asymmetric GroEL/GroES/ADP state shown in panel [1] is the ligand-accepting state, which rapidly binds ATP followed by non-native polypeptide ([1] and [2]) in its open ring. GroES collision with the mobilized apical domains follows ([3]), leading to further large apical movements and formation of the domed folding-active complex [4], in which polypeptide attempts to reach the native state in the cis ring. This is the longest-lived state of the reaction cycle. ATP hydrolysis then gates entry of ATP into the opposite ring, which allosterically discharges the ligands from the cis ring and sets up a folding-active complex in the newly bound ring.*

in a slower binding step by non-native substrate polypeptide binding to the apical domains (Tyagi et al., 2009). This order of binding reflects both the relative binding rate constants (Tyagi et al., 2009) and the large difference in concentration between ATP ($\sim$5 mM) and substrate polypeptides (<1 μM). ATP binding ([2]→[3]) brings about (or stabilizes) a 25° rotation of the collective of intermediate domains in the bound ring down onto the nucleotide pockets, associated with breakage of the collective of salt bridges between Glu386 in the intermediate domain of each subunit and Arg197 in the neighboring apical domain (Figure 10.2b; Ranson et al., 2001). The latter breakage of bonds frees the seven apical domains. During this process, ATP binding also results in the allosterically directed discharge of the ligands from the opposite ring ([3]; Rye et al., 1997). Initial GroES association [3] is followed by major rigid-body movements ([3]→[4]) of the apical domains,

which result in the release of the substrate protein from the hydrophobic substrate-binding surface into the now-encapsulated central cavity of the ring, where it commences folding.

The forceful movement that follows initial GroES association in the presence of ATP constitutes the power stroke of the chaperonin machine, driven by the energy of binding ATP and GroES (Chaudhry et al., 2003). Such a forceful movement is required for ejection of polypeptide off of the cavity wall, with ATP alone or ADP/GroES unable to accomplish this, that is, unable to rapidly and efficiently release "stringent," GroEL/GroES-dependent, substrate proteins into the chamber (Weissman et al., 1996; Motojima et al., 2004). Yet the action of ATP binding in producing an initial movement of the apical domains – that is, before GroES binds – is likely to contribute already to such a subsequent forceful release. For example,

ATP-directed movement is likely to differ from movement in ADP, both with respect to cooperative behavior (ADP is non-cooperative; Cliff et al., 1999), rate of movement of the apical domains, and potentially trajectory, but the latter points remain to be directly examined. Nevertheless, in the present discussion we refer to the power stroke as that phase of the cycle subsequent to where GroES has initially collided with the ATP-mobilized apical domains.

Polypeptide substrate is released during the rigid-body movements of the GroES/ATP-associated power stroke phase ([3]→[4]), and folding proceeds inside the *cis* chamber of the GroEL/GroES/ATP folding-active complex that is formed. The folding-active phase, the duty cycle of the machine, is the longest-lasting of the reaction cycle, with a $t_{1/2}$ ~10–15 s (Rye et al., 1999). The phase is ended by concerted ATP hydrolysis in the seven equatorial domains, which weakens the affinity for GroES and allosterically adjusts the equatorial domains of the *trans* ring to permit ATP binding (fifth panel; Rye et al., 1997; Ranson et al., 2001). ATP and non-native polypeptide bind (<1 s) to the *trans* ring (sixth panel), allosterically discharging the ligands, ADP, polypeptide, and GroES, from the *cis* ring (Rye et al., 1999). At this point, the non-native polypeptide may have folded to its native state and will no longer be a substrate; alternatively, it may not have folded and, after release, will remain a substrate for another round of binding and attempted folding (Todd et al., 1994; Weissman et al., 1994). Upon GroES encapsulation, what had been the *trans* ring now becomes folding-active. The GroEL system continues this alternating cycle, with ATP binding providing the driving force for the machine's movements, at once discharging the previously folding-active ring and nucleating a new one. ATP hydrolysis ensures the directionality of the cycle.

Details of the domain movements during this cycle, their relationship to polypeptide and GroES binding, and the routes of allosteric communication within the complex are discussed in more detail in the sections that follow.

III.2 ATP Binding
A cryo-EM study of GroEL in the ATP-bound state, using the D398A mutant to greatly reduce the rate of hydrolysis, yielded an approximate picture of the domain rotations in ATP (Ranson et al., 2001). However, it was clear that there was disorder in that data set, and that the resolution and interpretation were limited by the averaging of multiple dynamic states. This issue has been re-examined in more recent work, using a much larger data set and statistical methods for discriminating structural variations from orientational variations in single-particle data sets, in order to sort out multiple, coexisting structures (Elad et al., 2008; Orlova and Saibil, 2010). Multiple GroEL conformations were found to coexist in the prehydrolysis state (in the

absence of GroES), in accordance with kinetic data suggesting multiple kinetic intermediates (see later in this chapter). Three conformations were resolved with one ring occupied by ATP, and a further three with both rings occupied. The trajectory of motions can be deduced by placing these states in the order that gives the smallest steps in a series of progressive domain rotations (Clare, Vasishtan, Topf, Farr, Horwich, and Saibil, unpublished).

The initial event at the subunit level, occurring upon cooperative binding of ATP to a GroEL ring, is the closing of the long alpha helix of the intermediate domain, helix M, containing the catalytic base for ATP hydrolysis (D398), over the equatorial nucleotide binding pocket. This movement was predicted early by Karplus and coworkers, carrying out a normal-mode analysis (Ma and Karplus, 1998). The closing of the M helix brings with it rigid body movements of both the intermediate and apical domains. These movements break a salt bridge that interlocks the intermediate and apical domains of neighboring subunits of the unliganded machine and replace it with a new intermediate-to-equatorial contact. This switch contributes substantially to producing the shift of the ring from a "closed" T state (with low affinity for ATP) to an "open" R state (with high affinity for ATP). In particular, the salt bridge between E386 at the end of the M helix and R197 on the apical domain of the neighboring subunit (ccw when the ring is viewed from outside the complex) is replaced by a new contact from E386 to the top surface of the neighboring equatorial domain, likely at K80 (Figure 10.2b; Ranson et al, 2001). With breakage of the contact between apical residue R197 and intermediate E386, the apical domains are freed for a small rigid-body elevation towards a conformation that docks GroES.

The ATP-mediated movements of the apical domains could have several possible functional implications. On the one hand, given recent observations that polypeptide binding is favored by an ATP-bound open ring, the freeing of the apical domains and their excursion through a range of states of elevation and tilt could account for facilitated recognition of the ensemble of different non-native states that appears to be characteristic of the protein substrates that GroEL recognizes. Formally, however, the timescale of apical elevation and tilt is unknown, specifically whether it occurs during the first 20 ms of ATP binding or on a slower timescale of ~200 ms (see review of kinetic studies below). If apical movement occurs at the latter relatively slow rate, it could be possible that non-native states of polypeptide that become multi-valently bound prior to the full excursion of the apical domains are subject to stretching and unfolding action produced by the further elevation and tilting of the freed domains relative to each other. It can be noted that, in experiments where ATP is added to an already polypeptide-bound ring (the reverse of the physiologic order) a stretching of substrate has been measured by FRET (Lin et al., 2008; Sharma et al., 2008).

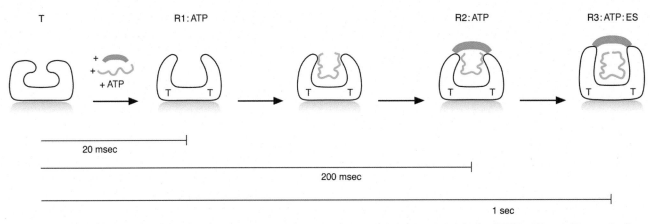

FIGURE 10.7: *Progression of a GroEL ring, as deduced from kinetic and EM studies, during the ATP/GroES reaction cycle, from unliganded (T state; opposite an ADP/GroES bound ring) to GroES-bound folding-active (R3) (see text). ATP binds rapidly to the ring (20 msec), associated with downward rotation of the intermediate domains onto the nucleotide pocket (R1). The apical domains of the ATP-bound ring undergo elevation as revealed by EM studies. The timescale of this movement may be rapid, given tryptophan fluorescence intensity changes that are as rapid as ATP binding, but may potentially occur over a longer time, corresponding to fluorescence changes occurring on the 200 msec scale, associated with non-native polypeptide binding (which is ~10 times slower than ATP binding). By ~200 msec, as indicated from fluorescence studies, GroES collides with the elevated apical domains (R2). Following the collision, further rigid body movement of the apical domains, corresponding to a further fluorescent phase, produces the domed folding-active cis cavity (R3). (Reproduced from Horwich and Fenton, 2009, with permission of Cambridge University Press.)*

At whatever point of apical elevation polypeptide binding occurs, it is indeed this step, prior to arrival of GroES, that has been the locus in the reaction cycle where release of substrate protein from a kinetically trapped state has been demonstrated. A variety of early studies have supported such an action (Zahn and Plückthun, 1994; Ranson et al., 1995; Bhutani and Udgaonkar, 2000). Such action has more recently been directly observed in an elegant study of Lin and Rye (2004), who placed fluorophores at either end of the primary structure of Rubisco and followed FRET signals first of the kinetically trapped protein while free in solution, then during binding by GroEL, and finally during GroES encapsulation. The investigators carried out their study at 4°C, where Rubisco free in solution did not reach the native state but also did not aggregate, that is, it became kinetically trapped as a monomer. They observed that, even at this low temperature, GroEL/GroES could efficiently bind and refold this kinetically trapped state to the native form. They observed in parallel FRET measurements that binding to GroEL was associated with rapid stretching of the kinetically trapped Rubisco monomer, on the seconds timescale that would correspond to an interval prior to GroES binding (see also Sharma et al., 2008). Concerning the conformational effect on substrate, hydrogen-deuterium studies of MDH, Rubisco, and DHFR in particular (Chen et al., 2001; Park et al., 2005; Horst et al., 2007) would indicate that non-native substrate free in solution lacks detectable stable secondary structure. Yet at the level of tertiary structural topology, substrates appear to be capable of becoming kinetically trapped, and the "stretching" observed constitutes a likely step of release from such a state.

The conformational changes of ATP binding are also transmitted to the less mobile equatorial domains – the presence of ATP in the binding pocket is signaled to the inter-ring interface through helix D, which, as mentioned earlier, runs directly from the site of the γ-phosphate to the ring-ring interface to make a homotypic contact with the same helix from a subunit of the opposite ring (Figure 10.4). This contact is distorted upon ATP binding, likely accounting for inhibiting ATP binding in the second ring (negative cooperativity) (Ranson et al, 2001).

III.3 Correlates of ATP-Directed Movement from Fluorescence and Mutational Studies

The foregoing domain movements of GroEL during ATP and GroES binding can be correlated with fluorescence changes reported by fluorescently labeled GroEL molecules. Because GroEL and GroES are devoid of tryptophan, substitution of this residue in innocuous positions in either the equatorial domain (W485) or apical domain (W231) can be used to monitor movements of the machine (Taniguchi et al., 2004; Cliff et al., 2006; see also W44 substitution, Yifrach and Horovitz, 1998). The studies of the Clarke and Kawata groups report, remarkably, the same kinetic phases from these distinct positions (summarized in Figure 10.7). The earliest phase associated with ATP binding occurs within 20 ms and involves an increase of fluorescence intensity. This phase is unaffected by the presence of GroES and occurs much faster than the recently measured binding of polypeptide. The state produced, termed R1, likely corresponds to the initial downward rotation of the intermediate domain onto the ATP binding pocket in the equatorial domain. The second fluorescent phase, over 200 ms, involves quenching of fluorescence, and this phase occurs on the timescale of polypeptide binding but precedes GroES binding. It produces a state, R2, that marks the point of initial GroES binding. That is, the amplitude of the R1-to-R2 transition is affected by GroES, and rates of two subsequent phases of fluorescence change are substantially affected by GroES. The nature of the R1-to-R2 transition remains unclear. It could conceivably correspond to the elevation and tilting of the free apical domains, although such a movement is not presently excluded from occurring at the T-to-R1 early transition.

The R2 state thus appears to correspond to a GroES collision state in which the elevated apical domains are recognized by the GroES mobile loops, docking the cochaperonin. Presumably, the loops can bind to the same apical domains as polypeptide, via helix H, if bound polypeptide substrate is localized more deeply in the cavity on helix I and the underlying segment. Alternatively, the mobile loops may only initially bind to apical domains that are not occupied with substrate protein (typically three or four consecutive domains). Further EM studies may be able to directly observe the GroES collision state as well as location of substrate, but it seems clear that this state, whatever its structural arrangement, will need to accommodate the efficient encapsulation of substrate protein without its "leakage" into the bulk solution.

III.4 Beyond Initial GroES Association – A Power Stroke of Apical Domain Movement Ejecting Substrate into the *cis* Folding Chamber

The initial contact of GroES with an ATP-bound GroEL ring is immediately followed by a further large rigid-body movement of the apical domains (Figure 10.6). The binding of GroES produces a subsequent forceful rigid body movement, a power stroke, that serves to completely remove the apical hydrophobic polypeptide binding sites from facing the central cavity, driving non-native polypeptide substrate off of the sites. This step of ejection allows the polypeptide to commence folding in the encapsulated *cis* chamber, whose surface is hydrophilic in character, in particular, rich in charged and polar side chains.

It was noted early that under so-called stringent conditions, i.e. where a substrate protein cannot fold without GroEL/GroES assistance, both ATP and GroES are absolutely required to support production of the native state. Interestingly, even though ADP, as well as ATPγS and AMP-PNP, permit GroES binding and substrate protein encapsulation in the *cis* cavity, these nucleotides do not lead to productive folding. Indeed, fluorescence anisotropy experiments showed that substrates remained bound to the cavity wall when ADP/GroES was added instead of ATP/GroES. For example, starting with binary complexes of pyrene-labeled rhodanese and GroEL, a rapid drop of fluorescence anisotropy was observed when ATP/GroES was added, but no change occurred when ADP/GroES was added (Weissman et al., 1996). Surprisingly, however, X-ray crystallography and cryo-EM reconstructions of the *cis* rings of GroEL/GroES/ADP indicated no differences from GroEL/GroES/ATP complexes (GroEL/GroES/ADP-AlF$_x$ had to suffice for an ATP analogous state in the X-ray studies). What, then, is the difference between these complexes that enables activity in release and protein folding of one but not the other?

One obvious difference is the presence of the γ-phosphate in ATP, which provides additional binding interaction with the nucleotide pocket as compared with, for example, ADP, and, hence, more available binding energy. The role of the γ-phosphate has been explored in detail, taking advantage of the observation that adding aluminum fluoride (or beryllium fluoride) to preformed complexes with ADP and GroES triggered productive folding (Chaudhry et al., 2003). The combination of such metal complexes with ADP has been observed in other systems to mimic ATP function, by simulating intermediate states along the trajectory of ATP hydrolysis. A crucial observation in the GroEL system was that addition of AlF$_x$ after formation of a stable, but folding-inactive, GroEL-GroES-substrate-ADP complex now triggered release of polypeptide from the cavity wall followed by productive folding, which proceeded at the same rate as observed when ATP and GroES were added to a GroEL-polypeptide complex. With such a sequence of additions, it was thus possible to estimate separately the energetic contributions of ADP, GroES, and the γ-phosphate to the power stroke (Chaudhry et al., 2003).

In particular, using the single-ring version of GroEL, SR1, to avoid the complications of the second ring, the binding constant of ADP was determined by isothermal titration calorimetry to be 32.7 μM, which gives a free energy of -43 kcal/mole GroEL rings for binding seven ADPs per ring. The binding constant of GroES to SR1 was measured in the presence of ADP by a Hummel-Dreyer experiment, giving 0.4 μM and an energetic contribution of −9 kcal/mole. Finally, a competitive binding assay using AlF$_x$ and labeled BeF$_x$ binding to SR1-GroES-ADP allowed an estimate of 16 μM for the AlF$_x$ binding constant and of −46 kcal/mole GroEL rings free energy, again for 7 metal complexes per ring. These results show clearly that the γ-phosphate contributes about half of the total binding energy available to drive the motion of the chaperonin's power stroke. This was reflected at the level of the nucleotide pocket in the X-ray structure of an ADP•AlF$_3$-bound GroEL/GroES complex where each metal complex formed an array of hydrogen bonding contacts with the surrounding pocket (see Figure 10.3). Yet the architecture of the GroEL-GroES-ADP•AlF$_x$ complex outside the nucleotide pocket was not significantly different from that determined for GroEL-GroES-ADP, for example, comparing the local structure of the apical domains or the GroES mobile loops. The question thus remains: If there are no large-scale structural differences between the ADP and ATP GroEL/GroES complexes, what is the contribution of the γ-phosphate to the power stroke that drives polypeptide release and initiates protein folding? Is it the force of movement, directed along the same trajectory as taken in ADP, or could it be that the extent of movement is greater, with ATP going beyond the end-state of ADP, even if only transiently, as a yet-to-be observed "overshoot"?

One answer to this question has come from observations of the rates of movement of the apical domains in response to ADP/GroES versus ATP/GroES binding. By labeling one residue in the relatively fixed equatorial domain and a second one on the outside of the mobile apical domain of a subunit with a pair of FRET reporters, Motojima et al. (2004) were able to carry out stopped-flow mixing experiments to measure the rates of apical domain movement, as well as its extent, under a variety of conditions. Binding of ATP and GroES to an empty GroEL led to a rate constant for movement of about 2.7 s^{-1}. With ADP and GroES, the rate was the same, and the extent appeared similar as well. However, when the same experiments were carried out with GroEL or SR1 that had a substrate polypeptide, either MDH or rhodanese, bound to the apical domains, large differences were observed. With ATP and GroES, the rates for GroEL with MDH or rhodanese bound were almost as fast as with unliganded GroEL (0.69 s^{-1} and 0.59 s^{-1}, respectively), with a similar extent of movement. With ADP and GroES, however, the rates of movement of the GroEL apical domains were considerably slower (0.074 s^{-1} and 0.033 s^{-1}, respectively), and the extent of movement was reduced (that is, the full change of FRET observed for GroEL in the absence of rhodanese or in the presence of rhodanese with ATP/GroES was not obtained with rhodanese and ADP/GroES). For SR1-MDH or SR1-rhodanese, the effects were even larger, with almost no apparent movement in ADP/GroES. Thus, the inability of ADP to support polypeptide release and folding is reflected in the slower movement of the apical domains and reduced extent of movement. This indicated that the bound non-native polypeptide is a physical load on the apical domains that requires the additional energy provided by the γ-phosphate of ATP to support a forceful movement of the domains, which is necessary to drive the release of the polypeptide into the folding chamber. In support of this conclusion, further experiments adding AlF$_x$ metal complex to preformed GroEL/GroES/ADP/polypeptide complexes showed an immediate further rapid movement of the apical domains, with FRET change now producing the same extent as achieved with ATP (Motojima et al., 2004), consistent with the ability of the metal complex to support folding under these conditions.

In sum, it seems clear that the power stroke of the GroEL machine brings about the large-scale movements of the apical domains detailed above, and does so at a rate and extent that achieves the release of the bound substrate polypeptide. It is likely that the difference in the rate of movement produced by ATP (or ADP•AlF$_x$) versus ADP accounts for the observation that ATP supports productive folding whereas ADP does not.

The question of what such forceful, rapid movement achieves relative to conformation of the bound substrate has been an important consideration. Is it required to "stretch" or otherwise change the structure of the bound polypeptide prior to its release into the cavity? Or is it more a matter of moving the apical binding sites away from the chamber faster than the non-native polypeptide can accommodate, so as to release it? Both hydrogen-deuterium exchange studies of substrates in binary complexes with GroEL (reviewed in Horwich and Fenton, 2009) and TROSY NMR studies of ^{15}N-labeled DHFR while bound to GroEL (Horst et al., 2005), have addressed the question of structure in bound substrate. No stable secondary structure was observed and a dynamic ensemble of states appeared to be present. Thus there is not an apparent stable secondary structure in the substrates studied to date that could report on significant stretching if it occurs at this phase. Concerning tertiary structural topology, single-molecule FRET measurements between reporters at various pairs of positions in several substrates studied to date have indicated that "stretching" is not observed during the power stroke (Lin and Rye, 2004; Sharma et al., 2008); rather, a "compaction" of substrates is observed. Thus, it does not seem that the power stroke phase of the reaction cycle is employed to directly stretch the non-native substrate protein before its release into the *cis* cavity. Rather, it seems that the power stroke is employed to accomplish efficient release of substrate protein from the apical binding sites so that substrate can collapse into a productive conformation.

III.5 *Cis* ATP Hydrolysis – A Molecular Timer that Allosterically Gates Entry of ATP and Substrate Protein into the Open *trans* Ring

The longest-lived state of the GroEL/GroES reaction cycle is the GroEL/GroES/ATP complex, with a t$_{1/2}$ ~10 s. This folding-active *cis* complex, whose formation was just described, exhibits only low affinity of its open *trans* ring for non-native substrate or ATP (Rye et al., 1999). The complex is stable relative to GroEL/GroES/ADP complexes, which are readily dissociated, for example, by gel filtration in the absence of nucleotide. The ATP complex is stable under such conditions, and is also stable against exposure to low concentrations of denaturant (e.g., 0.4 M guanidine HCl; Rye et al., 1997). The lifetime of the stable ATP complex thus determines the period for productive folding in the *cis* chamber. The measured half-time before the occurrence of what has been observed to be concerted ATP

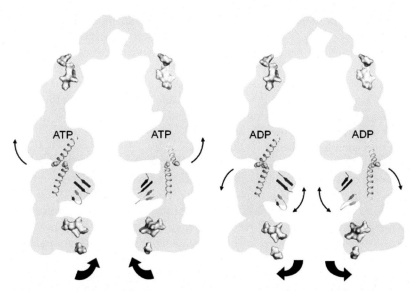

FIGURE 10.8: *Actions of* cis *ring ATP hydrolysis on the* trans *ring. Schematic illustration of sites of effects as derived from cryo-EM studies (Ranson et al., 2006). Hydrolysis in the* cis *ring produces an alignment of α-helices D between the subunits of the two rings (associated with pivoting on the other site of inter-ring contact [involving aa 461], not shown here). This is associated with tilting of the equatorial domains, leading to splitting of the 4-stranded beta sheet contact between subunits (purple and blue sheets) and with an altered attitude of the* trans *ring apical domains that enables binding of substrate protein. (Reproduced from Ranson et al., 2006, with permission of Nature Publishing Group.)*

hydrolysis (Yifrach and Horovitz, 1995) must represent an evolutionary compromise between the need to give even slowly folding substrate proteins sufficient time to reach native form and the timely release, in many cases to assemble into functional oligomeric proteins. Mechanistically, ATP hydrolysis, as a molecular timer, programs the dissociation of the *cis* complex and the establishment of a new complex on the opposite ring. In particular, at the level of the *cis* ring, ATP hydrolysis weakens the affinity of GroEL for GroES, and at the level of the *trans* ring, hydrolysis programs allosteric adjustments that gate the entry of ATP and substrate polypeptide. Relative to the *cis* ring, the entry of these ligands in *trans* serves to allosterically eject GroES and substrate protein (as well as ADP; note that free Pᵢ diffuses rapidly away [Jackson et al., 1993]), with ATP absolutely required and substrate polypeptide accelerating the process (Rye et al., 1999). With respect to the *trans* ring, the binding of ATP programs the rigid body movements that enable docking of GroES and favor substrate binding, and these events drive formation of a new folding-active complex, as described earlier.

Cryo-EM studies have been crucial to observing molecular events attendant to *cis* ATP hydrolysis, with information deriving from comparison of a hydrolysis-defective (D398A) GroEL/GroES/ATP asymmetric complex at 7.7 Å with a GroEL/GroES/ADP asymmetric complex at 8.7 Å (Ranson et al., 2006). The *cis* rings of the two complexes are practically superimposable, consistent with the notion that folding proceeds unimpeded in the *cis* cavity of the post-hydrolysis ADP complex despite its short lifetime (<1 s). By contrast, striking changes are observed in the *trans* ring. First, the ring-ring interface is altered. Whereas the 461 contact is preserved as a pivot-point (see Figure 10.4 for reference), the D helix contact is altered, effectively transiting from an offset arrangement in the ATP state to an aligned one in ADP (shown schematically in Figure 10.8). Presumably this adjustment accounts for gating of ATP into the *trans* ring, either by a structural shift or by a charge-based mechanism – for example, involving helix dipoles. The alignment of helices D is also associated with rotation of the *trans* ring equatorial domains (going from ATP to ADP) about the 461 pivot point. Such rotation in turn is associated with the splitting of a four-stranded beta sheet contact between neighboring subunits. In the unliganded and ATP-bound states, two antiparallel β-strands (aa 37–51) reach over from each subunit to stack on a two-stranded sheet formed by a short stretch of N-terminal and C-terminal sequence of the counterclockwise neighbor. This is a major inter-subunit contact in the oligomeric assembly. However, in the *cis* ADP complex, these pairs of strands become separated from each other through equatorial tilts, allowing rigid-body outward expansion of the entire *trans* ring, which opens out the apical domains (see Figure 10.8). This is the only state in which separation of the inter-subunit β-sheet contact is observed, and this conformation presumably accounts for, or contributes to, the destabilization of the *cis* ring and shift of the *trans* ring to a substrate acceptor state.

These structural adjustments begin to provide a mechanistic explanation of what happens upon *cis* hydrolysis, but there still remain a number of interesting questions. Why is the affinity for GroES weakened by ATP hydrolysis if the *cis* rings are effectively isomorphic? One must presume that the γ-phosphate contacts are supplying binding energy that transduces to stability of GroES binding. At the subsequent step, it is unclear how ligand binding in *trans* leads to the rapid eviction of the *cis* ligands, in <1 s. What allosteric movements in the *cis* ring are responsible for the departure of GroES? Are they counterclockwise apical rotational movements, effectively reversing direction from the nucleotide/GroES-bound state? The release of ADP following intermediate domain elevation off of the nucleotide pocket appears to be nearly as rapid as release of GroES, as recent experiments would indicate (Tyagi et al., 2010).

IV. OUTSTANDING QUESTIONS

IV.1 Effects on Substrate Proteins

Whereas much is known about the GroEL/GroES reaction cycle, there is still discussion about the actions that facilitate the folding of substrate proteins to native form. There seems to be general agreement that the step of polypeptide binding to an open ring leads to stretching of substrate proteins that can relieve kinetically trapped topologies (Lin and Rye, 2004; Sharma et al., 2008). It also seems clear that such binding as well as encapsulation by GroES to form the GroEL/GroES *cis* folding chamber further forestalls aggregation that could occur in free solution. Whether there are additional effects of the *cis* chamber that could favor productive folding remains an area of uncertainty. It has been suggested that close confinement in a chamber could produce entropic effects that might accelerate productive folding, as posited by Chakraborty et al. (2010) to account for the behavior of a mutant form of maltose binding protein. However, Apetri and Horwich (2008) presented evidence for behavior of the chaperonin as an infinite dilution chamber during folding of the same protein. Other studies, on human DHFR and bovine rhodanese, report no difference in folding rates inside the chamber versus free in solution (Horst et al., 2007; Hofmann et al., 2010). The latter study is of particular interest because it examined a GroEL/GroES-dependent substrate protein at the level of single-molecule fluorescence (FRET), following the kinetics of domain formation, observing that the overall rate of folding was the same at high dilution in free solution and inside a stable GroEL(SR1)/GroES folding chamber. Nevertheless, the folding of a fast-folding domain of rhodanese was slowed by half inside the chaperonin chamber, so it may be that the chamber walls have variable effects on

different domains and substrate proteins. Thus it seems that the chaperonin chamber consistently prevents aggregation of proteins but may exhibit variable effects from substrate to substrate on rates of recovery of the native state of at least some domains. Additional single-molecule studies should be able to further resolve the behavior of different substrate proteins.

IV.2 Proteins that are Not Assisted

At an extreme are substrate proteins that cannot be assisted by GroEL/GroES/ATP, despite encapsulation, such as actin and tubulin. In contrast, these proteins are efficiently folded by the type II chaperonin, CCT (chaperonin containing TCP1), which normally assists them in the Eukaryotic cytosol, in the presence of ATP (e.g. Tian et al, 1995). We have little understanding of what features of these proteins or of the GroEL machinery prevent the proteins from reaching the native state during cycling in the Bacterial type I system.

IV.3 Collision of GroES with GroEL

Although the recent study of ATP-mediated movements of the GroEL apical domains gives a clear indication of their trajectory, the point along the trajectory where GroES first contacts the domains and the topology of substrate proteins while still associated with the apical domains (before release into the *cis* cavity) remain unknown. We surmise, as discussed, that GroES only interacts with the apical domains after they have undergone a degree of elevation. Whether the mobile loops of GroES directly bind to substrate-associated apical domains or whether the loops can only associate with non-occupied apical domains remains unknown (see e.g., Miyazaki et al., 2002; Nojima et al., 2008). Can GroES directly compete for the same apical binding sites? However this may be, it seems that GroES binding occurs by a mechanism that prevents substrate proteins from escaping the cavity into free solution.

ACKNOWLEDGMENTS

We thank Wayne Fenton for critical reading of the manuscript and Dan Clare for assistance with preparation of figures. A.H. thanks HHMI and H.R.S. thanks the Wellcome Trust for generous support of this work.

REFERENCES

Anfinsen, C.B. (1973) Principles that govern the folding of protein chains. *Science* 181, 223–230.

Apetri, A.C., and Horwich, A.L. (2008) Chaperonin chamber accelerates protein folding through passive action of preventing aggregation. *Proc. Natl. Acad. Sci. USA* 105, 17351–17355.

Barraclough, R., and Ellis, R.J. (1980) Protein synthesis in chloropolasts. IX. Assembly of newly-synthesized large subunits into ribulose bisphosphate carboxylase in isolated intact pea chloroplasts. *Biochim. Biophys. Acta* 608, 19–31.

Bhutani, N., and Udgaonkar, J.B. (2000) A thermodynamic coupling mechanism can explain the GroEL-mediated acceleration of the folding of barstar. *J. Mol. Biol.* 297, 1037–1044.

Boisvert, D.C., Wang, J., Otwinowski, Z., Horwich, A.L., and Sigler, P.B. (1996) Structure of GroEL with bound ATPγS at 2.4 Å. *Nature Struct Biol* 3, 170–177.

Braig, K., Adams, P.D., and Brunger, A.T. (1995) Conformational variability in the refined structure of the chaperonin GroEL at 2.8 Å *resolution. Nature Struct. Biol.* 2, 1083–1094.

Braig, K., Otwinowski, Z., Hegde, R., Boisvert, D., Joachimiak, A., Horwich, A.L. and Sigler, P.B. (1994) The crystal structure of the bacterial chaperonin GroEL at 2.8 Å. *Nature* 371, 578–586.

Chakraborty, K., Chatila, M., Sinha, J., Shi, Q., Poschner, B.C., Sikor, M., Jiang, G., Lamb, D.C., Hartl, F.U., and Hayer-Hartl, M. (2010) Chaperonin-catalyzed rescue of kinetically trapped stats in protcin folding. *Cell* 142, 112–122.

Chandrasekhar, G.N., Tilly, K., Woolford, C., Hendrix, R., and Georgopoulos, C. (1986) Purification and properties of the groES morphogenetic protein of Escherichia coli. *J. Biol. Chem.* 261, 12414–12419.

Chaudhry, C., Farr, G.W., Todd, M.J., Rye, H.S., Brunger, A.T., Adams, P.D., Horwich, A.L., and Sigler, P.B. (2003) Role of the γ-phosphate of ATP in triggering protein folding by GroEL-GroES: Function, structure, and energetics. *EMBO J.* 22, 4877–4887.

Chen, J., Walter, S., Horwich, A.L., and Smith, D.L. (2001) Folding of malate dehydrogenase inside the GroEL-GroES cavity. *Nature Struct. Biol.* 8, 721–728.

Chen, S., Roseman, A.M., Hunter, A.S., Wood, S.P., Burston, S.G., Ranson, N.A., Clarke A.R., and Saibil, H.R. (1994) Location of a folding protein and shape changes in GroEL-GroES complexes imaged by cryo-electron microscopy. *Nature* 371, 261–264.

Cheng, M.-Y., Hartl, F.-U., Martin, J., Pollock, R.A., Kalousek, F., Neupert, W., Hallberg, E.M., Hallberg, R.L., and Horwich, A.L. (1989) Mitochondrial heat shock protein HSP60 is essential for assembly of proteins imported into yeast mitochondria. *Nature* 337, 620–625.

Clare, D.K., Bakkes, P.J., van Heerikhuizen, H., Van Der Vies, S.M., and Saibil, H.R. (2009) Chaperonin complex with a newly folded protein encapsulated in the folding chamber. *Nature* 457, 107–110.

Cliff, M.J., Kad, N.M., Hay, N., Lund, P.A., Webb, M.R., Burston, S.G., and Clarke A.R. (1999) A kinetic analysis of the nucleotide-induced allosteric transitions of GroEL. *J. Mol. Biol.* 293, 667–684.

Cliff, M.J., Limpkin, C., Cameron, A., Burston, S.G., and Clarke, A.R. (2006) Elucidation of steps in the capture of a protein substrate for efficient encapsulation by GroE. *J. Biol. Chem.* 281, 21266–21275.

Ditzel, L, Löwe, J., Stock, D., Stetter, K.O., Huber, H., Huber, R., Steinbacher, S. (1998) Crystal structure of the thermosome, the archaeal chaperonin and homolog of CCT. *Cell* 93, 125–138.

Eilers, M. and Schatz, G. (1986) Binding of a specific ligand inhibits import of a purified precursor protein into mitochondria. *Nature* 322, 228–232.

Elad, N., Clare, D.K., Saibil, H.R., and Orlova, E.V. (2008) Detection and separation of heterogeneity in molecular complexes by statistical analysis of their two-dimensional projections. *J. Struct. Biol.* 162, 108–120.

Elad, N., Farr, G.W., Clare, D.K., Orlova, E.V., Horwich, A.L. and Saibil, H.R. (2007) Topologies of a substrate protein bound to the chaperonin GroEL. *Mol Cell* 26, 415–426.

Epstein, C.J., Goldberger, R.F., and Anfinsen, C.B. (1963) The genetic control of tertiary protein structure: Studies with model systems. *Cold Spring Harb. Symp. Quant. Biol.* 28, 439–449.

Farr, G.W., Furtak, K., Rowland, M.C., Ranson, N.A., Saibil, H.R., Kirchhausen, T., and Horwich, A.L. (2000) Multivalent binding of non-native substrate proteins by the chaperonin GroEL. *Cell* 100, 561–573.

Fenton, W.A., Kashi, Y., Furtak, K., and Horwich, A.L. (1994) Residues in chaperonin GroEL required for polypeptide binding and release. *Nature* 371, 614–619.

Georgopoulos, C.P., Hendrix, R.W., Kaiser A.D., Wood, W.B. (1972) Role of the host cell in bacteriophage morphogenesis: Effects of a bacterial mutation on T4 head assembly. *Nature New Biol.* 239, 38–41.

Goloubinoff, P., Christeller, J.T., Gatenby, A.A., and Lorimer, G.H. (1989) Reconstitution of active dimeric ribulose bisphosphate carboxylase from an unfolded state depends on two chaperonin proteins and MgATP. *Nature* 342, 884–889.

Hemmingsen, S.M., Woolford, C., Van Der Vies, S.M., Tilly, K., Dennis, D.T., Georgopoulos, C.P., Hendrix, R.W., and Ellis R.J. (1988) Homologous plant and bacterial proteins chaperone oligomeric protein assembly. *Nature* 333, 330–334.

Hendrix R.W. (1979) Purification and properties of groE, a host protein involved in bacteriophage assembly. *J. Mol. Biol.* 129, 375–392.

Hofmann, H., Hillger, F., Pfeil, S.H., Hoffmann, A., Streich, D., Haenni, D., Nettels, D., Lipman, E.A., and Schuler B. (2010) Single-molecule spectroscopy of protein folding in a chaperonin cage. *Proc. Natl. Acad. Sci. USA* 107, 11793–11798.

Hohn, T., Hoh, B., Engel, A., Wurtz M., and Smith, P.R. (1979) Isolation and characterization of the host protein groE involved in bacteriophage lambda assembly. *J. Mol. Biol.* 129, 359–373.

Horst, R., Bertelsen, E.B., Fiaux, J., Wider, G., Horwich, A.L., and Wüthrich, K. (2005) Direct NMR observation of a substrate protein bound to the chaperonin GroEL. *Proc Natl Acad Sci USA* 102, 12748–12753.

Horst, R., Fenton, W.A., Englander, S.W., Wüthrich, K., and Horwich, A.L. (2007) Folding trajectories of human dihydrofolate reductase inside the GroEL-GroES chaperonin cavity and free in solution. *Proc. Natl. Acad. Sci. USA* 104, 20788–20792.

Horwich, A.L., Farr, G.W., and Fenton, W.A. (2006) GroEL-GroES-mediated protein folding. *Chem. Rev.* 106, 1917–1930.

Horwich, A.L. and Fenton, W.A. (2009) Chaperonin-mediated protein folding: using a central cavity to kinetically assist polypeptide chain folding. *Q. Rev. Biophys.* 42, 83–116.

Horwich, A.L., Fenton, W.A., Chapman, E., and Farr, G.W. (2007) Two families of chaperonin: Physiology and mechanism. *Annu. Rev. Cell Dev. Biol.* 23, 115–145.

Jackson, G.S., Staniforth, R.A., Halsall, D.J., Atkinson, T., Holbrook, J.J., Clarke, A.R., and Burston, S.G. (1993) Binding and hydrolysis of nucleotides in the chaperonin catalytic cycle: Implications for the mechanism of assisted protein folding. *Biochemistry* 32, 2554–2563.

Kiser, P.D., Lorimer, G.H., and Palczewski, K. (2010) Use of thallium to identify monovalent cation binding sites in GroEL. *Acta Cryst Sect F* 65, 967–971.

Lin, Z. and Rye, H.S. (2004) Expansion and compression of a protein folding intermediate by GroEL. *Mol. Cell* 16, 23–34.

Lin, Z., Madan, D., and Rye, H.S. (2008) GroEL stimulates protein folding through forced unfolding. *Nat. Struct. Mol. Biol.* 15, 303–311.

Liu C., Young, A.L., Starling-Windhof, A., Bracher, A., Saschenbrecker, S., Rao, B.V., Rao, K.V., Berninghausen O., Mielke, T., Hartl, F.U., Beckmann, R., and Hayer-Hartl, M. (2010) Coupled chaperone action in folding and assembly of hexadecameric Rubisco. *Nature* 463, 197–202.

Ma, J. and Karplus, M. (1998) The allosteric mechanism of the chaperonin GroEL: A dynamic analysis. *Proc. Natl. Acad. Sci. USA* 92, 8502–8507.

Martin, J., Langer, T., Boteva, R., Schramel, A., Horwich, A.L., and Hartl, F.-U. (1991) Chaperonin-mediated protein folding occurs at the surface of GroEL via a molten globule-like intermediate. *Nature* 352, 36–42.

Mayhew, M., da Silva, A.C.R., Martin, J., Erdjument-Bromage, H., Tempst, P. and Hartl, F.U. (1996) Protein folding in the central cavity of the GroEL-GroES chaperonin complex. *Nature* 379, 420–426.

McMullin, T.W. and Hallberg, R.L. (1988) A highly evolutionarily conserved mitochondrial protein is structurally related to the protein encoded by the Escherichia coli groEL gene. *Mol. Cell Biol.* 8, 371–380.

Mendoza, J.A., Lorimer, G.H., and Horowitz, P.M. (1991) Intermediates in the chaperonin-assisted refolding of rhodanese are trapped at low temperature and show a small stoichiometry. *J. Biol. Chem.* 266, 16973–16976.

Miyazaki, T., Yoshimi, T., Furutsu, Y., Hongo, K., Mizobata, T., Kanemori, M., and Kawata, Y. (2002) GroEL-substrate-GroES ternary complexes are an important transient intermediate of the chaperonin cycle. *J. Biol. Chem.* 277, 50621–50628.

Motojima, F., Chaudhry, C., Fenton, W.A., Farr, G.W., and Horwich, A.L. (2004) Substrate polypeptide presents a load on the apical domains of the chaperonin GroEL. *Proc. Natl. Acad. Sci. USA* 101, 15005–15012.

Nojima, T., Murayama, S., Yoshida, M., and Motojima, F. (2008) Determination of the number of active GroES subunits in the fused heptamer GroES required for interactions with GroEL. *J. Biol. Chem.* 283, 18385–18392.

Orlova, E. and Saibil, H.R. (2010) Methods for three-dimensional reconstruction of heterogeneous assemblies. *Meth. Enzymol.* 482, 321–341.

Park, E.S., Fenton, W.A., and Horwich, A.L. (2005) No evidence for a forced-unfolding mechanism during ATP/GroES binding to substrate-bound GroEL: No observable protection of metastable Rubisco intermediate, or GroEL-bound Rubisco from tritium exchange. *FEBS Lett.* 579, 1183–1186.

Pelham, H.R. (1986) Speculations on the functions of the major heat shock and glucose-regulated proteins. *Cell* 46, 959–961.

Ramakrishnan, C., Dani, V.S., and Ramasarma, T. (2002) A conformational analysis of Walker motif A [GXXXXGKT(S)] in nucleotide-binding and other proteins. *Protein Eng. Des. Sel.* 15, 783–798.

Ranson, N.A., Clare, D.K., Farr, G.W., Houldershaw, D., Horwich, A.L., and Saibil, H.R. (2006) Allosteric signaling of ATP hydrolysis in GroEL-GroES complexes. *Nat. Struct. Mol. Biol.* 13, 147–152.

Ranson, N.A., Dunster, N.J., Burston, S.G., and Clarke, A.R. (1995) Chaperonins can catalyse the reversal of early aggregation steps when a protein misfolds. *J. Mol. Biol.* 250, 581–586.

Ranson, N.A., Farr, G.W., Roseman, A.M., Gowen, B., Fenton, W.A., Horwich, A.L. and Saibil, H.R. (2001) ATP-bound states of GroEL captured by cryo-electron microscopy. *Cell* 107, 869–879.

Roseman, A.M., Chen, S., White, H., Braig, K., and Saibil, H.R. (1996) The chaperonin ATPase cycle: Mechanism of allosteric switching and movements of substrate-binding domains in GroEL. *Cell* 87, 241–251.

Rothman, J.E., and Kornberg R.D. (1986) An unfolding story of protein translocation. *Nature* 322, 209–210.

Rye, H.S., Burston, S.G., Fenton, W.A., Beechem, J.M., Xu, Z., Sigler, P.B., and Horwich, A.L. (1997) Distinct actions of *cis* and *trans* ATP within the double ring of the chaperonin GroEL. *Nature* 388, 792–798.

Rye, H.S., Roseman, A.M., Furtak, K., Fenton, W.A., Saibil, H.R., and Horwich, A.L. (1999) GroEL-GroES cycling: ATP and non-native polypeptide direct alternation of folding-active rings. *Cell* 97, 325–338.

Saibil, H.R., Zheng, D., Roseman, A.M., Hunter, A.S., Watson, G.M., Chen, S., Auf der Mauer, A., O'Hara, B.P., Wood, S.P., Mann, N.H., Barnett, L.K. and Ellis, R.J. (1993) ATP induces large quaternary rearrangements in a cage-like chaperonin structure. *Cur.Biol.*3, 265–273.

Sharma, S., Chakraborty, K., Müller, B.K., Astola, N., Tang, Y.C., Lamb, D.C., Hayer-Hartl, M., Hartl, F.U. (2008) Monitoring protein conformation along the pathway of chaperonin-assisted folding. *Cell* 133, 142–153.

Takano, T, Kakefuda, T. (1972) Involvement of a bacterial factor in morphogenesis of bacteriophage capsid. *Nature New Biol.* 239, 34–37

Taniguchi, M., Yoshimi, T., Hongo, K., Mizobata, T., and Kawata, Y. (2004) Stopped-flow fluorescente análisis of the conformational changes in the GroEL apical domain. *J. Biol. Chem.* 279, 16368–16376.

Tian, G., Vainberg, I.E., Tap, W.D., Lewis, S.A., and Cowan, N.J. (1995) Specificity in chaperonin-mediated protein folding. *Nature* 375, 250–253.

Tilly, K., Murialdo, H. & Georgopoulos, C. (1981) Identification of a second *Escherichia coli* groE gene whose product is necessary for bacteriophage morphogenesis. *Proc. Natl. Acad. Sci. USA* 78, 1629–33.

Todd, M.J., Viitanen, P.V., and Lorimer, G.H. (1993) Hydrolysis of adenosine 5'-triphosphate by Escherichia coli GroEL: effects of GroES and potassium ion. *Biochemistry* 32, 8560–8567.

Todd, M.J., Viitanen, P.V., and Lorimer, G.H. (1994) Dynamics of the chaperonin ATPase cycle: Implications for facilitated protein folding. *Science* 265, 659–666.

Tyagi, N., Fenton, W.A., and Horwich, A.L. (2009) GroEL/GroES cycling: ATP binds to an open ring before substrate protein favoring protein binding and production of the native state. *Proc. Natl. Acad. Sci. USA* 106, 20264–20269.

Tyagi, N.K., Fenton, W.A., and Horwich, A.L. (2010) ATP-triggered ADP release from the asymmetric chaperonin complex GroEL/GroES/ADP7 is not the rate-limiting step of the GroEL/GroES reaction cycle. *FEBS Lett.* 584, 951–953.

Viitanen, P.V., Lubben, T.H., Reed, J., Goloubinoff, P., O'Keefe, D.P., and Lorimer, G.H. (1990) Chaperonin-facilitated refolding of ribulosebisphosphate carboxylase and ATP hydrolysis by chaperonin 60 (groEL) are K$^+$ dependent. *Biochemistry* 29, 5665–5671.

Weissman, J.S., Hohl, C.M., Kovalenko, O., Chen, S., Braig, K., Saibil, H.R., Fenton, W.A., and Horwich, A.L. (1995) Mechanism of GroEL action: Productive release of polypeptide from a sequestered position under GroES. *Cell* 83, 577–587.

Weissman, J.S., Kashi, Y., Fenton, W.A., and Horwich, A.L. (1994) GroEL-mediated protein folding proceeds by multiple rounds of release and rebinding of non-native forms. *Cell* 78, 693–702.

Weissman, J.S., Rye, H.S., Fenton, W.A., Beechem, J.M., and Horwich, A.L. (1996) Characterization of the active intermediate of a GroEL-GroES-mediated protein folding reaction. *Cell* 84, 481–490.

White, F.H., and Anfinsen, C.B. (1959) Some relationships of structure to function in ribonuclease. *Ann. N.Y. Acad. Sci.* 81, 515–523.

Xu, Z., Horwich, A.L., and Sigler, P.B. (1997) The crystal structure of the asymmetric GroEL-GroES-(ADP)7 chaperonin complex. *Nature* 388, 741–751.

Yifrach, O. and Horovitz, A. (1995) Nested cooperativity in the ATPase activity in the oligomeric chaperonin GroEL. *Biochemistry* 34, 9716–9723.

Yifrach, O. and Horovitz, A. (1998) Transient kinetic analysis of adenosine 5′ triphosphate binding-induced conformational changes in the allosteric chaperonin GroEL. *Biochemistry* 37, 7083–7088.

Zahn, R. and Plückthun, A. (1994) Thermodynamic partitioning model for hydrophobic binding of polypeptides by GroEL: II. GroEL recognizes themally unfolded mature β-lactamase. *J. Mol. Biol.* 242, 165–174.

CHAPTER 11

ATP Synthase – A Paradigmatic Molecular Machine

Thomas Meier

José D. Faraldo-Gómez

Michael Börsch

I. ATP AND F_1F_O-ATP SYNTHASES

I.1 ATP – An Energy-Rich Compound with a Long History

Phosphorylation of ribose sugars is central to life in its present form as well as throughout evolution. This reaction chemically activates sugars and hence plays a major role in the transmission of information and energy conservation. Nature has chosen adenosine-5′-triphosphate (ATP) as a widely used energy source in a variety of cellular energy-converting processes. A few but important examples are the anabolic and catabolic biochemical pathways, solute and ion transport (osmotic work), and mechanical work (e.g., muscle contraction or cell motility).

ATP was first described by the German chemist Karl Lohmann in 1929, who isolated it from muscle and liver extracts (Langen and Hucho, 2008). The first chemical synthesis of ATP outside a living cell was performed by the Nobel Laureate Lord Alexander Robertus Todd in 1949 (Baddiley et al., 1949). Already in 1935, the Russian scientist Vladimir Engelhardt noted that muscle contraction requires ATP. Two years later, the Danish scientist Herman Moritz Kalckar established that ATP synthesis is linked with cell respiration and that ATP represents the final product of the catabolic reaction. In the years 1939–1941, Fritz Lipmann showed that ATP is the main bearer of chemical energy in the cell. He coined the phrase "energy-rich phosphate bonds" (Lipmann, 1941). The reason for this expression lies in the structure of ATP.

In the structure of ATP (Figure 11.1) three phosphates are bound in a row, of which the first one (α) forms a 5′-ester bond with the ribose. The second (β) and third (γ) bond chemically represent anhydride bonds. The hydrolysis of the γ anhydride bond is chemically slow in the absence of enzymes but is rapid in their presence, which makes the constellation of a triphosphate group connected with an adenosine group attractive for biochemical reactions (Westheimer, 1987). The cleavage of the third phosphate is exergonic; the energy released is considerably high ($\Delta G^{0\prime} = -32.3$ kJ/mol), and it is even higher under conditions found in living cells ($\Delta G^0 \approx -50$ kJ/mol).

I.2 The F_1F_O-ATP synthase

I.2.1 Discovery and Classification. Most of ATP produced in cells is synthesized by the enzyme F_1F_O-ATP synthase (other names: F-type ATP synthase, ATPase, Complex V, ion-driven rotary motor). The amount of ATP synthesis in a living creature can be impressively high. In a human body, for instance, each cell contains about a billion ATP molecules, and every day the body requires approximately its own weight of this compound for its basic life processes.

The discovery of the enzyme ATP synthase dates back to the 1960s. By inspecting sub-mitochondrial particles using transmission electron microscopy (Figure 11.2), Ephraim Racker and his research colleagues discovered lollipop-like shapes that were distributed along one side of the inner membranes of bovine heart mitochondria (Figure 11.2b).

The water-soluble components of the ATP synthase (today known as the F_1 sub-complex) were shown to participate in oxydative phosphorylation (Penefsky et al., 1960, Pullman et al., 1960). These particles of ~9 nm diameter (Figure 11.2a) were originally called elementary particles and were thought to contain the entire respiratory apparatus of the mitochondrion. The Racker lab was finally able to experimentally prove that this particle is involved with ATP hydrolysis activity, and suggested that it is also associated with ATP synthesis (Kagawa and Racker, 1966b). From these experiments, different types of ATPases were defined: The name F_1 historically goes back to a "soluble fraction" (Penefsky et al., 1960, Pullman et al., 1960), which was characterized as the ATP-hydrolyzing fraction. The term F_O was introduced on account of the biochemical sensitivity of the ATP-hydrolyzing fraction to the ATP synthase inhibitor *oligomycin* (Kagawa and Racker, 1966a).

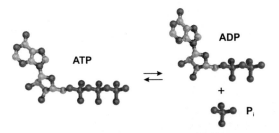

FIGURE 11.1: *Chemical structure of ATP (adenosine triphosphate) and the hydrolysis of ATP to ADP (adenosine diphosphate) and P$_i$ (phosphate). ATP hydrolysis requires an H$_2$O (not shown). The cleavage of the anhydride bond releases ΔG^0 ≈ − 50 kJ/mol under living cell conditions.*

To date (2011), several other ATPase families have been identified (AAA-/ A(anion)-/ A(archaea)-/ E(1–2)-/ F-/ P(1–5)-/ V-type), some of which are also ion-driven rotary motors similar to the F-type ATP synthases (A-/F-/ V-types): F-type ATPases/synthases are the classical ATP-generating enzymes using ion-motive forces (H$^+$ or Na$^+$). Built as molecular machines with a rotor-stator principle, they are able to perform both ATP synthesis and ATP hydrolysis (for reviews, see Section I.2.2). V-type ATPases are vacuolar ATPases that serve as proton pumps in the acidification of vacuoles in eukaryotes. Some V-ATPases are also found in bacterial systems. Their function is generally ATP hydrolysis (Forgac, 2007), but ATP synthesis has been shown to be also possible for some V-type ATPases of plants and yeast (Gambale et al., 1993; Kettner et al., 2003). Finally, A-type ATP synthases are found in Archaea. These are also rotor-stator molecular machines and share features of both the V-type and the F-type enzymes. Like the F-type, they are able to operate in ATP synthesis and hydrolysis direction, but they are more closely related to the V-type enzymes (Cross and Müller, 2004).

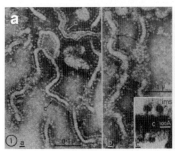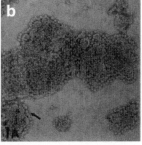

FIGURE 11.2: *Discovery of lollipop-like shape of ATP synthase. (a) Transmission electron microscopy (EM) images of inner membranes (cristae) from (rat liver) mitochondria. (b) Beef heart submitochondrial particles visualized by transmission EM and characterized as the enzyme catalyzing oxidative phosphorylation (Reproduced with permission from (a) Parsons, 1963; copyright AAAS and (b) Kagawa and Racker, 1966b).*

I.2.2 The F-ATP Synthase is an Ion-Driven Rotary Machine Composed of Two Motors.

The F$_1$F$_o$-ATP synthase is a nano-sized rotary molecular machine that comprises two opposing motors (for some recent reviews of different aspects of this enzyme, see Boyer, 1997a; Stock et al., 2000; Nakanishi-Matsui and Futai, 2006; Senior, 2007; Feniouk and Yoshida, 2008; Nakamoto et al., 2008; von Ballmoos et al., 2008; Junge et al., 2009; von Ballmoos et al., 2009; Hicks et al., 2010). One of the motors is the membrane-embedded F$_o$ complex and the other is the water-soluble F$_1$ complex. The two motors are driven by two different energy sources. When ATP is hydrolyzed by the F$_1$ motor, the released energy drives ions to be pumped through the F$_o$ complex, against the membrane gradient. When F$_1$ and F$_o$ are coupled in the operational holo-enzyme, the F$_1$F$_o$ complex functions as an ATP synthase that manufactures ATP powered by energy stored in a transmembrane electrochemical gradient (ion motive force) of either proton (Mitchell, 1961) or Na$^+$ (Dimroth, 1997). Depending on the physiological demand for ATP in the cell, the ATP synthase can reverse its operation direction (ATP synthesis or ATP hydrolysis).

The two motors, F$_1$ and F$_o$, are tightly coupled and can exchange energy with each other by a rotating the connecting structure (or "central stalk"), which acts as a camshaft. The two motors are also held together by a peripheral static connection or "outer stalk". Therefore, from a mechanical point of view, the enzyme can be divided into two distinct motor parts, a rotor and a stator (Capaldi and Aggeler, 2002).

The evolution of ATP synthases happened early in the history of life. Evolutionary considerations indicate that this family of enzymes originated from membrane-bound RNA translocases (Mulkidjanian et al., 2007). Apparently, the basic mechanistic principle of these molecular motors has been preserved since then. Figure 11.3 shows a cartoon representation of the composition of the bacterial form of this molecular machine, which is the most simplified version of the enzyme.

The subunit composition and stoichiometry of the water-soluble F$_1$ portion in the functional core is universally α$_3$β$_3$γδε. The molecular masses of the single subunits (e.g., in the bacterium *Escherichia coli*) are 55.3 kDa (α), 50.3 kDa (β), 31.6 kDa (γ), 19.3 (δ), and 15.0 (ε) (Walker et al., 1984). F$_1$ harbors three catalytic sites for ATP synthesis, located within three β-subunits (see Section I.4). F$_1$ interacts with the F$_o$ complex embedded in the cell membrane (bacterial plasma membrane; mitochondrial inner membrane, and chloroplast thylakoid membrane). In the most simplified version (bacterial type), F$_o$ comprises three subunits in the stoichiometry ab$_2$c$_n$ with the following molecular masses (*E. coli*): 30.3 (a), 17.2 (b) and 8.3 (c) (Kanazawa et al., 1981; Walker et al., 1984). The enzyme's rotor consists of subunits γεc$_n$ and the stator comprises

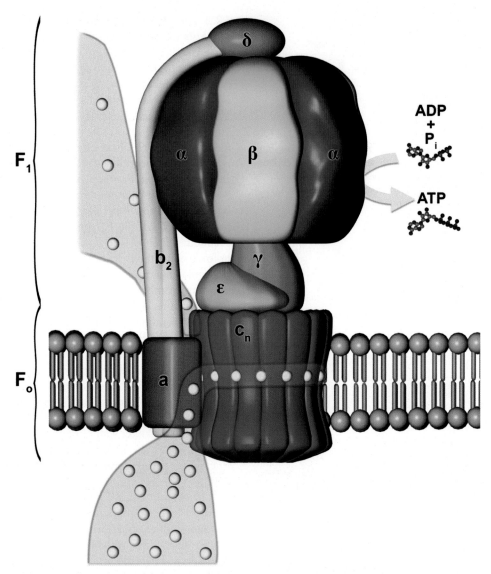

FIGURE 11.3: Schematic structural organization of the molecular machine F_1F_o-ATP synthase. The ATP synthase consists of the water-soluble F_1 and the membrane-embedded F_o complex. The F_o and F_1 complexes each represent a molecular motor, which are able to exchange energy by a rotational coupling mechanism. The F_1 domain, with the three catalytic sites on the three β-subunits (light green), is connected via the γ and ε subunits (blue, central stalk) to the c-oligomer (blue, in the membrane) of the F_o domain, and via the outer stalk δ (orange) and b_2 (yellow) subunits to subunit a (red). Mechanistically the enzyme can be divided into the rotor, which consists of subunits $c_{8-15}\gamma\varepsilon$ (in blue), and the stator, assembled from subunits $ab_2\alpha_3\beta_3\delta$ (green and orange/yellow). During ATP synthesis, H^+ (or Na^+ ions) pass the F_o complex, causing the rotor to rotate. Rotation is transduced into the F_1 headpiece ($\alpha_3\beta_3$) by the inherently asymmetric γ-subunit and causes conformational changes within the β-subunits. This process finally leads to the formation (and release) of ATP from ADP and P_i. Taken together, in the ATP synthesis mode, this molecular machine converts electrical (ion gradient) to mechanical (rotor) to chemical (ATP) energy. The enzyme is also able to operate in opposite direction, to act as an uphill ion pump by the use of the energy released by hydrolysis of ATP. The figure was created by Paolo Lastrico, Max-Planck-Institute of Biophysics.

Table 11.1: Subunit Composition, Stoichiometry, and Nomenclature of F_1F_o-ATP Synthases Found in Bacteria, Mitochondria, and Chloroplasts (adapted from Nakamoto et al., 2008)

Complex, Function / Origin	Bacteria	Mitochondria	Chloroplast
F_1 stator	α_3	α_3	α_3
	β_3	β_3	β_3
F_1 stator, peripheral stalk	δ	OSCP*	δ
F_1 rotor, central stalk	γ	γ	γ
	ε	δ	ε
		ε	
F_o stator	a	a ($= 6$), e, f, g, A6L ($= 8$, A6)	a ($=$ IV)
F_o stator, peripheral stalk	b_2 or bb′	b_2, d, F6	bb′ ($=$ I)
F_o rotor, central stalk	c_n**	c_n (or 9_n)	c_n ($=$ III$_n$)

*oligomycin sensitivity conferral protein (Kagawa and Racker, 1966a)
**$n = 8$–15 (see Section I.5.3.3)

subunits $\alpha_3\beta_3\delta ab_2$. The overall stoichiometry of a fully assembled bacterial enzyme can therefore be given as $\alpha_3\beta_3\gamma\delta\varepsilon ab_2c_n$ (8 different proteins) forming a minimum of 22 proteins per complex (if $c_n = c_{10}$, for example in *E. coli*; see Section I.5.3.3).

The subunit composition in different organisms can be more complex than that found in Bacteria. Furthermore, in complex organisms such as Eukarya (including fungi, plants, and animals), the subunit nomenclature can vary for historical reasons. Table 11.1 gives an overview of F-type ATP synthase subunits with respect to biochemical (F_1 and F_o) and functional (rotor and stator) differentiation.

I.3 Binding Change Mechanism

Years before any structural details of the F_1F_o-ATP synthases were revealed, Paul Boyer formulated a hypothetical mechanism for ATP synthesis (or hydrolysis) by the F_1 complex. This hypothesis is named the *binding change mechanism*, also known as the *alternating site mechanism* (Figure 11.4). The hypothesis proposed a conformational cycling of each of the three $(\alpha\beta)$ protomers through three different states, named *open* (O), *loose* (L), and *tight* (T), which finally leads to the synthesis of ATP (Cross, 1981; Boyer, 1993). This notion was based on unusual experimental properties of the F_1 complex first reported in (Boyer et al., 1973): the three catalytic sites displayed strong positive cooperativity for substrate binding and, at the same time, strong positive cooperativity for enzymatic activity (Weber et al., 1994).

Boyer proposed that three different catalytic sites are present at the same time in the F_1 complex. The O site has the lowest affinity for ADP and P_i, the L site binds the substrates loosely, and the T site binds them tightly. Torque from ion translocation in the F_o complex is used to inter-convert these three states by physical rotation of the central stalk, i.e. the rod-shaped, asymmetric γ-subunit. The γ-phosphoanhydride bond of ATP is formed in the T

state, and ADP and P_i can only be released in the O state. Hence the actual energy-requiring step is not the formation of ATP from ADP and P_i, but rather the release of ATP from the T state by formation of the O state (Boyer, 1997a). The hypothesis includes the notion of the so-called bi-site activation, meaning that ATP release is only possible when ADP and P_i are bound to one of the other subunits (catalytic cooperativity of F_1).

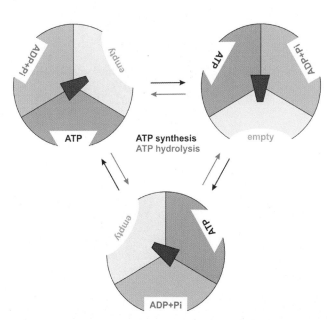

FIGURE 11.4: *The binding change mechanism of the F_1F_o-ATP synthase. F_1 harbors three conformationally different $\alpha\beta$ protomers in open (O), loose (L) and tight (T) conformation, each shown in a different color. O provides the very low-affinity binding site for ADP and P_i; L represents the state with loose binding to the ligands and is catalytically inactive; T is the catalytically active site and harbors the converted ligands (ATP). The energy which drives the ATP synthesis process is introduced by the inherently asymmetric γ-subunit. This central stalk synchronizes the three different conformations at the three different protomers at any time point by rotation around its own axis. Drawn after (Cross, 1981, Boyer, 1993).*

I.4 Structure of the F₁ Complex

In 1994, the structure of a large part of the F_1 subcomplex ($\alpha_3\beta_3\gamma'$) from bovine heart mitochondria was determined to 2.8 Å by X-ray crystallography (Abrahams et al., 1994). The structure demonstrated that the ATP synthase contains three inter-convertible catalytic sites (in the β-subunits), which are in three different states at any time point, under the control of the central, rotating stalk. Later, the structure of the $\alpha_3\beta_3\gamma\delta\varepsilon$ complex was solved (Gibbons et al., 2000) (Figure 11.5), also including most of the hydrophilic part of the outer stalk (Rees et al., 2009).

The F_1-δ complex consists of a hexameric assembly of alternating α- and β-subunits, forming the head-piece around the protruding γ-subunit, with an overall dimension of ~125 Å in height and a diameter of ~115 Å. The γ-subunit forms an asymmetric coiled-coil of α-helices (right-handed, see Section I.5.2 and Figure 11.7c). The α- and β-subunits share a high similarity both in the primary sequence and their tertiary structure. Each of the three α- and β-subunits is composed of three protein domains, an N-terminal six-stranded β-barrel, a central α – helical-β – barrel domain, and a C-terminal α-helical domain. The β-subunits accommodate the catalytically active centers of the enzyme. In the crystal structure, each of the three β-subunits could be resolved in three different states, depending on the position of the

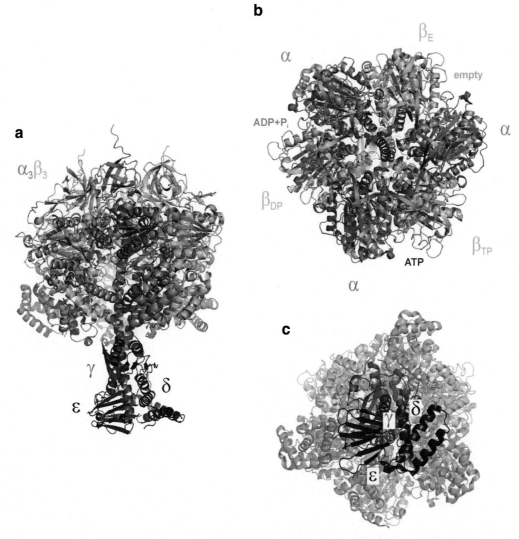

FIGURE 11.5: *Structure of the F₁-ATPase from bovine heart mitochondria. Cartoon representation mode showing the nucleotides (ADP and the non-hydrolyzable ATP analog AMPNP) in sphere representation. (a) Side view showing subunits α and β in light and dark green, respectively, and subunits γ, δ and ε in different blue tones. (b) View from the inner membrane space (cytoplasm in bacteria) on the enzyme. (c) View from the membrane on the foot of the central stalk of the enzyme. Three pairs of (αβ) protomers are arranged around subunit γ. The β-subunits are present in different conformations depending on the bound nucleotide (empty, ADP or AMPPNP). The figure shows the structure from Gibbons et al., (2000), Protein Data Bank ID entry 1E79.*

asymmetric γ-subunit, resulting in an empty (β_E), loose (β_{DP}, containing ADP and P_i), and tight (β_{TP}, containing AMPPNP, a non-hydrolyzable ATP analog) conformation (Lutter et al., 1993). The nucleotide- binding site of the β-subunits is characterized by a common nucleotide-binding fold, the so-called Walker A (GxxxxGK(T/S)) and Walker B motif (R/KxxxGxxxL/VhhhhD) (Walker et al., 1982b). These motifs, together with a set of other typical nucleotide-binding sequences (Davidson and Chen, 2004), are prominent in many ATPases, for example, ABC transporters. Furthermore, a conserved amino acid motif in the β-subunits, the DELSEED-motif (Hara et al., 2001), is involved in contacting the ε-subunit for ATP synthase activity regulation (see discussion later in this chapter).

These structural features are completely in accord with Paul Boyer's earlier prediction of the *binding change mechanism* (see Section I.3): the β-subunits of the enzyme, at any time point of operation, are seen in three different conformations, with the γ-subunit, an asymmetric rotating entity within the $\alpha_3\beta_3$ headpiece, causing these conformations to change from one to another (Capaldi and Aggeler, 2002). The N-terminal β-barrel region is in close contact with subunit γ (Lötscher et al., 1984; Dallmann et al., 1992; Aggeler et al., 1995) and with the c-subunit oligomer from the F_o sector (Watts et al., 1996; Zhang and Fillingame, 1995). "For their elucidation of the enzymatic mechanism underlying the synthesis of adenosine triphosphate (ATP)" the Nobel prize in Chemistry was awarded to John E. Walker and Paul D. Boyer in 1997 (Boyer, 1997b; Walker, 1997).

The role of the ε-subunit in the F_1 complex of ATP synthase is yet to be established but it is believed to be involved in the regulation of the enzyme activity (Feniouk and Yoshida, 2008). Structurally, this subunit is composed of two domains: the N-terminal domain, consisting of a 10-stranded, flat motif of β-barrels, and the C-terminal domain forming an α-helical hairpin. The latter can adopt two conformations referred to as the up and down state (Tsunoda et al., 2001b). In the up state, which is promoted by a higher $\Delta\mu H^+$ or by ADP binding, the hairpin motif was shown to electrostatically interact with the conserved DELSEED motif of the β-subunit. This conformation supports ATP synthesis but blocks ATP hydrolysis.

The structure of the bacterial δ-subunit has not been resolved in complex with F_1 so far. However, NMR spectroscopy revealed that this subunit forms a bundle of six α-helices (Wilkens et al., 1997). Cross-link data furthermore suggested that this subunit is located atop of the $\alpha_3\beta_3$ headpiece, forming contacts with both the α- and β-subunits (Wilkens et al., 2005) as well as with the outer b_2-dimer stalk (McLachlin and Dunn, 1997; McLachlin et al., 1998). Besides a structural role connecting these two parts of the enzyme, the δ-subunit has no other known function.

I.5 The Membrane-Embedded F_o Complex

The F_o complex of the ATP synthase consists, in its minimal prokaryotic composition, of subunits a, b, and c. The F_o portion has been shown to interact with the F_1 complex (Walker et al., 1982a), but a high-resolution structure of the complete membrane-embedded F_o sector is not available so far. Models of the structural arrangement of the F_o complex subunits (and their stoichiometry) were first suggested in the 1980s. Various possible positions of the a_1b_2 complex with respect to the c-oligomer were considered, in which a_1b_2 is either inside (Cox et al., 1984; Cox et al., 1986) or outside a ring of c-subunits (Hoppe and Sebald, 1986; Schneider and Altendorf, 1987; Fillingame, 1992). Electron microscopy and image analysis provided a low-resolution view of the shape and mass distribution of F_o. Two b-subunits and one a-subunit were found to be located outside of the c-subunit complex (Birkenhäger et al., 1995). Later, more focus on the oligomeric c-subunits confirmed that they form an annular structure (Birkenhäger et al., 1995; Singh et al., 1996; Takeyasu et al., 1996).

The main role of the F_o domain is H^+ (or Na^+) translocation, either to generate a torque, in ATP synthesis mode, or to energize the membrane, in ATP hydrolysis mode. The ion pathway itself is thought to involve subunit a and c (Junge et al., 1997; Vik and Antonio, 1994). The b-subunits provide the static connection between F_o and the F_1 headpiece (see section "outer stalk"). From biochemical experiments, the presence of all three F_o subunits (a, b, and c) is required for a functional enzyme complex (Schneider and Altendorf, 1984; Schneider and Altendorf, 1985; Wehrle et al., 2002a; Greie et al., 2004).

I.5.1 The a-subunit. The a-subunit is formed by the largest polypeptide chain in the F_o sector, consisting of 271 amino acids in *E. coli*. The a-subunit occurs in one copy per ATP synthase (Figure 11.3). It likely consists of five membrane-spanning α-helices (Zhang and Vik, 2003) with its amino terminus in the periplasm and carboxy-terminus in the cytoplasm (Jäger et al., 1998; Valiyaveetil and Fillingame, 1998; Wada et al., 1999; Zhang and Vik, 2003; Daley et al., 2005). The a-subunit is proposed to form two proton- (Na^+ ion) accessible half-channels that are offset relative to each other, which allow ion exchange between the environment and binding sites within the c-subunits at the a/c-ring interface (Vik and Antonio, 1994; Junge et al., 1997). The exact location of these half-channels is unclear, because no atomic structure of this subunit is available. However, recent work suggests that the periplasmic half-channel is located within the a-subunit, and that the cytoplasmic half-channel is at the interface of the a-subunit and the c-oligomer (Angevine et al., 2007; Moore et al., 2008; Steed and Fillingame, 2009).

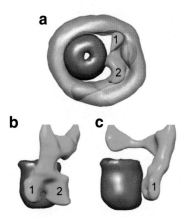

FIGURE 11.6: A first glimpse on the rotor-stator interface of the ATP synthase by electron microscopy. The picture shows three different views on the V_o/A_o motor of the ATP synthase from the bacterium Thermus thermophilus. (a) View from cytoplasm on the membrane. (b) and (c) Two views from the membrane plane. The number 1 and 2 marks two (small and large) membrane-embedded domains of the a-subunit, respectively. Color code: Pink: rotor ring; grey: detergent micelle; green: I-subunit (=a-subunit in F-type). Note that the micelle (in gray) is not part of the F_o complex structure. Courtesy of John Rubinstein and with kind permission (Lau and Rubinstein, 2010).

Numerous cross-links between the penultimate α-helix of the a-subunit and the C-terminal α-helix of the c-subunit have been found, suggesting that the two C-terminal α-helices from the a-subunit participate in coupling ion translocation by direct interaction with the c-subunits (Jiang and Fillingame, 1998; Vorburger et al., 2008). The conservation of several amino acids in these C-terminal helices indeed indicates that they are key for the function of the a-subunit. In particular a highly conserved amino acid sequence (RLFGN) appears on the penultimate α-helix in many of the a-subunit sequences. The arginine found in this sequence, aR210 in *E. coli*, is fully conserved throughout all species, essential for ATP-driven proton translocation through F_o (Lightowlers et al., 1987; Cain and Simoni, 1989; Eya et al., 1991) and plays a key role in the ion translocation process (Pogoryelov et al., 2010) (see also Section III). In addition to this important residue, site-directed mutagenesis has also revealed E219 and H245 to be important in the *E. coli* a-subunit (Cain and Simoni, 1986; Cain and Simoni, 1988; Lightowlers et al., 1988; Howitt et al., 1990; Eya et al., 1991; Hartzog and Cain, 1994).

Recently, a low-resolution view of the overall architecture of the V-type/A-type ATPase from *Thermus thermophilus* by cryo-EM has become available (Lau and Rubinstein, 2010), providing an idea of how this stator subunit may interact with the rotor ring in the membrane (Figure 11.6). Surprisingly, this interaction seems to be sustained only by a very small contact area between the rotor ring and the a-subunit, approximately at the level of the rotor ion-binding sites – that is, close to the

middle of the membrane. Experimental cross-linking data (Valiyaveetil and Fillingame, 1998) suggest that this region could possibly encompass the RLFGN motif of the fourth α-helix. Furthermore, hydrophilic access pathways from both sides, from the periplasm to the c-subunit ion-binding site through the a-subunit, and from the binding site to the cytoplasm via the a/c-ring interface, have been suggested on the basis of biochemical accessibility studies (Angevine et al., 2007; Moore et al., 2008; Steed and Fillingame, 2009). The low-resolution density map is, in principle, consistent with the two-half-channel model for ion translocation, but a detailed microscopic mechanism for torque generation by ion exchange at the a/c-ring interface requires further clarification (see Section III).

I.5.2 The b-subunit, the Outer Stalk of the ATP Synthase. The b-subunits constitute the major part of the outer peripheral stalk of the ATP synthase. From the first DNA sequence of this subunit (from *E. coli*), it was obvious that its N-terminus is highly hydrophobic, and that it is highly charged in the remainder of the structure (Walker et al., 1982a). This led to the correct assumption that the b-subunits are N-terminally anchored within the membrane and could play a role in ion conduction. From various biochemical work, it became clear that the polar domain of the b-subunit forms an elongated dimer, and that it interacts with the F_1 sector via the δ- and the α-subunit (Dunn, 1992; Rodgers et al., 1997; Sawada et al., 1997; Dunn and Chandler, 1998; McLachlin et al., 1998; Rodgers and Capaldi, 1998). However, the exact position of the b-subunit was not clear until the outer stalk of the ATP synthase was visualized for the first time by electron microscopy (Wilkens and Capaldi, 1998). Whereas in F-type ATP synthases only one outer stalk has been described, two are present in the V-type ATPase from *Thermus thermophilus* (Bernal and Stock, 2004) and the A-type ATP synthase from *Pyrococcus furiosus* (Vonck et al., 2009); remarkably, three outer stalks have recently reported for the V-type ATPase found in the caterpillar *Manduca sexta* (Muench et al., 2009).

In bacterial systems, two b-subunits form a dimer. The amphipathic protein consists of ~160 amino acids. The N-terminal 30 amino acids are hydrophobic and form an α-helical secondary structure providing a dimeric membrane anchor (Figure 11.3). The much longer and highly hydrophilic C-terminal part extends from the membrane surface and binds to subunit α presumably via the δ-subunit (Weber et al., 2003). Detailed structural information about the soluble part of the outer stalk has recently become available for the more complex bovine mitochondrial (Rees et al., 2009) as well as the bacterial ATP synthase (Lee et al., 2010) (Figure 11.7). Whereas in both cases the α-helical secondary structures are dominant, the outer stalk of the bacterial ATP synthase seems straighter than its

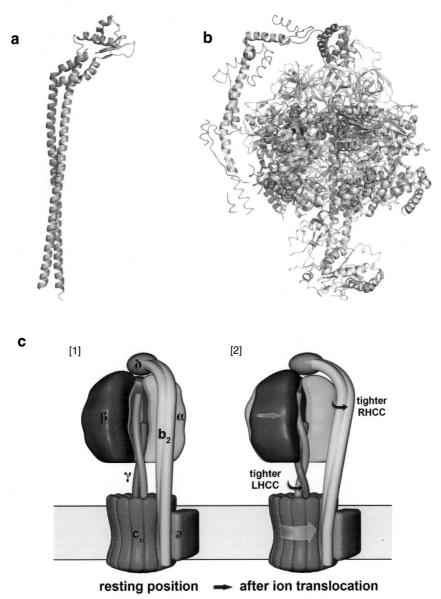

FIGURE 11.7: *The outer stalk of the ATP synthase. (a) shows the outer stalk subunits G and H of the* **Thermus thermophilus** *ATP synthase (V-type/A-type) (modified from Lee et al., (2010). (b) shows the outer stalk of the bovine ATP synthase (modified from Rees et al., (2009). The stalk is shown in color relative to the remaining F_1 subunits from bovine heart mitochondrial ATP synthase. (c) Proposal for the significance of right-handed coiled-coils (RHCCs) in rotational catalysis of ATP synthases redrawn from (Del Rizzo et al., 2006). The counter-clockwise rotation of the rotor during ATP synthesis induces twisting of the γ-subunit. An RHCC outer stalk (but not a left-handed CC) b_2 dimer helps counteracting the induced torque by winding up more ([1] to [2]). The figure does not intend to show real conformations of subunits γ and b_2 but is meant to illustrate the distribution of forces during enzyme operation.*

bovine counterpart. The b-subunits are connected with the δ-subunit at the top headpiece part of the F_1 complex (Wilkens et al., 2005). In the membrane F_o complex, the b-subunit has been shown to interact with the a-subunit, but it has also been proposed to be proximal to the rotor ring (c-ring) based on cross-linking experiments (Jones et al., 2000).

In functional terms, the outer stalk in the ATP synthases is believed to act as a stabilizing part of the enzyme as it connects the stator F_1 headgroup ($\alpha_3\beta_3$) with the membrane-embedded motor part (a/c-ring), and presumably also helps keeping the a-subunit in a stable position at the outer surface of the rotating c-ring. Biochemical data suggests that the two b-subunit α-helices form a long coiled-coil with a right-handed super-helical twist (Del Rizzo et al., 2002; Del Rizzo et al., 2006) . Novel single-molecule elastic compliance measurements support a rigid b_2 structure (Wächter et al., 2011). Deletions within the b-subunit were shown to disrupt the enzyme function but not enzyme assembly (Cipriano et al., 2006). The prediction of a right-handed coiled-coil (RHCC) has been recently confirmed by the X-ray structure of the b-subunit homologs (subunits E and H) in the ATPase/synthase (V-type/A-type) of the bacterium *Thermus thermophilus* (Lee et al., 2010). This outer stalk consists of an E/H heterodimer (Figure 11.7a). Together they form a globular head group, which is in contact with the water-soluble complex F_1 head group on top of the enzyme, and a long (140 Å) RHCC, which forms the actual peripheral stalk. The structure strongly emphasizes the significant role of RHCCs in rotational catalysis of ATP synthases (Del Rizzo et al., 2006) by counteracting the oppositely coiled left-handed CCs of the γ-subunit, which helps prevent bowing of the outer stalk during ATP synthesis (Figure 11.7c). Hence this structural constellation increases lateral rigidity of the peripheral stalk (Del Rizzo et al., 2006; Lee et al., 2010). Recent work rather confirms the stiff nature (Dickson et al., 2006) proposed for the eukaryotic ATP synthase outer stalk (Rees et al., 2009) (Figure 11.8b).

I.5.3 The c-subunit and the Rotor Ring.

The c-subunit is the best investigated subunit of the F_o sector. It is a small hydrophobic protein of approximately 70–90 amino acids, with a molecular mass between 7 and 9 kDa. It is also called a proteolipid because it can be easily dissolved in organic solvents. The discovery of the c-subunit dates back to 1951 when Folch and Lees described proteolipids as a new type of tissue lipoproteins, isolated from the white matter of the brain by organic-solvent (chloroform/methanol) extraction (Folch and Lees, 1951). Each c-subunit forms a building unit for the rotor ring, which binds and shuttles the coupling ions (H^+ or Na^+) across the membrane. The c-subunit's monomeric structure was first investigated by biochemical methods (Fillingame, 1990) and NMR spectroscopy (Girvin et al., 1998). Following the development of isolation procedures for the complex rotor rings (Meier et al., 2003), structural investigations using a variety of methodologies have proliferated (Table 11.2).

I.5.3.1 Monomeric Structures of the c-subunit.

The first insights into the c-subunit structure were gained by NMR spectroscopy of c-subunits dissolved in organic solvents. The first successful NMR structure (Girvin et al., 1998) showed that the c-subunit folds into a hairpin with two membrane spanning α-helices that are connected by a short polar loop with no secondary structure. Accordingly, the conserved carboxyl group (D61 in *E. coli*) located in the middle of the C-terminal helix would be positioned in the center of the membrane. This basic observation was

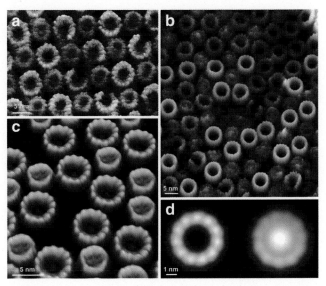

FIGURE 11.8: Atomic force microscopy of rotor rings. Examples are shown from (a) spinach chloroplasts c_{14} ring (Seelert et al., 2000), (b) Spirulina platensis c_{15} ring (Pogoryelov et al., 2005) and (c) Synechococcus elongatus SAG 89.79 c_{13} ring (Pogoryelov et al., 2007) and (d) Ilyobacter tartaricus c_{11} ring (Stahlberg et al., 2001), reproduced with kind permission. Image (b) courtesy of Daniel J. Müller and Adriana L. Klyszejko.

confirmed by subsequent NMR investigations at various pH of the solvent (chloroform/methanol/water) mixtures (Matthey et al., 1999; Rastogi and Girvin, 1999; Rivera-Torres et al., 2004). Nevertheless, biochemical analyses suggested that the monomeric c-subunit structure might not be in a native state, since reaction rates of inhibitor binding to isolated c-subunits varied by a factor of 400 compared to the native enzyme (Kluge and Dimroth, 1994). For these reasons, the native structure of the oligomeric c-subunit complex became the aim of many research groups.

I.5.3.2 The Rotor Rings Come into Focus.

A ring-like structure formed by several copies of c-subunits was the expected arrangement within the F_o part of the functional ATP synthase (Schneider and Altendorf, 1987; Fillingame, 1990; Groth and Walker, 1997). Its stoichiometry was estimated to be approximately 10 in accord with cross-linking studies and tryptophan substitutions (Jones et al., 1998; Schnick et al., 2000). In the early 1990s, however, the available data on the F_o complex only allowed estimating an approximate diameter (Soper et al., 1979; Fromme et al., 1987; Tsuprun et al., 1989; Gogol, 1994; Boekema et al., 1997). The first visualization of the F_o complex (*E. coli*) was due to atomic-force microscopy (Singh et al., 1996; Takeyasu et al., 1996) and electron microscopy (Birkenhäger et al., 1995), which confirmed the annular shape of c-rings previously anticipated (Cox et al., 1984; Cox et al., 1986).

Table 11.2: Subunit Stoichiometries of Rotor Rings from ATPases/synthases. The Number of c-subunits Varies Among Different Species but is Constant in a Species. Numbers in Bold Refer To The Most Appropriate Description

Species	c-ring Stoichiometry	Methods	Reference
Bos taurus mitochondria	8	X-ray crystallography	(Watt et al., 2010)
Saccharomyces cerevisiae mitochondria	10	X-ray crystallography	(Stock et al., 1999)
Escherichia coli	12, **10**	Genetic fusion technology	(Jones and Fillingame, 1998; Jiang et al., 2001)
Thermophilic *Bacillus* PS3	10	Genetic fusion technology	(Mitome et al., 2004)
Pyrococcus furiosus (A-type)	10	Mass spectrometry	(Vonck et al., 2009)
Enterococcus hirae (V-type)	7, **10** (2 hairpins per c)	Single particle analysis (7) X-ray crystallography (10)	(Murata et al., 2003, Murata et al., 2005)
Ilyobacter tartaricus	11	Atomic force microscopy Electron microscopy X-ray crystallography Mass spectrometry	(Stahlberg et al., 2001; Vonck et al., 2002; Meier et al., 2005a; Meier et al., 2007)
Propionigenium modestum	11	Electron microscopy	(Meier et al., 2003)
Clostridium paradoxum	11	Electron microscopy Mass spectrometry	(Meier et al., 2006, Meier et al., 2007)
Acetobacterium woodii	11 (9 hairpins + 1 duplicated hairpin)	Electron microscopy Atomic force microscopy, Mass spectrometry	(Fritz et al., 2008)
Thermus thermophilus (V-type/A-type)	12	Electron microscopy	(Toei et al., 2007)
Synechococcus elongatus SAG 89.79	13	Electrophoresis Atomic force microscopy Mass spectrometry	(Meier et al., 2007; Pogoryelov et al., 2007)
Bacillus sp. strain TA2.A1 (=*Caldalkalibacillus thermarum* strain TA2.A1)	13	Mass spectrometry Electron microscopy Atomic force microscopy	(Meier et al., 2007, Matthies et al., 2009)
Bacillus pseudofirmus OF4	13	X-ray crystallography	(Preiss et al., 2010)
Spinachia oleracea chloroplasts	14	Atomic force microscopy X-ray crystallography	(Seelert et al., 2000, Vollmar et al., 2009)
Synechocystis sp. PCC 6803 *Anabaena* sp. PCC 7120 *Synechococcus* sp. PCC 6716 *Synechococcus elongates* PCC 6301 *Synechococcus elongatus* PCC 7942	14	Electrophoresis	(Pogoryelov et al., 2007)
Spirulina platensis strain C1 (=*Arthrospira* sp. PCC 9438)	15	Atomic force microscopy Electrophoresis Mass spectrometry X-ray crystallography	(Pogoryelov et al., 2005; Meier et al., 2007; Pogoryelov et al., 2007; Pogoryelov et al., 2009)
Arthrospira PCC 9108 *Gloeobacter violaceus* PCC 7421	15	Electrophoresis	(Pogoryelov et al., 2007)

However, the stoichiometric composition of the c-ring and the exact helix packing was unclear and under discussion for a long time (Foster and Fillingame, 1982; Groth et al., 1998; Jones and Fillingame, 1998; Dmitriev et al., 1999; Schnick et al., 2000; Jiang et al., 2001). Construction of genetically-fused dimers and trimers of *E. coli* c-subunits (Table 11.2) yielded a functional ATPase, and the cross-linking of the fused c-subunits resulted in multimers up to c_{12}. Therefore a c_{12} ring was initially assumed (Jones and Fillingame, 1998) but the data was reinterpreted later to be a c_{10} oligomer (Jiang et al., 2001). Therefore, the first objective in rotor-ring investigations was to solve the question of the stoichiometries in c-rings.

I.5.3.3 Rotor Ring Stoichiometries: Constant for One but Variable Across Different Species. In the past years, the structures and stoichiometries of various c-rings have become available. The first visual determination of a c-ring stoichiometry was made with X-ray crystallography of the yeast mitochondrial F_1 c-ring complex, which showed electron densities of 10 individual c-subunits (Stock et al., 1999). This was followed by c-rings isolated from spinach chloroplast and crystallized in 2D (Seelert et al., 2000). Analysis of the 2D crystalline lattice by atomic force microscopy (AFM) revealed a stoichiometry of 14 c-subunits per ring (Figure 11.8).

The first rotor ring from a bacterial ATP synthase to be isolated was that from *Ilyobacter tartaricus* (Meier et al., 2003). In this case, the stoichiometry was shown to be 11 by both atomic-force microscopy (Figure 11.8) and by electron microscopy of 2D crystalline lattices (Stahlberg et al., 2001). A 3D density map of the *I. tartaricus* c_{11} rotor ring (Vonck et al., 2002) derived from electron microscopy of 2D crystals provided the first complete model of a bacterial rotor ring. Table 11.2 summarizes the sizes of rotor rings in various species studied since then, and the methodology used in each case.

The obvious result from these specific investigations is clearly that even though the rotor ring stoichiometry seems always constant within a selected species, it clearly can vary depending on what species they have been isolated from. The c-subunits from different organisms show highly conserved primary sequences (see e.g., Meier et al., 2009) but nevertheless assemble into rings of variable stoichiometries (Table 11.2).

The obvious question which therefore arises is whether these c-rings would occasionally change their stoichiometry depending on external living conditions, for example, by changing the carbon source as has been suggested for the c-ring from *E. coli* (Schemidt et al., 1995, Schemidt et al., 1998). This c-ring has been shown by several research groups to consist of 10 subunits in the native enzyme (Jiang et al., 2001; Ballhausen et al., 2009; Düser et al., 2009) but the stoichiometry has not yet been confirmed by structural approaches. The effect of a possible variability in the c-ring stoichiometry has been experimentally assessed by genetically engineering the c-ring size in *E. coli* from 10 to 12 hairpins. In both cases, an active ATP synthase was reported (Jones and Fillingame, 1998; Jiang et al., 2001). These data indicate a certain flexibility of this complex at least in the *E. coli* ATP synthase. In contrast, the number of c-subunits in the ring of the bacterium 'thermophilic *Bacillus* PS3' is restricted to 10, because only marginal ATP synthesis rates have been found with all ATP synthases containing c-rings deviating from the c_{10} oligomeric form (Mitome et al., 2004). In support of this finding, recent structural data of the c-rings suggest a fixed stoichiometry for a given species rather than a flexible one. AFM images of *I. tartaricus* c_{11} rings, plant chloroplast c_{14} rings, and c_{15} rings from *Arthrospira sp.* PCC 9438 showed that occasionally c-rings lack individual subunits. Instead of closing this gap immediately by collapsing to smaller rings, these incomplete oligomers retain the same shape and diameter, as observed for complete, intact c-rings (Müller et al., 2001; Pogoryelov et al., 2005). This observation indicates that the c_{11}, c_{14}, and c_{15} stoichiometries of these rings are intrinsic properties of the c-subunits, determined by the primary

structure of their protein (Müller et al., 2001). Observation of annular structures after isolation of monomeric c-*subunit*s from *E. coli* support these findings (Arechaga et al., 2002). Furthermore, subunits c from *I. tartaricus* and *Bacillus* sp. strain TA2.A1 when synthesized in *E. coli* were assembled correctly into c_{11} and c_{13} rings, respectively (Meier et al., 2005b; Matthies et al., 2009; Meier et al., 2007), despite of the preferred c_{10} stoichiometry of the native *E. coli* c-ring (Jiang et al., 2001). Therefore, the c-ring sizes are independent of external factors such as pH of the medium (Meier et al., 2007), host protein expression conditions, composition of the expression host´s lipid types (Meier and Dimroth, 2002, Meier et al., 2005b), source of the carbon used by the cell (Fritz et al., 2008), and rate of ATP synthesis (Krebstakies et al., 2008; Ballhausen et al., 2009).

Taken together, the data indicate that the size of the c-rings is defined entirely by the primary sequence of the c-subunit. Each species has a defined c-ring stoichiometry that is invariant within that species but variable across different species, ranging, as known so far, from 8 (Watt et al., 2010) to 15 (Pogoryelov et al., 2005) c-subunits. One cannot exclude the possibility that this known range might still be extended in the future.

I.5.3.4 The Variation of Rotor Ring Stoichiometries is an Adaptation to Bioenergetic Demands of the Cell.
The variability of rotor ring sizes has a profound impact on the ATP synthase motor function as one ion (H^+ or Na^+) is translocated across the membrane per c-subunit. This is the case at least under the premise of a tightly coupled enzyme and the fact that all ion binding sites have to be neutralized when facing the hydrophobic lipid membrane (see Section III for F_o-mechanistic details). From a kinetic point of view, per rotational cycle (360°), a c_n ring translocates n ions. In the F_1 complex, the three β-subunits constantly release three ATP molecules per rotation. Therefore, the ion-to-ATP ratio (= $n/β$ ratio = $n/3$) can be considered an important bioenergetic parameter of the ATP synthase, with implications in the energy balance of the whole cell (Ferguson, 2000; Ferguson, 2010). The next paragraphs will explain why this is the case.

Depending on the actual physiological function of the ATPase/ATP synthase, either ATP synthesis for ATP production or ATP hydrolysis for ion pumping, two opposed settings of the c-ring stoichiometry in principal exist: In the ATP synthesis mode, a c-ring with fewer ion binding sites (a lower c-ring stoichiometry, a "small c-ring") is advantageous for efficient ATP production, as fewer ions per ATP are transported across the membrane per cycle. The exactly opposite situation exists when the enzyme operates in the ATP hydrolysis direction: in this mode, it is more energy saving if more ions per ATP can be pumped across the membrane. Hence a "larger c-ring" with more ion binding sites (more c-subunits) is an advantage for the cell bioenergetics.

In this view, the variability of the c-ring stoichiometries therefore seems to represent a finely tuned, evolutionary adaptation of individual cells (species) to their individual cell-bioenergetic environment. This view, however, is challenged by biochemical data showing that the thermodynamic ion-to-ATP ratio of the ATP synthase can be constant for enzymes that are clearly different from their $n/3$ stoichiometric ratio (Turina et al., 2003; Krebstakies et al., 2008; Steigmiller et al., 2008).

An important question thus concerns the actual (evolutionary) determining factors for the rotor ring size in a given species. We will try to formulate a hypothesis here, based on the current knowledge of rotor ring stoichiometries and sizes. The driving force for ATP synthesis is the ion motive force (Δp) (Mitchell, 1963), which is composed of two components, the membrane potential ($\Delta \Psi$) and the transmembrane proton gradient (ΔpH, sometimes ΔpNa^+), as reflected by the equation $\Delta p = \Delta \Psi + (2.3RT/F)\Delta pH$ ($R =$ gas constant, $T =$ absolute temperature and $F =$ Faraday's constant). The ATP synthases from various sources differ in their ability to use these forces ($\Delta \Psi$, ΔpH, ΔpNa^+) (Kaim and Dimroth, 1999). Chloroplasts and cyanobacteria both use light-driven H^+ translocation of the photosystems to generate proton gradients. These systems are highly efficient and can generate considerably high ΔpH gradients up to 3 pH units across the (thylakoid) membranes (Kramer et al., 2003). The proton availability is high (low pH). Looking at the c-ring stoichiometries of these systems, the all consistently contain large c-rings with a large number of ion binding sites, as known today, ranging from 13 to 15 (Seelert et al., 2000, Pogoryelov et al., 2007). Hence, they have high ion-to-ATP ratios ($n/3 = 4.3$ to 5). In contrast, bovine or yeast mitochondria, as well as for example *E. coli* bacteria, ΔpH is the less dominant driving force for ATP synthesis (but certainly still required). In this case, a pH gradient of only approx. 0.5 pH units is available (1 pH unit is an equivalent of approximately 60 mV), although the overall electrochemical gradient is still typically 180 to 190 mV (e.g., in a respiring mitochondrion or in growing *E. coli* cells). Hence, the relative contribution of the $\Delta \Psi$ component to Δp is considerably higher in these cases than in light-driven systems; this fact was experimentally confirmed by detailed studies on the ATP synthase driving-force components in various species (Kaim and Dimroth, 1999). When looking at the c-ring stoichiometries of these species, they are indeed smaller. For example, c_8 in bovine (Watt et al., 2010) and c_{10} in yeast mitochondria (Stock et al., 1999) as well as in *E. coli* (Jiang et al., 2001) have been characterized. These cases represent the lower end of the c-ring stoichiometry scale and these c-rings contain less ion binding sites (8 and 10); their ATP synthesis is performed with a significantly lower ion-to-ATP ratio ($n/3 = 2.7$ to 3.3). It hence appears that

the c-ring stoichiometries, the number of ion binding sites, is influenced by the magnitudes of the components of the transmembrane electrochemical ion gradient in the corresponding species.

Further support of this hypothesis is given from the observed enlarged rotor rings in alkaliphilic bacteria: these bacteria are adapted to live in environments devoid of protons, and under alkaline conditions outside a cell, the electrochemical gradient is significantly lowered due to an inverted ΔpH, for example, pH 8.2 inside and pH 10.5 in *Bacillus pseudofirmus* OF4 (for a review, see Hicks et al., 2010). Rather surprising is hence the fact that these organisms still run ATP synthesis by a proton-dependent ATP synthase using the proton-motive force (Hicks and Krulwich, 1990; Hoffmann and Dimroth, 1990; Cook et al., 2003). At the thermodynamic equilibrium for ATP synthesis, the free energies of the ΔpH and the phosphorylation potential ($\Delta G_p = \Delta G^0$ for ATP, see Section I.1) are in balance and the equation $n/3 \times \Delta pH = \Delta G_p$ applies. At a low (or even inverted) ΔpH, a higher ion-to-ATP ratio and hence a higher c-ring stoichiometry (n) clearly seems advantageous for these bacteria. Indeed, alkaliphilic *Bacillus* species, beside having a set of several other unique adaptations to life at high pH (Hicks et al., 2010), harbour enlarged numbers of c-subunits in their c-rings (Meier et al., 2007; Matthies et al., 2009; Preiss et al., 2010), when compared with the closely related neutrophilic *Bacillus* (Mitome et al., 2004).

Taken together, the number of c-subunits (ion binding sites) is fixed within a species and determined by the primary structure of the protein. But the c-ring stoichiometries found in the various species vary in a certain range ($n = 8$ to 15, as known today, Table 11.2). These variations reflect different bioenergetic environments and challenges, which the F_1F_o-ATP synthases of these organisms/organelles have faced and adapted to throughout evolutionary processes.

I.5.3.5 Structures of the Rotor Ring Resemble a Turbine. Some of the rotor rings are highly stable protein complexes and can even resist incubation temperatures at 95°C for more than an hour without affecting its rotor complex composition (Meier and Dimroth, 2002). This high stability facilitated structural investigations of a number of these c-rings by various methods (Table 11.2). The first high-resolution structures of rotor rings from F-type ATP synthases have been solved by X-ray crystallography (Figure 11.9).

In these structures, the rotor rings share, in principal, a high overall structural similarity. The basic construction principle of a rotor ring is the following: every ring is composed of a number of c-subunits that assemble into a cylindric shape with a central pore. Sometimes the cylinder resembles an hourglass (Meier et al., 2005a; Pogoryelov et al., 2009), but in other cases (Preiss et al., 2010) the

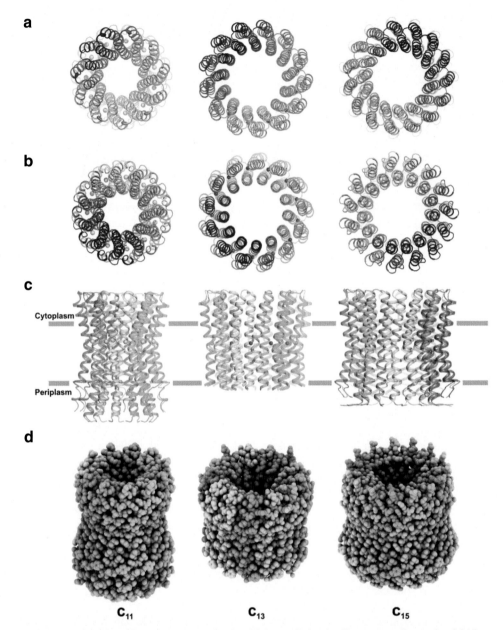

a

b

c

Cytoplasm

Periplasm

d

C_{11} C_{13} C_{15}

FIGURE 11.9: Rotor ring structures solved at high resolution by X-ray crystallography. (a) View from cytoplasm, (b) view from periplasm, (c) view parallel to the membrane (indicated by grey bars, membrane diameter = 35 Å) and (d) space-fill representation, tilted view from cytoplasm. The structures are from left to right: Ilyobacter tartaricus c_{11} ring (Meier et al., 2005a), Bacillus pseudofirmus OF4 c_{13} ring (Preiss et al., 2010) and Spirulina platensis c_{15} ring (Pogoryelov et al., 2009). The c-subunits are shown in different colors. The Na$^+$ ions in the I. tartaricus c_{11} ring and the H$_2$O in the ion binding site of the B. pseudofirmus OF4 c_{13} ring are indicated with yellow and red spheres, respectively. Protein Data Bank ID entries are 1YCE, 2X2V, and 2WIE for c_{11}, c_{13} and c_{15}, respectively.

shape is reminiscent of a tulip-beer glass. Each c-subunit spans the membrane twice, once with the N-terminal α-helix and once with the C-terminal α-helix. Both helices are connected by a short and highly conserved cytoplasmic loop (RQPE), and both the N- and the C-termini end in the periplasm of the (bacterial) cell. The N- and C-terminal α-helices of the c-subunits form an inner and outer ring,

respectively. The loop region of the c-ring forms the rigid interaction site with the rotor subunits γ and ε from the F$_1$ sub-complex (Pogoryelov et al., 2008). The overall construction of rotor rings is very tight (Figure 11.9d). Beside some especially stabilizing elements, such as, for example, Na$^+$ ion c-c-subunit cross-bridging effects (Meier and Dimroth, 2002), the solid arrangement of the rotor ring

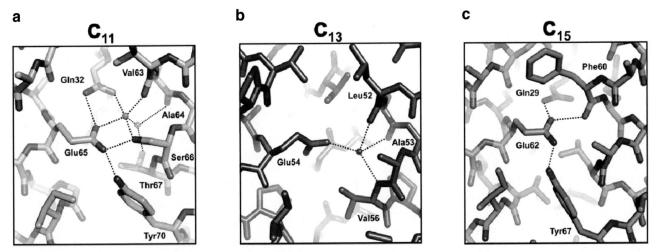

a C_{11}

Gln32 Val63 Ala64 Glu65 Ser66 Thr67 Tyr70

b C_{13}

Leu52 Ala53 Glu54 Val56

c C_{15}

Phe60 Gln29 Glu62 Tyr67

FIGURE 11.10: *Structural details of the ion binding sites of Na+ and H+ dependent ATP synthases. (a) The Na+ (+ H₂O) binding site in the I. tartaricus c_{11} ring. (b) The H+ (+ H₂O) binding site of B. pseudofirmus OF4 c_{13} ring. (c) The H+ binding site in the S. platensis c_{15} ring. Reproduced from (Preiss et al., 2010).*

helices explains the overall high stability of these biological turbines.

The inner pore of the c-rings has been shown to be filled with phospholipids (Meier et al., 2001), which originate from the bacterial membrane (Oberfeld et al., 2006). Such lipid plugs were visualized by atomic-force microscopy of single c-rings isolated from various species (Seelert et al., 2000; Stahlberg et al., 2001) (Figure 11.9). The distribution of large hydrophobic surfaces at the inner pore side of the c-ring is consistent with the presence of a lipid bilayer (of a few lipids) within the inner pore of the c-ring. In some cases (e.g., *I. tartaricus* and *S. platensis*), the two leaflets of the inner pore membrane are shifted toward the periplasmic space with respect to the membrane at the outside, but in others, like *Bacillus pseudofirmus* OF4 (Preiss et al., 2010) and *Bacillus sp.* strain TA2.A1 (Matthies et al., 2009), it is succinct to the level of the outer membrane. The lipid plug is thought to seal the inner pore from ion leakage through the membrane, which would be detrimental for any ion gradient and hence for the viability of the cell itself.

I.5.3.6 The Ion-Binding Site. The rotary mechanism of the ATP synthase is crucially dependent on the sequential loading and unloading of the ion-binding sites on the membrane rotor, the c-ring, during the reaction cycle. Collectively, these ion-binding sites can be seen as the culmination point of the energy conversion process; in that they provide a way to employ the transmembrane electrochemical gradients to bias the rotational motion of the c-ring in a particular direction. The ion-binding sites also confer specificity to the enzyme, either for H+ or Na+, and thus define whether the respiratory chain will be coupled to the proton or to the sodium motive force.

In the rotor rings of conventional F-type ATP synthases, the ion (Na+ or H+) is bound within the groove formed by

two neighboring c-subunits, near the outer surface of the c-ring. In these sites, the bound ion establishes a precise network of interactions with the protein (Figure 11.10). A strictly conserved carboxylate ligand is the key coordinating group for both Na+ and H+; in most organisms, this is provided by glutamate side chain, replaced sometimes by an aspartate – for example, in *E. coli*. This conserved side chain is also thought to be of crucial importance for the rotary mechanism of the enzyme for its presumed ability to interact with the conserved arginine side chain in the a-subunit (Lightowlers et al., 1987; Eya et al., 1991; Vik and Antonio, 1994; Junge et al., 1997; Jiang and Fillingame, 1998; Cain, 2000; Wehrle et al., 2002b; Langemeyer and Engelbrecht, 2007); it also prevents a short-circuit of ions by effectively blocking a direct path between the periplasmic and cytoplasmic reservoirs (Mitome et al., 2010). Nonetheless, other, more subtle and variable interactions are also important for modulating the binding properties of the c-subunits and to ensure the appropriate coordination of the ion regardless of the stoichiometry of the ring. For example, in the c_{15} ring from *S. platensis*, the proton that is covalently bound to the conserved glutamate finds another ligand in the backbone carbonyl of the adjacent c-subunit helix. In the c_{13} ring from *B. pseudofirmus*, in which the outer helices are further apart, a water molecule is involved to form the same interaction. Further stabilizing interactions of the protonated carboxylate side chain can, for instance, occur on a glutamine side chain in the c_{15} ring; but as the c_{13} ring illustrates, they appear to not be strictly required, and H+-binding may be sustained by sites that are mostly featureless, except for the glutamate/aspartate side chain. By contrast, Na+ coordination appears to be quite elaborate and it involves up to five ligands from the inner and outer helices, from both side chains and the protein backbone. Water molecules may again be found bridging ion-protein interactions, as shown for the c_{11} ring from *I. tartaricus*. This rotor also reveals how side chains beyond the first ion-coordination shell can influence the binding site – for example, a peripheral tyrosine side chain, which donates a hydrogen bond to the coordinating carboxylate and thus appears to stabilize its conformation in a way favorable for Na^+ binding.

II. MONITORING ROTATION

II.1 Observation of Subunit Rotation

The concept of rotational catalysis by Paul Boyer (Section I.3) was based on the observation of strong catalytic cooperativity of the three catalytic sites and the role of the γ-subunit on turnover rates by mutational studies. Evidence for subunit movements was provided by low-resolution electron microscopy images of *E. coli* F_1 in 1990 (Gogol et al., 1990). Depending on the nucleotides used, the central γ-subunit in F_1 was found to be located non-symmetrically with respect to the immunostained α-subunits. Three distinct positions for γ were identified. The crystal structure of bovine MF_1 in 1994 (Abrahams et al., 1994) showed the asymmetric orientation of γ at atomic resolution and the three different conformations of the three catalytic sites.

Inspired by the structural data on the mitochondrial F_1, elegant cross-linking experiments were designed by Thomas Duncan and Richard Cross in 1995 (Duncan et al., 1995; Zhou et al., 1997). Using a homology model for the *E. coli* F_1 motor, cysteines were introduced specifically to the β- and γ-subunits. These cysteines could be cross-linked by oxidation, which resulted in the loss of catalytic activity. Upon reduction, the activity was restored. By re-assembling the F_1 motor from individual subunits, it was possible to create F_1 complexes with two radioactive-labeled β-subunits plus one non-radioactive β. Starting with F_1 comprising the γ-subunit specifically cross-linked to the non-radioactive β-subunit, subsequent reduction of this cross-link and addition of Mg^{2+} ATP resulted in random contacts between γ- and β-subunits due to rotation. After reforming the disulfide bridge by oxidation, the γ-subunit was found to be cross-linked to about 66% of radioactive β-subunits, that is, to the other two β-subunits. Thus, the association of γ with one of the three β-subunits is being randomized during ATP hydrolysis.

Using the same approach to detect ATP-driven γ-subunit rotation, rotation of γ during ATP synthesis could be revealed. Everted membrane vesicles from *E. coli* were treated with EDTA to promote the dissociation of the F_1 parts from F_o. Afterward, radioactive-modified F_1 with γ-β cross-link was re-bound to obtain a fully functional F_1F_o. ATP synthesis was induced by addition of NADH, which generated the necessary ΔpH and membrane potential. Following reduction of the cross-link, initiation of catalytic turnover and subsequent oxidation

to reform the disulfide bridge yielded the expected random association of γ and β, indicating that the rotary motion of γ had taken place. In the presence of the ATP synthase inhibitor dicyclohexylcarbodiimide (DCCD), which binds to the c-subunit and prevents proton translocation through the F_o part, the randomization of γ-β cross-links was suppressed for both ATP hydrolysis and ATP synthesis. In conclusion, these static association/cross-linking experiments of radioactively tagged subunits provided the proof-of-principle for the rotation of the γ-subunit during catalysis.

Direct monitoring of γ-subunit rotation, specifically to measure the speed of rotation in real time, was accessible by spectroscopic experiments in the group of Wolfgang Junge (Sabbert et al., 1996). The F_1 part of the chloroplast enzyme contains native cysteines for attaching an eosin chromophore. The high absorbance anisotropy for eosin on CF_1 indicated a restricted mobility of the dye in its local environment near the C-terminus of the γ-subunit. Thus, the transition dipole moment of the attached dye served as a reporter of the orientation of the γ-subunit. When immobilized CF_1 was excited with a linearly polarized laser pulse at high power, a small fraction of those eosin molecules with dipole moments oriented parallel to the laser were photobleached. The fraction of photobleached chromophores resulted in a change of mean eosin absorption. Absorption was measured simultaneously by a second linearly polarized laser perpendicular to the bleaching laser, and it was measured in two orthogonal polarizations. The eosin absorbance loss yielded a larger change in absorption using parallel polarization with respect to the bleaching laser pulse. In the presence of the non-hydrolyzable AMPPNP that was expected to block rotation of the γ-subunit in CF_1, the ratio of polarized absorption (i.e., the anisotropy parameter) remained constant. In contrast, during ATP hydrolysis, the anisotropy decayed monoexponentially within 100 ms, indicating rotation of γ. The ATP hydrolysis rate of the CF_1 preparation was 14 ± 7 s^{-1} in good accordance with the anisotropy decay rate.

Using a small chromophore as the probe to detect subunit rotation allowed for measurement of the rotary dynamics without impairment by mechanical drag forces. The accessible time scale for rotation was in the millisecond time range. However, the problem of an ensemble measurement averaging over active and inactive F_1 parts remained. The best possible and most convincing demonstration of rotational catalysis was the direct visualization of γ-subunit rotation by video microscopy in real time.

II.2 First Single-Molecule Demonstration of Rotation During ATP Hydrolysis

In 1997, Hiroyuki Noji, Ryohei Yasuda, Masasuke Yoshida, and Kazuhiko Kinosita, Jr. reported the mechanical rotation of a fluorescent pointer on γ, that is, an actin filament that was mounted on the γ-subunit of a single F_1 part comprising subunits $\alpha_3\beta_3\gamma$ (Noji et al., 1997). They used the enzyme from thermophilic *Bacillus* PS3 with engineered his-tags on the β-subunits for oriented binding of F_1 to a Ni-NTA-modified cover glass. A single cysteine introduced to the γ-subunit was reacted with biotin-maleimide. After the addition of streptavidin with its four biotin binding sites, the fluorescent and also biotinylated actin were bound (Figure 11.11a). The length of the actin filament varied between 1 to 3 μm. Due to the large number of fluorophores per actin filament, an intensified CCD camera mounted on a conventional fluorescence microscope allowed for imaging the γ-subunit rotation with video rate (~30 frames per second).

ATP hydrolysis at millimolar [ATP] caused unidirectional rotation of the actin filaments as found for 90 F_1 molecules (Figure 11.11b). The direction of rotation was counter-clockwise when viewed from the membrane side of the enzymes, that is, from the F_o part. In the crystal structure of the mitochondrial F_1, this direction of rotation corresponds to sequential interactions of the γ-subunit with the β-subunits with catalytic binding sites in the order of β-empty (β_E) → β ADP-bound (β_{DP}) → β AMPPNP-bound (β_{TP}) (see Figure 11.4 and Figure 11.5). For a catalytic binding site on one β-subunit, its status would change in the order of $\beta_E \rightarrow \beta_{TP} \rightarrow \beta_{DP} \rightarrow \beta_E$, which is the expected sequence for ATP hydrolysis. The direction of the rotating actin filament was the direct proof of the concept of rotational catalysis.

The biological rotary motor with 5 nm radius produced a torque of about 40 pN nm when rotating one μm-long actin filament with 4 rounds per second (r.p.s.) during ATP hydrolysis. Using the curvature analysis of the bent actin filament in subsequent experiments, the torque was refined and about 56 pN nm was computed (Junge et al., 2001; Pänke et al., 2001).

From biochemical measurements of the ATP hydrolysis rate, the expected speed of rotation would have been 17 r.p.s. without the viscous drag by the attached actin filament. However, only about 1 out of 70 filaments per field-of-view rotated, while most of the actin filaments did not move. In rare cases, the anticipated 120° step size was observed for the rotating actin filament on γ. Lower ATP concentrations were required to clearly resolve the 120° stepping of the γ-subunit in F_1 (Kinosita et al., 1998; Yasuda et al., 1998). In the absence of ATP (or in the presence of azide, which inhibits ATP hydrolysis) moving actin filaments did not rotate continuously but fluctuated back and forth due to Brownian motion.

Adopting this video microscopy approach, the co-rotation of the ε subunit was demonstrated (Kato-Yamada et al., 1998). The direct visualization of a single rotating ε-subunit required a stable enzyme with non-dissociating ε-subunits at nanomolar concentrations. Thus, F_1 from PS3 was suitable to confirm the assignment of the ε subunit to the rotor part of the enzyme, which had been concluded from cross-linking experiments of the γ- and ε-subunits (Aggeler et al., 1997). Subsequently Masamitsu Futai and coworkers showed that the membrane-embedded part of the rotor subunits, that is, the c-ring of the *E. coli* F_1F_o-ATP synthase, rotated during ATP hydrolysis (Sambongi et al., 1999) with respect to the stator subunits ab_2-$\delta\alpha_3\beta_3$. This observation was confirmed afterward (Pänke et al., 2000; Tanabe et al., 2001; Ueno et al., 2005, Ishmukhametov et al., 2010).

The hydrodynamic friction of the rotation marker – that is, the actin filament or streptavidin-coated polystyrene beads with 250 to 500 nm in diameter – hampered rotation and limited the maximum achievable speed at high ATP concentrations. Improvements for drag-free measurements were made using a single nanometer-sized fluorophore as a pointer of γ-subunit orientation. Wolfgang Junge and coworkers refined their spectroscopic approach of monitoring subunit rotation by single-fluorophore excitation anisotropy (Hasler et al., 1998). Immobilizing single CF_1 on a cover glass containing a γ-subunit specifically labeled with tetramethylrhodamine (TMR) allowed confocal detection of fluorescence anisotropy. Using linearly polarized excitation of the TMR fluorophore, distinct fluorescence intensity levels corresponded to different orientations of the transition dipole moments of the fluorophore. During ATP hydrolysis, a sequence of changing intensity levels indicated the stepwise rotation of TMR with transition time faster than 10 ms. Accordingly, the γ-subunit moved with respect to the non-rotating $\alpha_3\beta_3$ subunits in CF_1, which were attached to the surface. TMR bound to the ε-subunit also showed these sequential fluorescence intensity fluctuations. On the level of single fluorophores, intensity changes might also reflect variations of the photophysical properties in the local protein environment in the millisecond time range. Therefore, measuring only constant intensity levels of TMR on the non-rotating δ-subunit was taken as the negative control experiment to exclude such photophysical artifacts.

Fluorescence anisotropy changes of single cyanine dyes (Cy3) attached to the c-subunits were analyzed to demonstrate rotation in membrane-embedded Na^+-driven F_1F_o-ATP synthases from *Propionigenium modestum*. The F_1F_o-ATP synthase was attached to the glass surface via his-tags on the β-subunits (Figure 11.11c). Circularly polarized laser light ensured the excitation of all Cy3 chromophores independent from their actual orientation on the c-subunits. Instead, fluorescence of Cy3 was split into two orthogonal polarized detection channels in the confocal microscope. Localization of the individual F_1F_o-ATP synthase was achieved using piezo-driven sample scanning of the surface. Subsequently, the fluorescence anisotropy time trajectories of Cy3 were recorded for selected F_1F_o. At low [ATP], stepwise changes of the fluorescence could

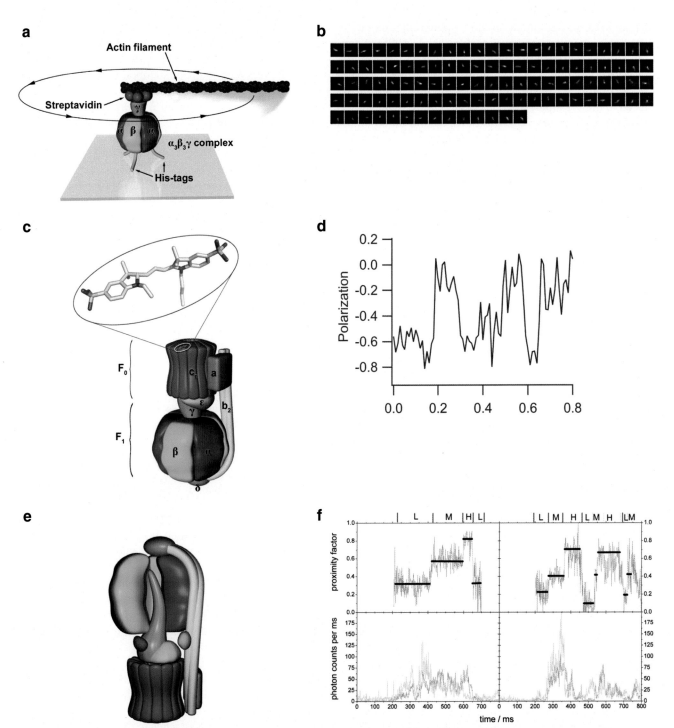

FIGURE 11.11: *Observing ATP-driven subunit rotation in single F_1 and F_1F_o-ATP synthases. (a) F_1 comprising subunits $\alpha_3\beta_3\gamma$ is attached to the glass surface via His-tags, and γ-rotation is monitored by a biotin-streptavidin connected, 1 to 3 μm-long fluorescent actin filament. (b) Image sequence of the actin filament showing counter-clockwise rotary movement in the presence of ATP (three full rotations). (c) F_1F_o in detergent or reconstituted into a liposome is attached to the glass surface via His-tags and a single Cy3 fluorophore reports c-subunit rotation by polarization-resolved fluorescence. (d) Time trajectory of fluorescence polarization, showing stepwise large fluctuations in the presence of ATP. (e) measuring rotation using intramolecular distance measurements based on fluorescence resonance energy transfer, FRET, between two fluorophores on a single liposome-reconstituted freely diffusing F_1F_o. The donor fluorophor TMR (green sphere) is bound to the static b-subunits and the acceptor fluorophore Cy5 (red sphere) to the rotating γ-subunit. (f) During ATP hydrolysis the ratio of fluorescence intensities of TMR (green traces) and Cy5 (red traces) changes sequentially in three steps indicating fluorophore distance changes (or proximity changes, blue trace) caused by γ-subunit rotation in 120°. (Figures reproduced from Noji et al., (1997) [a, b], Kaim et al., (2002) [c, d] and Börsch et al., (2002) [e, f], with kind permission.)*

be detected (Figure 11.11d). Dwell times for constant anisotropy levels in the presence of 0.5 μM ATP were found between 100 and 500 ms, corresponding to ATP hydrolysis rates in the range <10 s^{-1}, which is in agreement with the expected turnover for this ATP concentration. At high [ATP], the limited number of photons per time interval did not allow the steps to be resolved. Instead, the fast fluctuations of the anisotropies were analyzed by its autocorrelation function (ACF). Oscillations of the ACFs in the range of 20 to 40 ms indicated time constants 12.5 to 25 s^{-1} for the movement of the Cy3 chromophore attached to the c-subunits. The yield of rotating c-subunits in F_1F_o was found in the range of 70%, revealing the benefits of minimal sterical interference between the marker of rotation and the c-subunits. Use of small probes on the rotor subunit and reconstitution of the F_1F_o-ATP synthase into a lipid vesicle also enabled the measurement of subunit rotation during Na$^+$-driven ATP synthesis (see discussion later in this chapter).

A different approach to monitor rotation with single fluorophores was reported simultaneously from Peter Gräber and coworkers (Börsch et al., 2002). Here, rotation of the γ-subunit in membrane-embedded F_1F_o-ATP synthase was detected by changes in intra-molecular distances between one fluorophore on the rotating γ-subunit and a second fluorophore on the static b-subunits (Figure 11.11e). The distance measurement is based on Förster-type fluorescence resonance energy transfer (FRET) between the two fluorophores. As the rotor subunit moves, the distances are expected to change stepwise (Figure 11.11f). The intra-molecular distance measurement does not require a surface immobilization. Thus, F_1F_o-ATP synthase can diffuse and rotate within the lipid bilayer, and ATP synthesis can be induced by applying a pH gradient across the membrane through rapid buffer mixing. The time resolution for each internal FRET measurement was 1 ms per data point. Single-molecule FRET measurements resulted in a number of key findings: a three-stepped rotation of the γ- and ε-subunit at high [ATP] with dwell times between 13 and 19 ms, asymmetry of the three catalytic sites on the β-subunits depending on the attachment of the stator (Zimmermann et al., 2005), three-dimensional localization of the proton-translocating a-subunit (Düser et al., 2008), subunit contributions to the binding affinities of the F_1 part to the F_o part (Krebstakies et al., 2005), action mode of an non-competitive inhibitor of ATP hydrolysis aurovertin B (Johnson et al., 2009), and inversion in the direction of rotation during ATP synthesis compared to ATP hydrolysis (Diez et al., 2004). The transition time for one 120° step was estimated by single-molecule FRET to be in the range of 200 μs (Zimmermann et al., 2006). However, this approach with freely diffusing proteoliposomes limited the observation time of rotation in F_1F_o-ATP synthase to several hundred milliseconds. Thus, rotation measurements were possible only at high [ATP] for

ATP hydrolysis, or at large driving forces for ATP synthesis, namely high ΔpH values plus high electric potential differences.

Observation times for FRET-based rotation analysis could be prolonged by immobilizing F_1-ATPase to the surface via His-tags (Shimabukuro et al., 2003). At low [ATP], the identification of the relative γ-subunit orientation revealed that the MF_1 crystal structure was trapped in the catalytic state. In addition, single-molecule FRET measurements proved that the ε-subunit is changing its C-terminal conformation in an up/down fashion in response to the regulatory function to prevent ATP hydrolysis (Iino et al., 2005; Saita et al., 2010).

To resolve the details of rotary motion under drag-free conditions and for long observation times at low [ATP], a small and non-photobleachable probe had to be attached to the γ-subunit. Using a 40-nm gold bead as a light scatterer in laser dark field microscopy, Itoh and coworkers resolved sub-steps of γ-subunit rotation with a high-speed camera (Yasuda et al., 2001). At 8,000 frames per second, they discriminated the ATP waiting orientation of γ at an angle of 0° (or 120° or 240°) from the catalytic orientation for ATP hydrolysis at an angle of 80° to 90° (or 200/210° or 320/330°). The transition time at high [ATP] as well as low [ATP] was faster and in the range of 0.25 ms. From global fitting the rise and decay times in the dwell time histograms at several ATP concentrations, at least two different processes were found to be associated with the catalytic dwell, that is, the release of ADP and of phosphate as the products of ATP hydrolysis.

Subsequent work resolved the concerted participation of each β-subunit in the rotary reaction cycle. Correlating ATP binding to one catalytic site with a γ angle of 0°, this ATP is hydrolyzed after rotation of γ to 200°, which required sequential ATP hydrolysis on the two other calatytic sites (Ariga et al., 2007). Using fluorescent ATP derivatives in combination with a bead on γ as the marker of rotation, the timing of phosphate and ADP release with respect to γ-subunit rotation was unravelled (Adachi et al., 2007). Accordingly, the 40° rotation sub-step of γ corresponds to the release of phosphate. The conformational changes of the β-subunits enforcing γ rotation were identified by single-fluorophore anisotropy measurements in combination with the simultaneous bead-based monitoring of rotation (Masaike et al., 2008).

The time resolution for the rotary movements was improved further by the use of gold nanorods with a length of 40 to 90 nm as a light scatterer. Wayne Frasch and coworkers used the wavelength dependence of the scattered light from gold nanorods to measure rotation at 400,000 data points per second, that is, with 2.5 μs time resolution (Spetzler et al., 2006). The transition time for the 120° step of γ in *E. coli* F_1 was 270 μs, with a catalytic dwell time of 8.3 ms. The torque profile for the 120° movement was flat. Subsequently the torque was measured at different viscosities using the gold nanorod as marker, and found to be 63 pN nm (Hornung et al., 2008).

II.3 First Single-Molecule Demonstration of Rotation During ATP Synthesis

To observe subunit rotation in F_1F_o-ATP synthase during ATP synthesis requires a functional enzyme with all subunits present. ATP synthesis is driven by a pH difference (or Na^+ concentration difference, respectively) plus an electric potential across the lipid membrane. Accordingly F_1F_o-ATP synthase has to be reconstituted into lipid vesicles. A convenient way to identify subunit rotation during ATP synthesis is to cross-link those rotor subunits that had been shown to rotate during ATP hydrolysis (see earlier discussion). When F_1F_o-ATP synthase from E. coli was reconstituted in liposomes, Rod Capaldi and coworkers showed that the complete rotor – that is, the γ-, ε- and c-subunits – can be cross-linked simultaneously without affecting ATP hydrolysis nor ATP synthesis (Tsunoda et al., 2001a). The cross-linked enzyme retained its full inhibitory sensitivity, which was partly missing in detergent-solubilized F_1F_o-ATP synthase.

Thomas Duncan and Richard Cross applied selective radioactive labeling of β-subunits and cross-linking with γ in F_1 (Zhou et al., 1997). F_1 consisted of γ cross-linked to an unlabeled β- and two radioactive-flagged β-subunits. F_1 was reassembled on F_o in inverted E. coli membrane particles. Building up a ΔpH plus electric potential was achieved by addition of NADH. In the presence of ADP and phosphate, ATP synthesis resulted in a randomization of γ-β contacts after reduction of the cross-link, followed by several turnovers and the final cross-linking in the presence of oxidating reagents. Initiation of rotation during ATP synthesis could be inhibited by the addition of DCCD, or by omitting ADP and phosphate, respectively.

The first single fluorophore-based detection of c-ring rotation during ATP synthesis was reported in 2002 using the anisotropy of the dye (Kaim et al., 2002). Na^+-driven F_1F_o-ATP synthase was attached to a glass surface via his-tags and reconstituted into liposomes (Figure 11.12a). ATP synthesis was initiated by the addition of Na^+. The fluctuation of the fluorescence anisotropy was too fast to allow identification of steps in c-ring rotation. However, the autocorrelation of the anisotropy time trajectories showed oscillations with time periods of 20 to 40 ms, that is, repeating time periods for the re-orientations of the transition dipole of the dye on the rotating c-ring, which were abolished in the presence of DCCD blocking rotation (Figure 11.12b).

In solution, subunit rotation during ATP synthesis was shown by several types of single-molecule FRET approaches (Diez et al., 2004; Zimmermann et al., 2005; Düser et al., 2009). The FRET measurements of rotation revealed the opposite directions of γ- and ε-subunit rotation as compared to ATP hydrolysis. Three-stepped rotation of γ and ε was detected at maximum driving forces. In contrast, rotation of the proton-driven c-ring during ATP synthesis occurred in smaller steps (Figure 11.12c, d), which corresponded to the translocation mode of one proton at a time, or 36° step size, respectively (Düser et al., 2009).

As an alternative, forced mechanical rotation of the γ-subunit in F_1 in a sealed femtoliter-sized volume resulted in ATP synthesis (Figure 11.12e). A magnetic bead attached to γ was rotated by external field in the clockwise direction in the presence of ADP and phosphate. Synthesized ATP was detected biochemically using luciferin/luciferase (Itoh et al., 2004) or by video microscopy of the rotating bead (Figure 11.12f) during hydrolysis of the synthesized ATP after releasing the magnetic forces (Rondelez et al., 2005).

II.4 How Fast is Subunit Rotation in ATP Synthase?

Single-molecule determination of rotary motion in E. coli F_1 ATPase showed a broad distribution of rotational speeds. With more than 700 rounds per second, some F_1 rotated about 10 times faster than expected from biochemical ensemble assays for those conditions (Nakanishi-Matsui et al., 2006). For a single F_1, the rotational speed changes over time, indicating the stochastic nature of the rotation. Given these rotational speed limits, the question is raised about the most appropriate method to measure the rotational speed. The highest angular resolution of rotary step sizes is achieved with strong light-scattering probes like polystyrene beads or long actin filaments. Problems of unknown surface properties, charges, and obstacles due to surface roughness in the vicinity of the 10-nanometer-sized F_1-ATPase with the attached marker remain with this approach. However, to overcome the drawback of viscous drag, one has to use smaller gold nanoparticles (beads or rods) with sizes around 50 nm in combination with strong laser excitations to obtain the required signal-to-noise ratio (Spetzler et al., 2009). Using fluctuation theorem–based analysis of the probabilities of forward and backward steps improves the estimation of torque for an unknown viscosity on the surface (Hayashi et al., 2010).

In contrast, fluorophore-based measurements of rotation are not affected by viscous drag as the marker size is in the range 1 nm. Whereas fluorescence anisotropy measurements require surface attachment of the enzyme, intramolecular FRET measurements can be carried out using liposome-reconstituted enzymes. Hence rotation measurements during ATP synthesis are easily possible. However, photophysics and photochemistry (i.e., photobleaching) of the dyes limit the applicability of FRET measurements to observation times of milliseconds to seconds, thus demanding fast turnover at high driving forces. New ultraphotostable fluorophores based on nanocrystals such as fluorescent nanodiamonds (Tisler et al., 2009) with sizes below

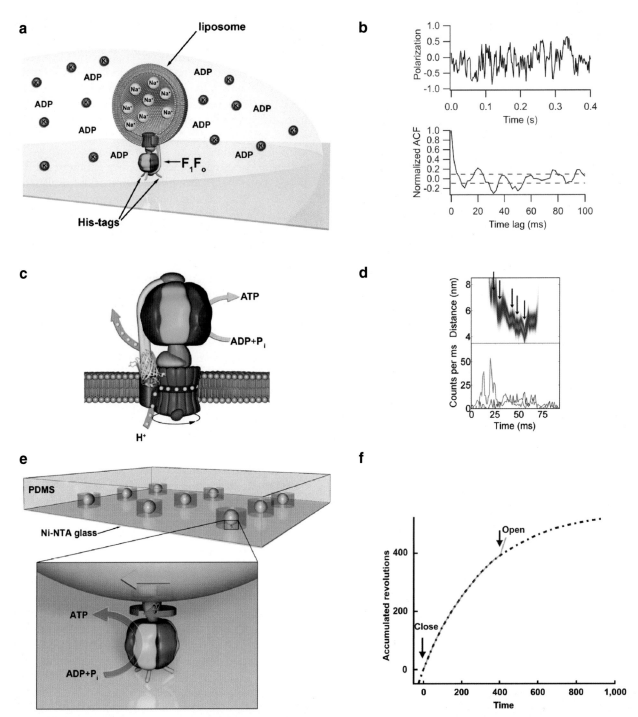

FIGURE 11.12: *Observing subunit rotation in single F_1F_o-ATP synthases during ion-driven ATP synthesis, and mechanically enforced in F_1 parts. (a) F_1F_o (from* Propionigenium modestum*) in liposomes is attached to the glass surface* via *His-tags, and ATP synthesis is started by generating a sodium gradient across the membrane. (b) Rotation of the c-subunit is measured by fast fluorescence polarization changes of a single Cy3 fluorophore on (c) (red sphere). Rotational speed is calculated from the autocorrelation function (ACF) of the fluorescence polarization time trajectories (right curve). (c) FRET-based rotation measurement of c-subunits during proton-driven ATP synthesis in single liposome-reconstituted freely diffusing F_1F_o (E. coli). The donor fluorophor EGFP is fused to the static a-subunit and the acceptor fluorophore Cy5 to the rotating c-subunit. (d) The ratio of fluorescence intensities of EGFP (green traces) and Cy5 (red traces) changes sequentially in multiple steps indicating smaller fluorophore distance changes than 120°. (e) Mechanically driven rotation of a magnetic bead on γ in a His-tag-attached single F_1 comprising subunits $\alpha_3\beta_3\gamma$. Clockwise-forced rotation of γ results in ATP synthesis. ATP is stored in a sealed femtoliter-sized reaction chamber. (f) After release of the external magnetic fields, detection of synthesized ATP is achieved by recording ATP hydrolysis-driven rotation of the magnetic bead. Concentration of synthesized ATP is correlated to rotational speed or time-dependent accumulated revolutions, respectively. (Figures reproduced from Kaim et al., (2002) [a, b], Düser et al., (2009) [c, d], and Rondelez et al., (2005) [e, f], with kind permission.)*

5 nm may overcome the photobleaching constraint. Still, the low signal-to-background ratio and the small number of detected photons to assign an orientation of the rotor subunit make the use of autocorrelation analysis for faster dynamics necessary. Alternatively, the application of Hidden Markov Models (McKinney et al., 2006) to find the transitions in rotary orientations and to determine the dwell times will improve the time resolution to sub-millisecond kinetics.

Finally, rotation measurements under physiological conditions and in vivo – that is, in living bacterial cells – are required. So far, the reported rotation approaches used either fragments of the $F_1 F_o$-ATP synthase or have applied artificial conditions (ATP hydrolysis, or high ΔpH or large ion gradients with unknown electric potentials during ATP synthesis) for monitoring the dynamics. The in vivo conditions with a small ΔpH but high electric potential across the membrane might results in a completely different rotational speed or behavior. Small fluorophores reporting rotation might allow these measurements, but the questions of specific labeling methods in living cells have to be solved.

III. MECHANISM OF F_o ACTION

The F_o motor supplies ATP synthases with the energy required for ATP production and release. This energy derives from the transmembrane gradients of protons or Na^+ created by neighboring respiratory-chain complexes. To be able to harvest this energy, and to transduce it to the F_1 sector, the F_o motor provides a translocation pathway for ions across the membrane and couples their movement to a mechanism that ultimately causes the physical rotation of the central stalk, thereby driving the functional rotation of the catalytic subunits. The microscopic details of the rotary mechanism of ion transport by the F_o domain remain to be elucidated. Nonetheless, significant advancements have been made over the years. In this section, we will focus on recent progress in understanding several key aspects of this ion-transport mechanism, resulting from the synergistic combination of experimental and theoretical approaches.

III.1 Ion Selectivity

A crucial feature of the rotary mechanism of ion transport is its specificity. Clearly, the fact that ATP synthesis is coupled to a unique electrochemical ion gradient, which at the same time is consistent with other components of the respiratory chain, enables the cell to regulate ATP synthases with precision. In most organisms, including all eukaryotes, ATP synthesis is coupled to H^+ transport (Mitchell and Moyle, 1967). Only in a few selected anaerobic bacteria is synthesis coupled to translocation of Na^+ ions (Laubinger and Dimroth, 1987; Heise et al., 1991; Neumann et al., 1998; Ferguson et al., 2006). No other cations, and no anions, are employed by ATP synthases to power synthesis.

How is this ion selectivity accomplished at the molecular level? The answer to this question lies within the structure of the c-subunit ring. As explained in the previous section I.5.3, the c-ring contains a series of binding sites that allow it to sequester ions from its environment. Analysis of the amino-acid composition of these binding sites reveals that a carboxylate side-chain – mostly Glu, sometimes Asp – is strictly conserved across the entire family of ATP synthases. Polar and hydrophobic amino acids are also found, but other charged side chains are excluded, such as Lys, Arg, and so on.

The presence of the carboxylate group has very important consequences with regard to ion selectivity. The most evident of these is that it excludes all anions as a possible energy source, owing to the prohibitive electrostatic repulsion they would experience on account of the carboxylate side-chain. The Glu/Asp side chain also serves to exclude the larger cations. This is not so much because these would not fit in the space of the binding sites (although this may be the case in some rotors), but because the strong electrostatic field created by the charged carboxylate favors smaller cations in general, such as Na^+ (Noskov and Roux, 2008). Lastly, the Glu/Asp side chain provides a means to bind H^+ covalently, through the protonation of its carboxylate group. Based on these considerations, the selectivity question becomes more focused, namely How does the c-ring discriminate between Na^+ and H^+? The answer lies again in the amino-acid composition of the binding site, but also in the physiological context of the ATP synthase (Meier et al., 2009; Krah et al., 2010).

Strictly speaking, all ATP-synthase c-rings are proton-selective, including those that in a physiological setting are coupled to Na^+ gradients. To a first approximation, this reflects the fact that the carboxylate group has intrinsically a greater affinity for protons ($pK_a \sim 4$) than for Na^+ ions (monosodium glutamate is a dissociable salt). Nevertheless, the selectivity against Na^+ of the c-rings known to be H^+ coupled is several orders of magnitude greater than, for instance, that of glutamate in solution. Thus, the selectivity cannot be attributed to the carboxylate group alone. What confers these c-rings their extreme proton selectivity (see e.g., Feniouk et al., 2004) is that the other amino acid side chains in the binding site are primarily hydrophobic, and therefore cannot sustain the coordination of a cation such as Na^+, and thus compensate for the dehydration cost. In some cases, the size of these binding sites may also be restricted by the proximity of the c-subunits, or by bulky side chains, further enhancing their selectivity for protons, and against Na^+.

Conversely, the intrinsic H^+ selectivity of the carboxylate side chains is too strong to explain the prevalence of Na^+ over H^+ coupling, even though Na^+ is present in great excess over H^+ under most physiological conditions. Thus, those c-rings that are functionally coupled to Na^+ gradients do so because their H^+ selectivity is weaker than what the carboxylate group confers. This selectivity is achieved by the decoration of the binding sites with additional polar side chains that can coordinate Na^+, such as Gln or Ser.

In summary, the strictly conserved Glu/Asp side chain is employed by c-rings to sequester H^+ and Na^+, at the expense of larger cations and all anions. This side chain also confers the binding site with a baseline H^+ selectivity. In c-rings coupled to the proton-motive force, this selectivity is enhanced by an arrangement in which the carboxylate group is surrounded with hydrophobic side chains. Conversely, rotor rings that are coupled to the sodium-motive force instead reduce this default H^+ selectivity by providing additional polar side chains; under physiological conditions of excess Na^+, these weakly proton-selective rings will thus be coupled to Na^+ gradients (Krah et al., 2010).

III.2 Rotary Mechanism

In addition to their role in selective ion binding, the c-rings of ATP synthases provide the key rotary mechanism to these enzymes. Their ability to rotate within the membrane is intrinsic to their symmetric structure – that is, it owes to the fact that multiple, energetically equivalent interfaces with the a-subunit can be formed as the ring rotates. Assuming the energetic barrier between these equivalent states is not too large, the c-ring will therefore rotate stochastically as a result of thermal fluctuations, without a preferred directionality.

The presence of transmembrane electrochemical gradients, however, imposes a net directionality on this stochastic motion, provided that ion access pathways exist that extend from within the a/c interface toward the bulk solution, on either side of the membrane (Figure 11.13). Excess of ATP over ADP, and the corresponding switch into hydrolysis-mode, also biases the rotational motion of the ring, but in the opposite direction – thus creating an electrochemical gradient where there was none. The transport properties of the proton-driven F_o motor are characterized by an ohmic conductance behavior of 10 fS, a high specificity of a H^+/Na^+ ratio larger than 10^7, and only a very weak pH dependence for proton translocation (Feniouk et al., 2004).

The key questions with respect to the rotary mechanism are therefore the following: What is the structure of the interface of c-ring and a-subunit? By which mechanism does the c-ring and a-subunit alternate between equivalent states of this interface? And what is the nature of the ion access pathways into the binding sites? Unfortunately, the answers to these questions are not known in molecular

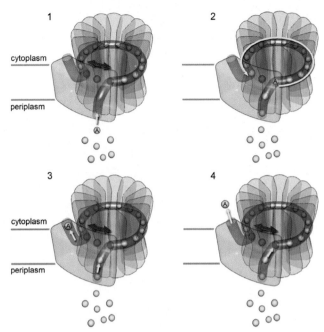

FIGURE 11.13: *Torque generation in the F_o complex. Four images showing torque generation by Brownian rotary motion and directed ion flow. The images show a c-ring (barrel with spots showing protonated sites in yellow and non-protonated and negatively charged sites in red) facing subunit a (transparent red, on the left) with its two laterally offset proton access and exit channels. The path of the proton is indicated by a yellow arrow. In step 1, the c-ring rotates back and forth in the plane of the membrane (indicated by the red arrow), as the charged site is not protonated and so avoids contact with the hydrophobic centre of the membrane. In step 2, after protonation has occurred, unidirectional rotation by one step has taken place (ATP synthesis direction, yellow arrow). Steps 3 and 4 show the proton in the exit channel of subunit a and a return to the initial state by further rotation of the c-ring. Redrawn from Junge et al., (2009).*

detail, owing to the limited structural information that is available on both the a-subunit and its interface with the c-ring.

This lack of understanding notwithstanding, it is generally accepted that there exist two independent, disconnected access pathways – one at the interface of c-ring and a-subunit; the other within the latter – which allow H^+ and Na^+ to be exchanged with the surrounding solution (Vik and Antonio, 1994; Junge et al., 1997; Angevine and Fillingame, 2003; Steed and Fillingame, 2008; Steed and Fillingame, 2009). Once an ion binds to the ring, it would be shuttled from one access channel to the other as the c-ring rotates in the membrane, effectively completing a pathway across the membrane (Vik and Antonio, 1994; Junge et al., 1997).

The principle that underlies this mechanism is that the occupancy of the c-subunit binding sites is dependent on the environment they face at a given time. As ions are shuttled within the membrane by the rotating

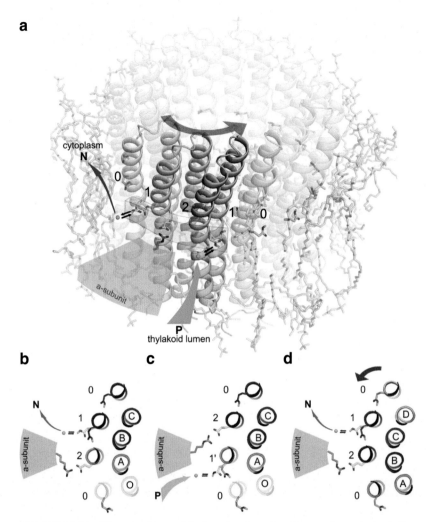

FIGURE 11.14: *Microscopic model of proton translocation and coupled rotation in the F_o complex from S. platensis. (a) Schematic representation of the different states envisaged by the binding sites (0-1-2-1'-0), as the c-subunits (green) face the lipid membrane or the hydrated interface (blue) of the a-subunit (grey). The view of the interface is tilted from the membrane plane ~20° toward the reader. (b, c, and d) Specific sequence of steps in the proposed mechanism; these are (b) opening and deprotonation; (c) rearrangement of Glu/Arg ion pair and reprotonation; (d) rotation. The c-subunits (colors) are shown from the cytoplasmic side at the level of the ion-binding sites. The a-subunit is schematically depicted (grey sector), highlighting the conserved arginine). The direction of the rotation in the figure is that in the ATP synthesis mode, but the model is evidently reversible. (Reproduced from Pogoryelov et al., (2010) with kind permission.)*

ring, they remain safely bound to the c-ring. However, the environment of the a-subunit somehow enables these sites to load and release ions from/to the access channels.

Recent structural and theoretical work focused on the H^+-coupled rings has confirmed the notion that the membrane provides a stable environment for the bound ion (Pogoryelov et al., 2010). The lack of hydration in the vicinity of the c-subunit binding sites implies that the pK_a of the conserved carboxylate side chain is strongly shifted upward. Hence, protons will remain tightly bound even under severely alkaline conditions.

The influence of the a-subunit interface is comparatively less clear. As mentioned earlier, the high-resolution structure of this interface has been elusive. Nonetheless, biochemical experiments have established that at this interface, the a-subunit displays a strictly conserved arginine side chain – which conceivably interacts with the also strictly conserved glutamate/aspartate side chain in the c-subunit binding site. This arginine is also believed to act as an electrostatic separator between the access pathways that allow ions to reach, or be released from, the ring (Vik and Antonio, 1994; Junge et al., 1997). A second crucial feature of this interface is that it is locally hydrated (Steed and Fillingame, 2009). This property is entirely consistent with the notion of access pathways, as hydration is required to facilitate ion permeation into the membrane, in the absence of direct protein coordination. Hydration would also be required to facilitate the sequential formation and dissociation of the proposed salt bridge between the a-subunit arginine and the c-subunit glutamate/aspartate, as the ring rotates within the membrane.

In consideration of these two important features, subsequent structural and theoretical work (Figure 11.14) has indicated that the influence of the a-subunit interface on the c-ring is two-fold (Pogoryelov et al., 2010). First, local hydration promotes a conformation of the binding site in which the conserved Glu/Asp adopts an alternate rotamer configuration, thus projecting toward the interface. Hydration of the side chain also restores its intrinsic pK_a in part, so that deprotonation may occur. Engagement by this carboxylate of the a-subunit arginine stabilizes the deprotonated state and contributes to rotating the ring by a one-subunit step. Meanwhile, the preceding subunit, now disengaged from the arginine side chain, is able to capture a proton from the second access pathway and fold back into the binding site, preparing the site to enter the membrane interface once again.

The detailed molecular model outlined above is reversible, which makes it applicable to both ATP synthesis and hydrolysis. A distinct characteristic of this recently proposed mechanism is the limited range of the structural changes in the c-subunit. This feature is consistent with the high-resolution structural data, in view of which large-scale changes in the c-ring structure are difficult to envisage. This tentative model is also consistent with the aforementioned notion that the a-subunit arginine prevents a short-circuit of the proton-transport mechanism (Mitome et al., 2010) by occluding a direct pathway between either side of the membrane. Lastly, an important characteristic of this model is that it would apply to the rotor rings of the V-type (and A-type) ATPases/synthases. In these rotors, more complex subunit architectures sometimes cause the ion-binding sites to be further apart from each other than in the classical c-rings found in F-type ATP synthases. This greater spacing in principle imposes a greater kinetic barrier for the subsequent dissociation and formation of the Glu/Asp-Arg salt-bridge pairs, but the presence of hydration would greatly facilitate such exchange. Similarly, the model can be generalized to Na^+-driven rotors.

This progress notwithstanding, high-resolution structures of the complete F_o complex are required to validate or refute this and other proposals, as well as to address further mechanistic questions, such as the potential role of conformational changes within the a-subunit, possibly coupled to the ring rotation, and the nature of the ion access pathways of this marvelous molecular machine.

ACKNOWLEDGEMENTS

TM: I want to express my gratitude to Peter Dimroth for many years of support and inspiration and to Werner Kühlbrandt for generous support and advice to build up my research at the Max-Planck-Institute of Biophysics, Frankfurt. I am very thankful to all my team members for their motivated and dedicated work and for sharing my fascination about the structure and function of the ATP synthase. I also wish to acknowledge the Cluster of Excellence 'Macriomolecular Complexes,' the Collaborative Research Center (SFB) 807 of the German Research Foundation (DFG) and the European Science Foundation (ESF, EuroSYNBIO) for financial support.

JFG: I am deeply thankful to Alexander Krah and Vanessa Leone for their contributions to the theoretical studies of the structure, selectivity and mechanism of F_o carried out in my group at the Max-Planck-Institute of Biophysics, and the Jülich Supercomputing Centre for the resources made available to us. I also wish to acknowledge the financial support from the Cluster of Excellence 'Macromolecular Complexes' and the Max-Planck Society.

MB: I want to thank all my coworkers at the 3rd Institute of Physics (Stuttgart) carefully performing and continuously improving the actual biophysical studies on F_1F_o-ATP synthase, namely M. G. Düser, N. Zarrabi, S. Ernst, T. Rendler, and A. Zappe, and my former coworkers at the Institute for Physical Chemistry (Freiburg), namely M. Diez, F. M. Boldt, S. Steigmiller, and B. Zimmermann, for developing the single-molecule FRET experiments. I am grateful for the enduring collaborations with S. D. Dunn (University of Western Ontario, London, Canada) and G. D. Glick (University of Michigan, Ann Arbor) for the strong biochemical support, and for the ongoing generous support by my mentors J. Wrachtrup (Stuttgart) and P. Gräber (Freiburg).

All authors thank Wolfgang Junge for his critical reading of this manuscript and his helpful suggestions. We are very grateful to Paolo Lastrico for creating the figures with the ATP synthase models (Figures 3, 7c, 11abc, 12abc and 13).

REFERENCES

Abrahams, J. P., Leslie, A. G. W., Lutter, R. & Walker, J. E. (1994) Structure at 2.8 Å resolution of F_1-ATPase from bovine heart mitochondria. *Nature*, 370, 621–628.

Adachi, K., Oiwa, K., Nishizaka, T., Furuike, S., Noji, H., Itoh, H., Yoshida, M. & Kinosita, K., Jr. (2007) Coupling of rotation and catalysis in F_1-ATPase revealed by single-molecule imaging and manipulation. *Cell*, 130, 309–321.

Aggeler, R., Ogilvie, I. & Capaldi, R. A. (1997) Rotation of a γ-ε subunit domain in the *Escherichia coli* F_1F_o-ATP synthase complex. The γ-ε subunits are essentially randomly distributed relative to the $\alpha_3\beta_3\delta$ domain in the intact complex. *J. Biol. Chem.*, 272, 19621–19624.

Aggeler, R., Weinreich, F. & Capaldi, R. A. (1995) Arrangement of the ε subunit in the *Escherichia coli* ATP synthase from the reactivity of cysteine residues introduced at different positions in this subunit. *Biochim. Biophys. Acta*, 1230, 62–68.

Angevine, C. M. & Fillingame, R. H. (2003) Aqueous access channels in subunit a of rotary ATP synthase. *J. Biol. Chem.*, 278, 6066–6074.

Angevine, C. M., Herold, K. A., Vincent, O. D. & Fillingame, R. H. (2007) Aqueous access pathways in ATP synthase subunit a. Reactivity of cysteine substituted into transmembrane helices 1, 3, and 5. *J. Biol. Chem.*, 282, 9001–9007.

Arechaga, I., Butler, P. J. & Walker, J. E. (2002) Self-assembly of ATP synthase subunit c rings. *FEBS Lett.*, 515, 189–193.

Ariga, T., Muneyuki, E. & Yoshida, M. (2007) F_1-ATPase rotates by an asymmetric, sequential mechanism using all three catalytic subunits. *Nat. Struct. Mol. Biol.*, 14, 841–846.

Baddiley, J., Michelson, A. M. & Todd, A. R. (1949) Synthesis of adenosine triphosphate. *Nature*, 161, 761.

Ballhausen, B., Altendorf, K. & Deckers-Hebestreit, G. (2009) Constant c_{10} ring stoichiometry in the *Escherichia coli* ATP synthase analyzed by cross-linking. *J. Bacteriol.*, 191, 2400–2404.

Bernal, R. A. & Stock, D. (2004) Three-dimensional structure of the intact *Thermus thermophilus* H^+-ATPase/synthase by electron microscopy. *Structure*, 12, 1789–1798.

Birkenhäger, R., Hoppert, M., Deckers-Hebestreit, G., Mayer, F. & Altendorf, K. (1995) The F_o complex of the *Escherichia coli* ATP synthase. Investigation by electron spectroscopic imaging and immunoelectron microscopy. *Eur. J. Biochem.*, 230, 58–67.

Boekema, E. J., Ubbink-Kok, T., Lolkema, J. S., Brisson, A. & Konings, W. N. (1997) Visualisation of the peripheral stalk in V-type ATPase: Evidence for a stator structure essential to rotational catalysis. *Proc. Natl. Acad. Sci. U.S.A.*, 94, 14291–14293.

Börsch, M., Diez, M., Zimmermann, B., Reuter, R. & Gräber, P. (2002) Stepwise rotation of the γ-subunit of EF_oF_1-ATP synthase observed by intramolecular single-molecule fluorescence resonance energy transfer. *FEBS Lett.*, 527, 147–152.

Boyer, P. D. (1993) The binding change mechanism for ATP synthase – some probabilities and possibilities. *Biochim. Biophys. Acta*, 1140, 215–250.

Boyer, P. D. (1997a) The ATP synthase – a splendid molecular machine. *Annu. Rev. Biochem.*, 66, 717–749.

Boyer, P. D. (1997b) Energy, Life, and ATP. *Nobel Lecture, December 8, 1997*.

Boyer, P. D., Cross, R. L. & Momsen, W. (1973) A new concept for energy coupling in oxidative phosphorylation based on a molecular explanation of the oxygen exchange reactions. *Proc. Natl. Acad. Sci. U.S.A.*, 70, 2837–2839.

Cain, B. D. (2000) Mutagenic analysis of the F_o stator subunits. *J. Bioenerg. Biomembr.*, 32, 365–3671.

Cain, B. D. & Simoni, R. D. (1986) Impaired proton conductivity resulting from mutations in the a subunit of F_1F_o ATPase in *Escherichia coli*. *J. Biol. Chem.*, 261, 10043–10050.

Cain, B. D. & Simoni, R. D. (1988) Interaction between Glu-219 and His-245 within the a subunit of F_1F_o-ATPase in *Escherichia coli*. *J. Biol. Chem.*, 263, 6606–6612.

Cain, B. D. & Simoni, R. D. (1989) Proton translocation by the F_1F_o ATPase of *Escherichia coli*. Mutagenic analysis of the a subunit. *J. Biol. Chem.*, 264, 3292–3300.

Capaldi, R. A. & Aggeler, R. (2002) Mechanism of the F_1F_o-type ATP synthase, a biological rotary motor. *Trends Biochem. Sci.*, 27, 154–160.

Cipriano, D. J., Wood, K. S., Bi, Y. & Dunn, S. D. (2006) Mutations in the dimerization domain of the b subunit from the *Escherichia coli* ATP synthase. Deletions disrupt function but not enzyme assembly. *J. Biol. Chem.*, 281, 12408–12413.

Cook, G. M., Keis, S., Morgan, H. W., von Ballmoos, C., Matthey, U., Kaim, G. & Dimroth, P. (2003) Purification and biochemical characterization of the F_1F_o-ATP synthase from thermophilic *Bacillus* sp. strain TA2.A1. *J. Bacteriol.*, 185, 4442–4449.

Cox, G. B., Fimmel, A. L., Gibson, F. & Hatch, L. (1986) The mechanism of ATP synthase: a reassessment of the functions of the b and a subunits. *Biochim. Biophys. Acta*, 849, 62–69.

Cox, G. B., Jans, D. A., Fimmel, A. L., Gibson, F. & Hatch, L. (1984) Hypothesis. The mechanism of ATP synthase. Conformational change by rotation of the β-subunit. *Biochim. Biophys. Acta*, 768, 201–208.

Cross, R. L. (1981) The mechanism and regulation of ATP synthesis by F_1-ATPases. *Annu. Rev. Biochem.*, 50, 681–714.

Cross, R. L. & Müller, V. (2004) The evolution of A-, F-, and V-type ATP synthases and ATPases: reversals in function and changes in the H^+/ATP coupling ratio. *FEBS Lett.*, 576, 1–4.

Daley, D. O., Rapp, M., Granseth, E., Melen, K., Drew, D. & von Heijne, G. (2005) Global topology analysis of the *Escherichia coli* inner membrane proteome. *Science*, 308, 1321–1323.

Dallmann, H. G., Flynn, T. G. & Dunn, S. D. (1992) Determination of the 1-ethyl-3-[(3-dimethylamino)propyl]-carbodiimide-induced cross-link between the β and ε subunits of *Escherichia coli* F_1-ATPase. *J. Biol. Chem.*, 267, 18953–18960.

Davidson, A. L. & Chen, J. (2004) ATP-binding cassette transporters in bacteria. *Annu. Rev. Biochem.*, 73, 241–268.

Del Rizzo, P. A., Bi, Y. & Dunn, S. D. (2006) ATP synthase b subunit dimerization domain: a right-handed coiled coil with offset helices. *J. Mol. Biol.*, 364, 735–746.

Del Rizzo, P. A., Bi, Y., Dunn, S. D. & Shilton, B. H. (2002) The "second stalk" of *Escherichia coli* ATP synthase: structure of the isolated dimerization domain. *Biochemistry*, 41, 6875–68784.

Dickson, V. K., Silvester, J. A., Fearnley, I. M., Leslie, A. G. & Walker, J. E. (2006) On the structure of the stator of the mitochondrial ATP synthase. *EMBO J.*, 25, 2911–2918.

Diez, M., Zimmermann, B., Börsch, M., König, M., Schweinberger, E., Steigmiller, S., Reuter, R., Felekyan, S., Kudryavtsev, V., Seidel, C. A. & Gräber, P. (2004) Proton-powered subunit rotation in single membrane-bound F_oF_1-ATP synthase. *Nat. Struct. Mol. Biol.*, 11, 135–141.

Dimroth, P. (1997) Primary sodium ion translocating enzymes. *Biochim. Biophys. Acta*, 1318, 11–51.

Dmitriev, O., Jones, P. C., Jiang, W. & Fillingame, R. H. (1999) Structure of the membrane domain of subunit b of the *Escherichia coli* F_oF_1 ATP synthase. *J. Biol. Chem.*, 274, 15598–15604.

Duncan, T. M., Bulygin, V. V., Zhou, Y., Hutcheon, M. L. & Cross, R. L. (1995) Rotation of subunits during catalysis by *Escherichia coli* F_1-ATPase. *Proc. Natl. Acad. Sci. U.S.A.*, 92, 10964–10968.

Dunn, S. D. (1992) The polar domain of the b subunit of *Escherichia coli* F_1F_o-ATPase forms an elongated dimer that interacts with the F_1 sector. *J. Biol. Chem.*, 267, 7630–7636.

Dunn, S. D. & Chandler, J. (1998) Characterization of a $b_2\delta$ complex from *Escherichia coli* ATP synthase. *J. Biol. Chem.*, 273, 8646–8651.

Düser, M. G., Bi, Y., Zarrabi, N., Dunn, S. D. & Börsch, M. (2008) The proton-translocating a subunit of F_oF_1-ATP synthase is allocated asymmetrically to the peripheral stalk. *J. Biol. Chem.*, 283, 33602–33610.

Düser, M. G., Zarrabi, N., Cipriano, D. J., Ernst, S., Glick, G. D., Dunn, S. D. & Börsch, M. (2009) 36 degrees step size of proton-driven c-ring rotation in F_oF_1-ATP synthase. *EMBO J.*, 28, 2689–2696.

Eya, S., Maeda, M. & Futai, M. (1991) Role of the carboxy terminal region of H^+-ATPase (F_oF_1) a subunit from *Escherichia coli. Arch. Biochem. Biophys.*, 284, 71–77.

Feniouk, B. A., Kozlova, M. A., Knorre, D. A., Cherepanov, D. A., Mulkidjanian, A. Y. & Junge, W. (2004) The of ATP synthase: ohmic conductance (10 fS), and absence of voltage gating. *Biophys. J.*, 86, 4094–4109.

Feniouk, B. A. & Yoshida, M. (2008) Regulatory mechanisms of proton-translocating F_oF_1-ATP synthase. *Results Probl. Cell Differ.*, 45, 279–308.

Ferguson, S. A., Keis, S. & Cook, G. M. (2006) Biochemical and molecular characterization of a Na^+-translocating F_1F_o-ATPase from the thermoalkaliphilic bacterium *Clostridium paradoxum. J. Bacteriol.*, 188, 5045–5054.

Ferguson, S. J. (2000) ATP synthase: what dictates the size of a ring? *Curr. Biol.*, 10, R804–R808.

Ferguson, S. J. (2010) ATP synthase: From sequence to ring size to the P/O ratio. *Proc. Natl. Acad. Sci. U.S.A.*, 107, 16755–16756.

Fillingame, R. H. (1990) Molecular mechanism of ATP synthesis by F_1F_o- Type H^+-transporting ATP synthases. *The Bacteria*, XII, 345–391.

Fillingame, R. H. (1992) H^+ transport and coupling by the F_o sector of the ATP synthase: insights into the molecular mechanism of function. *J. Bioenerg. Biomembr.*, 24, 485–491.

Folch, J. & Lees, M. (1951) Proteolipides, a new type of tissue lipoproteins; their isolation from brain. *J. Biol. Chem.*, 191, 807–817.

Forgac, M. (2007) Vacuolar ATPases: rotary proton pumps in physiology and pathophysiology. *Nat. Rev. Mol. Cell Biol.*, 8, 917–929.

Foster, D. L. & Fillingame, R. H. (1982) Stoichiometry of subunits in the H^+-ATPase complex of *Escherichia coli. J. Biol. Chem.*, 257, 2009–2015.

Fritz, M., Klyszejko, A. L., Morgner, N., Vonck, J., Brutschy, B., Müller, D. J., Meier, T. & Müller, V. (2008) An intermediate step in the evolution of ATPases: a hybrid F_o-V_0 rotor in a bacterial Na^+ F_1F_o ATP synthase. *FEBS J.*, 275, 1999–2007.

Fromme, P., Boekema, E. J. & Gräber, P. (1987) Isolation and characterization of a supramolecular complex of subunit III of the ATP-synthase from chloroplasts. *Z. Naturforsch.*, 42c, 1239–1245.

Gambale, F., Kolb, A., Cantù, A. M. & Hedrich, R. (1993) The voltage-dependent H^+-ATPase of the sugar beet vacuole is reversible. *Eur. Biophys. J.*, 22, 399–403.

Gibbons, C., Montgomery, M. G., Leslie, A. G. & Walker, J. E. (2000) The structure of the central stalk in bovine F_1-ATPase at 2.4 Å resolution. *Nat. Struct. Biol.*, 7, 1055–1061.

Girvin, M. E., Rastogi, V. K., Albildgaard, F., Markley, J. L. & Fillingame, R. H. (1998) Solution structure of the transmembrane H^+-transporting subunit c of the F_1F_o ATP synthase. *Biochemistry*, 37, 8817–8824.

Gogol, E. P. (1994) Electron microscopy of the F_1F_o ATP synthase: from structure to function. *Microsc. Res. Tech.*, 27, 294–306.

Gogol, E. P., Johnston, E., Aggeler, R. & Capaldi, R. A. (1990) Ligand-dependent structural variations in *Escherichia coli* F_1 ATPase revealed by cryoelectron microscopy. *Proc. Natl. Acad. Sci. U.S.A.*, 87, 9585–9589.

Greie, J. C., Heitkamp, T. & Altendorf, K. (2004) The transmembrane domain of subunit b of the *Escherichia coli* F_1F_O ATP synthase is sufficient for H^+-translocating activity together with subunits a and c. *Eur. J. Biochem.*, 271, 3036–3042.

Groth, G., Tilg, Y. & Schirwitz, K. (1998) Molecular architecture of the c-subunit oligomer in the membrane domain of F-ATPases probed by tryptophan substitution mutagenesis. *J. Mol. Biol.*, 281, 49–59.

Groth, G. & Walker, J. E. (1997) Model of the c-subunit oligomer in the membrane domain of F-ATPases. *FEBS Lett.*, 410, 117–123.

Hara, K. Y., Kato-Yamada, Y., Kikuchi, Y., Hisabori, T. & Yoshida, M. (2001) The role of the βDELSEED motif of F_1-ATPase: propagation of the inhibitory effect of the ε-subunit. *J. Biol. Chem.*, 276, 23969–23973.

Hartzog, P. E. & Cain, B. D. (1994) Second-site supressor mutations at Glycine 218 and Histidine 245 in the a subunit of F_1F_o ATP synthase in *Escherichia coli. J. Biol. Chem.*, 269, 32313–32317.

Hasler, K., Engelbrecht, S. & Junge, W. (1998) Three-stepped rotation of subunits γ and ε in single molecules of F-ATPase as revealed by polarized, confocal fluorometry. *FEBS Lett.*, 426, 301–304.

Hayashi, S., Ueno, H., Iino, R. & Noji, H. (2010) Fluctuation theorem applied to F_1-ATPase. *Phys. Rev. Lett.*, 104, 1–4.

Heise, R., Reidlinger, J., Müller, V. & Gottschalk, G. (1991) A sodium-stimulated ATP synthase in the acetogenic bacterium *Acetobacterium woodii. FEBS Lett.*, 295, 119–122.

Hicks, D. B. & Krulwich, T. A. (1990) Purification and reconstitution of the F_1F_o-ATP synthase from alkaliphilic *Bacillus firmus* OF4. Evidence that the enzyme translocates H^+ but not Na^+. *J. Biol. Chem.*, 265, 20547–20554.

Hicks, D. B., Liu, J., Fujisawa, M. & Krulwich, T. A. (2010) F_1F_o-ATP synthases of alkaliphilic bacteria: lessons from their adaptations. *Biochim. Biophys. Acta*, 1797, 1362–1377.

Hoffmann, A. & Dimroth, P. (1990) The ATPase of *Bacillus alkalophilus*. Purification and properties of the enzyme. *Eur. J. Biochem.*, 194, 423–430.

Hoppe, J. & Sebald, W. (1986) Topological studies suggest that the pathway of the protons through F_o is provided by amino acid residues accessible from the lipid phase. *Biochimie*, 68, 427–434.

Hornung, T., Ishmukhametov, R., Spetzler, D., Martin, J. & Frasch, W. D. (2008) Determination of torque generation from the power stroke of *Escherichia coli* F_1-ATPase. *Biochim. Biophys. Acta*, 1777, 579–582.

Howitt, S. M., Lightowlers, R. N., Gibson, F. & Cox, G. B. (1990) Mutational analysis of the function of the a-subunit of the F_0F_1-ATPase of *Escherichia coli*. *Biochim. Biophys. Acta*, 1015, 264–268.

Iino, R., Rondelez, Y., Yoshida, M. & Noji, H. (2005) Chemomechanical coupling in single-molecule F-type ATP synthase. *J. Bioenerg. Biomembr.*, 37, 451–454.

Ishmukhametov, R., Hornung, T., Spetzler, D., & Frasch, W. D. (2010) Direct observation of stepped proteolipid ring rotation in *E. coli* F_0F_1-ATP synthase. *EMBO J.*, 29, 3911–3923.

Itoh, H., Takahashi, A., Adachi, K., Noji, H., Yasuda, R., Yoshida, M. & Kinosita, K. (2004) Mechanically driven ATP synthesis by F_1-ATPase. *Nature*, 427, 465–468.

Jäger, H., Birkenhäger, R., Stalz, W. D., Altendorf, K. & Deckers-Hebestreit, G. (1998) Topology of subunit a of the *Escherichia coli* ATP synthase. *Eur. J. Biochem.*, 251, 122–132.

Jiang, W. & Fillingame, R. H. (1998) Interacting helical faces of subunits a and c in the F_1F_0 ATP synthase of *Escherichia coli* defined by disulfide cross-linking. *Proc. Natl. Acad. Sci. U.S.A.*, 95, 6607–6612.

Jiang, W., Hermolin, J. & Fillingame, R. H. (2001) The preferred stoichiometry of c subunits in the rotary sector of *Escherichia coli* ATP synthase is 10. *Proc. Natl. Acad. Sci. U.S.A.*, 98, 4966–4971.

Johnson, K. M., Swenson, L., Opipari, A. W., Jr., Reuter, R., Zarrabi, N., Fierke, C. A., Börsch, M. & Glick, G. D. (2009) Mechanistic basis for differential inhibition of the F_1F_0-ATPase by aurovertin. *Biopolymers*, 91, 830–840.

Jones, P. C. & Fillingame, R. H. (1998) Genetic fusions of subunit c in the F_0 sector of the H^+-transporting ATP synthase. Functional dimers and trimers and determination of stoichiometry by crosslinking analysis. *J. Biol. Chem.*, 273, 29701–29705.

Jones, P. C., Hermolin, J., Jiang, W. & Fillingame, R. H. (2000) Insights into the rotary catalytic mechanism of F_0F_1 ATP synthase from the cross-linking of subunits b and c in the *Escherichia coli* enzyme. *J. Biol. Chem.*, 275, 31340–31346.

Jones, P. C., Jiang, W. & Fillingame, R. H. (1998) Arrangement of the multicopy H^+-translocating subunit c in the membrane sector of the *Escherichia coli* F_1F_0 ATP synthase. *J. Biol. Chem.*, 273, 17178–17185.

Junge, W., Lill, H. & Engelbrecht, S. (1997) ATP synthase: An electrochemical transducer with rotatory mechanics. *Trends Biochem. Sci.*, 22, 420–423.

Junge, W., Pänke, O., Cherepanov, D. A., Gumbiowski, K., Müller, M. & Engelbrecht, S. (2001) Inter-subunit rotation and elastic power transmission in F_0F_1-ATPase. *FEBS Lett.*, 504, 152–160.

Junge, W., Sielaff, H. & Engelbrecht, S. (2009) Torque generation and elastic power transmission in the rotary F_0F_1-ATPase. *Nature*, 459, 364–370.

Kagawa, Y. & Racker, E. (1966a) Partial resolution of the enzymes catalyzing oxidative phosphorylation. VIII. Properties of a factor conferring oligomycin sensitivity on mitochondrial adenosine triphosphatase. *J. Biol. Chem.*, 241, 2461–2466.

Kagawa, Y. & Racker, E. (1966b) Partial resolution of the enzymes catalyzing oxidative phosphorylation. X. Correlation of morphology and function in submitochondrial particles. *J. Biol. Chem.*, 241, 2475–2482.

Kaim, G. & Dimroth, P. (1999) ATP synthesis by F-type ATP synthase is obligatorily dependent on the transmembrane voltage. *EMBO J.*, 18, 4118–4127.

Kaim, G., Prummer, M., Sick, B., Zumofen, G., Renn, A., Wild, U. P. & Dimroth, P. (2002) Coupled rotation within single F_0F_1 enzyme complexes during ATP synthesis or hydrolysis. *FEBS Lett.*, 525, 156–163.

Kanazawa, H., Mabuchi, K., Kayano, T., Noumi, T., Sekiya, T. & Futai, M. (1981) Nucleotide sequence of the genes for F_0 components of the proton-translocating ATPase from *Escherichia coli*: prediction of the primary structure of F_0 subunits. *Biochem. Biophys. Res. Commun.*, 103, 613–620.

Kato-Yamada, Y., Noji, H., Yasuda, R., Kinosita, K. & Yoshida, M. (1998) Direct observation of the rotation of the ε subunit in F_1-ATPase. *J. Biol. Chem.*, 273, 19375–19377.

Kettner, C., Bertl, A., Obermeyer, G., Slayman, C. & Bihler, H. (2003) Electrophysiological analysis of the yeast V-type proton pump: variable coupling ratio and proton shunt. *Biophys. J.*, 85, 3730–3738.

Kinosita, K., Jr., Yasuda, R., Noji, H., Ishiwata, S. & Yoshida, M. (1998) F_1-ATPase: a rotary motor made of a single molecule. *Cell*, 93, 21–24.

Kluge, C. & Dimroth, P. (1994) Modification of isolated subunit c of the F_1F_0-ATPase from *Propionigenium modestum* by dicyclohexylcarbodiimide. *FEBS Lett.*, 340, 245–248.

Krah, A., Pogoryelov, D., Langer, J. D., Bond, P. J., Meier, T. & Faraldo-Gómez, J. D. (2010) Structural and energetic basis for H^+ versus Na^+ binding selectivity in ATP synthase F_0 rotors. *Biochim. Biophys. Acta*, 1797, 763–772.

Kramer, D. M., Cruz, J. A. & Kanazawa, A. (2003) Balancing the central roles of the thylakoid proton gradient. *Trends Plant Sci.*, 8, 27–32.

Krebstakies, T., Aldag, I., Altendorf, K., Greie, J. C. & Deckers-Hebestreit, G. (2008) The stoichiometry of subunit c of *Escherichia coli* ATP synthase is independent of its rate of synthesis. *Biochemistry*, 47, 6907–6916.

Krebstakies, T., Zimmermann, B., Gräber, P., Altendorf, K., Börsch, M. & Greie, J. C. (2005) Both rotor and stator subunits are necessary for efficient binding of F_1 to F_0 in functionally assembled *Escherichia coli* ATP synthase. *J. Biol. Chem.*, 280, 33338–33345.

Langemeyer, L. & Engelbrecht, S. (2007) Essential arginine in subunit a and aspartate in subunit c of F_0F_1 ATP synthase: effect of repositioning within helix 4 of subunit a and helix 2 of subunit c. *Biochim. Biophys. Acta*, 1767, 998–1005.

Langen, P. & Hucho, F. (2008) Karl Lohmann and the discovery of ATP. *Angew. Chem. Int. Ed. Engl.*, 47, 1824–1827.

Lau, W. C. & Rubinstein, J. L. (2010) Structure of intact *Thermus thermophilus* V-ATPase by cryo-EM reveals organization of the membrane-bound V_O motor. *Proc. Natl. Acad. Sci. U.S.A.*, 107, 1367–1372.

Laubinger, W. & Dimroth, P. (1987) Characterization of the Na^+-stimulated ATPase of *Propionigenium modestum* as an enzyme of the F_1F_0 type. *Eur. J. Biochem.*, 168, 475–480.

Lee, L. K., Stewart, A. G., Donohoe, M., Bernal, R. A. & Stock, D. (2010) The structure of the peripheral stalk of *Thermus thermophilus* H^+-ATPase/synthase. *Nat. Struct. Mol. Biol.*, 17, 373–378.

Lightowlers, R. N., Howitt, S. M., Hatch, L., Gibson, F. & Cox, G. (1988) The proton pore in the *Escherichia coli* F_oF_1-ATPase: Substitution of glutamate by glutamine at position 219 of the a-subunit prevents F_o-mediated proton permeability. *Biochim. Biophys. Acta*, 933, 241–248.

Lightowlers, R. N., Howitt, S. M., Hatch, L., Gibson, F. & Cox, G. B. (1987) The proton pore in *Escherichia coli* F_oF_1- ATPase: A requirement of arginine at position 210 of the a-subunit. *Biochim. Biophys. Acta*, 894, 399–406.

Lipmann, F. (1941) Metabolic generation and utilization of phosphate bond energy. *Adv. Enzymol.*, 1, 99–162.

Lötscher, H. R., Dejong, C. & Capaldi, R. A. (1984) Inhibition of the adenosinetriphosphatase activity of *Escherichia coli* F_1 by the water-soluble carbodiimide 1-ethyl-3-[3-(dimethylamino)propyl]carbodiimide is due to modification of several carboxyls in the β subunit. *Biochemistry*, 23, 4134–4140.

Lutter, R., Abrahams, J. P., Van Raaij, M. J., Todd, R. J., Lundqvist, T., Buchanan, S. K., Leslie, A. G. & Walker, J. E. (1993) Crystallization of F_1-ATPase from bovine heart mitochondria. *J. Mol. Biol.*, 229, 787–790.

Masaike, T., Koyama-Horibe, F., Oiwa, K., Yoshida, M. & Nishizaka, T. (2008) Cooperative three-step motions in catalytic subunits of F_1-ATPase correlate with 80 degrees and 40 degrees substep rotations. *Nat. Struct. Mol. Biol.*, 15, 1326–1333.

Matthey, U., Kaim, G., Braun, D., Wüthrich, K. & Dimroth, P. (1999) NMR studies of subunit c of the ATP synthase from *Propionigenium modestum* in dodecylsulfate micelles. *Eur. J. Biochem.*, 261, 459–467.

Matthies, D., Preiss, L., Klyszejko, A. L., Müller, D. J., Cook, G. M., Vonck, J. & Meier, T. (2009) The c_{13} ring from a thermoalkaliphilic ATP synthase reveals an extended diameter due to a special structural region. *J. Mol. Biol.*, 388, 611–618.

Mckinney, S. A., Joo, C. & Ha, T. (2006) Analysis of single-molecule FRET trajectories using hidden Markov modeling. *Biophys. J.*, 91, 1941–1951.

Mclachlin, D. T., Bestard, J. A. & Dunn, S. D. (1998) The b and δ subunits of the *Escherichia coli* ATP synthase interact via residues in their C-terminal regions. *J. Biol. Chem.*, 273, 15162–15168.

Mclachlin, D. T. & Dunn, S. D. (1997) Dimerization interactions of the b-subunit of the *Escherichia coli* F_1F_o-ATPase. *J. Biol. Chem.*, 272, 21233–21239.

Meier, T. & Dimroth, P. (2002) Intersubunit bridging by Na^+ ions as a rationale for the unusual stability of the c-rings of Na^+-translocating F_1F_o ATP synthases. *EMBO Rep.*, 3, 1094–1098.

Meier, T., Ferguson, S. A., Cook, G. M., Dimroth, P. & Vonck, J. (2006) Structural investigations of the membrane-embedded rotor ring of the F-ATPase from *Clostridium paradoxum*. *J. Bacteriol.*, 188, 7759–7764.

Meier, T., Krah, A., Bond, P. J., Pogoryelov, D., Diederichs, K. & Faraldo-Gómez, J. D. (2009) Complete ion-coordination structure in the rotor ring of Na^+-dependent F-ATP synthases. *J. Mol. Biol.*, 391, 498–507.

Meier, T., Matthey, U., Henzen, F., Dimroth, P. & Müller, D. J. (2001) The central plug in the reconstituted undecameric c cylinder of a bacterial ATP synthase consists of phospholipids. *FEBS Lett.*, 505, 353–356.

Meier, T., Matthey, U., von Ballmoos, C., Vonck, J., Krug von Nidda, T., Kühlbrandt, W. & Dimroth, P. (2003) Evidence for structural integrity in the undecameric c-rings isolated from sodium ATP synthases. *J. Mol. Biol.*, 325, 389–397.

Meier, T., Morgner, N., Matthies, D., Pogoryelov, D., Keis, S., Cook, G. M., Dimroth, P. & Brutschy, B. (2007) A tridecameric c ring of the adenosine triphosphate (ATP) synthase from the thermoalkaliphilic *Bacillus* sp. strain TA2.A1 facilitates ATP synthesis at low electrochemical proton potential. *Mol. Microbiol.*, 65, 1181–1192.

Meier, T., Polzer, P., Diederichs, K., Welte, W. & Dimroth, P. (2005a) Structure of the rotor ring of F-type Na^+-ATPase from *Ilyobacter tartaricus*. *Science*, 308, 659–662.

Meier, T., Yu, J., Raschle, T., Henzen, F., Dimroth, P. & Müller, D. J. (2005b) Structural evidence for a constant c_{11} ring stoichiometry in the sodium F-ATP synthase. *FEBS J.*, 272, 5474–5483.

Mitchell, P. (1961) Coupling of phosphorylation to electron and hydrogen transfer by a chemi-osmotic type of mechanism. *Nature*, 191, 144–148.

Mitchell, P. & Moyle, J. (1967) Chemiosmotic hypothesis of oxidative phosphorylation. *Nature*, 213, 137–139.

Mitome, N., Ono, S., Sato, H., Suzuki, T., Sone, N. & Yoshida, M. (2010) Essential arginine residue of the F_o-a subunit in F_oF_1-ATP synthase has a role to prevent the proton shortcut without c-ring rotation in the Fo proton channel. *Biochem. J.*, 430, 171–177.

Mitome, N., Suzuki, T., Hayashi, S. & Yoshida, M. (2004) Thermophilic ATP synthase has a decamer c-ring: indication of noninteger 10:3 H^+/ATP ratio and permissive elastic coupling. *Proc. Natl. Acad. Sci. U.S.A.*, 101, 12159–12164.

Moore, K. J., Angevine, C. M., Vincent, O. D., Schwem, B. E. & Fillingame, R. H. (2008) The cytoplasmic loops of subunit a of *Escherichia coli* ATP synthase may participate in the proton translocating mechanism. *J. Biol. Chem.*, 283, 13044–13052.

Muench, S. P., Huss, M., Song, C. F., Phillips, C., Wieczorek, H., Trinick, J. & Harrison, M. A. (2009) Cryo-electron microscopy of the vacuolar ATPase motor reveals its mechanical and regulatory complexity. *J. Mol. Biol.*, 386, 989–999.

Mulkidjanian, A. Y., Makarova, K. S., Galperin, M. Y. & Koonin, E. V. (2007) Inventing the dynamo machine: the evolution of the F-type and V-type ATPases. *Nat. Rev. Microbiol.*, 5, 892–899.

Müller, D. J., Dencher, N. A., Meier, T., Dimroth, P., Suda, K., Stahlberg, H., Engel, A., Seelert, H. & Matthey, U. (2001) ATP synthase: constrained stoichiometry of the transmembrane rotor. *FEBS Lett.*, 504, 219–222.

Murata, T., Arechaga, I., Fearnley, I. M., Kakinuma, Y., Yamato, I. & Walker, J. E. (2003) The membrane domain of the Na^+-motive V-ATPase from *Enterococcus hirae* contains a heptameric rotor. *J. Biol. Chem.*, 278, 21162–21167.

Murata, T., Yamato, I., Kakinuma, Y., Leslie, A. G. & Walker, J. E. (2005) Structure of the rotor of the V-Type Na^+-ATPase from *Enterococcus hirae*. *Science*, 308, 654–659.

Nakamoto, R. K., Baylis Scanlon, J. A. & Al-Shawi, M. K. (2008) The rotary mechanism of the ATP synthase. *Arch. Biochem. Biophys.*, 476, 43–50.

Nakanishi-Matsui, M. & Futai, M. (2006) Stochastic proton pumping ATPases: from single molecules to diverse physiological roles. *IUBMB Life*, 58, 318–322.

Nakanishi-Matsui, M., Kashiwagi, S., Hosokawa, H., Cipriano, D. J., Dunn, S. D., Wada, Y. & Futai, M. (2006) Stochastic high-speed rotation of *Escherichia coli* ATP synthase F_1 sector: the ε subunit-sensitive rotation. *J. Biol. Chem.*, 281, 4126–4131.

Neumann, S., Matthey, U., Kaim, G. & Dimroth, P. (1998) Purification and properties of the F_1F_o ATPase of *Ilyobacter tartaricus*, a sodium ion pump. *J. Bacteriol.*, 180, 3312–3316.

Noji, H., Yasuda, R., Yoshida, M. & Kinosita, K. (1997) Direct observation of the rotation of F_1-ATPase. *Nature*, 386, 299–302.

Noskov, S. Y. & Roux, B. (2008) Control of ion selectivity in LeuT: two Na^+ binding sites with two different mechanisms. *J. Mol. Biol.*, 377, 804–818.

Oberfeld, B., Brunner, J. & Dimroth, P. (2006) Phospholipids occupy the internal lumen of the c ring of the ATP synthase of *Escherichia coli*. *Biochemistry*, 45, 1841–1851.

Pänke, O., Cherepanov, D. A., Gumbiowski, K., Engelbrecht, S. & Junge, W. (2001) Viscoelastic dynamics of actin filaments coupled to rotary F-ATPase: angular torque profile of the enzyme. *Biophys. J.*, 81, 1220–1233.

Pänke, O., Gumbiowski, K., Junge, W. & Engelbrecht, S. (2000) F-ATPase: specific observation of the rotating c subunit oligomer of EF_oEF_1. *FEBS Lett.*, 472, 34–38.

Parsons, D. F. (1963) Mitochondrial structure: two types of subunits on negatively stained mitochondrial membranes. *Science*, 140, 985–987.

Penefsky, H. S., Pullman, M. E., Datta, A. & Racker, E. (1960) Partial resolution of the enzymes catalyzing oxidative phosphorylation. II. Participation of a soluble adenosine tolphosphatase in oxidative phosphorylation. *J. Biol. Chem.*, 235, 3330–3336.

Pogoryelov, D., Krah, A., Langer, J. D., Yildiz, Ö., Faraldo-Gomez, J. D. & Meier, T. (2010) Microscopic rotary mechanism of ion translocation in the F_o complex of ATP synthases. *Nat. Chem. Biol.*, 6, 891–899.

Pogoryelov, D., Nikolaev, Y., Schlattner, U., Pervushin, K., Dimroth, P. & Meier, T. (2008) Probing the rotor subunit interface of the ATP synthase from *Ilyobacter tartaricus*. *FEBS J.*, 275, 4850–4862.

Pogoryelov, D., Reichen, C., Klyszejko, A. L., Brunisholz, R., Müller, D. J., Dimroth, P. & Meier, T. (2007) The oligomeric state of c rings from cyanobacterial F-ATP synthases varies from 13 to 15. *J. Bacteriol.*, 189, 5895–5902.

Pogoryelov, D., Yildiz, Ö., Faraldo-Gómez, J. D. & Meier, T. (2009) High-resolution structure of the rotor ring of a proton-dependent ATP synthase. *Nat. Struct. Mol. Biol.*, 16, 1068–1073.

Pogoryelov, D., Yu, J., Meier, T., Vonck, J., Dimroth, P. & Müller, D. J. (2005) The c_{15} ring of the *Spirulina platensis* F-ATP synthase: F_1/F_o symmetry mismatch is not obligatory. *EMBO Rep.*, 6, 1040–1044.

Preiss, L., Yildiz, Ö., Hicks, D. B., Krulwich, T. A. & Meier, T. (2010) A new type of proton coordination in an F_1F_o-ATP synthase rotor ring. *PLoS Biology*, 8, e000443.

Pullman, M. E., Penefsky, H. S., Datta, A. & Racker, E. (1960) Partial resolution of the enzymes catalyzing oxidative phosphorylation. I. Purification and properties of soluble

dinitrophenol-stimulated adenosine triphosphatase. *J. Biol. Chem.*, 235, 3322–3329.

Rastogi, V. K. & Girvin, M. E. (1999) Structural changes linked to proton translocation by subunit c of the ATP synthase. *Nature*, 402, 263–268.

Rees, D. M., Leslie, A. G. & Walker, J. E. (2009) The structure of the membrane extrinsic region of bovine ATP synthase. *Proc. Natl. Acad. Sci. U.S.A.*, 106, 21597–21601.

Rivera-Torres, I. O., Krueger-Koplin, R. D., Hicks, D. B., Cahill, S. M., Krulwich, T. A. & Girvin, M. E. (2004) pK_a of the essential Glu54 and backbone conformation for subunit c from the H^+-coupled F_1F_o ATP synthase from an alkaliphilic *Bacillus*. *FEBS Lett.*, 575, 131–135.

Rodgers, A. J. & Capaldi, R. A. (1998) The second stalk composed of the b- and δ-subunits connects F_o to F_1 via an a-subunit in the *Escherichia coli* ATP synthase. *J. Biol. Chem.*, 273, 29406–29410.

Rodgers, A. J., Wilkens, S., Aggeler, R., Morris, M. B., Howitt, S. M. & Capaldi, R. A. (1997) The subunit δ-subunit b domain of the *Escherichia coli* F_1F_o ATPase. The b subunits interact with F_1 as a dimer and through the δ subunit. *J. Biol. Chem.*, 272, 31058–31064.

Rondelez, Y., Tresset, G., Nakashima, T., Kato-Yamada, Y., Fujita, H., Takeuchi, S. & Noji, H. (2005) Highly coupled ATP synthesis by F_1-ATPase single molecules. *Nature*, 433, 773–777.

Sabbert, D., Engelbrecht, S. & Junge, W. (1996) Intersubunit rotation in active F-ATPase. *Nature*, 381, 623–625.

Saita, E., Iino, R., Suzuki, T., Feniouk, B. A., Kinosita, K., Jr. & Yoshida, M. (2010) Activation and stiffness of the inhibited states of F_1-ATPase probed by single-molecule manipulation. *J. Biol. Chem.*, 285, 11411–11417.

Sambongi, Y., Iko, Y., Tanabe, M., Omote, H., Iwamoto-Kihara, A., Ueda, I., Yanagida, T., Wada, Y. & Futai, M. (1999) Mechanical rotation of the c subunit oligomer in ATP synthase (F_oF_1): direct observation. *Science*, 286, 1722–1724.

Sawada, K., Kuroda, N., Watanabe, H., Moritani-Otsuka, C. & Kanazawa, H. (1997) Interaction of the δ and b subunits contributes to F_1 and F_o interaction in *Escherichia coli* F_1F_o-ATPase. *J. Biol. Chem.*, 272, 30047–30053.

Schemidt, R. A., Hsu, D. K., Deckers-Hebestreit, G., Altendorf, K. & Brusilow, W. S. (1995) The effects of an *atpE* ribosome-binding site mutation on the stoichiometry of the c subunit in the F_1F_o ATPase of *Escherichia coli*. *Arch. Biochem. Biophys.*, 323, 423–428.

Schemidt, R. A., QU, J., Williams, J. R. & Brusilow, W. S. (1998) Effects of carbon source on expression of F_o genes and on the stoichiometry of the c subunit in the F_1F_o ATPase of *Escherichia coli*. *J. Bacteriol.*, 180, 3205–3208.

Schneider, E. & Altendorf, K. (1984) Subunit b of the membrane moiety (F_o) of ATP synthase (F_1F_o) from *Escherichia coli* is indispensable for H^+ translocation and binding of the water-soluble F_1 moiety. *Proc. Natl. Acad. Sci. U.S.A.*, 81, 7279–7283.

Schneider, E. & Altendorf, K. (1985) All three subunits are required for the reconstitution of an active proton channel (F_o) of *Escherichia coli* ATP synthase (F_1F_o). *EMBO J.*, 4(2), 515–518.

Schneider, E. & Altendorf, K. (1987) Bacterial adenosine 5'-triphosphate synthase (F_1F_o): Purification and reconstitution of F_o complexes and biochemical and functional characterization of their subunits. *Microbiol. Rev.*, 51, 477–497.

Schnick, C., Forrest, L. R., Sansom, M. S. & Groth, G. (2000) Molecular contacts in the transmembrane c-subunit oligomer of F-ATPases identified by tryptophan substitution mutagenesis. *Biochim. Biophys. Acta*, 1459, 49–60.

Seelert, H., Poetsch, A., Dencher, N. A., Engel, A., Stahlberg, H. & Müller, D. J. (2000) Proton-powered turbine of a plant motor. *Nature*, 405, 418–419.

Senior, A. E. (2007) ATP synthase: motoring to the finish line. *Cell*, 130, 220–221.

Shimabukuro, K., Yasuda, R., Muneyuki, E., Hara, K. Y., Kinosita, K., Jr. & Yoshida, M. (2003) Catalysis and rotation of F_1 motor: cleavage of ATP at the catalytic site occurs in 1 ms before 40 degree substep rotation. *Proc. Natl. Acad. Sci. U.S.A.*, 100, 14731–14736.

Singh, S., Turina, P., Bustamante, C. J., Keller, D. J. & Capaldi, R. A. (1996) Topographical structure of membrane-bound *Escherichia coli* F_1F_o ATP synthase in aqueous buffer. *FEBS Lett.*, 397, 30–34.

Soper, J. W., Decker, G. L. & Pedersen, P. L. (1979) Mitochondrial ATPase complex. A dispersed, cytochrome-deficient, oligomycin-sensitive preparation from rat liver containing molecules with a tripartite structural arrangement. *J. Biol. Chem.*, 254, 11170–11176.

Spetzler, D., Ishmukhametov, R., Hornung, T., Day, L. J., Martin, J. & Frasch, W. D. (2009) Single molecule measurements of F_1-ATPase reveal an interdependence between the power stroke and the dwell duration. *Biochemistry*, 48, 7979–7985.

Spetzler, D., York, J., Daniel, D., Fromme, R., Lowry, D. & Frasch, W. (2006) Microsecond time scale rotation measurements of single F_1-ATPase molecules. *Biochemistry*, 45, 3117–3124.

Stahlberg, H., Müller, D. J., Suda, K., Fotiadis, D., Engel, A., Meier, T., Matthey, U. & Dimroth, P. (2001) Bacterial Na^+-ATP synthase has an undecameric rotor. *EMBO Rep.*, 2, 229–233.

Steed, P. R. & Fillingame, R. H. (2008) Subunit a facilitates aqueous access to a membrane-embedded region of subunit c in *Escherichia coli* F_1F_o ATP synthase. *J. Biol. Chem.*, 283, 12365–12372.

Steed, P. R. & Fillingame, R. H. (2009) Aqueous accessibility to the transmembrane regions of subunit c of the *Escherichia coli* F_1F_o ATP synthase. *J. Biol. Chem.*, 284, 23243–23250.

Steigmiller, S., Turina, P. & Gräber, P. (2008) The thermodynamic H^+/ATP ratios of the H^+-ATPsynthases from chloroplasts and *Escherichia coli*. *Proc. Natl. Acad. Sci. U.S.A.*, 105, 3745–3750.

Stock, D., Gibbons, C., Arechaga, I., Leslie, A. G. & Walker, J. E. (2000) The rotary mechanism of ATP synthase. *Curr. Opin. Struct. Biol.*, 10, 672–679.

Stock, D., Leslie, A. G. W. & Walker, J. E. (1999) Molecular architecture of the rotary motor in ATP synthase. *Science*, 286, 1700–1705.

Takeyasu, K., Omote, H., Nettikadan, S., Tokumasu, F., Iwamotu-Kihara, A. & Futai, M. (1996) Molecular imaging of *Escherichia coli* F_1F_o-ATPase in reconstituted membranes using atomic force microscopy. *FEBS Lett.*, 392, 110–113.

Tanabe, M., Nishio, K., Iko, Y., Sambongi, Y., Iwamoto-Kihara, A., Wada, Y. & Futai, M. (2001) Rotation of a complex of the γ subunit and c ring of *Escherichia coli* ATP synthase. The rotor and stator are interchangeable. *J. Biol. Chem.*, 276, 15269–15274.

Tisler, J., Balasubramanian, G., Naydenov, B., Kolesov, R., Grotz, B., Reuter, R., Boudou, J. P., Curmi, P. A., Sennour, M., Thorel, A., Börsch, M., Aulenbacher, K., Erdmann, R., Hemmer, P. R., Jelezko, F. & Wrachtrup, J. (2009) Fluorescence and spin properties of defects in single digit nanodiamonds. *ACS Nano*, 3, 1959–1965.

Toei, M., Gerle, C., Nakano, M., Tani, K., Gyobu, N., Tamakoshi, M., Sone, N., Yoshida, M., Fujiyoshi, Y., Mitsuoka, K. & Yokoyama, K. (2007) Dodecamer rotor ring defines H^+/ATP ratio for ATP synthesis of prokaryotic V-ATPase from *Thermus thermophilus*. *Proc. Natl. Acad. Sci. U.S.A.*, 104, 20256–20261.

Tsunoda, S. P., Aggeler, R., Yoshida, M. & Capaldi, R. (2001a) Rotation of the c subunit oligomer in fully functional F_1F_o ATP synthase. *Proc. Natl. Acad. Sci. U.S.A.*, 98, 898–902.

Tsunoda, S. P., Rodgers, A. J., Aggeler, R., Wilce, M. C., Yoshida, M. & Capaldi, R. A. (2001b) Large conformational changes of the ε-subunit in the bacterial F_1F_o ATP synthase provide a ratchet action to regulate this rotary motor enzyme. *Proc. Natl. Acad. Sci. USA*, 98, 6560–6564.

Tsuprun, V. L., Orlova, E. V. & Mesyanzhinova, I. V. (1989) Structure of the ATP-synthase studied by electron microscopy and image processing. *FEBS Lett.*, 244, 279–282.

Turina, P., Samoray, D. & Gräber, P. (2003) H^+/ATP ratio of proton transport-coupled ATP synthesis and hydrolysis catalysed by CF_oF_1-liposomes. *EMBO J.*, 22, 418–426.

Ueno, H., Suzuki, T., Kinosita, K., Jr. & Yoshida, M. (2005) ATP-driven stepwise rotation of F_oF_1-ATP synthase. *Proc. Natl. Acad. Sci. U.S.A.*, 102, 1333–1338.

Valiyaveetil, F. I. & Fillingame, R. H. (1998) Transmembrane topography of subunit a in the *Escherichia coli* F_1F_o ATP synthase. *J. Biol. Chem.*, 273, 16241–16247.

Vik, S. B. & Antonio, B. J. (1994) A mechanism of proton translocation by F_1F_o ATP synthases by double mutants of the a subunit. *J. Biol. Chem.*, 269, 30364–30369.

Vollmar, M., Schlieper, D., Winn, M., Buchner, C. & Groth, G. (2009) Structure of the c_{14} rotor ring of the proton translocating chloroplast ATP synthase. *J. Biol. Chem.*, 284, 18228–18235.

von Ballmoos, C., Cook, G. M. & Dimroth, P. (2008) Unique rotary ATP synthase and its biological diversity. *Annu. Rev. Biophys.*, 37, 43–64.

von Ballmoos, C., Wiedenmann, A. & Dimroth, P. (2009) Essentials for ATP synthesis by F_1F_o ATP synthases. *Annu. Rev. Biochem.*, 78, 649–672.

Vonck, J., Pisa, K. Y., Morgner, N., Brutschy, B. & Müller, V. (2009) Three-dimensional structure of A_1A_o ATP synthase from the hyperthermophilic archaeon *Pyrococcus furiosus* by electron microscopy. *J. Biol. Chem.*, 284, 10110–10119.

Vonck, J., Krug von Nidda, T. K., Meier, T., Matthey, U., Mills, D. J., Kühlbrandt, W. & Dimroth, P. (2002) Molecular architecture of the undecameric rotor of a bacterial Na^+-ATP synthase. *J. Mol. Biol.*, 321, 307–316.

Vorburger, T., Ebneter, J. Z., Wiedenmann, A., Morger, D., Weber, G., Diederichs, K., Dimroth, P. & von Ballmoos, C. (2008) Arginine-induced conformational change in the c-ring/a-subunit interface of ATP synthase. *FEBS J.*, 275, 2137–2150.

Wächter, A., Bi, Y., Dunn, S. D., Cain, B. D., Sielaff, H., Wintermann, F., Engelbrecht, S. & Junge, W. (2011) Two rotary motors in F-ATP synthase are elastically coupled by a flexible rotor and a stiff stator stalk. *Proc. Natl. Acad. Sci. U.S.A.*, 108, 3924–3929.

Wada, W., Long, J. C., Zhang, D. & Vik, S. B. (1999) A novel labeling approach supports the five-transmembrane model of subunit a of the *Escherichia coli* ATP synthase. *J. Biol. Chem.*, 274, 17353–17357.

Walker, J. E. (1997) ATP synthesis by rotary catalysis. *Nobel Lecture, December 8, 1997.*

Walker, J. E., Saraste, M. & Gay, N. J. (1982a) *E. coli* F_1-ATPase interacts with a membrane protein component of a proton channel. *Nature*, 298, 867–869.

Walker, J. E., Saraste, M. & Gay, N. J. (1984) The *unc* operon. Nucleotide sequence, regulation and structure of ATP-synthase. *Biochim. Biophys. Acta*, 768, 164–200.

Walker, J. E., Saraste, M., Runswick, M. J. & Gay, N. J. (1982b) Distantly related sequences in the alpha- and beta-subunits of ATP synthase, myosin, kinases and other ATP-requiring enzymes and a common nucleotide binding fold. *EMBO J.*, 1, 945–951.

Watt, I. N., Montgomery, M. G., Runswick, M. J., Leslie, A. G. & Walker, J. E. (2010) Bioenergetic cost of making an adenosine triphosphate molecule in animal mitochondria. *Proc. Natl. Acad. Sci. U.S.A.*, 107, 16823–16827.

Watts, S. D., Tang, C. & Capaldi, R. A. (1996) The stalk region of the *Escherichia coli* ATP synthase. Tyrosine 205 of the γ subunit is in the interface between the F_1 and F_o parts and can interact with both the ε and c oligomer. *J. Biol. Chem.*, 271, 28341–28347.

Weber, J., Muharemagic, A., Wilke-Mounts, S. & Senior, A. E. (2003) F_1F_o-ATP synthase. Binding of d subunit to a 22-residue peptide mimicking the N-terminal region of a subunit. *J. Biol. Chem.*, 278, 13623–13626.

Weber, J., Wilke-Mounts, S. & Senior, A. E. (1994) Cooperativity and stoichiometry of substrate binding to the catalytic sites of *Escherichia coli* F_1-ATPase. Effects of magnesium, inhibitors, and mutation. *J. Biol. Chem.*, 269, 20462–20467.

Wehrle, F., Appoldt, Y., Kaim, G. & Dimroth, P. (2002a) Reconstitution of F_o of the sodium ion translocating ATP synthase of *Propionigenium modestum* from its heterologously expressed and purified subunits. *Eur. J. Biochem.*, 269, 2567–2573.

Wehrle, F., Kaim, G. & Dimroth, P. (2002b) Molecular mechanism of the ATP synthase's F_o motor probed by mutational analyses of subunit a. *J. Mol. Biol.*, 322, 369–381.

Westheimer, F. H. (1987) Why nature chose phosphates. *Science*, 235, 1173–1178.

Wilkens, S., Borchardt, D., Weber, J. & Senior, A. E. (2005) Structural characterization of the interaction of the δ and a subunits of the *Escherichia coli* F_1F_o-ATP synthase by NMR spectroscopy. *Biochemistry*, 44, 11786–11794.

Wilkens, S. & Capaldi, R. A. (1998) ATP synthase's second stalk comes into focus. *Nature*, 393, 29.

Wilkens, S., Dunn, S. D., Chandler, J., Dahlquist, F. W. & Capaldi, R. A. (1997) Solution structure of the N-terminal domain of the δ subunit of the *E. coli* ATP synthase. *Nat. Struct. Biol.*, 4, 198–201.

Yasuda, R., Noji, H., Kinosita, K., Jr. & Yoshida, M. (1998) F_1-ATPase is a highly efficient molecular motor that rotates with discrete 120 degree steps. *Cell*, 93, 1117–1124.

Yasuda, R., Noji, H., Yoshida, M., Kinosita, K., Jr. & Itoh, H. (2001) Resolution of distinct rotational substeps by submillisecond kinetic analysis of F_1-ATPase. *Nature*, 410, 898–904.

Zhang, D. & Vik, S. B. (2003) Helix packing in subunit a of the *Escherichia coli* ATP synthase as determined by chemical labeling and proteolysis of the cysteine-substituted protein. *Biochemistry*, 42, 331–337.

Zhang, Y. & Fillingame, R. H. (1995) Subunits coupling H^+ transport and ATP synthesis in the *Escherichia coli* ATP synthase. Cys-Cys cross-linking of F_1 subunit ε to the polar loop of F_o subunit c. *J. Biol. Chem.*, 270, 24609–24614.

Zhou, Y., Duncan, T. M. & Cross, R. L. (1997) Subunit rotation in *Escherichia coli* F_oF_1-ATP synthase during oxidative phosphorylation. *Proc. Natl. Acad. Sci. U.S.A.*, 94, 10583–10587.

Zimmermann, B., Diez, M., Börsch, M. & Gräber, P. (2006) Subunit movements in membrane-integrated EF_oF_1 during ATP synthesis detected by single-molecule spectroscopy. *Biochim. Biophys. Acta*, 1757, 311–319.

Zimmermann, B., Diez, M., Zarrabi, N., Gräber, P. & Börsch, M. (2005) Movements of the ε-subunit during catalysis and activation in single membrane-bound H^+-ATP synthase. *EMBO J.*, 24, 2053–2063.

CHAPTER 12

ATP-Dependent Proteases: The Cell's Degradation Machines

Sucharita Bhattacharyya

Shameika R. Wilmington

Andreas Matouschek

I. INTRODUCTION

Protein concentration in the cell is a function of the rates of protein synthesis and destruction, and the regulation of both processes is necessary for a properly functioning cell. The degradation of proteins is mainly performed by a small number of ATP-dependent cellular proteases. ATP-dependent proteases are molecular motors that degrade substrates by translocating along the substrates' polypeptide chain. Degradation is directional, highly processive, and requires energy from ATP hydrolysis (Kim et al., 2000; Lee et al., 2001; Reid et al., 2001; Kenniston et al., 2003; Aubin-Tam et al., 2011; Maillard et al., 2011). In this manner, these proteases control the concentrations of hundreds of regulatory proteins involved in processes such as the cell cycle, transcription, and signal transduction and play an important housekeeping role by destroying misfolded and damaged proteins (Ciechanover, 1994; Glickman and Ciechanover, 2002; Goldberg, 2003; Collins and Tansey, 2006). Despite this wide range of substrates, proteases have to act specifically to avoid the unintended degradation of the rest of the cellular proteins. ATP-dependent proteases in Bacteria, Archaea, and Eukaryotes have evolved a similar way of solving this problem: their proteolytic sites are encapsulated within the protease structure where they are inaccessible to folded proteins (Baumeister et al., 1998). Substrates are targeted to the proteases via specific degradation signals, to be unraveled and translocated into the proteolytic chamber (Baker and Sauer, 2006; Schrader et al., 2009). The unfolding and translocation of substrates is accelerated by ATP hydrolysis and is catalyzed by ATPase domains or subunits that flank the proteolytic barrel and pull at the substrates' polypeptide chains (Prakash and Matouschek, 2004, Sauer et al., 2004; Aubin-Tam et al., 2011; Maillard et al., 2011). In this chapter, we will introduce the main ATP-dependent proteases in Bacteria, Archaea, and Eukaryotes and attempt to describe the common mechanisms through which they recognize and degrade their substrates.

II. THE MAIN GROUPS OF ATP-DEPENDENT PROTEASES

Most ATP-dependent proteases share the same overall architecture: They have a cylindrical structure in which one set of subunits forms a central proteolytic core particle. This core is flanked by a second set of subunits that contain the ATPase activity. The ATPases function as regulators of degradation by recognizing substrates, gating the proteolytic core particle, and unfolding and translocating substrate proteins to the protease active site (Ogura and Wilkinson, 2001).

II.1. Bacterial Proteases

ClpAP, ClpXP, HslUV (ClpYQ), FtsH, and Lon make up the main ATP-dependent proteases in *Escherichia coli*. Other Bacterial organisms have varied collections of proteases. For instance, the Gram-positive bacterium *Bacillus subtilis* encodes the ClpCP and ClpEP proteases while lacking ClpAP (Kress et al., 2009). This section will focus on the *Escherichia coli* proteases because they have been characterized in the most detail.

In ClpAP, ClpXP, and HslUV, the ATPase domains and proteolytic domains are separated into two different polypeptide chains (Gottesman, 2003). ClpP and HslV (ClpQ) are the proteolytic domains and they assemble as two rings of seven (ClpP) or six (HslV) subunits stacked on top of each other (Bochtler et al., 1997; Rohrwild et al., 1997; Wang et al., 1997). The ATPase subunits (ClpA, ClpX, or HslU) belong to the Clp/Hsp100 family of chaperone proteins and form hexameric rings that cap the proteolytic cores (Wawrzynow et al., 1995; Schirmer et al., 1996; Hoskins et al., 1998; Bochtler et al., 2000; Sousa et al., 2000). This arrangement protects the proteolytic sites in a central chamber which is accessible only through a narrow channel. The ATPase subunits sit at the entrance to the channel where they control access by recognizing and unfolding substrate proteins that contain the correct targeting signals (Grimaud et al., 1998; Baker and Sauer, 2006; Mogk et al., 2008). Interestingly, ClpA contains two ATPase domains per subunit, but the functional importance of this duplication is not known (Hanson and Whiteheart, 2005).

The remaining Bacterial proteases, Lon and FtsH, contain the ATPase and proteolytic domains within the same polypeptide chain, but their overall architecture seems to follow the same blueprint: The subunits form a hexameric-ringed structure with the protease active sites sequestered within a ring of protease domains (Ito and Akiyama, 2005; Tsilibaris et al., 2006). FtsH is a membrane-bound protease and contains an N-terminal transmembrane segment in addition to the cytosolic C-terminal ATPase and protease domains (Ito and Akiyama, 2005). Presumably, many substrates of FtsH will also be membrane proteins.

Some Bacteria contain an ATP-dependent protease that, while similar in design to the Clp and Hsl proteases, also appears to be homologous to the Archaeal and Eukaryotic proteasome. This Bacterial proteasome is present in the actinobacterial lineage and is composed of an ATPase called ARC (*A*TPase forming *r*ing-shaped *c*omplexes) and a homolog of the proteolytic particle from Archaea and Eukaryotes (Wolf et al., 1998).

II.2. Eukaryotic Proteasome

Unlike Bacteria, Eukaryotes have only one cytosolic ATP-dependent protease: the 26S proteasome (the organelles of Eukaryotic cells contain homologs of the Bacterial ATP-dependent proteases). Architecturally, the Eukaryotic 26S proteasome follows the same design as the multiple-ringed Clp proteases: a 20S core particle that contains the protease active sites and is capped on one or both ends by a 19S regulatory particle that contains the ATPase sites. However, the composition of these particles is much more complicated than in the Bacterial systems (Voges et al., 1999). The 20S core particle is composed of a stack of four heptameric rings, two central rings of β subunits, and one ring of α subunits on each side (yielding the pattern αββα). The seven α and seven β subunits within each ring are homologous but not identical, and three of the β subunits contain the proteolytic sites (Groll et al., 1997).

The 19S regulatory particle is an even more complex structure, of at least 19 distinct proteins arranged in what is referred to as the base and lid (Glickman et al., 1998). The base contains a hexameric ring of homologous but not identical ATPases and four non-ATPase proteins. The lid contains nine known proteins, but only one of them has a characterized enzymatic function (Finley, 2009). The 19S regulatory particle also contains some non-stoichiometric subunits that associate with the structure more loosely (Finley, 2009).

Assembly of the 26S proteasome is a complicated process. Each α and β subunit of the 20S core particle and distinct ATPase of the 19S occupies a specific position in their respective rings (Groll et al., 1997; Rubin et al., 1998). Assembly of the 20S core and 19S regulatory particle is facilitated by several chaperone proteins that appear not to be present in the final ATPase-protease structure (Funakoshi et al., 2009; Kaneko et al., 2009; Bedford et al., 2010; Tomko et al., 2010).

The 20S core particle can associate with a few other activating caps, the best characterized of which is probably the 11S regulator, also known as PA28 or REG (Finley, 2009). The 11S regulator activates the 20S particle in an ATP-independent fashion, thus it cannot degrade stably folded proteins, and it appears to participate in the degradation and presentation of peptide antigens involved in the immune response (Whitby et al., 2000; Förster et al., 2005; Demartino and Gillette, 2007). In mammals, this cap assembles with a specialized core particle, in which the catalytic β-subunits are replaced by counterparts with different cleavage preferences, and this assembled particle is often referred to as the immuno-proteasome. The synthesis of both the altered catalytic β-subunits and 11S regulator is induced by interferon-γ, a cytokine involved in adaptive and innate immunity (Rock and Goldberg, 1999). The altered cleavage specificities allow the production of antigenic peptides that are presented to MHC class I molecules (Rock and Goldberg, 1999). Other caps, such as PA200/Blm10, function in other ubiquitin-independent degradation processes (Finley, 2009).

The canonical image of the Eukaryotic proteasome is that of a 20S core particle flanked by a 19S regulatory particle on both ends. However, the 20S core particle can also be capped by a 19S regulatory particle on one end and the 11S regulator on the other, perhaps mediating the different functional requirements of the proteasome in Eukaryotic cells (Groettrup et al., 1996; Hendil et al., 1998; Tanahashi et al., 2000; Cascio et al., 2002). Additionally, free 20S core particles may also be found in cells, but their physiological function is somewhat unclear (Brooks et al., 2000; Tanahashi et al., 2000).

II.3. Proteasome in Archaea

The ATP-dependent protease in Archaea, called PAN-20S (*p*roteasome-*a*ctivating *n*ucleotidase), shares the simple architecture of the Bacterial proteases; however, the protease subunits are homologous to the Eukaryotic proteasome, thus suggesting the existence of a bridge between the Bacterial and Eukaryotic proteolytic systems (Zwickl et al., 1999). The PAN-ATPase complex is formed by six copies of the same ATPase subunit. Intriguingly, the individual ATPase subunits of the Eukaryotic 19S cap are more homologous to the PAN ATPase subunit than they are to each other (Smith et al., 2006). The Archaeal 20S core particle contains α and β subunits, arranged in four heptameric rings, that are also homologous to the subunits of the Eukaryotic core particle (Löwe et al., 1995). However, unlike the Eukaryotic 20S, all of the seven α subunits and seven β subunits are identical (Benaroudj and Goldberg, 2000). Because of its relative simplicity, the Archaeal proteasome has served as a useful experimental system to explore proteasome mechanism (Smith et al., 2006).

II.4. Catalytic Chemistries and Cleavage Sites

The ATPase subunits of ATP-dependent proteases all belong to the same family of proteins and share many structural and sequence motifs that will be discussed in a later section. In contrast, the proteolytic particles appear to be unrelated to one another. This is most evident in the fact that the proteolytic active sites have different catalytic residues. The Bacterial ClpP is a serine protease, and HslV has a unique N-terminal threonine that acts as the catalytic residue (Maurizi et al., 1990). This active-site catalytic residue is homologous to the 20S protease in Archaea and Eukaryotes, and thus HslV is thought to be the Bacterial predecessor of the 20S proteolytic core particle (Löwe et al., 1995; Bochtler et al., 1997). In contrast, FtsH is a zinc-dependent metalloprotease (Ito and Akiyama, 2005), and the active site of Lon contains a serine-lysine dyad (Botos et al., 2004).

The cleavage site preferences of the different proteases are typically not strongly pronounced but the Eukaryotic proteasome's catalytic subunits have distinct preferences for residues adjacent to the cleavage site. One β subunit has a chymotrypsin-like, one subunit has trypsin-like, and one subunit has caspase-like cleavage preference (Nussbaum et al., 1998; Finley, 2009). Despite differences that may exist with respect to cleavage preferences, a consistent feature among the proteases is that degradation is processive and generally yields peptide fragments that are too small to harbor residual biological activity.

II.5. Atomic-Level Structure Determination

As for other molecular machines, high-resolution structures have provided insight into the function of ATP-dependent proteases (Schmidt et al., 1999) and some of these will be described in detail in the following sections. The proteolytic core particles of ClpP, HslV, and the Archaeal and Eukaryotic 20S proteasome, as well as the proteolytic core fragments of both Lon and FtsH have been crystallized and their structures revealed that the proteolytic active sites are located in internal cavities, that substrates access these sites through a narrow channel that excludes folded structures, and that gates control access into the channel, even for peptides (Löwe et al., 1995; Bochtler et al., 1997; Groll et al., 1997; Wang et al., 1997; Unno et al., 2002; Botos et al., 2004; Bieniossek et al., 2006; Suno et al.,

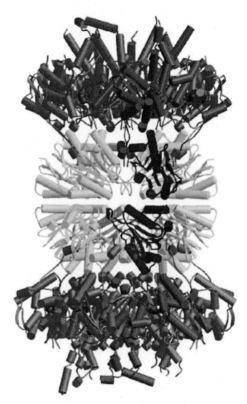

FIGURE 12.1: Ribbon diagram of the HslUV structure (Protein Data Bank accession code 1G3I; Sousa et al., 2000). The HslU ATPase is shown in pink, the HslV protease is shown in yellow, while individual subunits from both HslU and HslV are highlighed in blue and red, respectively. Reprinted from Crystal and Solution Structures of an HslUV, Vol. 103, Sousa et al., pp. 633–43, Copyright 2000, with permission from Elsevier.

2006). Likewise, atomic-level structures of the ATPase-subunits have firmly established that they are functional as hexamers, revealed how they associate with their proteolytic counterparts through specific interactions to gate the degradation channel, and where the regions are located that may interact with substrates for degradation (Bochtler et al., 2000; Sousa et al., 2000; Wang et al., 2001a; Guo et al., 2002; Kim and Kim, 2003; Bieniossek et al., 2006; Suno et al., 2006; Djuranovic et al., 2009; Glynn et al., 2009; Zhang et al., 2009a). So far, atomic-level structures for complete particles are only available for HslUV (Figure 12.1) and for the Lon protease, whose ATPase and proteolytic domains are contained within one polypeptide chain (Bochtler et al., 2000; Sousa et al., 2000; Wang et al., 2001a; Cha et al., 2010).

Atomic-level structural information also exists for one complete proteasome, the PA26–20S complex, PA26 is the *Trypanosoma brucei* homolog of the PA28 or 11S regulator and an ATP-independent activator for the 20S proteasome (Whitby et al., 2000) (Figure 12.2). This structure provided important insights into the gating of the 20S core particle.

The difficulty in obtaining atomic-resolution structures of complete ATP-dependent proteases highlights the fact

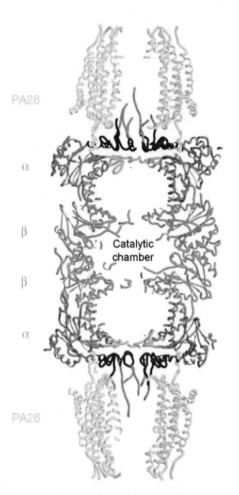

FIGURE 12.2: *Crystal structure of the PA26-20S proteasome complex. Ribbon diagram of the yeast 20S proteasome bound to the* Trypanosoma brucei *ATP-independent activator, PA26, which is a homolog of the yeast 11S regulator. (Protein Data Bank accession code 1FNT; Whitby et al., 2000). Reprinted with permission from Macmillan Publishers Ltd: Nature, Whitby et al. (2000), Copyright 2000.*

that these protein degradation machines are complex dynamic multi-subunit particles. Obtaining an atomic structure is further complicated by the fact that the composition of the entire complex is somewhat less well defined, especially for the Eukaryotic proteasome: some subunits appear to associate only transiently or bind the regulatory particle with a hierarchy of affinities, and some can be lost during isolation and purification of the holoenzyme (Finley, 2009; Förster et al., 2010). For the Eukaryotic proteasome, cryo-electron microscopic methods have been particularly useful, yielding medium-resolution structural information about the organization of these giant particles. Structures for the 26S proteasomes from various organisms have been solved using electron microscopy of negatively stained or frozen-hydrated single particles, and these structures correlate well with existing atomic-level data for the yeast 20S core particle (Walz et al., 1998; da

Fonseca and Morris, 2008; Nickell et al., 2009). Several approaches including cryo-EM, protein-protein crosslinking, and atomic modeling are currently being used in combination to obtain a detailed structural description of the entire Eukaryotic proteasome (Nickell et al., 2009; Förster et al., 2010). Recently, cryo-EM and computational methods have yielded a 9Å resolution structure of the Eukaryotic proteasome from *Saccharomyces pombe* (Figure 12.3), making it the highest-resolution structure of the entire 26S proteasome that exists to date (Bohn et al., 2010).

III. ATPASE DOMAINS

Substrate recognition, unfolding, and translocation are mediated by the regulatory ATPase caps that are composed of individual ATPases which oligomerize into a ring. In addition, the ATPase rings also activate the proteolytic core particle by gating, or opening the channel that leads to the proteolytic sites (Glickman et al., 1998; Grimaud et al., 1998; Wang et al., 2001a). These ATPases all belong to the same family of proteins and share several structural

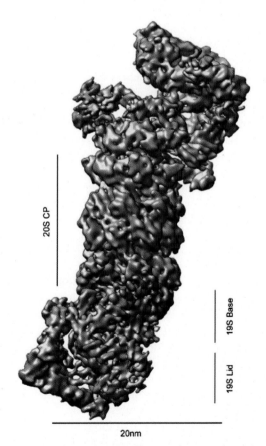

FIGURE 12.3: *Structure of the 26S proteasome from* S. pombe. *Structure of the 26S proteasome obtained by cryo-EM at 9Å resolution (Bohn et al., 2010). Reproduced from Bohn et al., (2010) with permission from Proceedings of the National Academy of Sciences USA Copyright 2010.*

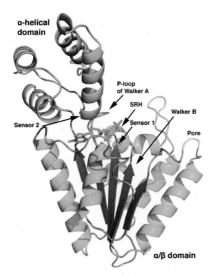

FIGURE 12.4: *Crystal structure of a representative AAA$^+$ domain from NSF (N-ethylmaleimide-sensitive fusion protein) showing the approximate locations of conserved regions that characterize the AAA$^+$ family. (Protein Data Bank accession code 1D2N, Lenzen et al., 1998) The ATPase domain can be divided into two smaller domains: an N-terminal α/β domain and a smaller C-terminal α-helical domain. SRH: Second Region of Homology. Pore: Pore-1 loop. Nucleotide is shown green. Adapted from Hanson and Whiteheart, 2005.*

features that appear to contribute to a similar function (Figure 12.4).

III.1. AAA$^+$ ATP-Binding Domain

The regulatory ATPase subunits of ATP-dependent proteases belong to the AAA$^+$, or *A*TPases *a*ssociated with diverse cellular *a*ctivities, family of proteins. AAA$^+$ proteins are classified as P-loop type NTP (nucleotide triphosphate)-binding proteins and share common structural features that allow the binding of adenine nucleotides (Ogura and Wilkinson, 2001). AAA$^+$ protein functions range from the regulation of transcription and RNA processing to membrane transport and protein destruction, and a hallmark of many ATPases from this family is that they use the energy from ATP hydrolysis to perform some kind of mechanical work (Neuwald et al., 1999; Ogura and Wilkinson, 2001; Lupas and Martin, 2002).

Each ATPase subunit contains one (or two in ClpA) 200–250 residue-long ATP-binding domain(s) that can be divided into two sub-domains: an N-terminal RecA-like α/β fold domain and a C-terminal α-helical bundle (Ogura and Wilkinson, 2001). They contain the widely conserved Walker A and Walker B motifs of P-loop ATPases, in addition to several loops and motifs that perform functions related to nucleotide hydrolysis, such as "sensing" the nucleotide-bound states of neighboring ATPases in the ring. The P-loop of the Walker A motif is required

for nucleotide binding, whereas the Walker B motif is important for nucleotide hydrolysis (Iyer et al., 2004). The Walker A and B motifs are followed by several other conserved residues that lie in a region of the ATP-binding domain referred to as the *s*econd *r*egion of *h*omology, or SRH. While the SRH is not strictly conserved in all AAA$^+$ proteases, a few key residues are consistently present within this region in all of these ATPases (Weibezahn et al., 2003; Hanson and Whiteheart, 2005). In particular, a residue at the N-terminus of the SRH known as "Sensor 1" and an arginine residue at the C-terminus of the SRH form contacts with the gamma phosphate of ATP to facilitate ATP hydrolysis and nucleotide-dependent conformational changes in the ATPase ring (Karata et al., 1999; Lupas and Martin, 2002).

The C-terminal α-helical bundle is called the *s*ensor and *s*ubstrate *d*iscrimination domain, or SSD in ATP-dependent proteases. It contains a conserved arginine residue, referred to as "Sensor 2," which makes contacts with the neighboring ATPase domain and may be required to transmit conformational changes during the ATP hydrolysis cycle (Schmidt et al., 1999; Smith et al., 1999; Ogura and Wilkinson, 2001). Additionally, the SSD could also play a role in recognizing substrates for degradation (Smith et al., 1999).

AAA$^+$ proteases also share conserved residues that are directly involved in recognizing substrates for degradation. As discussed earlier, the proteases share a common geometry, which includes a central pore in the ATPase rings through which substrates are translocated during degradation. The pore is lined by loops, and one of these, the pore-1 loop (labeled "pore" in Figure 12.4), contains the motif Ar-φG (an aromatic residue, Ar, followed by a hydrophobic residue, φ, and glycine). This motif is conserved in all AAA$^+$ proteases and is required for the recognition of substrates for degradation (Wang et al., 2001a; Yamada-Inagawa et al., 2003; Siddiqui et al., 2004; Hinnerwisch et al., 2005). It will be discussed in detail later.

Finally, the ATPase domains of some ATP-dependent proteases such as ClpX and ClpA contain an additional N-terminal domain upstream of the ATP-binding domain, which plays a role in recognizing certain substrates for degradation, but their deletion does not entirely abolish degradation (Sauer et al., 2004). The Bacterial N-terminal domains associate with, or bind to, adaptor proteins that recognize and deliver substrates to the protease complex, while an ATPase in the yeast 26S proteasome, Rpt5, can bind to ubiquitinated proteins, possibly via its N-terminus (Lam et al., 2002; Guo et al., 2002b; Zeth et al., 2002; Ishikawa et al., 2004; Thibault et al., 2006; Djuranovic et al., 2009). Furthermore, structural studies designed to follow the path of a substrate through ClpXP and ClpAP suggest that proteins initially bind at the very N-terminus of the ATPase particles as they are shuttled to the proteolytic active site (Ortega et al., 2000; Ishikawa et al., 2001; Kolygo et al., 2009).

III.2. ATP Hydrolysis

The energy of ATP hydrolysis drives the unfolding and translocation of substrates to the proteolytic sites and may also be required during substrate recognition (Sauer et al., 2004; Peth et al., 2010). The ATP hydrolysis cycle during degradation is perhaps best characterized for ClpX yet, how ATP hydrolysis is coupled to mechanical work remains a key question (Kenniston et al., 2003; Sauer et al., 2004; Martin et al., 2005; Baker and Sauer, 2006). The subunits in the ClpX hexamer do not hydrolyze ATP in a strictly ordered manner; rather, hydrolysis can occur probabilistically (Hersch et al., 2005; Martin et al., 2005). Remarkably, not all of the subunits within the ATPase ring need to be functional for degradation to occur and ClpXP can degrade proteins even when all but one ATPase subunit have been inactivated by mutagenesis (Martin et al., 2005). These findings do not rule out that some coordination between subunits occurs. In the mitochondrial FtsH homolog, Yta10/12, there is evidence that ATP hydrolysis is normally coordinated such that ATP binding in one subunit of Yta10/12 affects the nucleotide state of the neighboring subunit in the hexamer and may impose some directionality for the ATPase cycle (Augustin et al., 2009). Ordered hydrolysis has also been proposed for the PAN ATPases in Archaea and, by analogy, for the Eukaryotic proteasome (Smith et al., 2011). Whether hydrolysis is coordinated in only a subset of subunits, or around the entire ring, still remains unclear. Furthermore, the extent of coordination of ATP hydrolysis may differ among the various proteases.

Structural studies have shown that not all ATPase subunits in the HslU and ClpX hexamers are in the same nucleotide-bound state, a result which fits with the biochemical experiments arguing against a synchronous mechanism for ATP hydrolysis. Indeed, some of the hexamer structures of HslU and ClpX reveal that no more than four nucleotides are simultaneously bound in the ring (Bochtler et al., 2000; Wang et al., 2001b; Glynn et al., 2009). Each ClpX monomer shares the same sequence and domain folds, but the orientations of the α/β and α-helical domains relative to each other within each subunit vary in the hexamer, resulting in some subunits that are unable to bind ATP. This arrangement makes the ClpX ring asymmetric and nucleotide binding can result in rigid-body motions between domains in the ATPases, which allows conformational changes to be propagated around the ring (Glynn et al., 2009). These conformational changes result in the movement of substrates which are in contact with the ATPases through the pore loops mentioned above (and again later in more detail) (Martin et al., 2005; Glynn et al., 2009). The subunits in the HslU and PAN rings also fall into different functional classes and this arrangement

may provide clues to the Eukaryotic proteasome's ATPase cycle, where the individual ATPases also differ in sequence (Bochtler et al., 2000; Hersch et al., 2005; Yakamavich et al., 2008; Glynn et al., 2009; Smith et al., 2011).

IV. GATING

In the absence of the regulatory ATPase subunits, the entrance to the proteolytic core particle in Bacteria, Archaea, and Eukaryotes is generally gated shut to prevent unregulated degradation of cellular proteins. Gate opening appears to be a consequence of the docking of the regulatory ring to the protease particle (Figure 12.5). Structural studies have shown that the N-termini of α-subunits in the Archaeal and Eukaryotic 20S core particle block access to the channel through which substrates reach the proteolytic sites, and association with the ATPase or regulatory ring opens the gate, thus activating protein degradation (Groll et al., 2000; Whitby et al., 2000).

Many of the ATPase subunits dock onto the proteolytic core, through conserved sequence motifs in their C-terminal tails, in a way that allows them to interact with the N-termini of the alpha subunits. For example, the C-terminal tails of three of the six ATPase subunits of the 19S regulatory particle and also the Archaeal PAN harbor the sequence HbYX, where a hydrophobic amino acid is followed by a tyrosine residue (Smith et al., 2007; Gillette et al., 2008). Cryo-EM studies with Archaeal PAN suggest that this tail functions in a "key in a lock" manner upon docking to the 20S particle, causing the N-terminal regions of the α subunits of the 20S core particle to shift position (Rabl et al., 2008). Of the three ATPases in the yeast 19S particle that share the conserved HbYX motif, only the C-termini of Rpt5 and Rpt2 have been shown to induce gate opening in the 20S core particle (Smith et al., 2007; Gillette et al., 2008). This is consistent with genetic data that have implicated the yeast ATPase Rpt2 in gating (Köhler et al., 2001). It remains to be shown how the other ATPases of the Eukaryotic 26S proteasome contribute to gating the Eukaryotic 20S particle.

The 11S or PA28 regulator in Eukaryotes gates the 20S core particle by a different mechanism from that of the 19S ATPases. Binding to the 20S core requires association with the 11S C-termini, but gate opening requires an additional "activation loop" in each of the seven subunits of the 11S regulator to effectively displace the N-terminal regions of the α subunits (Zhang et al., 1998; Whitby et al., 2000) (Figure 12.5).

The HbYX motif does not appear to be conserved in the Bacterial proteases, where gating is instead mediated by loops in the ATPases that contain an IGF/L sequence motif, as illustrated by ClpAP and ClpXP (Kim et al., 2001; Singh et al., 2001; Joshi et al., 2004; Martin et al., 2007). The structure of isolated ClpP resembles a closed gate in which the N-terminal residues of ClpP block the axial channel to the protease active sites, much as the N-terminal

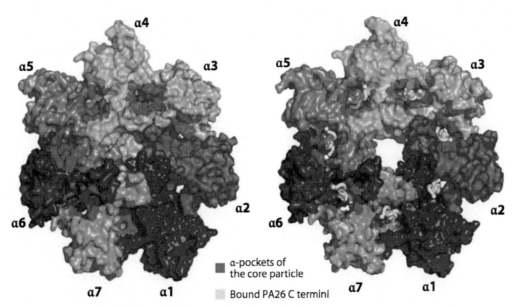

FIGURE 12.5: *The proteolytic particle is gated in ATP-dependent proteases. In the absence of a regulatory particle, the 20S core particle in Eukaryotes is generally gated shut (left, viewed as looking into the degradation chamber at the regulator-protease interface of the 20S core particle). PA28, or its homolog from* Trypanosoma brucei *PA26, opens the gate by inserting "activation" loops from each of its subunits into pockets in the 20S ring to move the N-termini out of the way (right). PA26 C-termini shown in yellow. Recognition and Processing of Ubiquitin-Protein Conjugates by the Proteasome (Volume 78) by Daniel Finley. Copyright 2009 by Annual Reviews Inc. Reproduced with permission of Annual Reviews Inc via Copyright Clearance Center.*

residues of the α subunits of the 20S proteasome block the channel in Eukaryotes (Effantin et al., 2010). A recent structural study in *Bacillus subtilis* showed that a small molecule inhibitor of ClpP deregulates the peptidase by disrupting the interaction between ClpP and its ATPase partners and thus activates gate opening in ClpP (Lee et al., 2010). This activation of ClpP in the absence of its ATPase ultimately leads to unregulated protein destruction (Kirstein et al., 2009). Presumably, the IGF/L motifs of the Clp ATPases activate ClpP by a similar mechanism as the inhibitor, as they both appear to interact with the same regions of ClpP (Lee et al., 2010).

Finally, a few studies of the Eukaryotic proteasome have suggested that the binding of substrate proteins that are specifically recognized by the 19S regulatory particle further enhances the proteolytic activity of the proteasome, perhaps by affecting 20S core particle gating. Substrates bearing these specific signals are recognized by the ATPases or other proteins that are present in the 19S cap, and this may result in a further opening of the gate for optimal degradation (Bech-Otschir et al., 2009; Li and Demartino, 2009; Peth et al., 2009).

V. SUBSTRATE TARGETING AND RECOGNITION

At first glance, it would appear that Bacterial and Eukaryotic proteolytic processes differ most in the way substrates are recognized by the proteases. For example, in Eukaryotes, substrates are generally targeted to the proteasome through post-translational modification by a small protein called ubiquitin. In contrast, Bacteria do not encode ubiquitin and instead use either small peptide tags or adaptor proteins to generally recognize their substrates. Despite these differences, the Bacterial and Eukaryotic mechanisms of substrate recognition and targeting bear many similarities, which are highlighted in the next few sections (Baker and Sauer, 2006; Schrader et al., 2009).

V.1. Targeting in Eukaryotes

In Eukaryotes, the degradation signal, or *degron*, that targets substrates to the 26S proteasome has two parts: a binding tag that is recognized by specific subunits of the proteasome, and an unstructured sequence that serves as the initiation region at which the proteasome engages the substrate and begins the proteolysis (Prakash et al., 2004; Takeuchi et al., 2007) (Figure 12.6). Intriguingly, the two components of the degron can also function in trans, when separated onto different polypeptides in a protein complex (Prakash et al., 2009).

V.2. The Binding Tag

The best-characterized proteasome-binding tag is the polyubiquitin chain (Weissman, 2001). Ubiquitin is a small protein of 76 amino acids that is covalently attached to substrates through the actions of a cascade of enzymes (Pickart, 2001). In the first step, ubiquitin is "activated" by

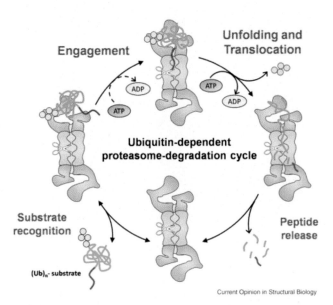

FIGURE 12.6: The 26S proteasome degradation cycle. Two components are required for proteasomal degradation; a proteasome-binding tag, such as polyubiquitin, and an initiation region. Polyubiquitinated proteins are recognized by ubiquitin receptors in the 19S regulatory particle of the proteasome. Once the ubiquitinated protein binds to the proteasome, degradation then starts at the initiation region (red region of protein). ATPases in the 19S regulatory particle unfold and translocate the protein to the proteolytic chamber where small peptides are released. Reprinted from Protein targeting to ATP-dependent proteases, Volume 18, Inobe et al., pages 43–51, Copyright 2008, with permission from Elsevier.

an enzyme called E1. The activated ubiquitin is then transferred to an enzyme called E2, or ubiquitin-conjugating enzyme, to form a covalent ubiquitin-E2 adduct. The final step of ubiquitin conjugation requires the action of an E3 enzyme, known as ubiquitin ligase, which transfers the ubiquitin to a substrate protein, although a few E2 proteins can transfer ubiquitin directly to the substrate protein (Hochstrasser, 2006). Different Eukaryotic organisms can have varied numbers of the E1, E2, and E3 enzymes. For instance, in mammalian organisms, there are only two known E1s, approximately forty E2s, and hundreds of E3s (Li et al., 2008; Deshaies and Joazeiro, 2009). The genome of yeast Saccharomyces cerevisiae only encodes a single E1, a few E2s and hundreds of E3s (Ciechanover, 1998; Pickart, 2004). There are two major classes of E3 enzymes: HECT and RING domain E3s. HECT-domain E3 proteins form a covalent ubiquitin-E3 intermediate as part of the reaction mechanism, whereas RING-domain E3s facilitate the ubiquitin transfer from the E2 to the substrate without an intermediate (Pickart, 2001; Hochstrasser, 2006). E3s work with specific E2s and confer specificity to the ubiquitination process by recognizing the substrates. Ubiquitin-conjugating enzymes have been found in both the nucleus and cytosol and in some cases also associated with the proteasome (Pickart, 2001; Deshaies and Joazeiro, 2009;

Finley, 2009). Additionally, a subset of E3 enzymes, sometimes referred to as E4s, can attach ubiquitin to a substrate that has already been ubiquitinated, thus extending the ubiquitin chain (Hoppe, 2005).

Ubiquitin is generally attached to a substrate protein through the epsilon amino group of lysine residues, but attachment of ubiquitin to the alpha-amino group of a protein, or to the side chains of cysteine, serine, and threonine residues has also been observed (Ciechanover and Ben-Saadon, 2004; Cadwell and Coscoy, 2005; Wang et al., 2007). The ubiquitination reaction continues after the first ubiquitin is attached to the substrate, with additional ubiquitins attached to lysine residues in the first ubiquitin, and effective proteasome degrons contain polyubiquitin chains of at least four ubiquitins (Thrower et al., 2000). Ubiquitin itself has seven lysine residues to which polyubiquitin chains can be linked. Polyubiquitin chains linked through K11, K29, and K48 are thought to route proteins to the proteasome for degradation physiologically (Chau et al., 1989; Koegl et al., 1999; Jin et al., 2008), whereas polyubiquitin chains linked through K63 may be primarily involved in other processes, such as DNA repair, trafficking, and signal transduction (Pickart and Fushman, 2004; Sun and Chen, 2004). However, it is not clear how well pronounced these rules are, as K63-linked polyubiquitin chains can serve as a proteasome-targeting signal in vitro (Pickart and Fushman, 2004; Saeki et al., 2009).

V.3. Ubiquitinated Substrates and Adaptor Proteins

The proteasome can recognize ubiquitinated substrates directly through subunits that serve as ubiquitin receptors: Rpn10 (Deveraux et al., 1994), Rpn13 (Husnjak et al., 2008), and Rpt5 (Lam et al., 2002). In other instances, the ubiquitinated proteins are shuttled to the proteasome with the help of adaptor proteins (Elsasser et al., 2004, Kim et al., 2004, Verma et al., 2004). Three proteasome adaptors have been identified in yeast: Rad23, Dsk2, and Ddi1 (Wilkinson et al., 2001; Chen and Madura, 2002; Rao and Sastry, 2002). These adaptor proteins each contain a ubiquitin-like (UbL) domain, which is recognized by the Rpn1 and Rpn13 proteasome subunits in yeast (as well as Rpn10/S5A in mammals but not yeast), and one or two ubiquitin-associated (UBA) domains that interact with the substrate (Elsasser et al., 2002; Husnjak et al., 2008). There may be even more unidentified substrate receptors, because deletion of all five receptors in yeast is not lethal (Husnjak et al., 2008). The reason for the relatively large number of receptors remains unclear, but presumably this redundancy makes degradation more robust.

V.4. Deubiquitinating Enzymes

Whereas substrates must generally be ubiquitinated to be recognized for degradation by the proteasome, the ubiquitins themselves are not degraded. Eukaryotic cells contain deubiquitinating enzymes (DUBs) that cleave the ubiquitin chains from substrate proteins (Wilkinson, 1997; D'Andrea and Pellman, 1998; Amerik and Hochstrasser, 2004; Wilkinson, 2009). Three DUBs are associated with the proteasome and one of them, Rpn11/POH1, is a stoichiometric component of the proteasome cap required for protein degradation (Amerik and Hochstrasser, 2004). Rpn11/POH1 is a ubiquitin-specific metalloprotease, but its deubiquitination activity depends on ATP hydrolysis by the 19S ATPases (Verma et al., 2002; Yao and Cohen, 2002). It cleaves the entire polyubiquitin chain from substrate proteins committed to degradation (Verma et al., 2002; Yao and Cohen, 2002). Two other deubiquitinating enzymes in yeast, Ubp6 and Uch37, associate with the proteasome (Lam et al., 1997; Borodovsky et al., 2001; Leggett et al., 2002; Hanna et al., 2006; Yao et al., 2006). They trim ubiquitin from the distal end of the chain, gradually releasing individual ubiquitins in the case of Uch37, or di- or tri-ubiquitins in the case of Ubp6 (Lam et al., 1997; Hanna et al., 2006). The action of these two enzymes weakens a substrate's affinity for the proteasome progressively and thus delays its degradation (Yao and Cohen, 2002; Hanna et al., 2006).

The action of the ubiquitin-trimming enzymes is balanced by E3s that can also associate with the proteasome, which makes the structure of ubiquitin tags dynamic. This fine-tuning or editing of ubiquitin tags is best understood for the Ubp6/Hul5 deubiquitinase/E3 pair in yeast (Crosas et al., 2006). The editing of the tag may function as a degradation timer for fine-tuning the targeting of a substrate, and it probably protects the proteasome from becoming clogged up by indigestible substrates (Crosas et al., 2006).

V.5. Ubiquitin-Independent Targeting

It appears that some substrates can be degraded by the proteasome even though they are not ubiquitinated (Verma and Deshaies, 2000; Baugh et al., 2009). Ornithine decarboxylase (ODC) was the first example of ubiquitin-independent degradation (Murakami et al., 1992), and it still is the best characterized. Degradation begins at the C-terminal tail of ODC and requires the protein antizyme as a cofactor, which may act by inducing a conformational change in ODC to expose its C-terminal tail and increases the affinity of ODC to the proteasome (Zhang et al., 2003; Zhang et al., 2004). Ubiquitin-independent degradation has been described for a range of other proteins, and there are some reports proposing that it may be a widespread mechanism catalyzed by the 20S proteasome lacking ATPase caps (Jariel-Encontre et al., 2008). Another possible mechanism for ubiquitin-independent degradation is that these proteins may contain particularly good proteasome initiation sites that can function by themselves like the Bacterial degrons (see Section V.6.). Finally, it is

also possible that some proteins are targeted for degradation by yet to be discovered proteasome adaptors that function in a manner similar to the Ubl-Uba adaptors (Elsasser et al., 2002; Husnjak et al., 2008; Prakash et al., 2009).

V.6. Substrate Targeting in Bacteria

Bacterial proteases commonly recognize degrons that are short stretches of amino acids within the substrate, often found at either the N- or C-terminus (Figure 12.7). For example, deleting the last four amino acids of the tetrameric transposase, MuA, protects it from degradation by ClpXP (Levchenko et al., 1995), the first 30 residues of the UmuD protein are required for it to be degraded by the Lon protease, and deletion of the first 15 amino acids of RecA makes it unable to be recognized by ClpAP (Gonzalez et al., 1998; Hoskins et al., 2000a; Hoskins et al., 2000b). These stretches of amino acids form unstructured tails and can simultaneously serve as the protease recognition sites and as initiation sites for degradation (Sauer et al., 2004). In some cases, Bacterial degrons can be found in regions other than the termini of a protein (Hoskins et al., 2002; Sauer et al., 2004). Degrons can be recognized internally by ClpXP and ClpAP, or, as in the case for the LexA repressor, a cleavage or processing event is required for a degron to be revealed (Hoskins et al., 2002). ClpXP recognizes the LexA repressor only after an autocleavage event in LexA has unmasked recognition signals from the intact protein (Flynn et al., 2003; Neher et al., 2003).

The amino acid sequences of these tags are well defined in some instances, such as the ssrA-tag discussed later in the chapter, but they can be degenerate in other instances (Keiler et al., 1996; Flynn et al., 2003). One particularly intriguing case is the N-end rule, where the degron is formed by the amino acid residue at the very N-terminus of the substrate protein (Bachmair et al., 1986; Varshavsky, 1992; Ravid and Hochstrasser, 2008). This residue can be recognized directly by ClpAP or by adaptor proteins, such as ClpS, that simultaneously bind the substrate and protease (see Section V.8.) (Dougan et al., 2002b; Erbse et al., 2006; Varshavsky, 2008). Interestingly, an analogous system exists in Eukaryotes, where specific N-terminal amino acids are recognized by E3 enzymes called N-recognins, which target these proteins to the proteasome by ubiquitinating them (Varshavsky, 1996; Mogk et al., 2007) (Figure 12.8a). ClpS and N-recognins share sequence homology in regions that recognize the N-end rule degron, suggesting that the two systems may be evolutionarily related (Erbse et al., 2006; Wang et al., 2008) (Figure 12.8).

V.7. Transferable Degradation Tags in Bacteria

There are two well-established examples of Bacterial degradation tags that are not initially encoded in the genes of the substrate proteins but are attached later. The

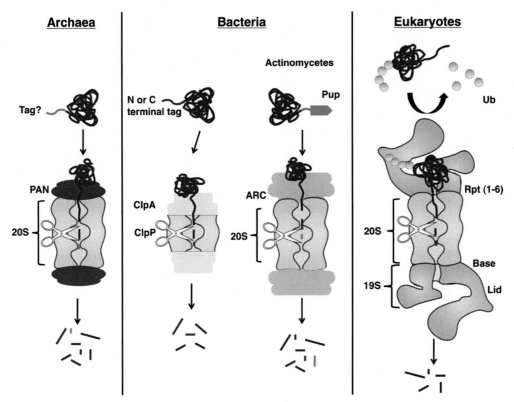

FIGURE 12.7: *. Substrate targeting to ATP-dependent proteases in Archaea, Bacteria, and Eukaryotes. The PAN component of Archaeal proteasomes can recognize substrates containing a terminal peptide tag. ClpAP, a Bacterial protease, recognizes substrate proteins with an N- or C-terminal peptide tag. ARC-proteasomes, present in the actinomycetes order of Bacteria, recognizes pupylated proteins. The Eukaryotic 26S proteasome contains the 19S regulatory particle which recognizes ubiquitinated proteins. The ubiquitins are cleaved from the substrate prior to degradation and recycled. Adapted from Streibel et al., (2009).*

best-understood peptide tag is probably the ssrA tag that is added to the C-terminus of proteins (Keiler et al., 1996). In a process called trans-translation, ssrA tags are added co-translationally to proteins when ribosomes are stalled at damaged mRNAs (Valle et al., 2003; Fu et al., 2010). An RNA construct called tmRNA containing a tRNA-like domain and an open reading frame (ORF) encoding the ssrA tag followed by a stop codon, enters the A site of the ribosome. The ribosome then switches transcript to the tmRNA ORF, allowing protein synthesis to finish and yielding a chimeric polypeptide chain that ends in the ssrA tag sequence. The released protein can be recognized through its ssrA-tag by the ClpAP, ClpXP, or FtsH proteases and is efficiently degraded (Keiler et al., 1996; Gottesman et al., 1998; Herman et al., 1998).

Actinobacteria and a few other bacteria use a protein tag that resembles ubiquitin in its mechanism, although there does not appear to be any clear sequence homology (Iyer et al., 2008; Striebel et al., 2009). The tag is a 64-amino-acid-long protein called prokaryotic ubiquitin-like modifier protein, or Pup, which is able to target proteins to the Bacterial proteasome (Pearce et al., 2008; Burns et al., 2009) (Figure 12.7). Much like ubiquitin, Pup is covalently

attached to the lysine residue of its target substrate (Burns et al., 2009). Pupylated proteins are then recognized by the ARC proteasomes in Bacteria where, unlike in the case of ubiquitin, which is cleaved from the substrate and recycled, degradation begins at the N-terminal tail of the Pup tag (Striebel et al., 2010).

V.8. Adaptor Proteins

The Bacterial degradation systems also include adaptors that can bring substrates to the proteases for degradation by a mechanism similar to that seen in Eukaryotes. Two in particular, SspB and ClpS, have been well characterized (Baker and Sauer, 2006; Kirstein et al., 2009). The adaptor protein SspB binds ssrA-tagged proteins and tethers them to the regulatory particle on the ClpXP protease. Although SspB is not required for ClpXP recognition of ssrA-tagged substrates, it increases the efficiency of their degradation, presumably by increasing the local concentration of ssrA-tagged protein on the protease (Levchenko et al., 2000). Just as the Eukaryotic adaptor proteins, SspB itself escapes from degradation (Sauer et al., 2004). As mentioned above, ClpS recognizes N-end rule substrates and delivers them to

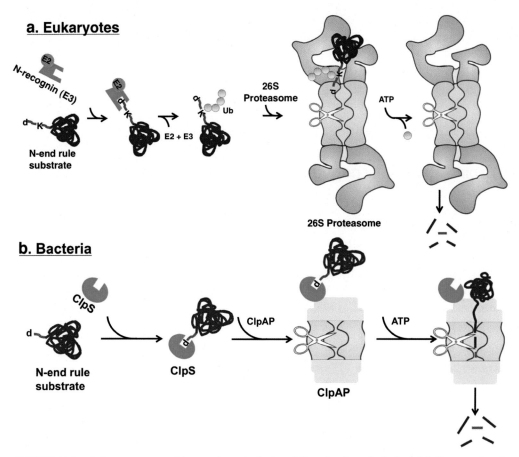

FIGURE 12.8: *Substrate recognition and proteolysis of N-end rule substrates. (a) N-recognins in Eukaryotes contain a direct binding site for N-end rule substrates and recognize destabilizing residues (d) at the N-termini. N-recognins associate with a specialized E2 and ubiquitinate substrates at an internal Lys residue. These substrates are then recognized by the proteasome where they are degraded. (b) In Bacteria, the adaptor protein ClpS recognizes destabilizing residues (d) of N-end rule substrates and shuttles them to the ClpAP protease where they are degraded. Adapted from Mogk et al., (2007).*

the ClpAP protease by binding to the N-terminal domain of ClpA (Dougan et al., 2002b; Erbse et al., 2006; Wang et al., 2008) (Figure 12.8).

Adaptor proteins can also re-target proteins from one protease to the other, shifting the flow of substrates (Dougan et al., 2002a) (Figure 12.8b). For example, SspB binding to ssrA tags masks sequences in the tag that are required for recognition by ClpA and thus redirects ssrA-tagged substrate to ClpXP (Flynn et al., 2001). ClpS in turn switches ClpAP's substrate preferences to enhance its degradation of N-end rule substrates and reduce its degradation of ssrA-tagged proteins (Erbse et al., 2006; Hou et al., 2008; Wang et al., 2008).

V.9. Substrate Targeting in Archaea

In Archaea, substrate targeting appears to have features of both the Bacterial and Eukaryotic systems (Figure 12.7). For example, similar to Bacterial proteases, Archaea can recognize and degrade proteins containing linear peptide

signals, as in the ssrA-tagged proteins (Benaroudj and Goldberg, 2000). However, ubiquitin-like modifiers have been found in Archaea that appear to be conjugated to proteins. Moreover, these ubiquitin-like modifiers are attached to proteins in Archaea through a similar chemistry as the ubiquitin-substrate conjugates characteristic in Eukaryotes (Humbard et al., 2010). These ubiquitin-like modifiers appear to be widespread among Archaea and may play a role in substrate recognition by Archaeal PAN, although this has yet to be directly shown (Iyer et al., 2006; Humbard et al., 2010).

V.10. Targeting by Substrate Localization

The localization of misfolded and damaged proteins in the cell may also play a role in their clearance and toxicity. For example, in Eukaryotic cells, misfolded and aggregated proteins are sequestered in either one of two quality control compartments, determined by their ubiquitination state and solubility (Kaganovich et al., 2008). The more toxic, non-soluble aggregates are sequestered away from the compartment containing high concentrations of cellular chaperones and the proteasome. Interestingly, protein aggregates in Bacteria also appear to be sequestered at foci near the cellular poles, sometimes colocalized with chaperones or chaperone-proteases, hence suggesting that the sequestration of aggregated proteins may be a conserved cellular response (Kirstein et al., 2008; Winkler et al., 2010)

VI. UNFOLDING AND TRANSLOCATION

Once substrates have been recognized by their cognate proteases, they must be translocated to the proteolytic sites. The dimensions of the degradation channel in all ATP-dependent proteases require that substrates are unfolded during their translocation. Structural studies have shown that the diameter of the entry pore can be as narrow as 10Å, which is several times smaller than the diameter of the average folded protein (Johnston et al., 1995; Wenzel and Baumeister, 1995; Groll et al., 1997; Wang et al., 1997).

VI.1. How Substrates are Unfolded by the ATPases

ATP-dependent proteases degrade their substrates sequentially by translocating along the substrates' polypeptide chains (Ishikawa et al., 2001; Lee et al., 2001; Reid et al., 2001; Kenniston et al., 2003). In doing so, they can unravel proteins much faster than the proteins would unfold spontaneously. For example, ClpA can unfold green fluorescent protein (GFP) in a matter of minutes, whereas spontaneous denaturation of the same substrate can take several years (Weber-Ban et al., 1999; Hoskins et al., 2000b; Singh et al., 2000). Interestingly, the resistance of proteins to unfolding and degradation by these proteases does not depend primarily on the global stability of the proteins but rather on the local stability of the structure first encountered by the protease (Lee et al., 2001; Kenniston et al., 2003). For instance, the resistance of a protein to degradation by a small non-specific protease such as Proteinase K or thermolysin correlates with the stability of the protein against thermal or solvent unfolding (Imoto et al., 1986; Park and Marqusee, 2005), whereas resistence to unfolding by ATP-dependent proteases correlates better with their stability against mechanical unfolding by atomic force microscopy (AFM) (Kenniston et al., 2003; Prakash and Matouschek, 2004; Wilcox et al., 2005). Earlier studies on protein unfolding during mitochondrial import had shown that coupling protein denaturation to translocation into a narrow channel can catalyze unfolding by changing the unfolding pathway (Huang et al., 1999; Huang et al., 2000). Similarly, AFM along with mutational studies have shown that simply pulling on a substrate protein can change its unfolding pathway (Lee et al., 2001; Brockwell et al., 2003; Carrion-Vazquez et al., 2003; Wilcox et al., 2005;

Martin et al., 2008b). Thus, a picture emerges in which the proteases translocate along their substrates' polypeptide chains, effectively pulling them into the degradation channel of the proteases. Substrate unfolding is then largely a manifestation of this translocation of the substrate into the protease core. Indeed, single-molecule optical trapping experiments on ClpXP have shown directly that the protease moves along its substrate in discrete steps, generating forces sufficient to unfold individual protein domains (Aubin-Tam et al., 2011; Maillard et al., 2011).

ATP-dependent proteases differ greatly (more than hundred-fold) in their ability to unfold proteins (Singh et al., 2000; Herman et al., 2003; Koodathingal et al., 2009), and at least for one protease, the unfolding strength can be regulated allosterically (Gur and Sauer, 2009). Because Bacterial ATP-dependent proteases recognize substrates with overlapping specificities, the differences in unfolding strength could contribute to substrate selection (Sauer et al., 2004; Koodathingal et al., 2009). For example, proteases with weak unfolding activities and broad sequence preferences could be particularly well suited to function as a stress response protease. The broad sequence preferences could allow it to recognize unstructured tails exposed by denaturation or damage of a wide range of proteins, and the weak unfolding strength would prevent degradation of proteins in their native structure even if they contain natively disordered regions.

VI.2. Getting a Grip on The Substrate

ATP-dependent proteases deal with a broad range of substrates, especially in Eukaryotes where a single protease is responsible for the large majority of protein turnover in the cytosol and nucleus. Even in Bacteria, only a handful of proteases deal with presumably hundreds of different regulatory proteins and even more damaged or denatured substrates. After the recognition step, the proteases have to be able to translocate all of these unrelated amino acid sequences through the degradation channel to the proteolytic sites. Translocation starts at the central pore formed by the ATPase subunits (Ortega et al., 2000; Ishikawa et al., 2001). Crosslinking and mutagenesis studies have provided key clues about how loops lining the pore exert a translocating force on a substrate's polypeptide chains (Wang et al., 2001a; Yamada-Inagawa et al., 2003; Siddiqui et al., 2004; Hinnerwisch et al., 2005; Martin et al., 2008a). Again, this translocation process is best understood in the ClpX and ClpA systems.

Each ClpX subunit provides three loops in the channel, and substrates interact with these loops sequentially, starting with the loop closest to the outside (Siddiqui et al., 2004; Martin et al., 2007; Glynn et al., 2009). The central loop, which appears to play the most important role in translocation, is characterized by the sequence motif

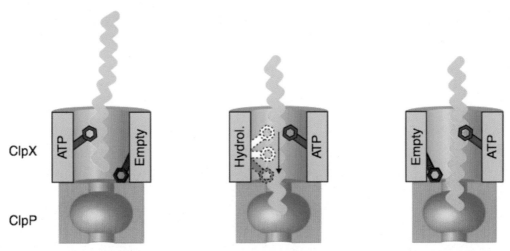

FIGURE 12.9: *Hypothetical mechanism of translocation. The Ar-ɸ motif in loops of the ClpX ATPases face the interior of the translocation channel and play a key role in translocating substrates to the proteolytic sites. The Ar-ɸ loops of two ClpX subunits are shown in conformations that depend on the nucleotide-bound state. Conformational changes allow the Ar-ɸ loops to move the substrates along the axial channel of the protease. Reprinted by permission from Macmillan Publishers Ltd:* **Nature Structural & Molecular Biology,** *Martin et al., (2008). Copyright 2008.*

GYVG. Whereas the exact sequence of the motif varies slightly among the ATPases, the aromatic and hydrophobic residue are conserved in almost all ATP-dependent proteases, and this pair is referred to as the aromatic or Ar-ɸ loop ('Ar' aromatic residue, 'ɸ' hydrophobic residue) (Wang et al., 2001a; Hinnerwisch et al., 2005). The Ar-ɸ loop is located in the pore-1 loop that connects the Walker A and Walker B motifs and its conformation is therefore likely to change with the ATP-hydrolysis cycle (Figure 12.9) (Martin et al., 2008a). Indeed, distinct conformations of the Ar-ɸ in individual subunits have been observed in crystal structures of HslU, ClpX, and FtsH depending on the nucleotide-bound state of the subunit (Wang et al., 2001b; Suno et al., 2006; Martin et al., 2008a; Bieniossek et al., 2009; Glynn et al., 2009).

A number of studies, particularly in the Bacterial proteases and Archaeal PAN, have suggested that this loop plays a critical role in translocation. In ClpX and HslU, mutation of the aromatic residue within the GYVG loops slows or abolishes degradation, while mutation of the hydrophobic residue in FtsH appears to abolish the recognition of certain substrates (Yamada-Inagawa et al., 2003; Siddiqui et al., 2004; Park et al., 2005; Martin et al., 2008a). Mutation of the other residues in this loop (and in other loops lining the channel in ClpA and ClpX) slows the degradation rate of folded substrates substantially and makes translocation less efficient (Hinnerwisch et al., 2005). The Ar-ɸ loop is also conserved in each of the Eukaryotic 19S ATPases and Archaeal PAN. Mutants of the Ar-ɸ loops of PAN show severe defects in unfolding and degradation activity (Zhang et al., 2009b). Whereas much less is known about the actions of the loop in Eukaryotes,

mutation of the Ar-ɸ sequence in the ATPases appears to result in the accumulation of ubiquitin-protein conjugates in yeast, thus indicating a defect in degradation of these substrates (Zhang et al., 2009b). A recent structural study of the 26S proteasome from *Saccharomyces pombe* concluded that the Ar-ɸ loops in the Eukaryotic ATPase subunits are indeed flexible and that they likely operate by a similar mechanism to the Bacterial ATPases (Bohn et al., 2010).

Thus, a model emerges in which ATP-dependent conformational changes cause the aromatic residue in the Ar-ɸ loop to move up and down. This perhaps allows the Ar-ɸ loop to act as a paddle, whose motions cause a substrate to be pulled into the axial channel of the protease for degradation (Baker and Sauer, 2006; Martin et al., 2008a; Glynn et al., 2009; Koga et al., 2009) (Figure 12.9).

VI.3. Unfolding and Translocation Require Large Amounts of Energy

Unfolding and translocation of substrate proteins to the proteolytic sites is driven by the chemical energy of ATP hydrolysis. However, the coupling between ATP-hydrolysis and translocation may not be tight, in the sense that ATP is hydrolyzed even without translocation and unfolding occurring. For example, ClpXP hydrolyzes approximately 600 ATP molecules during the degradation of one molecule of a small model substrate (the immunoglobulin-like domain 27 of titin) and approximately 100–150 molecules of ATP during the degradation of an unfolded variant of the same protein (Kenniston et al., 2003). The protease hydrolyzes ATP rapidly as it

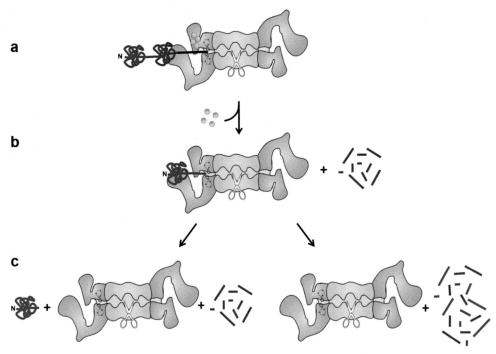

FIGURE 12.10: *Sequential degradation of a multi-domain protein by the 26S proteasome. The protein is recognized by the proteasome and degradation starts at the initiation site (a). The domain closest to the initiation site is degraded first (red domain) (b). As the proteasome proceeds along the polypeptide chain, two fates exist for the second domain (blue). The blue domain can be released if the proteasome is unable to degrade it, resulting in a partially degraded product. Alternatively, it can be degraded after the red domain, which leads to complete degradation of the protein (c).*

translocates along a substrate, and more slowly when it is stalled as it attempts to unravel a folded domain in a substrate. However, stalling also increases the overall time required for degradation so that more ATP is hydrolyzed during the degradation of a folded protein than during degradation of an unstructured protein (Kenniston et al., 2003; Sauer et al., 2004).

Other ATP-dependent proteases use similarly large amounts of ATP during degradation. For example, FtsH protease degradation of the heat shock factor σ^{32} requires the hydrolysis of about 140 ATP molecules (Okuno et al., 2004; Ito and Akiyama, 2005). The Archaeal proteasome seems to require 300–400 molecules of ATP hydrolyzed regardless of the folded state of its substrate (Benaroudj et al., 2003). While it is clear that a more tightly folded protein requires more ATP during degradation compared to a destabilized or unfolded protein, there is some debate as to whether translocation requires ATP hydrolysis. Studies with ClpXP and unfolded substrates suggest that translocation on its own requires a substantial amount of energy (Kenniston et al., 2003). However, other models for the coupling of ATP-hydrolysis and degradation have also been proposed, and studies with the Archaeal PAN-20S and 26S proteasome have suggested that translocation of unfolded substrates occurs by facilitated diffusion and that it is only dependent on ATP-binding

and not on hydrolysis (Smith et al., 2005; Liu et al., 2006).

VII. DEGRADATION

Once a substrate has been targeted, unfolded, and translocated to the proteolytic active sites, the final step of peptide bond cleavage can occur. In the Eukaryotic proteasome, substrates are degraded to small peptides that are, on average, about eight amino acids in length (Nussbaum et al., 1998). In Bacteria, ClpAP appears to degrade proteins to peptides of similar average lengths as seen for the proteasome (Choi and Licht, 2005). These peptide products are too small to form a functional protein and the cell recycles the remaining amino acids Glickman and Ciechanover, 2002.

In some instances substrates are not completely degraded, and larger, partially degraded protein fragments can be released by the protease. The Eukaryotic 26S proteasome is known to partially degrade a handful of physiological proteins, all of which are transcription factors. For example, the p105 subunit of the NF-κB transcription factor is trapped in the cytosol, however it is released into the nucleus when its C-terminal half is degraded by the proteasome. The released fragment is known as the p50 subunit of NFκB (Palombella et al., 1994). Similarly, the

Gli3 transcription factor and its *Drosophila* homolog *Cubitus interruptus* (Ci) are partially degraded by the proteasome in some cells in response to changes in Hedgehog signaling (Aza-Blanc et al., 1997). In this example, the full-length proteins and fragments have opposite biological activities: Gli3/Ci are transcriptional activators that are converted into fragments containing the DNA-binding domains, but not the transactivation domains, and therefore acting as competitive repressors of transcription (Aza-Blanc et al., 1997). Finally, two yeast transcription factors, Spt23 and Mga2 are inactive because they are prevented from entering the nucleus through C-terminal domains that serve as membrane anchors (Hoppe et al., 2000). During their activation, they are released when the proteasome degrades their C termini but leaves the remainder of the protein intact (Hoppe et al., 2000).

It has been proposed that partial degradation occurs when the proteasome stalls and releases its substrate as it translocates along the substrate's polypeptide chain (Lee et al., 2001; Tian et al., 2005) (Figure 12.10). The stalling could be caused as the proteasome encounters a tightly folded domain in the substrate and its association with the substrate at this moment is weakened by some property of the substrate's amino acid sequence (Kenniston et al., 2005; Tian et al., 2005). Although stalling has been shown to occur in the Bacterial proteases in vitro, there is as yet no known physiological role for partial degradation in Bacteria (Kenniston et al., 2005).

VIII. CONCLUSION

ATP-dependent proteases are large protein degradation machines that play an important role in the cell. They are required for the proper functioning of many regulatory processes by controlling the concentrations of regulatory proteins such as transcription factors, signaling proteins, and cell cycle regulators. As such they become the target of drug development (Sakamoto et al., 2001; Navon and Ciechanover, 2009; Goldenberg et al., 2010). For example, a proteasome inhibitor has seen significant success in the treatment of multiple myeloma (Shah and Orlowski, 2009).

ATP-dependent proteases are also a central factor in the cell's response to stress and environmental damage because they remove misfolded and damaged proteins from the cell (Wickner et al., 1999). This role of ATP-dependent proteases connects them to human disease (Goldberg, 2003; Schwartz and Ciechanover, 2009). Many neurodegenerative diseases are linked to the accumulation of protein aggregates (Ross and Poirier, 2004). Often these aggregates contain ubiquitin, in many cases attached to the protein linked to the disease, suggesting that the proteasome somehow fails to clear the aggregates (Morishima-Kawashima et al., 1993; Kalchman et al., 1996; Hasegawa et al., 2002). In Eukaryotes, pharmaceutical approaches aim at interfering in the pathways that target proteins to degradation (Colland, 2010). Thus, understanding the mechanism by which these related ATP-dependent proteases operate may allow for better drug development in the treatment of these diseases.

ACKNOWLEDGMENTS

We would like to thank Janine Kirstein-Miles, Erin Kennedy Schrader, and members of the Matouschek lab (Northwestern University) for a critical reading of the manuscript. This work was supported by grant R01GM64003 from the U.S. National Institutes of Health, and by grant U54CA143869 from the National Cancer Institute as well as the Robert H. Lurie Comprehensive Cancer Center at Northwestern University. S.R.W. was supported by The Cellular and Molecular Basis of Disease Training Program grant T32 NIH T32 GM08061 at Northwestern University funded by the National Institutes of Health, Institute of General Medical Sciences.

REFERENCES

Amerik, A. Y. & Hochstrasser, M. (2004) Mechanism and function of deubiquitinating enzymes. *Biochim Biophys Acta*, 1695, 189–207.

Aubin-Tam, M. E., Olivares, A. O., Sauer, R. T., Baker, T. A. & Lang, M. J. (2011) Single-molecule protein unfolding and translocation by an ATP-fueled proteolytic machine. *Cell*, 145, 257–67.

Augustin, S., Gerdes, F., Lee, S., Tsai, F. T. F., Langer, T. & Tatsuta, T. (2009) An intersubunit signaling network coordinates ATP hydrolysis by m-AAA proteases. *Mol Cell*, 35, 574–85.

Aza-Blanc, P., Ramirez-Weber, F. A., Laget, M. P., Schwartz, C. & Kornberg, T. B. (1997) Proteolysis that is inhibited by hedgehog targets Cubitus interruptus protein to the nucleus and converts it to a repressor. *Cell*, 89, 1043–53.

Bachmair, A., Finley, D. & Varshavsky, A. (1986) In vivo half-life of a protein is a function of its amino-terminal residue. *Science*, 234, 179–86.

Baker, T. A. & Sauer, R. T. (2006) ATP-dependent proteases of bacteria: recognition logic and operating principles. *Trends Biochem Sci*, 31, 647–53.

Baugh, J. M., Viktorova, E. G. & Pilipenko, E. V. (2009) Proteasomes can degrade a significant proportion of cellular proteins independent of ubiquitination. *J Mol Biol*, 386, 814–27.

Baumeister, W., Walz, J., Zühl, F. & Seemüller, E. (1998) The proteasome: paradigm of a self-compartmentalizing protease. *Cell*, 92, 367–80.

Bech-Otschir, D., Helfrich, A., Enenkel, C., Consiglieri, G., Seeger, M., Holzhütter, H.-G., Dahlmann, B. & Kloetzel, P.-M. (2009) Polyubiquitin substrates allosterically activate their own degradation by the 26S proteasome. *Nat Struct Mol Biol*, 16, 219–25.

Bedford, L., Paine, S., Sheppard, P. W., Mayer, R. J. & Roelofs, J. (2010) Assembly, structure, and function of the 26S proteasome. *Trends Cell Biol*, 20, 391–401.

Benaroudj, N. & Goldberg, A. L. (2000) PAN, the proteasome-activating nucleotidase from archaebacteria, is a protein-unfolding molecular chaperone. *Nat Cell Biol*, 2, 833–9.

Benaroudj, N., Zwickl, P., Seemüller, E., Baumeister, W. & Goldberg, A. L. (2003) ATP hydrolysis by the proteasome regulatory complex PAN serves multiple functions in protein degradation. *Mol Cell*, 11, 69–78.

Bieniossek, C., Schalch, T., Bumann, M., Meister, M., Meier, R. & Baumann, U. (2006) The molecular architecture of the metalloprotease FtsH. *Proc Natl Acad Sci USA*, 103, 3066–71.

Bieniossek, C., Niederhauser, B. & Baumann, U. M. (2009) The crystal structure of apo-FtsH reveals domain movements necessary for substrate unfolding and translocation. *Proc Natl Acad Sci USA*, 106, 21579–84.

Bochtler, M., Ditzel, L., Groll, M. & Huber, R. (1997) Crystal structure of heat shock locus V (HslV) from *Escherichia coli*. *Proc Natl Acad Sci USA*, 94, 6070–4.

Bochtler, M., Hartmann, C., Song, H. K., Bourenkov, G. P., Bartunik, H. D. & Huber, R. (2000) The structures of HsIU and the ATP-dependent protease HsIU-HsIV. *Nature*, 403, 800–5.

Bohn, S., Beck, F., Sakata, E., Walzthoeni, T., Beck, M., Aebersold, R., Förster, F., Baumeister, W. & Nickell, S. (2010) From the cover: structure of the 26S proteasome from *Schizosaccharomyces pombe* at subnanometer resolution. *Proc Natl Acad Sci USA*, 107, 20992–7.

Borodovsky, A., Kessler, B. M., Casagrande, R., Overkleeft, H. S., Wilkinson, K. D. & Ploegh, H. L. (2001) A novel active site-directed probe specific for deubiquitylating enzymes reveals proteasome association of USP14. *EMBO J*, 20, 5187–96.

Botos, I., Melnikov, E. E., Cherry, S., Tropea, J. E., Khalatova, A. G., Rasulova, F., Dauter, Z., Maurizi, M. R., Rotanova, T. V., Wlodawer, A. & Gustchina, A. (2004) The catalytic domain of *Escherichia coli* Lon protease has a unique fold and a Ser-Lys dyad in the active site. *J Biol Chem*, 279, 8140–8.

Brockwell, D. J., Paci, E., Zinober, R. C., Beddard, G. S., Olmsted, P. D., Smith, D. A., Perham, R. N. & Radford, S. E. (2003) Pulling geometry defines the mechanical resistance of a β-sheet protein. *Nat Struct Biol*, 10, 731–7.

Brooks, P., Fuertes, G., Murray, R. Z., Bose, S., Knecht, E., Rechsteiner, M. C., Hendil, K. B., Tanaka, K., Dyson, J. & Rivett, J. (2000) Subcellular localization of proteasomes and their regulatory complexes in mammalian cells. *Biochem J*, 346 Pt 1, 155–61.

Burns, K. E., Liu, W.-T., Boshoff, H. I. M., Dorrestein, P. C. & Barry, C. E. (2009) Proteasomal protein degradation in Mycobacteria is dependent upon a prokaryotic ubiquitin-like protein. *J Biol Chem*, 284, 3069–75.

Cadwell, K. & Coscoy, L. (2005) Ubiquitination on nonlysine residues by a viral E3 ubiquitin ligase. *Science*, 309, 127–30.

Carrion-Vazquez, M., Li, H., Lu, H., Marszalek, P. E., Oberhauser, A. F. & Fernandez, J. M. (2003) The mechanical stability of ubiquitin is linkage dependent. *Nat Struct Biol*, 10, 738–43.

Cascio, P., Call, M., Petre, B. M., Walz, T. & Goldberg, A. L. (2002) Properties of the hybrid form of the 26S proteasome containing both 19S and PA28 complexes. *EMBO J*, 21, 2636–45.

Cha, S. S, An, Y. J., Lee, C. R., Lee, H. S., Kim, Y. G., Kim, S. J., Kwon, K. K., De Donatis, G. M., Lee, J. H., Maurizi, M. R. & Kang, S. G. (2010) Crystal structure of Lon protease:

molecular architecture of gated entry to a sequestered degradation chamber. *EMBO J*, 29, 3520–30.

Chau, V., Tobias, J. W., Bachmair, A., Marriott, D., Ecker, D. J., Gonda, D. K. & Varshavsky, A. (1989) A multiubiquitin chain is confined to specific lysine in a targeted short-lived protein. *Science*, 243, 1576–83.

Chen, L. & Madura, K. (2002) Rad23 promotes the targeting of proteolytic substrates to the proteasome. *Mol Cell Biol*, 22, 4902–13.

Choi, K.-H. & Licht, S. (2005) Control of peptide product sizes by the energy-dependent protease ClpAP. *Biochemistry*, 44, 13921–31.

Ciechanover, A. (1998) The ubiquitin-proteasome pathway: on protein death and cell life. *EMBO J*, 17, 7151–7160.

Ciechanover, A. & Ben-Saadon, R. (2004) N-terminal ubiquitination: more protein substrates join in. *Trends Cell Biol*, 14, 103–6.

Colland, F. (2010) The therapeutic potential of deubiquitinating enzyme inhibitors. *Biochem Soc Trans*, 38, 137–43.

Collins, G. A. & Tansey, W. P. (2006) The proteasome: a utility tool for transcription? *Curr Opin Genet Dev*, 16, 197–202.

Crosas, B., Hanna, J., Kirkpatrick, D. S., Zhang, D. P., Tone, Y., Hathaway, N. A., Buecker, C., Leggett, D. S., Schmidt, M., King, R. W., Gygi, S. P. & Finley, D. (2006) Ubiquitin chains are remodeled at the proteasome by opposing ubiquitin ligase and deubiquitinating activities. *Cell*, 127, 1401–13.

D'andrea, A. & Pellman, D. (1998) Deubiquitinating enzymes: a new class of biological regulators. *Crit Rev Biochem Mol Biol*, 33, 337–52.

DA Fonseca, P. C. A. & Morris, E. P. (2008) Structure of the human 26S proteasome: subunit radial displacements open the gate into the proteolytic core. *J Biol Chem*, 283, 23305–14.

Demartino, G. N. & Gillette, T. G. (2007) Proteasomes: machines for all reasons. *Cell*, 129, 659–62.

Deshaies, R. J. & Joazeiro, C. A. P. (2009) RING domain E3 ubiquitin ligases. *Annu Rev Biochem*, 78, 399–434.

Deveraux, Q., Ustrell, V., Pickart, C. & Rechsteiner, M. (1994) A 26 S protease subunit that binds ubiquitin conjugates. *J Biol Chem*, 269, 7059–61.

Djuranovic, S., Hartmann, M. D., Habeck, M., Ursinus, A., Zwickl, P., Martin, J., Lupas, A. N. & Zeth, K. (2009) Structure and activity of the N-terminal substrate recognition domains in proteasomal ATPases. *Mol Cell*, 34, 580–90.

Dougan, D. A., Mogk, A., Zeth, K., Turgay, K. & Bukau, B. (2002a) AAA$^+$ proteins and substrate recognition, it all depends on their partner in crime. *FEBS Lett*, 529, 6–10.

Dougan, D. A., Reid, B. G., Horwich, A. L. & Bukau, B. (2002b) ClpS, a substrate modulator of the ClpAP machine. *Mol Cell*, 9, 673–83.

Effantin, G., Maurizi, M. R. & Steven, A. C. (2010) Binding of the CLP-A unfoldase opens the axial gate of CLP-P peptidase. *J Biol Chem*, 269, 18201–8.

Elsasser, S., Chandler-Militello, D., Müller, B., Hanna, J. & Finley, D. (2004) Rad23 and Rpn10 serve as alternative ubiquitin receptors for the proteasome. *J Biol Chem*, 279, 26817–22.

Elsasser, S., Gali, R. R., Schwickart, M., Larsen, C. N., Leggett, D. S., Müller, B., Feng, M. T., Tübing, F., Dittmar, G. A. G. & Finley, D. (2002) Proteasome subunit Rpn1 binds ubiquitin-like protein domains. *Nat Cell Biol*, 4, 725–30.

Erbse, A., Schmidt, R., Bornemann, T., Schneider-Mergener, J., Mogk, A., Zahn, R., Dougan, D. A. & Bukau, B. (2006) ClpS is an essential component of the N-end rule pathway in *Escherichia coli. Nature*, 439, 753–6.

Finley, D. (2009) Recognition and processing of ubiquitin-protein conjugates by the proteasome. *Annu Rev Biochem*, 78, 477–513.

Flynn, J. M., Levchenko, I., Seidel, M., Wickner, S. H., Sauer, R. T. & Baker, T. A. (2001) Overlapping recognition determinants within the ssrA degradation tag allow modulation of proteolysis. *Proc Natl Acad Sci USA*, 98, 10584–9.

Flynn, J. M., Neher, S. B., Kim, Y. I., Sauer, R. T. & Baker, T. A. (2003) Proteomic discovery of cellular substrates of the ClpXP protease reveals five classes of ClpX-recognition signals. *Mol Cell*, 11, 671–83.

Förster, A., Masters, E. I., Whitby, F. G., Robinson, H. & Hill, C. P. (2005) The 1.*9 A structure of a proteasome-11S activator complex and implications for proteasome-PAN/PA700 interactions. Mol Cell*, 18, 589–99.

Förster, F., Lasker, K., Nickell, S., Sali, A. & Baumeister, W. (2010) Toward an integrated structural model of the 26S proteasome. *Mol Cell Proteomics*, 9, 1666–77.

Frank, J. (2002) Single-particle imaging of macromolecules by cryo-electron microscopy. *Annu Rev Biophys Biomol Struct*, 31, 303–19.

Fu, J., Hashem, Y., Wower, I., Lei, J., Liao, H. Y., Zwieb, C., Wower, J. & Frank, J. (2010) Visualizing the transfer-messenger RNA as the ribosome resumes translation. *EMBO J*, 29, 3819–25.

Funakoshi, M., Tomko, R. J., Kobayashi, H. & Hochstrasser, M. (2009) Multiple assembly chaperones govern biogenesis of the proteasome regulatory particle base. *Cell*, 137, 887–99.

Gillette, T. G., Kumar, B., Thompson, D., Slaughter, C. A. & Demartino, G. N. (2008) Differential roles of the COOH termini of AAA subunits of PA700 (19 S regulator) in asymmetric assembly and activation of the 26 S proteasome. *J Biol Chem*, 283, 31813–22.

Glickman, M. H. & Ciechanover, A. (2002) The ubiquitin-proteasome proteolytic pathway: destruction for the sake of construction. *Physiol Rev*, 82, 373–428.

Glickman, M. H., Rubin, D. M., Coux, O., Wefes, I., Pfeifer, G., Cjeka, Z., Baumeister, W., Fried, V. A. & Finley, D. (1998) A subcomplex of the proteasome regulatory particle required for ubiquitin-conjugate degradation and related to the COP9-signalosome and eIF3. *Cell*, 94, 615–23.

Glynn, S. E., Martin, A., Nager, A. R., Baker, T. A. & Sauer, R. T. (2009) Structures of asymmetric ClpX hexamers reveal nucleotide-dependent motions in a AAA$^+$ protein-unfolding machine. *Cell*, 139, 744–56.

Goldberg, A. L. (2003) Protein degradation and protection against misfolded or damaged proteins. *Nature*, 426, 895–9.

Goldenberg, S. J., Marblestone, J. G., Mattern, M. R. & Nicholson, B. (2010) Strategies for the identification of ubiquitin ligase inhibitors. *Biochem Soc Trans*, 38, 132–6.

Gonzalez, M., Frank, E. G., Levine, A. S. & Woodgate, R. (1998) Lon-mediated proteolysis of the *Escherichia coli* UmuD mutagenesis protein: in vitro degradation and identification of residues required for proteolysis. *Genes Dev*, 12, 3889–99.

Gottesman, S., Roche, E., Zhou, Y. & Sauer, R. T. (1998) The ClpXP and ClpAP proteases degrade proteins with carboxy-terminal peptide tails added by the SsrA-tagging system. *Genes Dev*, 12, 1338–47.

Gottesman, S. (2003) Proteolysis in bacterial regulatory circuits. *Annu. Rev.Cell Dev.Biol.*, 19, 565–587.

Grimaud, R., Kessel, M., Beuron, F., Steven, A. C. & Maurizi, M. R. (1998) Enzymatic and structural similarities between the *Escherichia coli* ATP-dependent proteases, ClpXP and ClpAP. *J Biol Chem*, 273, 12476–81.

Groettrup, M., Soza, A., Eggers, M., Kuehn, L., Dick, T. P., Schild, H., Rammensee, H. G., Koszinowski, U. H. & Kloetzel, P. M. (1996) A role for the proteasome regulator PA28alpha in antigen presentation. *Nature*, 381, 166–8.

Groll, M., Bajorek, M., Köhler, A., Moroder, L., Rubin, D. M., Huber, R., Glickman, M. H. & Finley, D. (2000) A gated channel into the proteasome core particle. *Nat Struct Biol*, 7, 1062–7.

Groll, M., Ditzel, L., Löwe, J., Stock, D., Bochtler, M., Bartunik, H. D. & Huber, R. (1997) Structure of 20S proteasome from yeast at 2.*4 Å resolution. Nature*, 386, 463–71.

Guo, F., Maurizi, M. R., Esser, L. & Xia, D. (2002a) Crystal structure of ClpA, an Hsp100 chaperone and regulator of ClpAP protease. *J Biol Chem*, 277, 46743–52.

Guo, F., Esser, L., Singh, S. K., Maurizi, M. R. & Xia, D. (2002b) Crystal structure of the heterodimeric complex of the adaptor, ClpS, with the N-domain of the AAA + chaperone ClpA. *J Biol Chem*, 277, 46753–46762.

Gur, E. & Sauer, R. T. (2009) Degrons in protein substrates program the speed and operating efficiency of the AAA Lon proteolytic machine. *Proc Natl Acad Sci USA*, 106, 18503–8.

Hanna, J., Hathaway, N. A., Tone, Y., Crosas, B., Elsasser, S., Kirkpatrick, D. S., Leggett, D. S., Gygi, S. P., King, R. W. & Finley, D. (2006) Deubiquitinating enzyme Ubp6 functions noncatalytically to delay proteasomal degradation. *Cell*, 127, 99–111.

Hanson, P. I. & Whiteheart, S. W. (2005) AAA$^+$ proteins: have engine, will work. *Nat Rev Mol Cell Biol*, 6, 519–29.

Hasegawa, M., Fujiwara, H., Nonaka, T., Wakabayashi, K., Takahashi, H., Lee, V. M.-Y., Trojanowski, J. Q., Mann, D. & Iwatsubo, T. (2002) Phosphorylated alpha-synuclein is ubiquitinated in alpha-synucleinopathy lesions. *J Biol Chem*, 277, 49071–6.

Hendil, K. B., Khan, S. & Tanaka, K. (1998) Simultaneous binding of PA28 and PA700 activators to 20 S proteasomes. *Biochem J*, 332 (Pt 3), 749–54.

Herman, C., Prakash, S., Lu, C. Z., Matouschek, A. & Gross, C. A. (2003) Lack of a robust unfoldase activity confers a unique level of substrate specificity to the universal AAA protease FtsH. *Mol Cell*, 11, 659–69.

Herman, C., Thévenet, D., Bouloc, P., Walker, G. C. & D'ari, R. (1998) Degradation of carboxy-terminal-tagged cytoplasmic proteins by the *Escherichia coli* protease HflB (FtsH). *Genes Dev*, 12, 1348–55.

Hersch, G. L., Burton, R. E., Bolon, D. N., Baker, T. A. & Sauer, R. T. (2005) Asymmetric interactions of ATP with the AAA$^+$ ClpX6 unfoldase: allosteric control of a protein machine. *Cell*, 121, 1017–27.

Hinnerwisch, J., Fenton, W. A., Furtak, K. J., Farr, G. W. & Horwich, A. L. (2005) Loops in the central channel of ClpA chaperone mediate protein binding, unfolding, and translocation. *Cell*, 121, 1029–41.

Hochstrasser, M. (2006) Lingering mysteries of ubiquitin-chain assembly. *Cell*, 124, 27–34.

Hoppe, T. (2005) Multiubiquitylation by E4 enzymes: 'one size' doesn't fit all. *Trends Biochem Sci*, 30, 183–7.

Hoppe, T., Matuschewski, K., Rape, M., Schlenker, S., Ulrich, H. D. & Jentsch, S. (2000) Activation of a membrane-bound transcription factor by regulated ubiquitin/proteasome-dependent processing. *Cell*, 102, 577–86.

Hoskins, J. R., Kim, S. Y. & Wickner, S. (2000a) Substrate recognition by the ClpA chaperone component of ClpAP protease. *J Biol Chem*, 275, 35361–7.

Hoskins, J. R., Pak, M., Maurizi, M. R. & Wickner, S. (1998) The role of the ClpA chaperone in proteolysis by ClpAP. *Proc Natl Acad Sci USA*, 95, 12135–40.

Hoskins, J. R., Singh, S. K., Maurizi, M. R. & Wickner, S. (2000b) Protein binding and unfolding by the chaperone ClpA and degradation by the protease ClpAP. *Proc Natl Acad Sci USA*, 97, 8892–7.

Hoskins, J., Yanagihara, K., Mizuuchi, K. & Wickner, S. (2002) ClpAP and ClpXP degrade proteins with tags located in the interior of the primary sequence. *Proc Natl Acad Sci USA*, 99, 11037–42.

Hou, J. Y., Sauer, R. T. & Baker, T. A. (2008) Distinct structural elements of the adaptor ClpS are required for regulating degradation by ClpAP. *Nat Struct Mol Biol*, 15, 288–94.

Humbard, M. A., Miranda, H. V., Lim, J.-M., Krause, D. J., Pritz, J. R., Zhou, G., Chen, S., Wells, L. & Maupin-Furlow, J. A. (2010) Ubiquitin-like small archaeal modifier proteins (SAMPs) in *Haloferax volcanii*. *Nature*, 463, 54–60.

Huang, S., Ratliff, K. S., Schwarz, M. P., Spenner, J. M. & Matouschek, A. (1999) Mitochondria unfold precursor proteins by unraveling them from their N-termini. *Nat Struct Biol*, 6, 1132–8.

Huang, S., Murphy, S. & Matouschek, A. (2000) Effect of the protein import machinery at the mitochondrial surface on precursor stability. *Proc Natl Acad Sci USA*, 97, 12991–6.

Husnjak, K., Elsasser, S., Zhang, N., Chen, X., Randles, L., Shi, Y., Hofmann, K., Walters, K. J., Finley, D. & Dikic, I. (2008) Proteasome subunit Rpn13 is a novel ubiquitin receptor. *Nature*, 453, 481–8.

Imoto, T., Yamada, H. & Ueda, T. (1986) Unfolding rates of globular proteins determined by kinetics of proteolysis. *J Mol Biol*, 190, 647–9.

Ishikawa, T., Beuron, F., Kessel, M., Wickner, S., Maurizi, M. R. & Steven, A. C. (2001) Translocation pathway of protein substrates in ClpAP protease. *Proc Natl Acad Sci USA*, 98, 4328–33.

Ishikawa, T., Maurizi, M. R. & Steven, A. C. (2004) The N-terminal substrate-binding domain of ClpA unfoldase is highly mobile and extends axially from the distal surface of ClpAP protease. *J Struct Biol*, 146, 180–8.

Ito, K. & Akiyama, Y. (2005) Cellular functions, mechanism of action, and regulation of FtsH protease. *Annu Rev Microbiol*, 59, 211–31.

Iyer, L. M., Burroughs, A. M. & Aravind, L. (2006) The prokaryotic antecedents of the ubiquitin-signaling system and the early evolution of ubiquitin-like beta-grasp domains. *Genome Biol*, 7, R60.

Iyer, L. M., Burroughs, A. M. & Aravind, L. (2008) Unraveling the biochemistry and provenance of pupylation: a prokaryotic analog of ubiquitination. *Biol Direct*, 3, 45.

Iyer, L. M., Leipe, D. D., Koonin, E. V. & Aravind, L. (2004) Evolutionary history and higher order classification of AAA$^+$ ATPases. *J Struct Biol*, 146, 11–31.

Jariel-Encontre, I., Bossis, G. & Piechaczyk, M. (2008) Ubiquitin-independent degradation of proteins by the proteasome. *Biochim Biophys Acta*, 1786, 153–77.

Jin, L., Williamson, A., Banerjee, S., Philipp, I. & Rape, M. (2008) Mechanism of ubiquitin-chain formation by the human anaphase-promoting complex. *Cell*, 133, 570–2.

Johnston, J. A., Johnson, E. S., Waller, P. R. & Varshavsky, A. (1995) Methotrexate inhibits proteolysis of dihydrofolate reductase by the N-end rule pathway. *J Biol Chem*, 270, 8172–8.

Joshi, S. A., Hersch, G. L., Baker, T. A. & Sauer, R. T. (2004) Communication between ClpX and ClpP during substrate processing and degradation. *Nat Struct Mol Biol*, 11, 404–11.

Kaganovich, D., Kopito, R. & Frydman, J. (2008) Misfolded proteins partition between two distinct quality control compartments. *Nature*, 454, 1088–95.

Kalchman, M. A., Graham, R. K., Xia, G., Koide, H. B., Hodgson, J. G., Graham, K. C., Goldberg, Y. P., Gietz, R. D., Pickart, C. M. & Hayden, M. R. (1996) Huntingtin is ubiquitinated and interacts with a specific ubiquitin-conjugating enzyme. *J Biol Chem*, 271, 19385–94.

Kaneko, T., Hamazaki, J., Iemura, S.-I., Sasaki, K., Furuyama, K., Natsume, T., Tanaka, K. & Murata, S. (2009) Assembly pathway of the Mammalian proteasome base subcomplex is mediated by multiple specific chaperones. *Cell*, 137, 914–25.

Karata, K., Inagawa, T., Wilkinson, A. J., Tatsuta, T. & Ogura, T. (1999) Dissecting the role of a conserved motif (the second region of homology) in the AAA family of ATPases. Site-directed mutagenesis of the ATP-dependent protease FtsH. *J Biol Chem*, 274, 26225–32.

Keiler, K. C., Waller, P. R. & Sauer, R. T. (1996) Role of a peptide tagging system in degradation of proteins synthesized from damaged messenger RNA. *Science*, 271, 990–3.

Kenniston, J. A., Baker, T. A., Fernandez, J. M. & Sauer, R. T. (2003) Linkage between ATP consumption and mechanical unfolding during the protein processing reactions of an AAA$^+$ degradation machine. *Cell*, 114, 511–20.

Kenniston, J. A., Baker, T. A. & Sauer, R. T. (2005) Partitioning between unfolding and release of native domains during ClpXP degradation determines substrate selectivity and partial processing. *Proc Natl Acad Sci USA*, 102, 1390–5.

Kim, Y. I., Burton, R. E., Burton, B. M., Sauer, R. T. & Baker, T. A. (2000) Dynamics of substrate denaturation and translocation by the ClpXP degradation machine. *Mol Cell*, 5, 639–48.

Kim, D. Y. & Kim, K. K. (2003) Crystal structure of ClpX molecular chaperone from Helicobacter pylori. *J Biol Chem*, 278, 50664–70.

Kim, I., Mi, K. & Rao, H. (2004) Multiple interactions of rad23 suggest a mechanism for ubiquitylated substrate delivery important in proteolysis. *Mol Biol Cell*, 15, 3357–65.

Kim, Y. I., Levchenko, I., Fraczkowska, K., Woodruff, R. V., Sauer, R. T. & Baker, T. A. (2001) Molecular determinants of complex formation between Clp/Hsp100 ATPases and the ClpP peptidase. *Nat Struct Biol*, 8, 230–3.

Kirstein, J., Hoffmann, A., Lilie, H., Schmidt, R., Rübsamen-Waigmann, H., Brötz-Oesterhelt, H., Mogk, A. & Turgay, K. (2009) The antibiotic ADEP reprogrammes ClpP, switching it from a regulated to an uncontrolled protease. *EMBO Mol Med*, 1, 37–49.

Kirstein, J., Molière, N., Dougan, D. A. & Turgay, K. (2009) Adapting the machine: adaptor proteins for Hsp100/Clp and AAA+ proteases. *Nat Rev Microbiol*, 7, 589–99.

Kirstein, J., Strahl, H., Molière, N., Hamoen, L. W. & Turgay, K. (2008) Localization of general and regulatory proteolysis in Bacillus subtilis cells. *Mol Microbiol*, 70, 682–94.

Koegl, M., Hoppe, T., Schlenker, S., Ulrich, H. D., Mayer, T. U. & Jentsch, S. (1999) A novel ubiquitination factor, E4, is involved in multiubiquitin chain assembly. *Cell*, 96, 635–44.

Koga, N., Kameda, T., Okazaki, K.-I. & Takada, S. (2009) Paddling mechanism for the substrate translocation by AAA+ motor revealed by multiscale molecular simulations. *Proc Natl Acad Sci USA*, 106, 18237–42.

Köhler, A., Cascio, P., Leggett, D. S., Woo, K. M., Goldberg, A. L. & Finley, D. (2001) The axial channel of the proteasome core particle is gated by the Rpt2 ATPase and controls both substrate entry and product release. *Mol Cell*, 7, 1143–52.

Kolygo, K., Ranjan, N., Kress, W., Striebel, F., Hollenstein, K., Neelsen, K., Steiner, M., Summer, H. & Weber-Ban, E. (2009) Studying chaperone-proteases using a real-time approach based on FRET. *J Struct Biol*, 168, 267–77.

Koodathingal, P., Jaffe, N. E., Kraut, D. A., Prakash, S., Fishbain, S., Herman, C. & Matouschek, A. (2009) ATP-dependent proteases differ substantially in their ability to unfold globular proteins. *J Biol Chem*, 284, 18674–84.

Kress, W., Maglica, Z. & Weber-Ban, E. (2009) Clp chaperone-proteases: structure and function. *Res Microbiol*, 160, 618–28.

Lam, Y. A., Lawson, T. G., Velayutham, M., Zweier, J. L. & Pickart, C. M. (2002) A proteasomal ATPase subunit recognizes the polyubiquitin degradation signal. *Nature*, 416, 763–7.

Lam, Y. A., Xu, W., Demartino, G. N. & Cohen, R. E. (1997) Editing of ubiquitin conjugates by an isopeptidase in the 26S proteasome. *Nature*, 385, 737–40.

Lee, B.-G., Park, E. Y., Lee, K.-E., Jeon, H., Sung, K. H., Paulsen, H., Rübsamen-Schaeff, H., Brötz-Oesterhelt, H. & Song, H. K. (2010) Structures of ClpP in complex with acyldepsipeptide antibiotics reveal its activation mechanism. *Nat Struct Mol Biol*, 17, 471–8.

Lee, C., Schwartz, M. P., Prakash, S., Iwakura, M. & Matouschek, A. (2001) ATP-dependent proteases degrade their substrates by processively unraveling them from the degradation signal. *Mol Cell*, 7, 627–37.

Leggett, D. S., Hanna, J., Borodovsky, A., Crosas, B., Schmidt, M., Baker, R. T., Walz, T., Ploegh, H. & Finley, D. (2002) Multiple associated proteins regulate proteasome structure and function. *Mol Cell*, 10, 495–507.

Lenzen, C. U., Steinmann, D., Whiteheart, S. W. & Weis, W. I. (1998) Crystal structure of the hexamerization domain of N-ethylmaleimide-sensitive fusion protein. Cell, 94, 525–36.

Levchenko, I., Luo, L. & Baker, T. A. (1995) Disassembly of the Mu transposase tetramer by the ClpX chaperone. *Genes Dev*, 9, 2399–408.

Levchenko, I., Seidel, M., Sauer, R. T. & Baker, T. A. (2000) A specificity-enhancing factor for the ClpXP degradation machine. *Science*, 289, 2354–6.

Li, W., Bengtson, M. H., Ulbrich, A., Matsuda, A., Reddy, V. A., Orth, A., Chanda, S. K., Batalov, S. & Joazeiro, C. A. P. (2008) Genome-wide and functional annotation of human E3 ubiquitin ligases identifies MULAN, a mitochondrial E3 that regulates the organelle's dynamics and signaling. *PLoS ONE* 3(1): e1487. doi:10.1371/journal.pone.0001487.

Li, X. & Demartino, G. N. (2009) Variably modulated gating of the 26S proteasome by ATP and polyubiquitin. *Biochem J*, 421, 397–404.

Liu, C.-W., Li, X., Thompson, D., Wooding, K., Chang, T.-L., Tang, Z., Yu, H., Thomas, P. J. & Demartino, G. N. (2006) ATP binding and ATP hydrolysis play distinct roles in the function of 26S proteasome. *Mol Cell*, 24, 39–50.

Löwe, J., Stock, D., Jap, B., Zwickl, P., Baumeister, W. & Huber, R. (1995) Crystal structure of the 20S proteasome from the archaeon *T. acidophilum* at 3.4 Å resolution. *Science*, 268, 533–9.

Lupas, A. N. & Martin, J. (2002) AAA proteins. *Curr Opin Struct Biol*, 12, 746–53.

Maillard, R. A., Chistol, G., Sen, M., Righini, M., Tan, J., Kaiser, C. M., Hodges, C., Martin, A. & Bustamante, C. (2011) ClpX(P) generates mechanical force to unfold and translocate its protein substrates. *Cell*, 145, 459–69.

Martin, A., Baker, T. A. & Sauer, R. T. (2005) Rebuilt AAA+ motors reveal operating principles for ATP-fuelled machines. *Nature*, 437, 1115–20.

Martin, A., Baker, T. A. & Sauer, R. T. (2007) Distinct static and dynamic interactions control ATPase-peptidase communication in a AAA+ protease. *Mol Cell*, 27, 41–52.

Martin, A., Baker, T. A. & Sauer, R. T. (2008a) Pore loops of the AAA+ ClpX machine grip substrates to drive translocation and unfolding. *Nat Struct Mol Biol*, 15, 1147–51.

Martin, A., Baker, T. A. & Sauer, R. T. (2008b) Protein unfolding by a AAA+ protease is dependent on ATP-hydrolysis rates and substrate energy landscapes. *Nat Struct Mol Biol*, 15, 139–45.

Maurizi, M. R., Clark, W. P., Kim, S. H. & Gottesman, S. (1990) Clp P represents a unique family of serine proteases. *J Biol Chem*, 265, 12546–52.

Mogk, A., Haslberger, T., Tessarz, P. & Bukau, B. (2008) Common and specific mechanisms of AAA+ proteins involved in protein quality control. *Biochem Soc Trans*, 36, 120–5.

Mogk, A., Schmidt, R. & Bukau, B. (2007) The N-end rule pathway for regulated proteolysis: prokaryotic and eukaryotic strategies. *Trends Cell Biol*, 17, 165–72.

Morishima-Kawashima, M., Hasegawa, M., Takio, K., Suzuki, M., Titani, K. & Ihara, Y. (1993) Ubiquitin is conjugated with amino-terminally processed tau in paired helical filaments. *Neuron*, 10, 1151–60.

Murakami, Y., Matsufuji, S., Kameji, T., Hayashi, S., Igarashi, K., Tamura, T., Tanaka, K. & Ichihara, A. (1992) Ornithine decarboxylase is degraded by the 26S proteasome without ubiquitination. *Nature*, 360, 597–9.

Navon, A. & Ciechanover, A. (2009) The 26 S proteasome: from basic mechanisms to drug targeting. *J Biol Chem*, 284, 33713–8.

Neher, S. B., Flynn, J. M., Sauer, R. T. & Baker, T. A. (2003) Latent ClpX-recognition signals ensure LexA destruction after DNA damage. *Genes Dev*, 17, 1084–9.

Neuwald, A. F., Aravind, L., Spouge, J. L. & Koonin, E. V. (1999) AAA⁺: A class of chaperone-like ATPases associated with the assembly, operation, and disassembly of protein complexes. *Genome Res*, 9, 27–43.

Nickell, S., Beck, F., Scheres, S. H. W., Korinek, A., Förster, F., Lasker, K., Mihalache, O., Sun, N., Nagy, I., Sali, A., Plitzko, J. M., Carazo, J.-M., Mann, M. & Baumeister, W. (2009) Insights into the molecular architecture of the 26S proteasome. *Proc Natl Acad Sci USA*, 106, 11943–7.

Nussbaum, A. K., Dick, T. P., Keilholz, W., Schirle, M., Stevanović, S., Dietz, K., Heinemeyer, W., Groll, M., Wolf, D. H., Huber, R., Rammensee, H. G. & Schild, H. (1998) Cleavage motifs of the yeast 20S proteasome beta subunits deduced from digests of enolase 1. *Proc Natl Acad Sci USA*, 95, 12504–9.

Ogura, T. & Wilkinson, A. J. (2001) AAA⁺ superfamily ATPases: common structure–diverse function. *Genes Cells*, 6, 575–97.

Okuno, T., Yamada-Inagawa, T., Karata, K., Yamanaka, K. & Ogura, T. (2004) Spectrometric analysis of degradation of a physiological substrate sigma32 by Escherichia coli AAA protease FtsH. *J Struct Biol*, 146, 148–54.

Ortega, J., Singh, S. K., Ishikawa, T., Maurizi, M. R. & Steven, A. C. (2000) Visualization of substrate binding and translocation by the ATP-dependent protease, ClpXP. *Mol Cell*, 6, 1515–21.

Palombella, V. J., Rando, O. J., Goldberg, A. L. & Maniatis, T. (1994) The ubiquitin-proteasome pathway is required for processing the NF-kappa B1 precursor protein and the activation of NF-kappa B. *Cell*, 78, 773–85.

Park, C. & Marqusee, S. (2005) Pulse proteolysis: a simple method for quantitative determination of protein stability and ligand binding. *Nat Methods*, 2, 207–12.

Park, E., Rho, Y. M., Koh, O.-J., Ahn, S. W., Seong, I. S., Song, J.-J., Bang, O., Seol, J. H., Wang, J., Eom, S. H. & Chung, C. H. (2005) Role of the GYVG pore motif of HslU ATPase in protein unfolding and translocation for degradation by HslV peptidase. *J Biol Chem*, 280, 22892–8.

Pearce, M. J., Mintseris, J., Ferreyra, J., Gygi, S. P. & Darwin, K. H. (2008) Ubiquitin-like protein involved in the proteasome pathway of *Mycobacterium tuberculosis*. *Science*, 322, 1104–7.

Peth, A., Besche, H. C. & Goldberg, A. L. (2009) Ubiquitinated proteins activate the proteasome by binding to Usp14/Ubp6, which causes 20S gate opening. *Mol Cell*, 36, 794–804.

Peth, A., Uchiki, T. & Goldberg, A. L. (2010) ATP-dependent steps in the binding of ubiquitin conjugates to the 26S proteasome that commit to degradation. *Mol Cell*, 40, 671–81.

Pickart, C. M. (2001) Mechanisms underlying ubiquitination. *Annu Rev Biochem*, 70, 503–33.

Pickart, C. M. (2004) Back to the future with ubiquitin. *Cell*, 116, 181–90.

Pickart, C. M. & Cohen, R. E. (2004) Proteasomes and their kin: proteases in the machine age. *Nat Rev Mol Cell Biol*, 5, 177–87.

Pickart, C. M. & Fushman, D. (2004) Polyubiquitin chains: polymeric protein signals. *Curr Opin Chem Biol*, 8, 610–6.

Prakash, S., Inobe, T., Hatch, A. J. & Matouschek, A. (2009) Substrate selection by the proteasome during degradation of protein complexes. *Nat Chem Biol*, 5, 29–36.

Prakash, S. & Matouschek, A. (2004) Protein unfolding in the cell. *Trends Biochem Sci*, 29, 593–600.

Prakash, S., Tian, L., Ratliff, K. S., Lehotzky, R. E. & Matouschek, A. (2004) An unstructured initiation site is required for efficient proteasome-mediated degradation. *Nat Struct Mol Biol*, 11, 830–7.

Rabl, J., Smith, D. M., Yu, Y., Chang, S.-C., Goldberg, A. L. & Cheng, Y. (2008) Mechanism of gate opening in the 20S proteasome by the proteasomal ATPases. *Mol Cell*, 30, 360–8.

Rao, H. & Sastry, A. (2002) Recognition of specific ubiquitin conjugates is important for the proteolytic functions of the ubiquitin-associated domain proteins Dsk2 and Rad23. *J Biol Chem*, 277, 11691–5.

Ravid, T. & Hochstrasser, M. (2008) Diversity of degradation signals in the ubiquitin-proteasome system. *Nat Rev Mol Cell Biol*, 9, 679–90.

Reid, B. G., Fenton, W. A., Horwich, A. L. & Weber-Ban, E. U. (2001) ClpA mediates directional translocation of substrate proteins into the ClpP protease. *Proc Natl Acad Sci USA*, 98, 3768–72.

Rock, K. L. & Goldberg, A. L. (1999) Degradation of cell proteins and the generation of MHC class I-presented peptides. *Annu Rev Immunol*, 17, 739–79.

Rohrwild, M., Pfeifer, G., Santarius, U., Müller, S. A., Huang, H. C., Engel, A., Baumeister, W. & Goldberg, A. L. (1997) The ATP-dependent HslVU protease from *Escherichia coli* is a four-ring structure resembling the proteasome. *Nat Struct Biol*, 4, 133–9.

Ross, C. A. & Poirier, M. A. (2004) Protein aggregation and neurodegenerative disease. *Nat Med*, 10 Suppl, S10–S17.

Rubin, D. M., Glickman, M. H., Larsen, C. N., Dhruvakumar, S. & Finley, D. (1998) Active site mutants in the six regulatory particle ATPases reveal multiple roles for ATP in the proteasome. *EMBO J*, 17, 4909–19.

Saeki, Y., Kudo, T., Sone, T., Kikuchi, Y., Yokosawa, H., Toh-E, A. & Tanaka, K. (2009) Lysine 63-linked polyubiquitin chain may serve as a targeting signal for the 26S proteasome. *EMBO J*, 28, 359–71.

Sakamoto, K. M., Kim, K. B., Kumagai, A., Mercurio, F., Crews, C. M. & Deshaies, R. J. (2001) Protacs: chimeric molecules that target proteins to the Skp1-Cullin-F box complex for ubiquitination and degradation. *Proc Natl Acad Sci USA*, 98, 8554–9.

Sauer, R. T., Bolon, D. N., Burton, B. M., Burton, R. E., Flynn, J. M., Grant, R. A., Hersch, G. L., Joshi, S. A., Kenniston, J. A., Levchenko, I., Neher, S. B., Oakes, E. S. C., Siddiqui, S. M., Wah, D. A. & Baker, T. A. (2004) Sculpting the proteome with AAA(⁺) proteases and disassembly machines. *Cell*, 119, 9–18.

Schirmer, E. C., Glover, J. R., Singer, M. A. & Lindquist, S. (1996) HSP100/Clp proteins: a common mechanism explains diverse functions. *Trends Biochem Sci*, 21, 289–96.

Schmidt, M., Lupas, A. N. & Finley, D. (1999) Structure and mechanism of ATP-dependent proteases. *Curr Opin Chem Biol*, 3, 584–91.

Schrader, E. K., Harstad, K. G. & Matouschek, A. (2009) Targeting proteins for degradation. *Nat Chem Biol*, 5, 815–22.

Schwartz, A. L. & Ciechanover, A. (2009) Targeting proteins for destruction by the ubiquitin system: implications for human pathobiology. *Annu Rev Pharmacol Toxicol*, 49, 73–96.

Shah, J. J. & Orlowski, R. Z. (2009) Proteasome inhibitors in the treatment of multiple myeloma. *Leukemia*, 23, 1964–79.

Siddiqui, S. M., Sauer, R. T. & Baker, T. A. (2004) Role of the processing pore of the ClpX AAA⁺ ATPase in the recognition and engagement of specific protein substrates. *Genes Dev*, 18, 369–74.

Singh, S. K., Grimaud, R., Hoskins, J. R., Wickner, S. & Maurizi, M. R. (2000) Unfolding and internalization of proteins by the ATP-dependent proteases ClpXP and ClpAP. *Proc Natl Acad Sci USA*, 97, 8898–903.

Singh, S. K., Rozycki, J., Ortega, J., Ishikawa, T., Lo, J., Steven, A. C. & Maurizi, M. R. (2001) Functional domains of the ClpA and ClpX molecular chaperones identified by limited proteolysis and deletion analysis. *J Biol Chem*, 276, 29420–9.

Smith, C. K., Baker, T. A. & Sauer, R. T. (1999) Lon and Clp family proteases and chaperones share homologous substrate-recognition domains. *Proc Natl Acad Sci USA*, 96, 6678–82.

Smith, D. M., Benaroudj, N. & Goldberg, A. (2006) Proteasomes and their associated ATPases: a destructive combination. *J Struct Biol*, 156, 72–83.

Smith, D. M., Chang, S.-C., Park, S., Finley, D., Cheng, Y. & Goldberg, A. L. (2007) Docking of the proteasomal ATPases' carboxyl termini in the 20S proteasome's alpha ring opens the gate for substrate entry. *Mol Cell*, 27, 731–44.

Smith, D. M., Kafri, G., Cheng, Y., Ng, D., Walz, T. & Goldberg, A. L. (2005) ATP binding to PAN or the 26S ATPases causes association with the 20S proteasome, gate opening, and translocation of unfolded proteins. *Mol Cell*, 20, 687–98.

Smith, D. M., Fraga, H., Reis, C., Kafri, G. & Goldberg, A. L. (2011) ATP binds to proteasomal ATPases in pairs with distinct functional effects, implying an ordered reaction cycle. *Cell*, 144, 526–38.

Sousa, M. C., Trame, C. B., Tsuruta, H., Wilbanks, S. M., Reddy, V. S. & Mckay, D. B. (2000) Crystal and solution structures of an HslUV protease-chaperone complex. *Cell*, 103, 633–43.

Striebel, F., Hunkeler, M., Summer, H. & Weber-Ban, E. (2010) The mycobacterial Mpa-proteasome unfolds and degrades pupylated substrates by engaging Pup's N-terminus. *EMBO J*, 29, 1262–71.

Striebel, F., Kress, W. & Weber-Ban, E. (2009) Controlled destruction: AAA⁺ ATPases in protein degradation from bacteria to eukaryotes. *Curr Opin Struct Biol*, 19, 209–17.

Sun, L. & Chen, Z. J. (2004) The novel functions of ubiquitination in signaling. *Curr Opin Cell Biol*, 16, 119–26.

Suno, R., Niwa, H., Tsuchiya, D., Zhang, X., Yoshida, M. & Morikawa, K. (2006) Structure of the whole cytosolic region of ATP-dependent protease FtsH. *Mol Cell*, 22, 575–85.

Takeuchi, J., Chen, H. & Coffino, P. (2007) Proteasome substrate degradation requires association plus extended peptide. *EMBO J*, 26, 123–31.

Tanahashi, N., Murakami, Y., Minami, Y., Shimbara, N., Hendil, K. B. & Tanaka, K. (2000) Hybrid proteasomes. Induction by interferon-gamma and contribution to ATP-dependent proteolysis. *J Biol Chem*, 275, 14336–45.

Thibault, G., Tsitrin, Y., Davidson, T., Gribun, A. & Houry, W. A. (2006) Large nucleotide-dependent movement of the N-terminal domain of the ClpX chaperone. *EMBO J*, 25, 3367–76.

Thrower, J. S., Hoffman, L., Rechsteiner, M. & Pickart, C. M. (2000) Recognition of the polyubiquitin proteolytic signal. *EMBO J*, 19, 94–102.

Tian, L., Holmgren, R. A. & Matouschek, A. (2005) A conserved processing mechanism regulates the activity of transcription factors *Cubitus interruptus* and NF-kappaB. *Nat Struct Mol Biol*, 12, 1045–53.

Tomko, R. J., Funakoshi, M., Schneider, K., Wang, J. & Hochstrasser, M. (2010) Heterohexameric ring arrangement of the eukaryotic proteasomal ATPases: implications for proteasome structure and assembly. *Mol Cell*, 38, 393–403.

Tomoyasu, T., Gamer, J., Bukau, B., Kanemori, M., Mori, H., Rutman, A. J., Oppenheim, A. B., Yura, T., Yamanaka, K. & Niki, H. (1995) Escherichia coli FtsH is a membrane-bound, ATP-dependent protease which degrades the heat-shock transcription factor sigma 32. *EMBO J*, 14, 2551–60.

Tsilibaris, V., Maenhaut-Michel, G. & Van Melderen, L. (2006) Biological roles of the Lon ATP-dependent protease. *Res Microbiol*, 157, 701–13.

Unno, M., Mizushima, T., Morimoto, Y., Tomisugi, Y., Tanaka, K., Yasuoka, N. & Tsukihara, T. (2002) The structure of the mammalian 20S proteasome at 2.75 Å resolution. *Structure*, 10, 609–18.

Valle, M., Gillet, R., Kaur, S., Henne, A., Ramakrishnan, V. & Frank, J. (2003) Visualizing tmRNA entry into a stalled ribosome. *Science*, 300, 127–30.

Varshavsky, A. (1992) The N-end rule. *Cell*, 69, 725–35.

Varshavsky, A. (1996) The N-end rule: Functions, mysteries, uses. *Proc Natl Acad Sci USA*, 93, 12142–49.

Varshavsky, A. (2008) The N-end rule at atomic resolution. *Nat Struct Mol Biol*, 15, 1238–40.

Verma, R., Aravind, L., Oania, R., Mcdonald, W. H., Yates, J. R., Koonin, E. V. & Deshaies, R. J. (2002) Role of Rpn11 metalloprotease in deubiquitination and degradation by the 26S proteasome. *Science*, 298, 611–5.

Verma, R. & Deshaies, R. J. (2000) A proteasome howdunit: the case of the missing signal. *Cell*, 101, 341–4.

Verma, R., Oania, R., Graumann, J. & Deshaies, R. J. (2004) Multiubiquitin chain receptors define a layer of substrate selectivity in the ubiquitin-proteasome system. *Cell*, 118, 99–110.

Voges, D., Zwickl, P. & Baumeister, W. (1999) The 26S proteasome: a molecular machine designed for controlled proteolysis. *Annu Rev Biochem*, 68, 1015–68.

Walz, J., Erdmann, A., Kania, M., Typke, D., Koster, A. J. & Baumeister, W. (1998) 26S proteasome structure revealed by three-dimensional electron microscopy. *J Struct Biol*, 121, 19–29.

Wang, J., Hartling, J. A. & Flanagan, J. M. (1997) The structure of ClpP at 2.3 *A resolution suggests a model for ATP-dependent proteolysis. Cell*, 91, 447–56.

Wang, J., Song, J. J., Franklin, M. C., Kamtekar, S., Im, Y. J., Rho, S. H., Seong, I. S., Lee, C. S., Chung, C. H. & Eom, S. H. (2001a) Crystal structures of the HslVU peptidase-ATPase complex reveal an ATP-dependent proteolysis mechanism. *Structure*, 9, 177–84.

Wang, J., Song, J. J., Seong, I. S., Franklin, M. C., Kamtekar, S., Eom, S. H. & Chung, C. H. (2001b) Nucleotide-dependent conformational changes in a protease-associated ATPase HsIU. *Structure*, 9, 1107–16.

Wang, K. H., Roman-Hernandez, G., Grant, R. A., Sauer, R. T. & Baker, T. A. (2008) The molecular basis of N-end rule recognition. *Mol Cell*, 32, 406–14.

Wang, X., Herr, R. A., Chua, W.-J., Lybarger, L., Wiertz, E. J. H. J. & Hansen, T. H. (2007) Ubiquitination of serine, threonine, or lysine residues on the cytoplasmic tail can induce ERAD of MHC-I by viral E3 ligase mK3. *J Cell Biol*, 177, 613–24.

Wawrzynow, A., Wojtkowiak, D., Marszalek, J., Banecki, B., Jonsen, M., Graves, B., Georgopoulos, C. & Zylicz, M. (1995) The ClpX heat-shock protein of *Escherichia coli*, the ATP-dependent substrate specificity component of the ClpP-ClpX protease, is a novel molecular chaperone. *EMBO J*, 14, 1867–77.

Weber-Ban, E. U., Reid, B. G., Miranker, A. D. & Horwich, A. L. (1999) Global unfolding of a substrate protein by the Hsp100 chaperone ClpA. *Nature*, 401, 90–3.

Weibezahn, J., Schlieker, C., Bukau, B. & Mogk, A. (2003) Characterization of a trap mutant of the AAA$^+$ chaperone ClpB. *J Biol Chem*, 278, 32608–17.

Weissman, A. M. (2001) Themes and variations on ubiquitylation. *Nat Rev Mol Cell Biol*, 2, 169–78.

Wenzel, T. & Baumeister, W. (1995) Conformational constraints in protein degradation by the 20S proteasome. Nat Struct Biol, 3, 199–204.

Whitby, F. G., Masters, E. I., Kramer, L., Knowlton, J. R., Yao, Y., Wang, C. C. & Hill, C. P. (2000) Structural basis for the activation of 20S proteasomes by 11S regulators. *Nature*, 408, 115–20.

Wickner, S., Maurizi, M. R. & Gottesman, S. (1999) Posttranslational quality control: folding, refolding, and degrading proteins. *Science*, 286, 1888–93.

Wilcox, A. J., Choy, J., Bustamante, C. & Matouschek, A. (2005) Effect of protein structure on mitochondrial import. *Proc Natl Acad Sci USA*, 102, 15435–40.

Wilkinson, C. R., Seeger, M., Hartmann-Petersen, R., Stone, M., Wallace, M., Semple, C. & Gordon, C. (2001) Proteins containing the UBA domain are able to bind to multi-ubiquitin chains. *Nat Cell Biol*, 3, 939–43.

Wilkinson, K. D. (1997) Regulation of ubiquitin-dependent processes by deubiquitinating enzymes. *FASEB J*, 11, 1245–56.

Wilkinson, K. D. (2009) DUBs at a glance. *J Cell Sci*, 122, 2325–29.

Winkler, J., Seybert, A., König, L., Pruggnaller, S., Haselmann, U., Sourjik, V., Weiss, M., Frangakis, A. S., Mogk, A. & Bukau, B. (2010) Quantitative and spatio-temporal features of protein aggregation in *Escherichia coli* and consequences on protein quality control and cellular ageing. *EMBO J*, 29, 910–23.

Wolf, S., Nagy, I., Lupas, A., Pfeifer, G., Cejka, Z., Müller, S. A., Engel, A., De Mot, R. & Baumeister, W. (1998) Characterization of ARC, a divergent member of the AAA ATPase family from *Rhodococcus erythropolis*. *J Mol Biol*, 277, 13–25.

Yakamavich, J. A., Baker, T. A. & Sauer, R. T. (2008) Asymmetric nucleotide transactions of the HslUV protease. *J Mol Biol*, 380, 946–57.

Yamada-Inagawa, T., Okuno, T., Karata, K., Yamanaka, K. & Ogura, T. (2003) Conserved pore residues in the AAA protease FtsH are important for proteolysis and its coupling to ATP hydrolysis. *J Biol Chem*, 278, 50182–7.

Yao, T. & Cohen, R. E. (2002) A cryptic protease couples deubiquitination and degradation by the proteasome. *Nature*, 419, 403–7.

Yao, T., Song, L., Xu, W., Demartino, G. N., Florens, L., Swanson, S. K., Washburn, M. P., Conaway, R. C., Conaway, J. W. & Cohen, R. E. (2006) Proteasome recruitment and activation of the Uch37 deubiquitinating enzyme by Adrm1. *Nat Cell Biol*, 8, 994–1002.

Zeth, K., Ravelli, R. B., Paal, K., Cusack, S., Bukau, B. & Dougan, D. A. (2002) Structural analysis of the adaptor protein ClpS in complex with the N-terminal domain of ClpA. *Nat Struct Biol*, 9, 906–11.

Zhang, F., Hu, M., Tian, G., Zhang, P., Finley, D., Jeffrey, P. D. & Shi, Y. (2009a) Structural insights into the regulatory particle of the proteasome from Methanocaldococcus jannaschii. *Mol Cell*, 34, 473–84.

Zhang, F., Wu, Z., Zhang, P., Tian, G., Finley, D. & Shi, Y. (2009b) Mechanism of substrate unfolding and translocation by the regulatory particle of the proteasome from *Methanocaldococcus jannaschii*. *Mol Cell*, 34, 485–96.

Zhang, M., Macdonald, A. I., Hoyt, M. A. & Coffino, P. (2004) Proteasomes begin ornithine decarboxylase digestion at the C terminus. *J Biol Chem*, 279, 20959–65.

Zhang, M., Pickart, C. M. & Coffino, P. (2003) Determinants of proteasome recognition of ornithine decarboxylase, a ubiquitin-independent substrate. *EMBO J*, 22, 1488–96.

Zhang, Z., Clawson, A., Realini, C., Jensen, C. C., Knowlton, J. R., Hill, C. P. & Rechsteiner, M. (1998) Identification of an activation region in the proteasome activator REGalpha. *Proc Natl Acad Sci USA*, 95, 2807–11.

Zwickl, P., Ng, D., Woo, K. M., Klenk, H. P. & Goldberg, A. L. (1999) An archaebacterial ATPase, homologous to ATPases in the eukaryotic 26 S proteasome, activates protein breakdown by 20 S proteasomes. *J Biol Chem*, 274, 26008–14.

Index